*James E. Turner*
**Atoms, Radiation, and
Radiation Protection**

## 1807–2007 Knowledge for Generations

Each generation has its unique needs and aspirations. When Charles Wiley first opened his small printing shop in lower Manhattan in 1807, it was a generation of boundless potential searching for an identity. And we were there, helping to define a new American literary tradition. Over half a century later, in the midst of the Second Industrial Revolution, it was a generation focused on building the future. Once again, we were there, supplying the critical scientific, technical, and engineering knowledge that helped frame the world. Throughout the 20th Century, and into the new millennium, nations began to reach out beyond their own borders and a new international community was born. Wiley was there, expanding its operations around the world to enable a global exchange of ideas, opinions, and know-how.

For 200 years, Wiley has been an integral part of each generation's journey, enabling the flow of information and understanding necessary to meet their needs and fulfill their aspirations. Today, bold new technologies are changing the way we live and learn. Wiley will be there, providing you the must-have knowledge you need to imagine new worlds, new possibilities, and new opportunities.

Generations come and go, but you can always count on Wiley to provide you the knowledge you need, when and where you need it!

*William J. Pesce*
President and Chief Executive Officer

*Peter Booth Wiley*
Chairman of the Board

*James E. Turner*

# Atoms, Radiation, and Radiation Protection

Third, Completely Revised and Enlarged Edition

WILEY-VCH Verlag GmbH & Co. KGaA

**The Author**

*J.E. Turner*
127 Windham Road
Oak Ridge, TN 37830
USA

1st Edition 2007
  1st Reprint 2009
  2nd Reprint 2010
  3rd Reprint 2012
  4th Reprint 2015
  5th Reprint 2016

**Library of Congress Card No.:**
applied for

**British Library Cataloguing-in-Publication Data**
A catalogue record for this book is available
from the British Library.

**Bibliographic information published by
the Deutsche Nationalbibliothek**
The Deutsche Nationalbibliothek lists this
publication in the Deutsche
Nationalbibliografie; detailed bibliographic
data are available in the Internet at
<http://dnb.d-nb.de>.

© 2007 WILEY-VCH Verlag GmbH & Co.
KGaA, Weinheim

**Typesetting**   VTEX, Vilnius, Lithuania
**Printing**   betz-druck GmbH, Darmstadt
**Binding**   Litges & Dopf GmbH, Heppenheim
**Wiley Bicentennial Logo**   Richard J. Pacifico

Printed in the Federal Republic of Germany
Printed on acid-free paper

**ISBN**   978-3-527-40606-7

*To Renate*

# Contents

*Atoms, Radiation, and Radiation Protection.* James E. Turner
Copyright © 2007 WILEY-VCH Verlag GmbH & Co. KGaA, Weinheim
ISBN: 978-3-527-40606-7

**Appendices**

# Preface to the First Edition

*Atoms, Radiation, and Radiation Protection* was written from material developed
by the author over a number of years of teaching courses in the Oak Ridge Res-
ident Graduate Program of the University of Tennessee's Evening School. The
courses dealt with introductory health physics, preparation for the American Board
of Health Physics certification examinations, and related specialized subjects such
as microdosimetry and the application of Monte Carlo techniques to radiation protection. As the title of the book is meant to imply, atomic and nuclear physics and
the interaction of ionizing radiation with matter are central themes. These subjects
are presented in their own right at the level of basic physics, and the discussions are
developed further into the areas of applied radiation protection. Radiation dosimetry, instrumentation, and external and internal radiation protection are extensively
treated. The chemical and biological effects of radiation are not dealt with at length,
but are presented in a summary chapter preceding the discussion of radiation-
protection criteria and standards. Non-ionizing radiation is not included. The book
is written at the senior or beginning graduate level as a text for a one-year course
in a curriculum of physics, nuclear engineering, environmental engineering, or an
allied discipline. A large number of examples are worked in the text. The traditional
units of radiation dosimetry are used in much of the book; SI units are employed in
discussing newer subjects, such as ICRP Publications 26 and 30. SI abbreviations
are used throughout. With the inclusion of formulas, tables, and specific physical
data, *Atoms, Radiation, and Radiation Protection* is also intended as a reference for
professionals in radiation protection.

I have tried to include some important material not readily available in textbooks
on radiation protection. For example, the description of the electronic structure
of isolated atoms, fundamental to understanding so much of radiation physics,
is further developed to explain the basic physics of "collective" electron behavior
in semiconductors and their special properties as radiation detectors. In another
area, under active research today, the details of charged-particle tracks in water are
described from the time of the initial physical, energy-depositing events through
the subsequent chemical changes that take place within a track. Such concepts are
basic for relating the biological effects of radiation to particle-track structure.

I am indebted to my students and a number of colleagues and organizations,
who contributed substantially to this book. Many individual contributions are ac-

knowledged in figure captions. In addition, I would like to thank J. H. Corbin and W. N. Drewery of Martin Marietta Energy Systems, Inc.; Joseph D. Eddleman of Pulcir, Inc.; Michael D. Shepherd of Eberline; and Morgan Cox of Victoreen for their interest and help. I am especially indebted to my former teacher, Myron F. Fair, from whom I learned many of the things found in this book in countless discussions since we first met at Vanderbilt University in 1952.

It has been a pleasure to work with the professional staff of Pergamon Press, to whom I express my gratitude for their untiring patience and efforts throughout the production of this volume.

The last, but greatest, thanks are reserved for my wife, Renate, to whom this book is dedicated. She typed the entire manuscript and the correspondence that went with it. Her constant encouragement, support, and work made the book a reality.

*Oak Ridge, Tennessee*  
November 20, 1985

*James E. Turner*

# Preface to the Second Edition

The second edition of *Atoms, Radiation, and Radiation Protection* has several important new features. SI units are employed throughout, the older units being defined but used sparingly. There are two new chapters. One is on statistics for health physics. It starts with the description of radioactive decay as a Bernoulli process and treats sample counting, propagation of error, limits of detection, type-I and type-II errors, instrument response, and Monte Carlo radiation-transport computations. The other new chapter resulted from the addition of material on environmental radioactivity, particularly concerning radon and radon daughters (not much in vogue when the first edition was prepared in the early 1980s). New material has also been added to several earlier chapters: a derivation of the stopping-power formula for heavy charged particles in the impulse approximation, a more detailed discussion of beta-particle track structure and penetration in matter, and a fuller description of the various interaction coefficients for photons. The chapter on chemical and biological effects of radiation from the first edition has been considerably expanded. New material is also included there, and the earlier topics are generally dealt with in greater depth than before (e.g., the discussion of data on human exposures). The radiation exposure limits from ICRP Publications 60 and 61 and NCRP Report No. 116 are presented and discussed. Annotated bibliographies have been added at the end of each chapter. A number of new worked examples are presented in the text, and additional problems are included at the ends of the chapters. These have been tested in the classroom since the 1986 first edition. Answers are now provided to about half of the problems. In summary, in its new edition, *Atoms, Radiation, and Radiation Protection* has been updated and expanded both in breadth and in depth of coverage. Most of the new material is written at a somewhat more advanced level than the original.

I am very fortunate in having students, colleagues, and teachers who care about the subjects in this book and who have shared their enthusiasm, knowledge, and talents. I would like to thank especially the following persons for help I have received in many ways: James S. Bogard, Wesley E. Bolch, Allen B. Brodsky, Darryl J. Downing, R. J. Michael Fry, Robert N. Hamm, Jerry B. Hunt, Patrick J. Papin, Herwig G. Paretzke, Tony A. Rhea, Robert W. Wood, Harvel A. Wright, and Jacquelyn Yanch. The continuing help and encouragement of my wife, Renate, are gratefully acknowledged. I would also like to thank the staff of John Wiley & Sons, with whom

I have enjoyed working, particularly Gregory T. Franklin, John P. Falcone, and Angioline Loredo.

*Oak Ridge, Tennessee*                                                                    *James E. Turner*
January 15, 1995

# Preface to the Third Edition

Since the preparation of the second edition (1995) of *Atoms, Radiation, and Radiation Protection*, many important developments have taken place that affect the profession of radiological health protection. The International Commission on Radiological Protection (ICRP) has issued new documents in a number of areas that are addressed in this third edition. These include updated and greatly expanded anatomical and physiological data that replace "reference man" and revised models of the human respiratory tract, alimentary tract, and skeleton. At this writing, the Main Commission has just adopted the Recommendations 2007, thus laying the foundation and framework for continuing work from an expanded contemporary agenda into future practice. *Dose constraints, dose limits,* and *optimization* are given roles as core concepts. Medical exposures, exclusion levels, and radiation protection of nonhuman species are encompassed. The National Council on Radiation Protection and Measurements (NCRP) in the United States has introduced new limiting criteria and provided extensive data for the design of structural shielding for medical X-ray imaging facilities. *Kerma* replaces the traditional *exposure* as the shielding design parameter. The Council also completed its shielding report for megavoltage X- and gamma-ray radiotherapy installations. In other areas, the National Research Council's Committee on the Biological Effects of Ionizing Radiation published the BEIR VI and BEIR VII Reports, respectively dealing with indoor radon and with health risks from low levels of radiation. The very successful completion of the DS02 dosimetry system and the continuing Life Span Study of the Japanese atomic-bomb survivors represent additional major accomplishments discussed here.

Rapid advances since the last edition of this text have been made in instrumentation for the detection, monitoring, and measurement of ionizing radiation. These have been driven by improvements in computers, computer interfacing, and, in no small part, by heightened concern for nuclear safeguards and home security. Chapter 10 on Methods of Radiation Detection required extensive revision and the addition of considerable new material.

As in the previous edition, the primary regulatory criteria used here for discussions and working problems follow those given in ICRP Publication 60 with limits on *effective dose* to an individual. These recommendations are the principal ones employed throughout the world today, except in the United States. The ICRP-60

*Atoms, Radiation, and Radiation Protection.* James E. Turner
Copyright © 2007 WILEY-VCH Verlag GmbH & Co. KGaA, Weinheim
ISBN: 978-3-527-40606-7

limits for individual effective dose, with which current NCRP recommendations are consistent, are also generally encompassed within the new ICRP Recommendations 2007. The earlier version of the protection system, limiting *effective dose equivalent* to an individual, is generally employed in the U.S. Some discussion and comparison of the two systems, which both adhere to the ALARA principle ("as low as reasonable achievable"), has been added in the present text. As a practical matter, both maintain a comparable degree of protection in operating experience.

It will be some time until the new model revisions and other recent work of the ICRP become fully integrated into unified general protocols for internal dosimetry. While there has been partial updating at this time, much of the formalism of ICRP Publication 30 remains in current use at the operating levels of health physics in many places. After some thought, this formalism continues to be the primary focus in Chapter 16 on Internal Dosimetry and Radiation Protection. To a considerable extent, the newer ICRP Publications follow the established format. They are described here in the text where appropriate, and their relationships to Publication 30 are discussed.

As evident from acknowledgements made throughout the book, I am indebted to many sources for material used in this third edition. I would like to express my gratitude particularly to the following persons for help during its preparation: M. I. Al-Jarallah, James S. Bogard, Rhonda S. Bogard, Wesley. E. Bolch, Roger J. Cloutier, Darryl J. Downing, Keith F. Eckerman, Joseph D. Eddlemon, Paul W. Frame, Peter Jacob, Cynthia G. Jones, Herwig G. Paretzke, Charles A. Potter, Robert C. Ricks, Joseph Rotunda, Richard E. Toohey, and Vaclav Vylet. Their interest and contributions are much appreciated. I would also like to thank the staff of John Wiley & Sons, particularly Esther Dörring, Anja Tschörtner, and Dagmar Kleemann, for their patience, understanding, and superb work during the production of this volume.

*Oak Ridge, Tennessee*                                                             *James E. Turner*
March 21, 2007

# 1
# About Atomic Physics and Radiation

## 1.1
## Classical Physics

As the nineteenth century drew to a close, man's physical understanding of the world appeared to rest on firm foundations. Newton's three laws accounted for the motion of objects as they exerted forces on one another, exchanging energy and momentum. The movements of the moon, planets, and other celestial bodies were explained by Newton's gravitation law. Classical mechanics was then over 200 years old, and experience showed that it worked well.

Early in the century Dalton's ideas revealed the atomic nature of matter, and in the 1860s Mendeleev proposed the periodic system of the chemical elements. The seemingly endless variety of matter in the world was reduced conceptually to the existence of a finite number of chemical elements, each consisting of identical smallest units, called atoms. Each element emitted and absorbed its own characteristic light, which could be analyzed in a spectrometer as a precise signature of the element.

Maxwell proposed a set of differential equations that explained known electric and magnetic phenomena and also predicted that an accelerated electric charge would radiate energy. In 1888 such radiated electromagnetic waves were generated and detected by Hertz, beautifully confirming Maxwell's theory.

In short, near the end of the nineteenth century man's insight into the nature of space, time, matter, and energy seemed to be fundamentally correct. While much exciting research in physics continued, the basic laws of the universe were generally considered to be known. Not many voices forecasted the complete upheaval in physics that would transform our perception of the universe into something undreamed of as the twentieth century began to unfold.

## 1.2
## Discovery of X Rays

The totally unexpected discovery of X rays by Roentgen on November 8, 1895 in Wuerzburg, Germany, is a convenient point to regard as marking the beginning of

*Atoms, Radiation, and Radiation Protection*. James E. Turner
Copyright © 2007 WILEY-VCH Verlag GmbH & Co. KGaA, Weinheim
ISBN: 978-3-527-40606-7

CATHODE

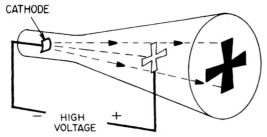

HIGH VOLTAGE

**Fig. 1.1** Schematic diagram of an early Crooke's, or cathode-ray, tube. A Maltese cross of mica placed in the path of the rays casts a shadow on the phosphorescent end of the tube.

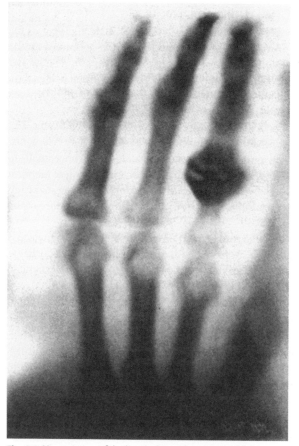

**Fig. 1.2** X-ray picture of the hand of Frau Roentgen made by Roentgen on December 22, 1895, and now on display at the Deutsches Museum. (Figure courtesy of Deutsches Museum, Munich, Germany.)

the story of ionizing radiation in modern physics. Roentgen was conducting experiments with a Crooke's tube—an evacuated glass enclosure, similar to a television picture tube, in which an electric current can be passed from one electrode to another through a high vacuum (Fig. 1.1). The current, which emanated from the cathode and was given the name cathode rays, was regarded by Crooke as a fourth state of matter. When the Crooke's tube was operated, fluorescence was excited in the residual gas inside and in the glass walls of the tube itself.

It was this fluorescence that Roentgen was studying when he made his discovery. By chance, he noticed in a darkened room that a small screen he was using fluoresced when the tube was turned on, even though it was some distance away. He soon recognized that he had discovered some previously unknown agent, to which he gave the name X rays.[1] Within a few days of intense work, Roentgen had observed the basic properties of X rays—their penetrating power in light materials such as paper and wood, their stronger absorption by aluminum and tin foil, and their differential absorption in equal thicknesses of glass that contained different amounts of lead. Figure 1.2 shows a picture that Roentgen made of a hand on December 22, 1895, contrasting the different degrees of absorption in soft tissue and bone. Roentgen demonstrated that, unlike cathode rays, X rays are not deflected by a magnetic field. He also found that the rays affect photographic plates and cause a charged electroscope to lose its charge. Unexplained by Roentgen, the latter phenomenon is due to the ability of X rays to ionize air molecules, leading to the neutralization of the electroscope's charge. He had discovered the first example of ionizing radiation.

## 1.3
## Some Important Dates in Atomic and Radiation Physics

Events moved rapidly following Roentgen's communication of his discovery and subsequent findings to the Physical–Medical Society at Wuerzburg in December 1895. In France, Becquerel studied a number of fluorescent and phosphorescent materials to see whether they might give rise to Roentgen's radiation, but to no avail. Using photographic plates and examining salts of uranium among other substances, he found that a strong penetrating radiation was given off, independently of whether the salt phosphoresced. The source of the radiation was the uranium metal itself. The radiation was emitted spontaneously in apparently undiminishing intensity and, like X rays, could also discharge an electroscope. Becquerel announced the discovery of radioactivity to the Academy of Sciences at Paris in February 1896.

1   That discovery favors the prepared mind is exemplified in the case of X rays. Several persons who noticed the fading of photographic film in the vicinity of a Crooke's tube either considered the film to be defective or sought other storage areas. An interesting account of the discovery and near-discoveries of X rays as well as the early history of radiation is given in the article by R. L. Kathren cited under "Suggested Reading" in Section 1.6.

The following tabulation highlights some of the important historical markers in the development of modern atomic and radiation physics.

1810    Dalton's atomic theory.

1859    Bunsen and Kirchhoff originate spectroscopy.

1869    Mendeleev's periodic system of the elements.

1873    Maxwell's theory of electromagnetic radiation.

1888    Hertz generates and detects electromagnetic waves.

1895    Lorentz theory of the electron.

1895    Roentgen discovers X rays.

1896    Becquerel discovers radioactivity.

1897    Thomson measures charge-to-mass ratio of cathode rays (electrons).

1898    Curies isolate polonium and radium.

1899    Rutherford finds two kinds of radiation, which he names "alpha" and "beta," emitted from uranium.

1900    Villard discovers gamma rays, emitted from radium.

1900    Thomson's "plum pudding" model of the atom.

1900    Planck's constant, $h = 6.63 \times 10^{-34}$ J s.

1901    First Nobel prize in physics awarded to Roentgen.

1902    Curies obtain 0.1 g pure $RaCl_2$ from several tons of pitchblend.

1905    Einstein's special theory of relativity ($E = mc^2$).

1905    Einstein's explanation of photoelectric effect, introducing light quanta (photons of energy $E = h\nu$).

1909    Millikan's oil drop experiment, yielding precise value of electronic charge, $e = 1.60 \times 10^{-19}$ C.

1910    Soddy establishes existence of isotopes.

1911    Rutherford discovers atomic nucleus.

1911    Wilson cloud chamber.

1912    von Laue demonstrates interference (wave nature) of X rays.

1912    Hess discovers cosmic rays.

1913    Bohr's theory of the H atom.

1913    Coolidge X-ray tube.

1914    Franck–Hertz experiment demonstrates discrete atomic energy levels in collisions with electrons.

1917    Rutherford produces first artificial nuclear transformation.

1922    Compton effect.

1924    de Broglie particle wavelength, $\lambda = h/$momentum.

1925    Uhlenbeck and Goudsmit ascribe electron with intrinsic spin $\hbar/2$.

1925    Pauli exclusion principle.

1925    Heisenberg's first paper on quantum mechanics.

1926    Schroedinger's wave mechanics.

1927    Heisenberg uncertainty principle.

1927    Mueller discovers that ionizing radiation produces genetic mutations.

1927    Birth of quantum electrodynamics, Dirac's paper on "The Quantum Theory of the Emission and Absorption of Radiation."

1928    Dirac's relativistic wave equation of the electron.

1930   Bethe quantum-mechanical stopping-power theory.
1930   Lawrence invents cyclotron.
1932   Anderson discovers positron.
1932   Chadwick discovers neutron.
1934   Joliot-Curie and Joliot produce artificial radioisotopes.
1935   Yukawa predicts the existence of mesons, responsible for short-range nuclear force.
1936   Gray's formalization of Bragg-Gray principle.
1937   Mesons found in cosmic radiation.
1938   Hahn and Strassmann observe nuclear fission.
1942   First man-made nuclear chain reaction, under Fermi's direction at University of Chicago.
1945   First atomic bomb.
1948   Transistor invented by Shockley, Bardeen, and Brattain.
1952   Explosion of first fusion device (hydrogen bomb).
1956   Discovery of nonconservation of parity by Lee and Yang.
1956   Reines and Cowen experimentally detect the neutrino.
1958   Discovery of Van Allen radiation belts.
1960   First successful laser.
1964   Gell-Mann and Zweig independently introduce quark model.
1965   Tomonaga, Schwinger, and Feynman receive Nobel Prize for fundamental work on quantum electrodynamics.
1967   Salam and Weinberg independently propose theories that unify weak and electromagnetic interactions.
1972   First beam of 200-GeV protons at Fermilab.
1978   Penzias and Wilson awarded Nobel Prize for 1965 discovery of 2.7 K microwave radiation permeating space, presumably remnant of "big bang" some 10–20 billion years ago.
1981   270 GeV proton–antiproton colliding-beam experiment at European Organization for Nuclear Research (CERN); 540 GeV center-of-mass energy equivalent to laboratory energy of 150,000 GeV.
1983   Electron–positron collisions show continuing validity of radiation theory up to energy exchanges of 100 GeV and more.
1984   Rubbia and van der Meer share Nobel Prize for discovery of field quanta for weak interaction.
1994   Brockhouse and Shull receive Nobel Prize for development of neutron spectroscopy and neutron diffraction.
2001   Cornell, Ketterle, and Wieman awarded Nobel Prize for Bose-Einstein condensation in dilute gases for alkali atoms.
2002   Antihydrogen atoms produced and measured at CERN.
2004   Nobel Prize presented to Gross, Politzer, and Wilczek for discovery of asymptotic freedom in development of quantum chromodynamics as the theory of the strong nuclear force.
2005   World Year of Physics 2005, commemorates Einstein's pioneering contributions of 1905 to relativity, Brownian motion, and the photoelectric effect (for which he won the Nobel Prize).

Figures 1.3 through 1.5 show how the complexity and size of particle accelerators have grown. Lawrence's first cyclotron (1930) measured just 4 in. in diameter. With it he produced an 80-keV beam of protons. The Fermi National Accelerator Laboratory (Fermilab) is large enough to accommodate a herd of buffalo and other wildlife on its grounds. The LEP (large electron-positron) storage ring at the European Organization for Nuclear Research (CERN) on the border between Switzerland and France, near Geneva, has a diameter of 8.6 km. The ring allowed electrons and positrons, circulating in opposite directions, to collide at very high energies for the study of elementary particles and forces in nature. The large size of the ring was needed to reduce the energy emitted as synchrotron radiation by the charged particles as they followed the circular trajectory. The energy loss per turn was made up by an accelerator system in the ring structure. The LEP was recently retired, and the tunnel is being used for the construction of the Large Hadron Collider (LHC), scheduled for completion in 2007. The LHC will collide head-on two beams of 7-TeV protons or other heavy ions.

In Lawrence's day experimental equipment was usually put together by the individual researcher, possibly with the help of one or two associates. The huge machines of today require hundreds of technically trained persons to operate. Earlier radiation-protection practices were much less formalized than today, with little public involvement.

**Fig. 1.3** E. O. Lawrence with his first cyclotron. (Photo by Watson Davis, Science Service; figure courtesy of American Institute of Physics Niels Bohr Library. Reprinted with permission from *Physics Today*, November 1981, p. 15. Copyright 1981 by the American Institute of Physics.)

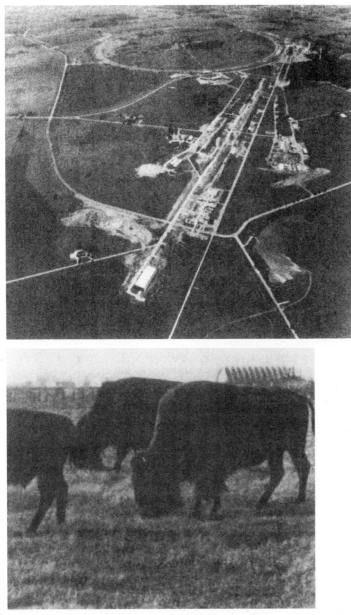

**Fig. 1.4** Fermi National Accelerator Laboratory, Batavia, Illinois. Buffalo and other wildlife live on the 6800 acre site. The 1000 GeV proton synchrotron (Tevatron) began operation in the late 1980s. (Figure courtesy of Fermi National Accelerator Laboratory. Reprinted with permission from *Physics Today*, November 1981, p. 23. Copyright 1981 by the American Institute of Physics.)

**Fig. 1.5** Photograph showing location of underground LEP ring with its 27 km circumference. The SPS (super proton synchrotron) is comparable to Fermilab. Geneva airport is in foreground. [Figure courtesy of the European Organization for Nuclear Research (CERN).]

## 1.4
### Important Dates in Radiation Protection

X rays quickly came into widespread medical use following their discovery. Although it was not immediately clear that large or repeated exposures might be harmful, mounting evidence during the first few years showed unequivocally that they could be. Reports of skin burns among X-ray dispensers and patients, for example, became common. Recognition of the need for measures and devices to protect patients and operators from unnecessary exposure represented the beginning of radiation health protection.

Early criteria for limiting exposures both to X rays and to radiation from radioactive sources were proposed by a number of individuals and groups. In time, organizations were founded to consider radiation problems and issue formal recommendations. Today, on the international scene, this role is fulfilled by the International Commission on Radiological Protection (ICRP) and, in the United States, by the National Council on Radiation Protection and Measurements (NCRP). The International Commission on Radiation Units and Measurements (ICRU) recommends radiation quantities and units, suitable measuring procedures, and numerical values for the physical data required. These organizations act as independent bodies composed of specialists in a number of disciplines—physics, medicine, biology, dosimetry, instrumentation, administration, and so forth. They are not government affiliated and they have no legal authority to impose their recommendations. The NCRP today is a nonprofit corporation chartered by the United States Congress.

Some important dates and events in the history of radiation protection follow.

1895    Roentgen discovers ionizing radiation.
1900    American Roentgen Ray Society (ARRS) founded.
1915    British Roentgen Society adopts X-ray protection resolution; believed to be the first organized step toward radiation protection.
1920    ARRS establishes standing committee for radiation protection.
1921    British X-Ray and Radium Protection Committee presents its first radiation protection rules.
1922    ARRS adopts British rules.
1922    American Registry of X-Ray Technicians founded.
1925    Mutscheller's "tolerance dose" for X rays.
1925    First International Congress of Radiology, London, establishes ICRU.
1928    ICRP established under auspices of the Second International Congress of Radiology, Stockholm.
1928    ICRU adopts the roentgen as unit of exposure.
1929    Advisory Committee on X-Ray and Radium Protection (ACXRP) formed in United States (forerunner of NCRP).
1931    The roentgen adopted as unit of X radiation.
1931    ACXRP publishes recommendations (*National Bureau of Standards Handbook* 15).
1934    ICRP recommends daily tolerance dose.
1941    ACXRP recommends first permissible body burden, for radium.
1942    Manhattan District begins to develop atomic bomb; beginning of health physics as a profession.
1946    U.S. Atomic Energy Commission created.
1946    NCRP formed as outgrowth of ACXRP.
1947    U.S. National Academy of Sciences establishes Atomic Bomb Casualty Commission (ABCC) to initiate long-term studies of A-bomb survivors in Hiroshima and Nagasaki.
1949    NCRP publishes recommendations and introduces risk/benefit concept.
1952    Radiation Research Society formed.
1953    ICRU introduces concept of absorbed dose.
1955    United Nations Scientific Committee on the Effects of Atomic Radiation (UNSCEAR) established.
1956    Health Physics Society founded.
1956    International Atomic Energy Agency organized under United Nations.
1957    NCRP introduces age proration for occupational doses and recommends nonoccupational exposure limits.
1957    U.S. Congressional Joint Committee on Atomic Energy begins series of hearings on radiation hazards, beginning with "The Nature of Radioactive Fallout and Its Effects on Man."

1958    United Nations Scientific Committee on the Effects of Atomic Radiation publishes study of exposure sources and biological hazards (first UNSCEAR Report).

1958    Society of Nuclear Medicine formed.

1959    ICRP recommends limitation of genetically significant dose to population.

1960    U.S. Congressional Joint Committee on Atomic Energy holds hearings on "Radiation Protection Criteria and Standards: Their Basis and Use."

1960    American Association of Physicists in Medicine formed.

1960    American Board of Health Physics begins certification of health physicists.

1964    International Radiation Protection Association (IRPA) formed.

1964    Act of Congress incorporates NCRP.

1969    Radiation in space. Man lands on moon.

1974    U.S. Nuclear Regulatory Commission (NRC) established.

1974    ICRP Publication 23, "Report of Task Group on Reference Man."

1975    ABCC replaced by binational Radiation Effects Research Foundation (RERF) to continue studies of Japanese survivors.

1977    ICRP Publication 26, "Recommendations of the ICRP."

1977    U.S. Department of Energy (DOE) created.

1978    ICRP Publication 30, "Limits for Intakes of Radionuclides by Workers."

1978    ICRP adopts "effective dose equivalent" terminology.

1986    Dosimetry System 1986 (DS86) developed by RERF for A-bomb survivors.

1986    Growing public concern over radon. U.S. Environmental Protection Agency publishes pamphlet, "A Citizen's Guide to Radon."

1987    NCRP Report No. 91, "Recommendations on Limits for Exposure to Ionizing Radiation."

1988    United Nations Scientific Committee on the Effects of Atomic Radiation, "Sources, Effects and Risks of Ionizing Radiation." Report to the General Assembly.

1988    U.S. National Academy of Sciences BEIR IV Report, "Health Risks of Radon and Other Internally Deposited Alpha Emitters—BEIR IV."

1990    U.S. National Academy of Sciences BEIR V Report, "Health Effects of Exposure to Low Levels of Ionizing Radiation—BEIR V."

1991    International Atomic Energy Agency report on health effects from the April 1986 Chernobyl accident.

1991    10 CFR Part 20, NRC.

1991    ICRP Publication 60, "1990 Recommendations of the International Commission on Radiological Protection."

1993    10 CFR Part 835, DOE.

1993    NCRP Report No. 115, "Risk Estimates for Radiation Protection."

1993    NCRP Report No. 116, "Limitation of Exposure to Ionizing Radiation."

1994    Protocols developed for joint U.S., Ukraine, Belarus 20-y study of thyroid disease in 85,000 children exposed to radioiodine following Chernobyl accident in 1986.

1994    ICRP Publication 66, "Human Respiratory Tract Model for Radiological Protection."

2000  UNSCEAR 2000 Report on sources of radiation exposure, radiation-associated cancer, and the Chernobyl accident.

2003  Dosimetry System 2002 (DS02) formally approved.

2005  ICRP proposes system of radiological protection consisting of dose constraints and dose limits, complimented by optimization.

2007  Final decision expected. ICRP 2007 Recommendations.

## 1.5
## Sources and Levels of Radiation Exposure

The United Nations Scientific Committee on the Effects of Atomic Radiation (UNSCEAR) has carried out a comprehensive study and analysis of the presence and effects of ionizing radiation in today's world. The UNSCEAR 2000 Report (see "Suggested Reading" at the end of the chapter) presents a broad review of the various sources and levels of radiation exposure worldwide and an assessment of the radiological consequences of the 1986 Chernobyl reactor accident.

Table 1.1, based on information from the Report, summarizes the contributions that comprise the average annual effective dose of about 2.8 mSv (see Chapter 14) to an individual. They do not necessarily pertain to any particular person, but

**Table 1.1** Annual per Capita Effective Doses in Year 2000 from Natural and Man-Made Sources of Ionizing Radiation Worldwide[*]

| Source | Annual Effective Dose (mSv) | Typical Range (mSv) |
|---|---|---|
| Natural Background | | |
| External | | |
| Cosmic rays | 0.4 | 0.3–1.0 |
| Terrestrial gamma rays | 0.5 | 0.3–0.6 |
| Internal | | |
| Inhalation (principally radon) | 1.2 | 0.2–10. |
| Ingestion | 0.3 | 0.2–0.8 |
| Total | 2.4 | 1–10 |
| Medical (primarily diagnostic X rays) | 0.4 | 0.04–1.0 |
| Man-Made Environmental | | |
| Atmospheric nuclear-weapons tests | 0.005 | Peak was 0.15 in 1963. |
| Chernobyl accident | 0.002 | Highest average was 0.04 in northern hemisphere in 1986. |
| Nuclear power production | 0.0002 | See paragraph 34 in Report for basis of estimate. |

[*] Based on UNSCEAR 2000 Report.

reflect averages from ranges given in the last column. Natural background radiation contributes the largest portion (~85%), followed by medical (~14%), and then man-made environmental (<1%). As noted in the table, background can vary greatly from place to place, due to amounts of radioactive minerals in soil, water, and rocks and to increased cosmic radiation at higher altitudes. Radon contributes roughly one-half of the average annual effective dose from natural background. Medical uses of radiation, particularly diagnostic X rays, result in the largest average annual effective dose from man-made sources. Depending on the level of healthcare, however, the average annual medical dose is very small in many parts of the world. The last three sources in Table 1.1 represent the relatively small contributions from man-made environmental radiation. Of all man's activities, atmospheric nuclear-weapons testing has resulted in the largest releases of radionuclides into the environment. According to the UNSCEAR Report, the annual effective dose from this source at its maximum in 1963 was about 7% as large as natural background. The Report also includes an analysis of occupational radiation exposures.

## 1.6
## Suggested Reading

1 Cropper, William H., *Great Physicists*, Oxford University Press, Oxford (2001). [Portrays the lives, personalities, and contributions of 29 scientists from Galileo to Stephen Hawkin.]

2 Glasstone, S., *Sourcebook on Atomic Energy*, 3d ed., D. Van Nostrand, Princeton, NJ (1967).

3 Kathren, R. L., "Historical Development of Radiation Measurement and Protection," pp. 13–52 in *Handbook of Radiation Protection and Measurement*, Section A, Vol. I, A. B. Brodsky, ed., CRC Press, Boca Raton, FL (1978). [An interesting and readable account of important discoveries and experience with radiation exposures, measurements, and protection. Contains bibliography.]

4 Kathren, R. L., and Ziemer, P. L., eds., *Health Physics: A Backward Glance*, Pergamon Press, Elmsford, NY (1980). [Thirteen original papers on the history of radiation protection.]

5 Meinhold, Charles B., "Lauriston S. Taylor Lecture: The Evolution of Radiation protection—from Erythema

to Genetic Risks to Risks of Cancer to . . .?," *Health Phys.* **87**, 241–248 (2004). [President Emeritus of the NCRP describes the evolution of radiation protection through the present-day ICRP, NCRP, and other organizations. This issue (Vol. 87, No. 3) contains the proceedings of the 2003 annual meeting of the NCRP, on the subject of radiation protection at the beginning of the 21st century.]

6 Moeller, Dade W., "Environmental Health Physics—50 Years of Progress," *Health Phys.* **87**, 337–357 (2004). [Review article, discussing sources of environmental radiation and the transport and monitoring of radioactive materials in the biosphere. Extensive bibliography.]

7 Morgan, K. Z., "History of Damage and Protection from Ionizing Radiation," Chapter 1 in *Principles of Radiation Protection*, K. Z. Morgan and J. E. Turner, eds., Wiley, New York (1967). [Morgan is one of the original eight health physicists of the Manhattan Project at the University of Chicago

(1942) and the first president of the Health Physics Society.]

8  National Research Council, *Health Effects of Exposures to Low Levels of Ionizing Radiation—BEIR V*, National Academy Press, Washington, DC (1990).

9  NCRP Report No. 93, *Ionizing Radiation Exposure of the Population of the United States*, National Council on Radiation Protection and Measurements, Bethesda, MD (1987).

10  Pais, Abraham, *Inward Bound*, Oxford University Press, Oxford (1986). [Subtitled *Of Matter and Forces in the Physical World*, this is a very readable account of what happened between 1895 and 1983 and the persons and personalities that played a role during that time.]

11  *Physics Today*, Vol. 34, No. 11 (Nov. 1981). [Fiftieth anniversary of the American Institute of Physics. Special issue devoted to "50 Years of Physics in America."]

12  *Physics Today*, Vol. 36, No. 7 (July 1983). [This issue features articles on physics in medicine to commemorate the twenty-fifth anniversary of the founding of the American Association of Physicists in Medicine.]

13  Ryan, Michael T., "Happy 100th Birthday to Dr. Lauriston S. Taylor," *Health Phys.* **82**, 773 (2002). [The many contributions of Taylor (1902–2004), the first President of the NCRP, are honored in this issue (Vol. 82, No. 6) of the journal.]

14  Segrè, Emilio, *From X-Rays to Quarks*, W. H. Freeman, San Francisco (1980).

[Describes physicists and their discoveries from 1895 to the present. Segrè received the Nobel Prize for the discovery of the antiproton.]

15  Stannard, J. N., *Radioactivity and Health*, National Technical Information Service, Springfield, VA (1988). [A comprehensive, detailed history (1963 pp.) of the age.]

16  Taylor, L. S., *Radiation Protection Standards*, CRC Press, Boca Raton, FL (1971). [The history of radiation protection as written by one of its leading international participants.]

17  Taylor, L. S., "Who Is the Father of Health Physics?" *Health Phys.* **42**, 91 (1982).

18  United Nations Scientific Committee on the Effects of Atomic Radiation, UNSCEAR 2000 Report to the General Assembly, with scientific annexes, Vol. I Sources, Vol. II Effects, United Nations Publications, New York, NY and Geneva, Switzerland (2000).

19  Weart, Spencer R. and Phillips, Melba, Eds., *History of Physics*, American Institute of Physics, New York, NY (1985). [Forty-seven articles of historical significance are reprinted from *Physics Today*. Included are personal accounts of scientific discoveries and developments in modern physics. One section, devoted to social issues in physics, deals with effects of the great depression in the 1930s, science and secrecy, development of the atomic bomb in World War II, federal funding, women in physics, and other subjects.]

The following Internet sources are available:

www.hps.org

www.icrp.org

www.icru.org

www.ncrponline.org

www.nobelprize.org/physics/laureates

# 2
# Atomic Structure and Atomic Radiation

## 2.1
## The Atomic Nature of Matter (ca. 1900)

The work of John Dalton in the early nineteenth century laid the foundation for modern analytic chemistry. Dalton formulated and interpreted the laws of definite, multiple, and equivalent proportions, based on the existence of identical atoms as the smallest indivisible unit of a chemical element. The law of definite proportions states that in every sample of a chemical compound, the proportion by mass or weight of the constituent elements is always the same. When two elements combine to form more than one compound, the law of multiple proportions says that the proportions by mass of the different elements are always in simple ratios to one another. When two elements react completely with a third, then the ratio of the masses of the two is the same, regardless of what the third element is, a fact expressed by the law of equivalent proportions. Dalton also assumed a rule of greatest simplicity—that elements forming only a single compound do so by means of a simple one-to-one combination of atoms. This rule does not always hold.

These ideas were supported by the work of Dalton's contemporary, Gay-Lussac, on the law of combining volumes of gases. This law states that the volumes of gases that enter into chemical combination with one another are in the ratio of simple whole numbers when all volumes are measured under the same conditions of pressure and temperature. Avogadro hypothesized that equal volumes of any gases at the same pressure and temperature contain the same number of molecules. Avogadro also suggested that the molecules of some gaseous elements could be composed of two or more atoms of that element.

Today we recognize that a gram atomic weight of any element contains Avogadro's number, $N_0 = 6.022 \times 10^{23}$, of atoms.[1] Furthermore, a gram molecular weight of any gas also contains $N_0$ molecules and occupies a volume of 22.41 L (liters) at standard temperature and pressure [STP, 0°C (=273 K on the absolute temperature scale) and 760 torr (1 torr = 1 mm Hg)]. The modem scale of atomic and molecular weights is set by stipulating that the gram atomic weight of the carbon isotope, $^{12}C$, is exactly 12.000... g. A periodic chart, giving atomic numbers,

1    See Appendices A and B for physical
     constants, units, and conversion factors.

*Atoms, Radiation, and Radiation Protection.* James E. Turner
Copyright © 2007 WILEY-VCH Verlag GmbH & Co. KGaA, Weinheim
ISBN: 978-3-527-40606-7

atomic weights, densities, and other information about the chemical elements, is shown in the back of this book.

*Example*

How many grams of oxygen combine with 2.3 g of carbon in the reaction $C + O_2 \rightarrow CO_2$? How many molecules of $CO_2$ are thus formed? How many liters of $CO_2$ are formed at 20°C and 752 torr?

*Solution*

In the given reaction, 1 atom of carbon combines with one molecule (2 atoms) of oxygen. From the atomic weights given on the periodic chart in the back of the book, it follows that 12.011 g of carbon reacts with $2 \times 15.9994 = 31.9988$ g of oxygen. Rounding off to three significant figures, letting $y$ represent the number of grams of oxygen asked for, and taking simple proportions, we have $y = (2.3/12.0) \times 32.0 = 6.13$ g. The number $N$ of molecules of $CO_2$ formed is equal to the number of atoms in 2.3 g of C, which is 2.3/12.0 times Avogadro's number: $N = (2.3/12.0) \times 6.02 \times 10^{23} = 1.15 \times 10^{23}$. Since Avogadro's number of molecules occupies 22.4 L at STP, the volume of $CO_2$ at STP is $(1.15 \times 10^{23}/6.02 \times 10^{23}) \times 22.4 = 4.28$ L. At the given higher temperature of 20°C = 293 K, the volume is larger by the ratio of the absolute temperatures, 293/273; the volume is also increased by the ratio of the pressures, 760/752. Therefore, the volume of $CO_2$ made from 2.3 g of C at 20°C and 752 torr is 4.28 (293/273) (760/752) = 4.64 L. This would also be the volume of oxygen consumed in the reaction under the same conditions of temperature and pressure, since 1 molecule of oxygen is used to form 1 molecule of carbon dioxide.

As mentioned in Chapter 1, mid-nineteenth century scientists could analyze light to identify the elements present in its source. Light entering an optical spectrometer is collimated by a lens and slit system, through which it is then directed toward an analyzer (e.g., a diffraction grating or prism). The analyzer disperses the light, changing its direction by an amount that depends on its wavelength. White light, for example, is spread out into the familiar rainbow of colors. Light that is dispersed at various angles with respect to the incident direction can be seen with the eye, photographed, or recorded electronically. Light from a single chemical element is observed as a series of discrete line images of the entrance slit that emerge at various angles from the analyzer. The spectrometer can be calibrated so that the angles at which the lines occur give the wavelengths of the light that appears there. Each chemical element produces its own unique, characteristic series of lines which identify it. The series is referred to as the optical, or line, spectrum of the element, or simply as the spectrum. When a number of elements are present in a light source, their spectra appear superimposed in the spectrometer, and the individual elemental spectra can be sorted out. Elements absorb light of the same wavelengths they emit.

Figure 2.1 shows the lines in the visible and near-ultraviolet spectrum of atomic hydrogen. [The wavelength of visible light is between about 4000 Å (violet) and 7500 Å (red).] In 1885 Balmer published an empirical formula that gives these

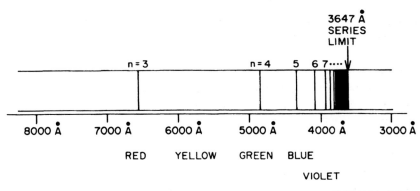

**Fig. 2.1** Balmer series of lines in the spectrum of atomic hydrogen.

observed wavelengths, $\lambda$, in the hydrogen spectrum. His formula is equivalent to the following:

$$\frac{1}{\lambda} = R_\infty \left( \frac{1}{2^2} - \frac{1}{n^2} \right), \tag{2.1}$$

where $R_\infty = 1.09737 \times 10^7$ m$^{-1}$ is called the Rydberg constant and $n = 3, 4, 5, \ldots$ represents any integer greater than 2. When $n = 3$, the formula gives $\lambda = 6562$ Å; when $n = 4$, $\lambda = 4861$ Å; and so on. The series of lines, which continue to get closer together as $n$ increases, converges to the limit $\lambda = 3647$ Å in the ultraviolet as $n \to \infty$. Balmer correctly speculated that other series might exist for hydrogen, which could be described by replacing the $2^2$ in Eq. (2.1) by the square of other integers. These other series, however, lie entirely in the ultraviolet or infrared portions of the electromagnetic spectrum. We shall see in Section 2.3 how the Balmer formula (2.1) was derived theoretically by Bohr in 1913.

As mentioned in Section 1.3, J. J. Thomson in 1897 measured the charge-to-mass ratio of cathode rays, which marked the experimental "discovery" of the electron as a particle of matter. The value he found for the ratio was about 1700 times that associated with the hydrogen atom in electrolysis. One concluded that the electron was less massive than the hydrogen atom by this factor. Thomson pictured atoms as containing a large number of the negatively charged electrons in a positively charged matrix filling the volume of the electrically neutral atom. When a gas was ionized by radiation, some electrons were knocked out of the atoms in the gas molecules, leaving behind positive ions of much greater mass. Thomson's concept of the structure of the atom is sometimes referred to as the "plum pudding" model.

## 2.2
## The Rutherford Nuclear Atom

The existence of alpha, beta, and gamma rays was known by 1900. With the discovery of these different kinds of radiation came their use as probes to study the structure of matter itself.

Rutherford and his students, Geiger and Marsden, investigated the penetration of alpha particles through matter. Because the range of these particles is small, an energetic source and thin layers of material were employed. In one set of experiments, 7.69-MeV collimated alpha particles from $^{214}_{84}$Po (RaC′) were directed at a $6 \times 10^{-5}$ cm thick gold foil. The relative number of particles leaving the foil at various angles with respect to the incident beam could be observed through a microscope on a scintillation screen. While most of the alpha particles passed through the foil with only slight deviation from their original direction, an occasional particle was scattered through a large angle, even backwards from the foil. About 1 in 8000 was deflected more than 90°. An enormously strong electric or magnetic field would be required to reverse the direction of the fast and relatively massive alpha particle. (In 1909 Rutherford conclusively established that alpha particles are doubly charged helium ions.) "It was about as credible as if you had fired a 15-in. shell at a piece of tissue paper and it came back and hit you," said Rutherford of this surprising discovery. He reasoned that the large-angle deflection of some alpha particles was evidence for the existence of a very small and massive nucleus, which was also the seat of the positive charge of an atom. The rare scattering of an alpha particle through a large angle could then be explained by the large repulsive force it experienced when it approached the tiny nucleus of a single atom almost head-on. Furthermore, the light electrons in an atom must move rapidly about the nucleus, filling the volume occupied by the atom. Indeed, atoms must be mostly empty space, allowing the majority of alpha particles to pass right through a foil with little or no scattering. Following these ideas, Rutherford calculated the distribution of scattering angles for the alpha particles and obtained quantitative agreement with the experimental data. In contrast to the plum pudding model. Rutherford's atom is sometimes called a planetary model, in analogy with the solar system.

Today we know that the radius of the nucleus of an atom of atomic mass number $A$ is given approximately by the formula

$$R \cong 1.3A^{1/3} \times 10^{-15} \text{ m.} \tag{2.2}$$

The radius of the gold nucleus is $1.3(197)^{1/3} \times 10^{-15} = 7.56 \times 10^{-15}$ m. The atomic radius of gold is $1.79 \times 10^{-10}$ m. The ratio of the two radii is $(7.56 \times 10^{-15}/1.79 \times 10^{-10}) = 4.22 \times 10^{-5}$. In physical extent, the massive nucleus is only a tiny speck at the center of the atom.

Nuclear size increases with atomic mass number $A$. Equation (2.2) indicates that the nuclear volume is proportional to $A$. The so-called strong, or nuclear, forces[2] that hold nucleons (protons and neutrons) together in the nucleus have short ranges ($\sim 10^{-15}$ m). Nuclear forces saturate; that is, a given nucleon interacts with only a few others. As a result, nuclear size is increased in proportion as more and more nucleons are merged to form heavier atoms. The size of all atoms, in contrast, is more or less the same. All electrons in an atom, no matter how many, are attracted to the nucleus and repelled by each other. Electric forces do not saturate—all pairs of charges interact with one another.

## 2.3
### Bohr's Theory of the Hydrogen Atom

An object that does not move uniformly in a straight line is accelerated, and an accelerated charge emits electromagnetic radiation. In view of these laws of classical physics, it was not understood how Rutherford's planetary atom could be stable. Electrons orbiting about the nucleus should lose energy by radiation and spiral into the nucleus.

In 1913 Bohr put forward a bold new hypothesis, at variance with classical laws, to explain atomic structure. His theory gave correct predictions for the observed spectra of the H atom and single-electron atomic ions, such as $He^+$, but gave wrong answers for other systems, such as He and $H_2^+$. The discovery of quantum mechanics in 1925 and its subsequent development has led to the modem mathematical theory of atomic and molecular structure. Although it proved to be inadequate, Bohr's theory gives useful insight into the quantum nature of matter. We shall see that a number of properties of atoms and radiation can be understood from its basic concepts and their logical extensions.

Bohr assumed that an atomic electron moves without radiating only in certain discrete orbits about the nucleus. He further assumed that the transition of the electron from one orbit to another must be accompanied by the emission or absorption of a photon of light, the photon energy being equal to the orbital energy lost or gained by the electron. In principle, Bohr's ideas thus account for the existence of discrete optical spectra that characterize an atom and for the fact that an element emits and absorbs photons of the same wavelengths.

Bohr discovered that the proper electronic energy levels, yielding the observed spectra, were obtained by requiring that the angular momentum of the electron about the nucleus be an integral multiple of Planck's constant $h$ divided by $2\pi$ ($\hbar = h/2\pi$). (Classically, any value of angular momentum is permissible.) For an

---

2  The four fundamental forces in nature are (1) gravitational, (2) electromagnetic, (3) strong (nuclear), and (4) weak (responsible for beta decay). The attractive nuclear force is strong enough to overcome the mutual Coulomb repulsion of protons in the nucleus (Section 3.1). The electromagnetic and weak forces are now recognized as a single, unified force.

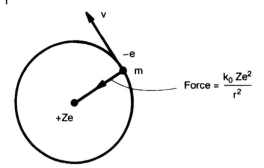

**Fig. 2.2** Schematic representation of electron (mass $m$, charge $-e$) in uniform circular motion (speed $v$, orbital radius $r$) about nucleus of charge $+Ze$.

electron of mass $m$ moving uniformly with speed $v$ in a circular orbit of radius $r$ (Fig. 2.2), we thus write

$$mvr = n\hbar, \tag{2.3}$$

where $n$ is a positive integer, called a quantum number ($n = 1, 2, 3, \ldots$). [Angular momentum, $mvr$, is defined in Appendix C; and $\hbar = 1.05457 \times 10^{-34}$ J s (Appendix A)]. If the electron changes from an initial orbit in which its energy is $E_i$ to a final orbit of lower energy $E_f$, then a photon of energy

$$h\nu = E_i - E_f \tag{2.4}$$

is emitted, where $\nu$ is the frequency of the photon. ($E_f > E_i$ if a photon is absorbed.) Equations (2.3) and (2.4) are two succinct statements that embody Bohr's ideas quantitatively. We now use them to derive the properties of single-electron atomic systems.

When an object moves with constant speed $v$ in a circle of radius $r$, it experiences an acceleration $v^2/r$, directed toward the center of the circle. By Newton's second law, the force on the object is $mv^2/r$, also directed toward the center (Problem 10). The force on the electron in Fig. 2.2 is supplied by the Coulomb attraction between the electronic and nuclear charges, $-e$ and $+Ze$. Therefore, we write for the equation of motion of the electron,

$$\frac{mv^2}{r} = \frac{k_0 Ze^2}{r^2}, \tag{2.5}$$

where $k_0 = 8.98755 \times 10^9$ N m$^2$ C$^{-2}$ (Appendix C). Solving for the radius gives

$$r = \frac{k_0 Ze^2}{mv^2}. \tag{2.6}$$

Solving Eq. (2.3) for $v$ and substituting into (2.6), we find for the radii $r_n$ of the allowed orbits

$$r_n = \frac{n^2 \hbar^2}{k_0 Ze^2 m}. \tag{2.7}$$

Substituting values of the constants from Appendix A, we obtain

$$r_n = \frac{n^2(1.05457 \times 10^{-34})^2}{(8.98755 \times 10^9 Z)(1.60218 \times 10^{-19})^2(9.10939 \times 10^{-31})}$$

$$= 5.29 \times 10^{-11}\frac{n^2}{Z}\text{ m.} \tag{2.8}$$

The innermost orbit ($n = 1$) in the hydrogen atom ($Z = 1$) thus has a radius of $5.29 \times 10^{-11}$ m $= 0.529$ Å, often referred to as the Bohr radius.

In similar fashion, eliminating $r$ between Eqs. (2.3) and (2.6) yields the orbital velocities

$$v_n = \frac{k_0 Z e^2}{n\hbar} = 2.19 \times 10^6 \frac{Z}{n}\text{ m s}^{-1}. \tag{2.9}$$

The velocity of the electron in the first Bohr orbit ($n = 1$) of hydrogen ($Z = 1$) is $2.19 \times 10^6$ m s$^{-1}$. In terms of the speed of light $c$, the quantity $v_1/c = k_0 e^2/\hbar c \cong 1/137$ is called the fine-structure constant. Usually denoted by $\alpha$, it determines the relativistic corrections to the Bohr energy levels, which give rise to a fine structure in the spectrum of hydrogen.

It follows that the kinetic and potential energies of the electron in the $n$th orbit are

$$KE_n = \frac{1}{2}mv_n^2 = \frac{k_0^2 Z^2 e^4 m}{2n^2\hbar^2} \tag{2.10}$$

and

$$PE_n = -\frac{k_0 Z e^2}{r_n} = -\frac{k_0^2 Z^2 e^4 m}{n^2\hbar^2}, \tag{2.11}$$

showing that the potential energy is twice as large in magnitude as the kinetic energy (virial theorem). The total energy of the electron in the $n$th orbit is therefore

$$E_n = KE_n + PE_n = -\frac{k_0^2 Z^2 e^4 m}{2n^2\hbar^2} = -\frac{13.6 Z^2}{n^2}\text{ eV.} \tag{2.12}$$

[The energy unit, electron volt (eV), given in Appendix B, is defined as the energy acquired by an electron in moving freely through a potential difference of 1 V: 1 eV $= 1.60 \times 10^{-19}$ J.] The lowest energy occurs when $n = 1$. For the H atom, this normal, or ground-state, energy is –13.6 eV; for He$^+$ ($Z = 2$) it is $-13.6 \times 4 = -54.4$ eV. The energy required to remove the electron from the ground state is called the ionization potential, which therefore is 13.6 eV for the H atom and 54.4 eV for the He$^+$ ion.

It remains to calculate the optical spectra for the single-electron systems based on Bohr's theory. Balmer's empirical formula (2.1) gives the wavelengths found in the visible spectrum of hydrogen. According to postulate (2.4), the energies of photons that can be emitted or absorbed are equal to the differences in the energy values given by Eq. (2.12). When the electron makes a transition from an initial orbit with quantum number $n_i$ to a final orbit of lower energy with quantum number $n_f$ (i.e., $n_i > n_f$), then from Eqs. (2.4) and (2.12) the energy of the emitted photon is

$$h\nu = \frac{hc}{\lambda} = \frac{k_0^2 Z^2 e^4 m}{2\hbar^2}\left(-\frac{1}{n_i^2} + \frac{1}{n_f^2}\right), \tag{2.13}$$

where $\lambda$ is the wavelength of the photon and $c$ is the speed of light. Substituting the numerical values[3] of the physical constants, one finds from Eq. (2.13) that

$$\frac{1}{\lambda} = 1.09737 \times 10^7 Z^2 \left(\frac{1}{n_f^2} - \frac{1}{n_i^2}\right)\, m^{-1}. \tag{2.14}$$

When $Z = 1$, the constant in front of the parentheses is equal to the Rydberg constant $R_\infty$ in Balmer's empirical formula (2.1). The integer 2 in the Balmer formula is interpretable from Bohr's theory as the quantum number of the orbit into which the electron falls when it emits the photon. Derivation of the Balmer formula and calculation of the Rydberg constant from the known values of $e$, $m$, $h$, and $c$ provided undeniable evidence for the validity of Bohr's postulates for single-electron atomic systems, although the postulates were totally foreign to classical physics.

Figure 2.3 shows a diagram of the energy levels of the hydrogen atom, calculated from Eq. (2.12), together with vertical lines that indicate the electron transitions that result in the emission of photons with the wavelengths shown. There are infinitely many orbits in which the electron has negative energy (bound states of the H atom). The orbital energies get closer together near the ionization threshold, 13.6 eV above the ground state. When an H atom becomes ionized, the electron is not bound and can have any positive energy. In addition to the Balmer series, Bohr's theory predicts other series, each corresponding to a different final-orbit quantum number $n_f$ and having an infinite number of lines. The set that results from transitions of electrons to the innermost orbit ($n_f = 1$, $n_i = 2, 3, 4, \ldots$) is called the Lyman series. The least energetic photon in this series has an energy

$$E = -13.6\left(\frac{1}{2^2} - \frac{1}{1^2}\right) = -13.6\left(-\frac{3}{4}\right) = 10.2\, eV, \tag{2.15}$$

as follows from Eqs. (2.4) and (2.12) with $Z = 1$. Its wavelength is 1216 Å. As $n_i$ increases, the Lyman lines get ever closer together, like those in the Balmer series, converging to the energy limit of 13.6 eV, the ionization potential of H. The photon wavelength at the Lyman series limit ($n_f = 1$, $n_i \to \infty$) is obtained from Eq. (2.14):

$$\frac{1}{\lambda} = 1.09737 \times 10^7 \left(\frac{1}{1^2} - \frac{1}{\infty^2}\right) = 1.09737 \times 10^7\, m^{-1}, \tag{2.16}$$

or $\lambda = 911$ Å. The Lyman series lies entirely in the ultraviolet region of the electromagnetic spectrum. The series with $n_f \geq 3$ lie in the infrared. The shortest wavelength in the Paschen series ($n_f = 3$) is given by $1/\lambda = (1.09737 \times 10^7)/9\, m^{-1}$, or $\lambda = 8.20 \times 10^{-7}\, m = 8200$ Å.

---

3   For high accuracy, the *reduced* mass of the
    electron must be used. See last paragraph in
    this section.

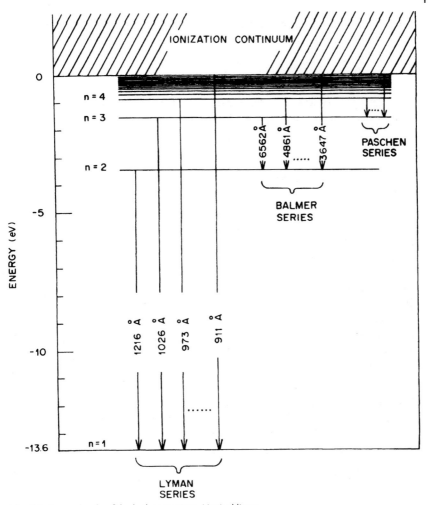

**Fig. 2.3** Energy levels of the hydrogen atom. Vertical lines represent transitions that the electron can make between various levels with the associated emitted photon wavelengths shown.

*Example*

Calculate the wavelength of the third line in the Balmer series in Fig. 2.1. What is the photon energy in eV?

*Solution*

We use Eq. (2.14) with $Z = 1$, $n_f = 2$, and $n_i = 5$:

$$\frac{1}{\lambda} = 1.09737 \times 10^7 \left( \frac{1}{4} - \frac{1}{25} \right) = 2.30448 \times 10^6 \text{ m}^{-1}. \tag{2.17}$$

Thus $\lambda = 4.34 \times 10^{-7}$ m $= 4340$ Å. A photon of this wavelength has an energy

$$E = \frac{hc}{\lambda} = \frac{6.63 \times 10^{-34} \times 3 \times 10^8}{4.34 \times 10^{-7}} = 4.58 \times 10^{-19} \text{ J,} \qquad (2.18)$$

or 2.86 eV. Alternatively, we can obtain the photon energy from Eq. (2.12). The energy levels involved in the electronic transition are $13.6/4 = 3.40$ eV and $13.6/25 = 0.544$ eV; their difference is 2.86 eV.

### Example

What is the largest quantum number of a state of the $Li^{2+}$ ion with an orbital radius less than 50 Å?

### Solution

The radii of the orbits are described by Eq. (2.8) with $Z = 3$. Setting $r_n = 50$ Å $= 5 \times 10^{-9}$ m and solving for $n$, we find that

$$n = \sqrt{\frac{r_n Z}{5.29 \times 10^{-11}}} = \sqrt{\frac{5 \times 10^{-9} \times 3}{5.29 \times 10^{-11}}} = 16.8. \qquad (2.19)$$

A nonintegral quantum number is not defined in the Bohr theory. Equation (2.19) tells us, though, that $r_n > 50$ Å when $n = 17$ and $r_n < 50$ Å when $n = 16$. Therefore, $n = 16$ is the desired answer.

### Example

Calculate the angular velocity of the electron in the ground state of $He^+$.

### Solution

With quantum number $n$, the angular velocity $\omega_n$ in radians s$^{-1}$ is equal to $2\pi f_n$, where $f_n$ is the frequency, or number of orbital revolutions of the electron about the nucleus per second. In general, $f_n = v_n/(2\pi r_n)$; and so $\omega_n = v_n/r_n$. With $n = 1$ and $Z = 2$, Eqs. (2.8) and (2.9) give $\omega_1 = v_1/r_1 = 1.66 \times 10^{17}$ s$^{-1}$, where the dimensionless angular unit, radian, is understood.

In deriving Eq. (2.14) it was tacitly assumed that an electron of mass $m$ orbits about a stationary nucleus. In reality, the electron and nucleus (mass $M$) orbit about their common center of mass. The energy levels are determined by the relative motion of the two, in which the effective mass is the reduced mass of the system (electron plus nucleus), given by

$$m_r = \frac{mM}{m+M}. \qquad (2.20)$$

For the hydrogen atom, $M = 1836m$, and so the reduced mass $m_r = 1836m/1837 = 0.9995m$ is nearly the same as the electron mass. The heavier the nucleus, the closer the reduced mass is to the electron mass. The symbol $R_\infty$ is used to denote the

Rydberg constant for a stationary (infinitely heavy) nucleus, with $m_r = m$. Then the Rydberg constant for ions with different nuclear masses $M$ is given by

$$R_M = \frac{R_\infty}{1 + m/M}. \tag{2.21}$$

Problem 29 shows an example in which the reduced mass plays a significant role. Problem 31 indicates how Eq. (2.20) can be derived for motion in one dimension.

## 2.4
## Semiclassical Mechanics, 1913–1925

The success of Bohr's theory for hydrogen and single-electron ions showed that atoms are "quantized" systems. They radiate photons with the properties described earlier by Planck and by Einstein. At the same time, the failure of the Bohr theory to give correct predictions for other systems led investigators to search for a more fundamental expression of the quantum nature of atoms and radiation.

Between Bohr's 1913 theory and Heisenberg's 1925 discovery of quantum mechanics, methods of semiclassical mechanics were explored in physics. A general quantization procedure was sought that would incorporate Bohr's rules for single-electron systems and would also be applicable to many-electron atoms and to molecules. Basically, as we did above with Eq. (2.5), one used classical equations of motion to describe an atomic system and then superimposed a quantum condition, such as Eq. (2.3).

A principle of "adiabatic invariance" was used to determine which variables of a system should be quantized. It was recognized that quantum transitions occur as a result of *sudden* perturbations on an atomic system, not as a result of gradual changes. For example, the rapidly varying electric field of a passing photon can result in an electronic transition with photon absorption by a hydrogen atom. On the other hand, the electron is unlikely to make a transition if the atom is simply placed in an external electric field that is slowly increased in strength. The principle thus asserted that those variables in a system that were invariant under slow, "adiabatic" changes were the ones that should be quantized.

A generalization of Bohr's original quantum rule (2.3) was also worked out (by Wilson and Sommerfeld, independently) that could be applied to pairs of variables, such as momentum and position. So-called phase integrals were used to quantize systems after the classical laws of motion were applied.

These semiclassical procedures had some successes. For example, elliptical orbits were introduced into Bohr's picture and relativistic equations were used in place of the nonrelativistic Eq. (2.5). The relativistic theory predicted a split in some atomic energy levels with the same quantum number, the magnitude of the energy difference depending on the fine-structure constant. The existence of the split gives rise to a fine structure in the spectrum of most elements in which some "lines" are observed under high resolution to be two closely separated lines. The well-known doublet in the sodium spectrum, consisting of two yellow lines at 5890 Å and 5896 Å, is due to transitions from two closely spaced energy levels, degenerate in nonrelativistic theory. In spite of its successes in some areas, semiclassical atomic theory did not work for many-electron atoms and for such simple systems as some diatomic molecules, for which it gave unambiguous but incorrect spectra.

As a guide for discovering quantum laws, Bohr in 1923 introduced his correspondence principle. This principle states that the predictions of quantum physics must be the same as those of classical physics in the limit of very large quantum number $n$. In addition, any relationships between states that are needed to obtain the classical results for large $n$ also hold for all $n$. The diagram of energy levels in Fig. 2.3 illustrates the approach of a quantum system to a classical one when the quantum numbers become very large. Classically, the electron in a bound state has continuous, rather than discrete, values of the energy. As $n \to \infty$, the bound-state energies of the H atom get arbitrarily close together.

Advances toward the discovery of quantum mechanics were also being made along other lines. The classical Maxwellian wave theory of electromagnetic radiation seemed to be at odds with the existence of Einstein's corpuscular photons of light. How could light act like waves in some experiments and like particles in others? The diffraction and interference of X rays was demonstrated in 1912 by von Laue, thus establishing their wave nature. The Braggs used X-ray diffraction from crystal layers of known separation to measure the wavelength of X rays. In 1922, discovery of the Compton effect (Section 8.4)—the scattering of X-ray photons from atoms with a decrease in photon energy—demonstrated their nonwave, or corpuscular, nature in still another way. The experimental results were explained by assuming that a photon of energy $E$ has a momentum $p = E/c = h\nu/c$, where $\nu$ is the photon frequency and $c$ is the speed of light. In 1924, de Broglie proposed that the wave/particle dualism recognized for photons was a characteristic of all fundamental particles of nature. An electron, for example, hitherto regarded as a particle, also might have wave properties associated with it. The universal formula that links the property of wavelength, $\lambda$, with the particle property of momentum, $p$, is that which applies to photons: $p = h\nu/c = h/\lambda$. Therefore, de Broglie proposed that the wavelength associated with a particle be given by the relation

$$\lambda = \frac{h}{p} = \frac{h}{\gamma m v},$$

(2.22)

where $m$ and $v$ are rest mass and speed of the particle and $\gamma$ is the relativistic factor defined in Appendix C.

Davisson and Germer in 1927 published the results of their experiments, which demonstrated that a beam of electrons incident on a single crystal of nickel is diffracted by the regularly spaced crystal layers of atoms. Just as the Braggs measured the wavelength of X rays from crystal diffraction, Davisson and Germer measured the wavelength for electrons. They found excellent agreement with Eq. (2.22). The year before this experimental confirmation of the existence of electron waves, Schroedinger had extended de Broglie's ideas and developed his wave equation for the new quantum mechanics, as described in the next section.

A convenient formula can be used to obtain the wavelength $\lambda$ of a non-relativistic electron in terms of its kinetic energy $T$. (An electron is nonrelativistic as long as $T$ is small compared with its rest energy, $mc^2 = 0.511$ MeV.) The nonrelativistic formula relating momentum $p$ and kinetic energy $T$ is $p = \sqrt{2mT}$ (Appendix C). It follows from Eq. (2.22) that

$$\lambda = \frac{h}{\sqrt{2mT}}. \tag{2.23}$$

It is often convenient to express the wavelength in Å and the energy in eV. Using these units for $\lambda$ and $T$ in Eq. (2.23), we write

$$\lambda_{\text{Å}} \times 10^{-10} = \frac{6.6261 \times 10^{-34}}{\sqrt{2 \times 9.1094 \times 10^{-31} \times T_{\text{eV}} \times 1.6022 \times 10^{-19}}}, \tag{2.24}$$

or

$$\lambda_{\text{Å}} = \frac{12.264}{\sqrt{T_{\text{eV}}}} = \frac{12.3}{\sqrt{T_{\text{eV}}}}. \qquad \text{(Nonrelativistic electrons)} \tag{2.25}$$

The subscripts indicate the units for $\lambda$ and $T$ when this formula is used.

An analogous expression can be derived for photons. Since the photon energy is given by $E = h\nu$, the wavelength is $\lambda = c/\nu = ch/E$. Analogously to Eq. (2.25), we find

$$\lambda_{\text{Å}} = \frac{12398}{E_{\text{eV}}} = \frac{12400}{E_{\text{eV}}}. \qquad \text{(Photons)} \tag{2.26}$$

*Example*

In some of their experiments, Davisson and Germer used electrons accelerated through a potential difference of 54 V. What is the de Broglie wavelength of these electrons?

*Solution*

The nonrelativistic formula (2.25) gives, with $T_{\text{eV}} = 54$ eV, $\lambda_{\text{Å}} = 12.3/\sqrt{54} = 1.67$ Å. Electron wavelengths much smaller than optical ones are readily obtainable. This is the basis for the vastly greater resolving power that electron microscopes have over optical microscopes (wavelengths $\gtrsim 4000$ Å).

*Example*

Calculate the de Broglie wavelength of a 10-MeV electron.

*Solution*

We must treat the problem relativistically. We thus use Eq. (2.22) after determining $\gamma$ and $v$. From Appendix C, with $T = 10$ MeV and $mc^2 = 0.511$ MeV, we have

$$10 = 0.511(\gamma - 1), \tag{2.27}$$

giving $\gamma = 20.6$. We can compute $v$ directly from $\gamma$. In this example, however, we know that $v$ is very nearly equal to $c$. Using $v = c$ in Eq. (2.22), we therefore write

$$\lambda = \frac{h}{\gamma mc} = \frac{6.63 \times 10^{-34}}{20.6 \times 9.11 \times 10^{-31} \times 3 \times 10^{8}} = 1.18 \times 10^{-13} \text{ m.} \tag{2.28}$$

For the last example one can, alternatively, derive the relativistic form of Eq. (2.25) for electrons. The result is (Problem 42)

$$\lambda_{\text{Å}} = \frac{12.264}{\sqrt{T_{\text{eV}}\left(1 + \frac{T_{\text{eV}}}{1.022 \times 10^{6}}\right)}}. \qquad \text{(Relativistic electrons)} \tag{2.29}$$

The last term in the denominator is $T_{\text{eV}}/(2mc^2)$ and is, therefore, not important when the electron's kinetic energy can be neglected compared with its rest energy. Equation (2.29) then becomes identical with Eq. (2.25).

Also, in the period just before the discovery of quantum mechanics, Pauli formulated his famous exclusion principle. This rule can be expressed by stating that no two electrons in an atom can have the same set of four quantum numbers. We shall discuss the Pauli principle in connection with the periodic system of the elements in Section 2.6.

## 2.5
## Quantum Mechanics

Quantum mechanics was discovered by Heisenberg in 1925 and, from a completely different point of view, independently by Schroedinger at about the same time. Heisenberg's formulation is termed matrix mechanics and Schroedinger's is called wave mechanics. Although they are entirely different in their mathematical formulation, Schroedinger showed in 1926 that the two systems are completely equivalent and lead to the same results. We shall discuss each in turn.

Heisenberg associated the failure of the Bohr theory with the fact that it was based on quantities that are not directly observable, like the classical position and speed of an electron in orbit about the nucleus. He proposed a system of mechanics based on observable quantities, notably the frequencies and intensities of the lines in the emission spectrum of atoms and molecules. He then represented dynamical variables (e.g., the position $x$ of an electron) in terms of observables and worked out rules for representing $x^2$ when the representation for $x$ is given. In so doing, Heisenberg found that certain pairs of variables did not commute multiplicatively (i.e., $xp \neq px$ when $x$ and $p$ represent position and momentum in the direction of $x$), a mathematical property of matrices recognized by others after Heisenberg's original formulation. Heisenberg's matrix mechanics was applied to various systems and gave results that agreed with those predicted by Bohr's theory where the latter was consistent with experiment. In other instances it gave new theoretical predictions that also agreed with observations. For example, Heisenberg explained the pattern of alternating strong and weak lines in the spectra of diatomic molecules, a problem in which Bohr's theory had failed. He showed that two forms of molecular hydrogen should exist, depending on the relative directions of the proton spin, and that the form with spins aligned (orthohydrogen) should be three times as abundant as the other with spins opposed (parahydrogen). This discovery was cited in the award of the 1932 Nobel Prize in physics to Heisenberg.

The concept of building an atomic theory on observables and its astounding successes let to a revolution in physics. In classical physics, objects move with certain endowed properties, such as position and velocity at every moment in time. If one knows these two quantities at any one instant and also the total force that an object experiences, then its motion is determined completely for all times by Newton's second law. Such concepts are applied in celestial mechanics, where the positions of the planets can be computed backwards and forwards in time for centuries. The same determinism holds for the motion of familiar objects in everyday life. However, on the atomic scale, things are inherently different. In an experiment that would measure the position and velocity of an electron in orbit about a nucleus, the act of measurement itself introduces uncontrollable perturbations that prevent one's obtaining all the data precisely. For example, photons of very short wavelength would be required to localize the position of an electron within an atomic dimension. Such photons impart high momentum in scattering from an electron, thereby making simultaneous knowledge of the electron's position and momentum imprecise.

In 1927 Heisenberg enunciated the uncertainty principle, which sets the limits within which certain pairs of quantities can be known simultaneously. For momentum $p$ and position $x$ (in the direction of the momentum) the uncertainty relation states that

$$\Delta p \, \Delta x \geq \hbar. \tag{2.30}$$

Here $\Delta p$ and $\Delta x$ are the uncertainties (probable errors) in these quantities, determined simultaneously; the product of the two can never be smaller than $\hbar$, which it can approach under optimum conditions. Another pair of variables consists of energy $E$ and time $t$, for which

$$\Delta E \, \Delta t \geq \hbar. \tag{2.31}$$

The energy of a system cannot be measured with arbitrary precision in a very short time interval. These uncertainties are not due to any shortcomings in our measuring ability. They are a result of the recognition that only observable quantities have an objective meaning in physics and that there are limits to making measurements on an atomic scale. The question of whether an electron "really" has a position and velocity simultaneously—whether or not we try to look—is metaphysical. Schools of philosophy differ on the fundamental nature of our universe and the role of the observer.

Whereas observation is immaterial to the future course of a system in classical physics, the observer's role is a basic feature of quantum mechanics, a formalism based on observables. The uncertainty relations rule out classical determinism for atomic systems. Knowledge obtained from one measurement, say, of an electron's orbital position, will not enable one to predict with certainty the result of a second measurement of the orbital position. Instead of this determinism, quantum mechanics enables one to predict only the *probabilities* of finding the electron in various positions when the second measurement is made. Operationally, such a probability distribution can be measured by performing an experiment a large number of times under identical conditions and compiling the frequency distribution of the different results. The laws of quantum mechanics are definite, but they are statistical, rather than deterministic, in nature. As an example of this distinction, consider a sample of $10^{16}$ atoms of a radioactive isotope that is decaying at an average rate of $10^4$ atoms per second. We cannot predict which particular atoms will decay during any given second nor can we say exactly how many will do so. However, we can predict with assurance the probability of obtaining any given number of counts (e.g., 10,132) in a given second, as can be checked by observation.

*Example*

What is the minimum uncertainty in the momentum of an electron that is localized within a distance $\Delta x = 1$ Å, approximately the diameter of the hydrogen atom? How large can the kinetic energy of the electron be, consistent with this uncertainty?

*Solution*

The relation (2.30) requires that the uncertainty in the momentum be at least as large as the amount

$$\Delta p \cong \frac{\hbar}{\Delta x} = \frac{1.05 \times 10^{-34} \text{ J s}}{10^{-10} \text{ m}} \sim 10^{-24} \text{ kg m s}^{-1}. \qquad (2.32)$$

To estimate how large the kinetic energy of the electron can be, we note that its momentum $p$ can be as large as $\Delta p$. With $p \sim \Delta p$, the kinetic energy $T$ of the electron (mass $m$) is

$$T = \frac{p^2}{2m} \sim \frac{(10^{-24})^2}{2 \times 9.11 \times 10^{-31}} \sim 5 \times 10^{-19} \text{ J}, \qquad (2.33)$$

or about 3 eV. This analysis indicates that an electron confined within a distance $\Delta x \sim 1$ Å will have a kinetic energy in the eV range. In the case of the H atom we saw that the electron's kinetic energy is 13.6 eV. The uncertainty principle implies that electrons confined to even smaller regions become more energetic, as the next example illustrates.

*Example*

If an electron is localized to within the dimensions of an atomic nucleus, $\Delta x \sim 10^{-15}$ m, estimate its kinetic energy.

*Solution*

In this case, we have $p \sim \Delta p \sim \hbar/\Delta x \sim 10^{-34}/10^{-15} \sim 10^{-19}$ kg m s$^{-1}$. Comparison with the last example ($p \sim 10^{-24}$ kg m s$^{-1}$) indicates that we must use the relativistic formula to find the kinetic energy from the momentum in this problem (Appendix C). The total energy $E_T$ of the electron is given by

$$E_T^2 = p^2 c^2 + m^2 c^4 \tag{2.34}$$

$$= (10^{-19})^2 (3 \times 10^8)^2 + (9 \times 10^{-31})^2 (3 \times 10^8)^4 \tag{2.35}$$

$$= 9 \times 10^{-22} \; \text{J}^2, \tag{2.36}$$

from which we obtain $E_T \sim 200$ MeV. It follows that electrons in atomic nuclei would have energies of hundreds of MeV. Before the discovery of the neutron in 1932 it was speculated that a nucleus of atomic number $Z$ and atomic mass number $A$ consists of $A$ protons and $A - Z$ electrons. The uncertainty principle argued against such a picture, since maximum beta-particle energies of only a few MeV are found. (In addition, some nuclear spins would be different from those observed if electrons existed in the nucleus.) The beta particle is created in the nucleus at the time of decay.

We turn now to Schroedinger's wave mechanics. Schroedinger began with de Broglie's hypothesis (Eq. 2.22) relating the momentum and wavelength of a particle. He introduced an associated oscillating quantity, $\psi$, and constructed a differential equation for it to satisfy. The coefficients in the equation involve the constants $h$ and the mass and charge of the particle. Equations describing waves are well known in physics. Schroedinger's wave equation is a linear differential equation, second order in the spatial coordinates and first order in time. It is linear, so that the sum of two or more solutions is also a solution. Linearity thus permits the superposition of solutions to produce interference effects and the construction of wave packets to represent particles. The wave-function solution $\psi$ must satisfy certain boundary conditions, which lead to discrete values, called eigenvalues, for the energies of bound atomic states. Applied to the hydrogen atom, Schroedinger's wave equation gave exactly the Bohr energy levels. It also gave correct results for the other systems to which it was applied. Today it is widely used to calculate the properties of many-electron atomic and molecular systems, usually by numerical solution on a computer.

A rough idea can be given of how wave mechanics replaces Bohr's picture of the H atom. Instead of the concept of the electron moving in discrete orbits about the nucleus, we envision the electron as being represented by an oscillating cloud. Furthermore, the electron cloud oscillates in such a way that it sets up a standing wave about the nucleus. A familiar example of standing waves is provided by a vibrating string of length $L$ stretched between two fixed points $P_1$ and $P_2$, as illustrated in Fig. 2.4. Standing waves are possible only with wavelengths $\lambda$ given by

$$L = n\frac{\lambda}{2}, \quad n = 1, 2, 3, \dots. \tag{2.37}$$

This relation describes a discrete set of wavelengths $\lambda$. In an analogous way, an electron standing-wave cloud in the H atom can be envisioned by requiring that an integral number of wavelengths $n\lambda$ fit exactly into a circumferential distance $2\pi r$ about the nucleus: $2\pi r = n\lambda$. Using the de Broglie relation (2.22) then implies

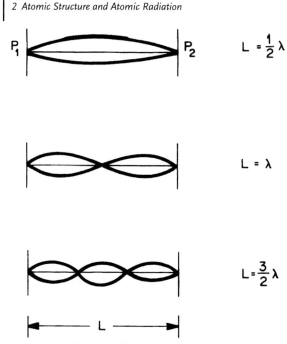

**Fig. 2.4** Examples of standing waves in string of length $L$ stretched between two fixed points $P_1$ and $P_2$. Such waves exist only with discrete wavelengths given by $\lambda = 2L/n$, where $n = 1, 2, 3, \ldots$.

nonrelativistically ($\gamma = 1$) that $2\pi r = nh/mv$, or $mvr = n\hbar$. One thus arrives at Bohr's original quantization law, Eq. (2.3).

Schroedinger's wave equation is nonrelativistic, and he proposed a modification of it in 1926 to meet the relativistic requirement for symmetry between space and time. As mentioned earlier, the Schroedinger differential equation is second order in space and first order in time variables. His relativistic equation, which contained the second derivative with respect to time, led to a fine structure in the hydrogen spectrum, but the detailed results were wrong. Taking a novel approach, Dirac proposed a wave equation that was first order in both the space and time variables. In 1928 Dirac showed that the new equation automatically contained the property of intrinsic angular momentum for the electron, rotating about its own axis. The predicted value of the electron's spin angular momentum was $\hbar/2$, the value ascribed experimentally in 1925 by Uhlenbeck and Goudsmit to account for the structure of the spectra of the alkali metals. Furthermore, the fine structure of the hydrogen-atom spectrum came out correctly from the Dirac equation. Dirac's equation also implied the existence of a positive electron, found later by Anderson, who discovered the positron in cosmic radiation in 1932. In 1927 Dirac also laid the foundation for quantum electrodynamics—the modern theory of the emission and absorption of electromagnetic radiation by atoms. The reader is referred to the historical outline in Section 1.3 for a chronology of events that occurred with the discovery of quantum mechanics.

## 2.6
## The Pauli Exclusion Principle

Originally based on the older, semiclassical quantum theory, the Pauli exclusion principle plays a vital role in modern quantum mechanics. The principle was developed for electrons in orbital states in atoms. It holds that no two electrons in an atom can be in the same state, characterized by four quantum numbers which we now define. The Pauli principle enables one to use atomic theory to account for the periodic system of the chemical elements.

In the Bohr theory with circular orbits, described in Section 2.3, only a single quantum number, $n$, was used. This is the first of the four quantum numbers, and we designate it as the principal quantum number. In Bohr's theory each value of $n$ gives an orbit at a given distance from the nucleus. For an atom with many electrons, we say that different values of $n$ correspond to different electron shells. Each shell can accommodate only a limited number of electrons. In developing the periodic system, we postulate that, as more and more electrons are present in atoms of increasing atomic number, they fill the innermost shells. The outer-shell electrons determine the gross chemical properties of an element; these properties are thus repeated successively after each shell is filled. When $n = 1$, the shell is called the K shell; $n = 2$ denotes the L shell; $n = 3$, the M shell; and so on.

The second quantum number arises in the following way. As mentioned in Section 2.4, elliptical orbits and relativistic mechanics were also considered in the older quantum theory of the hydrogen atom. In nonrelativistic mechanics, the mean energy of an electron is the same for all elliptical orbits having the same major axis. Furthermore, the mean energy is the same as that for a circular orbit with a diameter equal to the major axis. (The circle is the limiting case of an ellipse with equal major and minor axes.) Relativistically, the situation is different because of the increase in velocity and hence mass that an electron experiences in an elliptical orbit when it comes closest to the nucleus. In 1916 Sommerfeld extended Bohr's theory to include elliptical orbits with the nucleus at one focus. The formerly degenerate energies of different ellipses with the same major axis are slightly different relativistically, giving rise to the fine structure in the spectra of elements. The observed fine structure in the hydrogen spectrum was obtained by quantizing the ratio of the major and minor axes of the elliptical orbits, thus providing a second quantum number, called the azimuthal quantum number. In modern theory it amounts to the same thing as the orbital angular-momentum quantum number, $l$, with values $l = 0, 1, 2, \ldots, n-1$. Thus, for a given shell, the second quantum number can be any non-negative integer smaller than the principal quantum number. When $n = 1$, $l = 0$ is the only possible azimuthal quantum number; when $n = 2$, $l = 0$ and $l = 1$ are both possible.

The magnetic quantum number $m$ is the third. It was introduced to account for the splitting of spectral lines in a magnetic field (Zeeman effect). An electron orbiting a nucleus constitutes an electric current, which produces a magnetic field. When an atom is placed in an external magnetic field, its own orbital magnetic field lines up only in certain discrete directions with respect to the external field. The magnetic quantum number gives the component of the orbital angular momentum in the direction of the external field. Accordingly, $m$ can have any integral value between $+l$ and $-l$; viz., $m = 0, \pm 1, \pm 2, \ldots, \pm l$. With $l = 1$, for example, $m = -1, 0, 1$.

Although the Bohr–Sommerfeld theory explained a number of features of atomic spectra, problems still persisted. Unexplained was the fact that the alkali-metal spectra (e.g., Na) show a doublet structure even though these atoms have only a single valence electron in their outer shell (as we show in the next section). In addition, spectral lines do not split into a normal pattern in a weak magnetic field (anomalous Zeeman effect). These problems were cleared up when Pauli introduced a fourth quantum number of "two-valuedness," having no classical analogue. Then, in 1925, Uhlenbeck and Goudsmit proposed that the electron has an intrinsic angular momentum $\frac{1}{2}\hbar$ due to rotation about its own axis; thus the physical significance of Pauli's fourth quantum number was evident. The electron's intrinsic spin endows it with magnetic properties. The spin quantum number, $s$, has two values, $s = \pm\frac{1}{2}$. In an external magnetic field, the electron aligns itself either with "spin up" or "spin down" with respect to the field direction.

The Pauli exclusion principle states that no two electrons in an atom can occupy a state with the same set of four quantum numbers $n, l, m$, and $s$. The principle can also be expressed equivalently, but more generally, by saying that no two electrons in a system can have the same complete set of quantum numbers. Beyond atomic physics, the Pauli exclusion principle applies to all types of identical particles of half-integral spin (called fermions and having intrinsic angular momentum $\frac{1}{2}\hbar$, $\frac{3}{2}\hbar$, etc.). Such particles include positrons, protons, neutrons, muons, and others. Integral-spin particles (called bosons) do not obey the exclusion principle. These include photons, alpha particles, pions, and others.

We next apply the Pauli principle as a basis for understanding the periodic system of the elements.

## 2.7
## Atomic Theory of the Periodic System

The K shell, with $n = 1$, can contain at most two electrons, since $l = 0$, $m = 0$, and $s = \pm\frac{1}{2}$ are the only possible values of the other three quantum numbers. The two electrons in the K shell differ only in their spin directions. The element with atomic number $Z = 2$ is the noble gas helium. Like the other noble-gas atoms it has a completed outer shell and is chemically inert. The electron configurations of H and He are designated, respectively, as $1s^1$ and $1s^2$. The symbols in the configurations give the principal quantum number, a letter designating the azimuthal quantum number (s denotes $l = 0$; p denotes $l = 1$; d, $l = 2$; and f, $l = 3$), and a superscript giving the total number of electrons in the states with the given values of $n$ and $l$. The electron configurations of each element are shown in the periodic table in the back of this book, to which the reader is referred in this discussion. The first period contains only hydrogen and helium.

The next element, Li, has three electrons. Two occupy the full K shell and the third occupies a state in the L shell ($n = 2$). Electrons in this shell can have $l = 0$ (s states) or $l = 1$ (p states). The 2s state has lower energy than the 2p, and so the electron configuration of Li is $1s^2 2s^1$. With $Z = 4$ (Be), the other 2s state is occupied, and the configuration is $1s^2 2s^2$. No additional s electrons ($l = 0$) can be added in these two shells. However, the L shell can now accommodate electrons with $l = 1$ (p electrons) and with three values of $m$: $-1, 0, +1$. Since two electrons with opposite spins (spin quantum numbers $\pm\frac{1}{2}$) can occupy each state of given $n$, $l$, and $m$, there can be a total of six electrons in the 2p states. The configurations for the next six elements involve the successive filling of these states, from $Z = 5$ (B), $1s^2 2s^2 p^1$, to $Z = 10$ (Ne), $1s^2 2s^2 p^6$. The noble gas neon has the completed L shell. To save repeating the writing of the identical inner-shell configurations for other elements, one denotes the neon configuration by [Ne]. The second period of the table begins with Li and ends with Ne.

With the next element, sodium, the filling of the M shell begins. Sodium has a 3s electron and its configuration is $[Ne]3s^1$. Its single outer-shell electron gives it properties akin to those of lithium. One sees that the other alkali metals in the group IA of the periodic table are all characterized by having a single s electron in their outer shell. The third period ends with the filling of the $3s^2 p^6$ levels in the noble gas, Ar ($Z = 18$). The chemical and physical properties of the eight elements in the third period are similar to those of the eight elements in the second period with the same outer-shell electron configurations.

The configuration of Ar ($Z = 18$), which is $[Ne]3s^2 p^6$, is also designated as [Ar]. All of the states with $n = 1$, $l = 0$; $n = 2$, $l = 0, 1$; and $n = 3$, $l = 0, 1$ are occupied in Ar. However, the M shell is not yet filled, because d states ($l = 2$) are possible when $n = 3$. Because there are five values of $m$ when $l = 2$, there are five d states, which can accommodate a total of ten electrons (five pairs with opposite spin), which is the number needed to complete the M shell. It turns out that the 4s energy levels are lower than the 3d. Therefore, the next two elements, K and Ca, that follow Ar have the configurations $[Ar]4s^1$ and $[Ar]4s^2$. The next ten elements, from Sc ($Z = 21$) through Zn ($Z = 30$), are known as the transition metals. This series fills the 3d levels, sometimes in combination with $4s^1$ and sometimes with $4s^2$ electrons. The configuration of Zn is $[Ar]3d^{10}4s^2$, at which point the M shell ($n = 3$) is complete. The next six elements after Zn fill the six 4p states, ending with the noble gas, Kr, having the configuration $[Ar]3d^{10}4s^2 p^6$.

After Ar, the shells with a given principal quantum number do not get filled in order. Nevertheless, one can speak of the filling of certain subshells in order, such as the 4s subshell and then the 3d in the transition metals. The lanthanide series of rare-earth elements, from $Z = 58$ (Ce) to $Z = 71$ (Lu), occurs when the 4f subshell is being filled. For these states $l = 3$, and since $-3 \leq m \leq 3$, a total of $7 \times 2 = 14$ elements compose the series. Since it is an inner subshell that is being filled, these elements all have very nearly the same chemical properties. The situation is repeated with the actinide elements from $Z = 90$ (Th) to $Z = 103$ (Lr), in which the 5f subshell is being filled.

The picture given here is that of an independent-electron model of the atom, in which each electron independently occupies a given state. In reality, the atomic electrons are indistinguishable from one another and an atomic wave function is one in which any electron can occupy any state with the same probability as any other electron. Moreover, hybrid atomic states of mixed configurations are used to explain still other phenomena (e.g., the tetrahedral bonds in $CH_4$).

## 2.8
## Molecules

Quantum mechanics has also been very successful in areas other than atomic structure and spectroscopy. It has also explained the physics of molecules and condensed matter (liquids and solids). Indeed, the nature of the chemical bond between two atoms, of either the same or different elements, is itself quantum mechanical in nature, as we now describe.

Consider the formation of the $H_2$ molecule from two H atoms. Experimentally, it is known from the vibrational spectrum of $H_2$ that the two protons' separation oscillates about an equilibrium distance of 0.74 Å and the dissociation energy of the molecule is 4.7 eV. The two electrons move very rapidly about the two nuclei, which, by comparison, move slowly back and forth along the direction between their centers. When the nuclei approach each other, their Coulomb repulsion causes them to reverse their directions and move apart. The electrons more than keep pace and move so that the separating nuclei again reverse directions and approach one another. Since the electrons move so quickly, they make many passes about the nuclei during any time in which the latter move appreciably. Therefore, one can gain considerable insight into the structure of $H_2$ and other molecules by considering the electronic motion at different fixed separations of the nuclei (Born-Oppenheimer approximation).

To analyze $H_2$, we begin with the two protons separated by a large distance $R$, as indicated in Fig. 2.5(a). The lowest energy of the system will then occur when each electron is bound to one of the protons. Thus the ground state of the $H_2$ system at large nuclear separations is that in which the two hydrogen atoms, $H_A$ and $H_B$, are present in their ground states. We denote this structure by writing $(H_A1, H_B2)$, indicating that electron number one is bound in the hydrogen atom $H_A$ and electron number two in $H_B$. Another stable structure at large $R$ is an ionic one, $(H_A12^-, H_B^+)$, in which both electrons orbit one of the protons. This structure is shown in Fig. 2.5(b). Since 13.6 eV is required to remove an electron from H and its binding energy in the $H^-$ ion is only 0.80 eV, the ionic structure in Fig. 2.5(b) has less binding energy than the natural one in (a). In addition to the states shown in the figure, one can consider the same two structures in which the two electrons are interchanged, $(H_A2, H_B1)$ and $(H_A21^-, H_B^+)$.

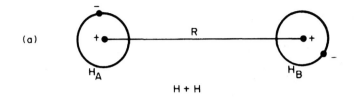

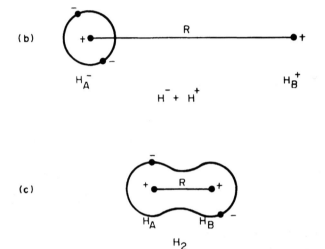

**Fig. 2.5** At large internuclear separation $R$, the structure of the $H_2$ molecule can approach (a) that of two neutral H atoms, H + H, or (b) that of two ions, $H^- + H^+$. These structures merge at close separations in (c). The indistinguishability of the two electrons gives stability to the bond formed through the quantum-mechanical phenomenon of resonance.

Next, we consider what happens when $R$ becomes smaller. When the nuclei move close together, as in Fig. 2.5(c), the electron wave functions associated with each nucleus overlap. Detailed calculations show that neither of the structures shown in Fig. 2.5(a) or (b) nor a combination of the two leads to the formation of a stable molecule. Instead, stability arises from the indistinguishable participation of both electrons. The neutral structure alone will bind the two atoms when, in place of either ($H_A1$, $H_B2$) or ($H_A2$, $H_B1$) alone, one uses the superposed structure ($H_A1$, $H_B2$) + ($H_A2$, $H_B1$). The need for the superposed structure is a purely quantum-mechanical concept, and is due to the fact that the two electrons are indistinguishable and their roles must be exchangeable without affecting observable quantities. The energy contributed to the molecular binding by the electron exchange is called the resonance energy, and its existence with the neutral structure in Fig. 2.5(c) accounts for ~80% of the binding energy of $H_2$. The type of electron-pair bond that

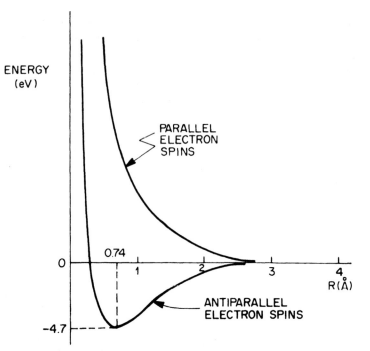

**Fig. 2.6** Total energy of the $H_2$ molecule as a function of internuclear separation R. A stable molecule is formed when the spins of the two electrons are antiparallel. A nonbonding energy level is formed when the spins are parallel. The two energies coincide at large R.

is thus formed by the exchange is called covalent. Resonance also occurs between the ionic structures ($H_A 12^-$, $H_B^+$) and ($H_A 21^-$, $H_B^+$) from Fig. 2.5(b) and contributes ~5% of the binding energy, giving the $H_2$ bond a small ionic character. The remaining 15% of the binding energy comes from other effects, such as deformation of the electron wave functions from the simple structures discussed here and from partial shielding of the nuclear charges by each electron from the other. In general, for covalent bonding to occur, the two atoms involved must have the same number of unpaired electrons, as is the case with hydrogen. However, the atoms need not be identical.

The character of the bond in HF and HCl, for example, is more ionic than in the homonuclear $H_2$ or $N_2$. The charge distribution in a heteronuclear diatomic molecule is not symmetric, and so the molecule has a permanent electric dipole moment. (The two types of bonds are called homopolar and heteropolar.)

Figure 2.6 shows the total energy of the $H_2$ molecule as a function of the internuclear separation R. (The total energy of the two H atoms at large R is taken as the reference level of zero energy.) The bound state has a minimum energy of –4.7 eV at the equilibrium separation of 0.74 Å, in agreement with the data given earlier in this section. In this state the spins of the bonding electron pair are antiparallel.

A second, nonbonding state is formed with parallel spins. These two energy levels are degenerate at large $R$, where the wave functions of the two atomic electrons do not overlap appreciably.

Molecular spectra are very complicated. Changes in the rotational motion of molecules accompany the emission or absorption of photons in the far infrared. Vibrational changes together with rotational ones usually produce spectra in the near infrared. Electronic transitions are associated with the visible and ultraviolet part of molecular spectra. Electronic molecular spectra have a fine structure due to the vibrational and rotational motions of the molecule. Molecular spectra also show isotopic structure. The presence of the naturally occurring $^{35}Cl$ and $^{37}Cl$ isotopes in chlorine, for example, gives rise to two sets of vibrational and rotational energy-level differences in the spectrum of HCl.

## 2.9
## Solids and Energy Bands

We briefly discuss the properties of solids and the origin of energy bands, which are essential for understanding how semiconductor materials can be used as radiation detectors (Chapter 10).

Solids can be crystalline or noncrystalline (e.g., plastics). Crystalline solids, of which semiconductors are an example, can be put into four groups according to the type of binding that exists between atoms. Crystals are characterized by regular, repeated atomic arrangements in a lattice.

In a *molecular* solid the bonds between molecules are formed by the weak, attractive van der Waals forces. Examples are the noble gases, $H_2$, $N_2$, and $O_2$, which are solids only at very low temperatures and can be easily deformed and compressed. All electrons are paired and hence molecular solids are poor electrical conductors.

In an *ionic* solid all electrons are also paired. A crystal of NaCl, for example, exists as alternating charged ions, $Na^+$ and $Cl^-$, in which all atomic shells are filled. These solids are also poor conductors. The electrostatic forces between the ions are very strong, and hence ionic solids are hard and have high melting points. They are generally transparent to visible light, because their electronic absorption frequencies are in the ultraviolet region and lattice vibration frequencies are in the infrared.

A *covalent* solid is one in which adjacent atoms are covalently bound by shared valence electrons. Such bonding is possible only with elements in Group IVB of the periodic system; diamond, silicon, and germanium are examples. In diamond, a carbon atom (electronic configuration $1s^2 2s^2 p^2$) shares one of its four L-shell electrons with each of four neighbors, which, in turn, donates one of its L electrons for sharing. Each carbon atom thus has its full complement of eight L-shell electrons through tight binding with its neighbors. Covalent solids are very hard and have high melting points. They have no free electrons, and are therefore poor conductors. Whereas diamond is an insulator, Si and Ge are semiconductors, as will be discussed in Chapter 10.

In a *metallic* solid (e.g., Cu, Au) the valence electrons in the outermost shells are weakly bound and shared by all of the atoms in the crystal. Vacancies in these shells permit electrons to move with ease through the crystal in response to the presence of an electric field. Metallic solids are good conductors of electricity and heat. Many of their properties can be understood by regarding some of the electrons in the solid as forming an "electron gas" moving about in a stationary lattice of positively charged ions. The electrons satisfy the Pauli exclusion principle and occupy a range of energies consistent with their temperature. This continuous range of energies is called a conduction band.

To describe the origin of energy bands in a solid, we refer to Fig. 2.6. We saw that the twofold exchange degeneracy between the electronic states of two widely separated hydrogen atoms was broken when their separation was reduced enough for their electron wave functions to overlap appreciably. The same twofold splitting occurs whenever any two identical atoms bind together. Moreover, the excited energy levels of isolated atoms also undergo a similar twofold splitting when the two atoms unite. Such splitting of exchange-degenerate energy levels is a general quantum-mechanical phenomenon. If three identical atoms are present, then the energy levels at large separations are triply degenerate and split into three different levels when the atoms are brought close together. In this case, the three levels all lie in about the same energy range as the first two if the interatomic distances are comparable. If N atoms are brought together in a regular arrangement, such as a crystal solid, then there are N levels in the energy interval.

Figure 2.7 illustrates the splitting of electronic levels for $N = 2$, 4, and 8 and the onset of band formation. Here the bound-state energies $E$ are plotted schematically as functions of the atomic separations $R$, with $R_0$ being the normal atomic spacing in the solid. When $N$ becomes very large, as in a crystal, the separate levels are "compressed together" into a band, within which an electron can have any energy. At a given separation, the band structure is most pronounced in the weakly bound states, in which the electron cloud extends over large distances. The low-lying levels, with tightly bound electrons, remain discrete and unperturbed by the presence of neighboring atoms. Just as for the discrete levels, electrons cannot exist in the solid with energies between the allowed bands. (The existence of energy bands also arises directly out of the quantum-mechanical treatment of the motion of electrons in a periodic lattice.)

While the above properties are those of "ideal" crystalline solids, the presence of impurities—even in trace amounts—often changes the properties markedly. We shall see in Chapter 10 how doping alters the behavior of intrinsic semiconductors and the scintillation characteristics of crystals.

## 2.10
### Continuous and Characteristic X Rays

Roentgen discovered that X rays are produced when a beam of electrons strikes a target. The electrons lose most of their energy in collisions with atomic electrons

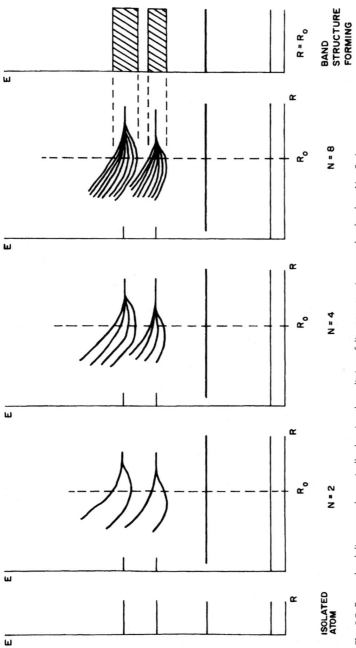

**Fig. 2.7** Energy-level diagram schematically showing the splitting of discrete atomic energy levels when $N = 2$, 4, and 8 atoms are brought together in a regular array. When $N$ is very large, continuous energy bands result, as indicated at the right of the figure.

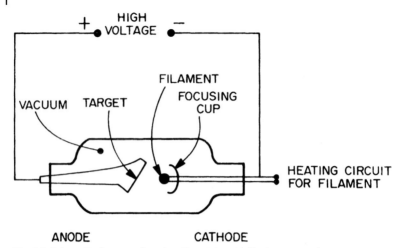

**Fig. 2.8** Schematic diagram of modern X-ray tube with fixed target anode.

in the target, causing the ionization and excitation of atoms. In addition, they can be sharply deflected in the vicinity of the atomic nuclei, thereby losing energy by irradiating X-ray photons. Heavy nuclei are much more efficient than light nuclei in producing the radiation because the deflections are stronger. A single electron can emit an X-ray photon having any energy up to its own kinetic energy. As a result, a monoenergetic beam of electrons produces a continuous spectrum of X rays with photon energies up to the value of the beam energy. The continuous X rays are also called bremsstrahlung, or "braking radiation."

A schematic diagram, showing the basic elements of a modern X-ray tube, is shown in Fig. 2.8. The tube has a cathode and anode sealed inside under high vacuum. The cathode assembly consists of a heated tungsten filament contained in a focusing cup. When the tube operates, the filament, heated white hot, "boils off" electrons, which are accelerated toward the anode in a strong electric field produced by a large potential difference (high voltage) between the cathode and anode. The focusing cup concentrates the electrons onto a focal spot on the anode, usually made of tungsten. There the electrons are abruptly brought to rest, emitting continuous X rays in all directions. Typically, less than 1% of the electrons' energy is converted into useful X rays that emerge through a window in the tube. The other 99+ % of the energy, lost in electronic collisions, is converted into heat, which must be removed from the anode. Anodes can be cooled by circulating oil or water. Rotating anodes are also used in X-ray tubes to keep the temperature lower.

Figure 2.9 shows typical continuous X-ray spectra generated from a tube operated at different voltages with the same current. The efficiency of bremsstrahlung production increases rapidly when the electron energy is raised. Therefore, the X-ray intensity increases considerably with tube voltage, even at constant current. The wavelength of an X-ray photon with maximum energy can be computed from Eq. (2.26). For the top curve in Fig. 2.9, we find $\lambda_{min} = 12400/50000 = 0.248$ Å, where

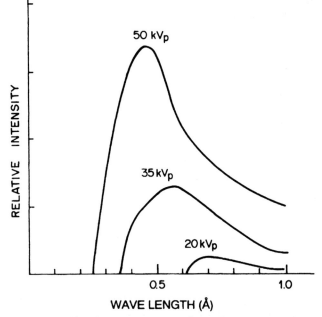

**Fig. 2.9** Typical continuous X-ray spectra from tube operating at three different peak voltages with the same current.

this curve intersects the abscissa. The X-ray energies are commonly referred to in terms of their peak voltages in kilovolts, denoted by kVp.

If the tube voltage is sufficient, electrons striking the target can eject electrons from the target atoms. (The K-shell binding energy is $E_K = 69.525$ keV for tungsten.) Discrete X rays are then also produced. These are emitted when electrons from higher shells fill the inner-shell vacancies. The photon energies are characteristic of the element of which the target is made, just as the optical spectra are in the visible range. Characteristic X rays appear superimposed on the continuous spectrum, as illustrated for tungsten in Fig. 2.10. They are designated $K_\alpha$, $K_\beta$, and so forth, when the K-shell vacancy is filled by an electron from the L shell, M shell, and so on. (In addition, when L-shell vacancies are filled, characteristic $L_\alpha$, $L_\beta$, and so forth, X rays are emitted. These have low energy and are usually absorbed in the tube housing.)

Because the electron energies in the other shells are not degenerate, the K X rays have a fine structure, not shown in Fig. 2.10. The L shell, for example, consists of three subshells, in which for tungsten the electron binding energies in keV are $E_{LI} = 12.098$, $E_{LII} = 11.541$, and $E_{LIII} = 10.204$. The transition LIII → K gives a $K_{\alpha 1}$ photon with energy $E_K - E_{LIII} = 69.525 - 10.204 = 59.321$ keV; the transition LII → K gives a $K_{\alpha 2}$ photon with energy 57.984 keV. The optical transition LI → K is quantum mechanically forbidden and does not occur.

The first systematic study of characteristic X rays was carried out in 1913 by the young British physicist, H. G. J. Moseley, working in Rutherford's laboratory.

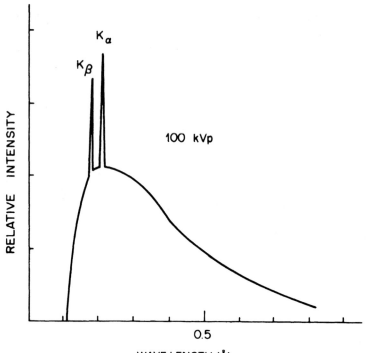

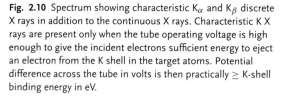

**Fig. 2.10** Spectrum showing characteristic $K_\alpha$ and $K_\beta$ discrete X rays in addition to the continuous X rays. Characteristic K X rays are present only when the tube operating voltage is high enough to give the incident electrons sufficient energy to eject an electron from the K shell in the target atoms. Potential difference across the tube in volts is then practically $\geq$ K-shell binding energy in eV.

The diffraction of X rays by crystals had been discovered by von Laue in 1912, and Moseley used this process to compare characteristic X-ray wavelengths. He found that the square root of the frequencies of corresponding lines (e.g., $K_{\alpha 1}$) in the characteristic X-ray spectra increases by an almost constant amount from element to element in the periodic system. Alpha-particle scattering indicated that the number of charge units on the nucleus is about half the atomic weight. Moseley concluded that the number of positive nuclear charges and the number of electrons both increase by one from element to element. Starting with $Z = 1$ for hydrogen, the number of charge units $Z$ determines the atomic number of an element, which gives its place in the periodic system.

The linear relationship between $\sqrt{\nu}$ and $Z$ would be predicted if the electrons in many-electron atoms occupied orbits like those predicted by Bohr's theory for single-electron systems. As seen from Eq. (2.13), the frequencies of the photons for a given transition i → f in different elements are proportional to $Z^2$.

In view of Moseley's findings, the positions of cobalt and nickel had to be reversed in the periodic system. Although Co has the larger atomic weight, 58.93 compared with 58.70, its atomic number is 27, while that of Ni is 28. Moseley also predicted the existence of a new element with $Z = 43$. Technetium, which has no stable form, was discovered after nuclear fission.

## 2.11
## Auger Electrons

An atom in which an L electron makes a transition to fill a vacancy in the K shell does not always emit a photon, particularly if it is an element of low $Z$. A different, nonoptical transition can occur in which an L electron is ejected from the atom, thereby leaving two vacancies in the L shell. The electron thus ejected from the atom is called an Auger electron.

The emission of an Auger electron is illustrated in Fig. 2.11. The downward arrow indicates the transition of an electron from the $L_I$ level into the K-shell vacancy,

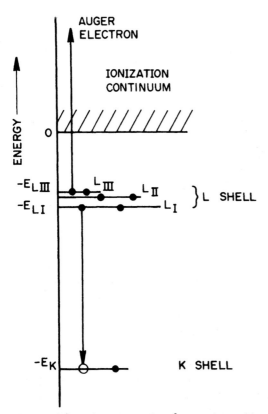

**Fig. 2.11** Schematic representation of an atomic transition that results in Auger-electron emission.

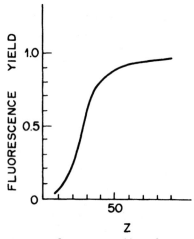

**Fig. 2.12** K fluorescence yield as a function of atomic number Z.

thus releasing an energy equal to the difference in binding energies, $E_K - E_{LI}$. As the alternative to photon emission, this energy can be transferred to an $L_{III}$ electron, ejecting it from the atom with a kinetic energy

$$T = E_K - E_{LI} - E_{LIII}. \tag{2.38}$$

Two L-shell vacancies are thus produced. The Auger effect can occur with other combinations of the three L-shell levels. Equations analogous to (2.38) provide the possible Auger-electron energies.

The Auger process is not one in which a photon is emitted by one atomic electron and absorbed by another. In fact, the $L_I \to K$ transition shown in Fig. 2.11 is optically forbidden.

The K fluorescence yield of an element is defined as the number of K X-ray photons emitted per vacancy in the K shell. Figure 2.12 shows how the K fluorescence yield varies from essentially zero for the low-Z elements to almost unity for high Z. Auger-electron emission is thus favored over photon emission for elements of low atomic number.

The original inner-shell vacancy in an Auger-electron emitter can be created by orbital electron capture, internal conversion, or photoelectric absorption of a photon from outside the atom. (These processes are described in Chapter 3.) As pointed out previously, emission of an Auger electron increases the number of vacancies in the atomic shells by one unit. Auger cascades can occur in relatively heavy atoms, as inner-shell vacancies are successively filled by the Auger process, with simultaneous ejections of the more loosely bound atomic electrons. An original, singly charged ion with one inner-shell vacancy can thus be converted into a highly charged ion by an Auger cascade. This phenomenon is being studied in radiation research and therapy. Auger emitters can be incorporated into DNA and other biological molecules. For example, [125]I decays by electron capture. The ensuing cascade can release some 20 electrons, depositing a large amount of energy ($\sim 1$ keV) within a few nanometers. A highly charged $^{125}$Te ion is left behind. A number of biological effects can be produced, such as DNA strand breaks, chromatid aberrations, mutations, bacteriophage inactivation, and cell killing.

## 2.12
## Suggested Reading

1 Alvarez, L. W., *Adventures of a Physicist*, Basic Books, New York, NY (1987). [Alvarez received the 1968 Nobel Prize for pioneering contributions in the field of high-energy physics and subatomic particles. This volume is part of the Alfred P. Sloan Foundation to encourage outstanding scientists to write personal accounts of their experiences for the general reader.]

2 Born, Max, *The Mechanics of the Atom*, Frederick Ungar Publ., New York, NY (1960). [An advanced text on semi-classical atomic theory. A number of atomic and molecular systems are solved explicitly using Hamilton–Jacobi theory and applying the older quantization procedures.]

3 Fano, U., and Fano, L., *Physics of Atoms and Molecules*, University of Chicago Press, Chicago, IL (1972). [An intermediate text on much of the material of this chapter and other subjects.]

4 Gamow, George, *Thirty Years that Shook Physics*, Dover Publications, New York, NY (1985). [The nonmathematical story of quantum mechanics as interestingly as related by a participant.]

5 Halliday, David; Resnick, Robert; and Walker, Jearl, *Fundamentals of Physics*, 7th ed., John Wiley, New York, NY (2004). [This excellent, time-tested text covers many of the subjects in this chapter.]

6 Hawking, Stephen, *A Brief History of Time*, Bantam Books, New York, NY (1988). [Describes the uncertainty principle, elementary particles, nature's forces, and cosmology. A readable, nontechnical book with only the single equation, $E = mc^2$.]

7 Hawking, Stephen, *A Brief History of Time: The Updated and Expanded Tenth Anniversary Edition*, Bantam Books, New York, NY (1998).

8 Heisenberg, Werner, *The Physical Principles of the Quantum Theory*, Dover Publications, New York, NY (1930). [A reprint of Heisenberg's 1929 lectures at the University of Chicago.]

9 Johns, H. E., and Cunningham, J. R., *The Physics of Radiology*, 4th ed., Charles C Thomas, Springfield, IL (1983). [A classic treatise. Chapter 2 is devoted to the fundamentals of diagnostic and therapeutic X-ray production.]

10 Pauling, Linus, *The Nature of the Chemical Bond*, 3rd ed., Cornell University Press, Ithaca, NY (1960). [The author received the Nobel Prize in Chemistry for his research into the nature of the chemical bond and its application to understanding the structure of complex substances.]

11 Pauling, L., and Haywood, R., *The Architecture of Molecules*, W. H. Freeman, San Francisco, CA (1964). [An introductory look at 57 pictured molecules. Chemical bonds are discussed.]

12 Ponomarev, L. I., *The Quantum Dice*, Institute of Physics Publ., Bristol, England (1993). ["A ray of sunlight over the cradle of a baby has always been a symbol of peace. But the ray carries not just caressing heat, it brings us rich information about the fiery storms and flares on the Sun, and about the elements it is composed of—you need only know how to read this information" (p. 25).]

13 Rozemtal, S., ed., *Niels Bohr*, North-Holland, Amsterdam, Netherlands (1967). [A collection of essays and personal accounts of the life and work of Bohr by his colleagues and contemporaries.]

14 Selman, Joseph, *The Fundamentals of Imaging Physics and Radiobiology*, 9th ed., Charles C Thomas, Springfield, IL (2000). [Previous (8th) edition, *The Fundamentals of X-Ray and Radium Physics*. An outstanding text for technicians that explains X-ray generation and utilization in radiology.]

15 Weisskopf, Victor, *The Privilege of Being a Physicist*, W. H. Freeman, New York, NY (1988). [The author knew most and worked with many of the leading physicists during the development of quantum mechanics and made many important contributions himself. This autobiography is another in the Alfred P. Sloan Foundation series (see Alvarez, above) to convey the excitement of science and discovery to general readers. This delightful book will also speak to those who are fond of classical music.]

16 Whittaker, E. T., *A History of the Theories of the Aether and Electricity*, Thomas Nelson and Sons, London, England (1953). [Volume 2, subtitled *The Modern Theories. 1900–1926*, begins with "The Age of Rutherford" (Chapter 1) and ends with "The Discovery of Wave Mechanics" (Chapter IX). Interesting, informative, and personal accounts of the discoveries of the time.]

## 2.13
## Problems

1. How many atoms are there in 3 L of $N_2$ at STP?

2. What is the volume occupied by 1 kg of methane ($CH_4$) at 23°C and 756 torr?

3. How many hydrogen atoms are there in the last problem?

4. What is the mass of a single atom of aluminum?

5. Estimate the number of atoms/cm$^2$ in an aluminum foil that is 1 mm thick.

6. Estimate the radius of a uranium nucleus. What is its cross-sectional area?

7. What is the density of the nucleus in a gold atom?

8. What was the minimum distance to which the 7.69-MeV alpha particles could approach the center of the gold nuclei in Rutherford's experiments?

9. How much energy would an alpha particle need in order to "just touch" the nuclear surface in a gold foil?

10. Figure 2.13 represents a particle moving with constant speed $v(= |\mathbf{v}_1| = |\mathbf{v}_2|)$ in a circular orbit of radius $r$. When the particle advances through a small angle $\Delta\theta$ about the center of the circle, the change in velocity $\Delta\mathbf{v} = \mathbf{v}_2 - \mathbf{v}_1$ is indicated. Denoting the elapsed time by $\Delta t$, one has $\Delta\theta = \omega\Delta t$, where $\omega = v/r$ is the angular velocity expressed in radians per unit time.

    (a) For small $\Delta t$, show that the acceleration of the particle is given by $a \cong \Delta v/\Delta t \cong v\Delta\theta/\Delta t$.

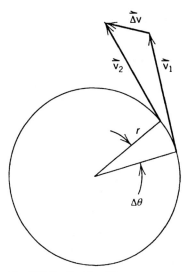

**Fig. 2.13** See Problem 10.

(b) In the limit $\Delta t \to 0$, show that the acceleration of the
particle is directed toward the center of the circle and has a
magnitude $a = \omega v = v^2/r$, as used in writing Eq. (2.5).

11. How much force acts on the electron in the ground state of the
hydrogen atom?

12. What is the angular momentum of the electron in the $n = 5$
state of the H atom?

13. How does the angular momentum of the electron in the $n = 3$
state of H compare with that in the $n = 3$ state of He$^+$?

14. Calculate the ionization potential of Li$^{2+}$.

15. Calculate the radius of the $n = 2$ electron orbit in the Bohr
hydrogen atom.

16. Calculate the orbital radius for the $n = 2$ state of Li$^{2+}$.

17. Do H and He$^+$ have any states with the same orbital radius?

18. What is the principal quantum number $n$ of the state of the H
atom with an orbital radius closest to that of the $n = 3$ state of
He$^+$?

19. What are the energies of the photons with the two longest
wavelengths in the Paschen series (Fig. 2.3)?

20. Calculate the wavelengths in the visible spectrum of the He$^+$
ion in which the electron makes transitions from higher states
to states with quantum number $n = 1, 2, 3,$ or 4.

21. How many energy levels of the He$^+$ ion lie below $-1$ eV?

22. Calculate the current of the electron in the ground state of the
hydrogen atom.

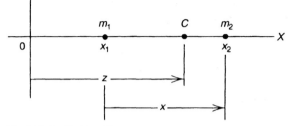

**Fig. 2.14** See Problem 31.

23. What is the lowest quantum number of an H-atom electron orbit with a radius of at least 1 cm?

24. According to Bohr theory, how many bound states of He$^+$ have energies equal to bound-state energies in H?

25. How much energy is needed to remove an electron from the $n = 5$ state of He$^+$?

26. **(a)** In the Balmer series of the hydrogen atom, what is the smallest value of the principal quantum number of the initial state for emission of a photon of wavelength less than 4200 Å?

    **(b)** What is the change in the angular momentum of the electron for this transition?

27. Calculate the reduced mass for the He$^+$ system.

28. What percentage error is made in the Rydberg constant for hydrogen if the electron mass is used instead of the reduced mass?

29. The negative muon is an elementary particle with a charge equal to that of the electron and a mass 207 times as large. A proton can capture a negative muon to form a hydrogen-like "mesic" atom. (The muon was formerly called the mu meson.) For such a system, calculate

    **(a)** the radius of the first Bohr orbit

    **(b)** the ionization potential.

    Do not assume a stationary nucleus.

30. What is the reduced mass for a system of two particles of equal mass, such as an electron and positron, orbiting about their center of mass?

31. Figure 2.14 shows two interacting particles, having masses $m_1$ and $m_2$ and positions $x_1$ and $x_2$. The particles are free to move only along the X-axis. Their total energy is
    $E = \frac{1}{2}m_1\dot{x}_1^2 + \frac{1}{2}m_2\dot{x}_2^2 + V(x)$, where $\dot{x}_1 = dx_1/dt$ and $\dot{x}_2 = dx_2/dt$ are the velocities, and the potential energy $V(x)$ depends only on the separation $x = x_2 - x_1$ of the particles. Let $z$ be the coordinate of the center of mass $C: m_1(z - x_1) = m_2(x_2 - z)$.

Show that $E = \frac{1}{2}(m_1 + m_2)\dot{z}^2 + \frac{1}{2}m_r\dot{x}^2 + V(x)$, where $m_r$, is the reduced mass, given by Eq. (2.20), and $\dot{x} = dx/dt$. The total energy is thus the sum of the translational kinetic energy of the motion of the total mass along the X-axis and the total energy associated with the relative motion (coordinate $x$ and mass $m_r$).

32. What is meant by the fine structure in the spectrum of hydrogen and what is its physical origin?

33. Calculate the momentum of an ultraviolet photon of wavelength 1000 Å.

34. What is the momentum of a photon of lowest energy in the Balmer series of hydrogen?

35. Calculate the de Broglie wavelength of the 7.69-MeV alpha particles used in Rutherford's experiment. Use nonrelativistic mechanics.

36. What is the energy of a proton that has the same momentum as a 1-MeV photon?

37. What is the energy of an electron having a wavelength of 0.123 Å?

38. Calculate the de Broglie wavelength of a 245-keV electron.

39. (a) What is the momentum of an electron with a de Broglie wavelength of 0.02 Å?
    (b) What is the momentum of a photon with a wavelength of 0.02 Å?

40. Calculate the kinetic energy of the electron and the energy of the photon in the last problem.

41. A microscope can resolve as distinct two objects or features that are no closer than the wavelength of the light or electrons used for the observation.
    (a) With an electron microscope, what energy is needed for a resolution of 0.4 Å?
    (b) What photon energy would be required of an optical microscope for the same resolution?

42. Show that Eq. (2.29) follows from Eq. (2.22) for relativistic electrons.

43. Estimate the uncertainty in the momentum of an electron whose location is uncertain by a distance of 2 Å. What is the uncertainty in the momentum of a proton under the same conditions?

44. What can one conclude about the relative velocities and energies of the electron and proton in the last problem? Are wave phenomena apt to be more apparent for light particles than for heavy ones?

45. The result given after Eq. (2.36) shows that an electron confined to nuclear dimensions, $\Delta x \sim 10^{-15}$ m, could be

expected to have a kinetic energy $T \sim 200$ MeV. What would be the value of $\Delta x$ for $T \sim 100$ eV?

46. (a) Write the electron configuration of carbon.
    (b) How many s electrons does the C atom have?
    (c) How many p electrons?

47. The configuration of boron is $1s^2 2s^2 p^1$.
    (a) How many electrons are in the L shell?
    (b) How many electrons have orbital angular-momentum quantum number $l = 0$?

48. How many electrons does the nickel atom have with azimuthal quantum number $l = 2$?

49. What is the electron configuration of the magnesium ion, $Mg^{2+}$?

50. What is incorrect in the electron configuration $1s^2 2s^2 p^6 3s^2 p^8 d^{10}$?

51. (a) What are the largest and the smallest values that the magnetic quantum number $m$ has in the Zn atom?
    (b) How many electrons have $m = 0$ in Zn?

52. Show that the total number of states available in a shell with principal quantum number $n$ is $2n^2$.

53. What is the wavelength of a photon of maximum energy from an X-ray tube operating at a peak voltage of 80 kV?

54. If the operating voltage of an X-ray tube is doubled, by what factor does the wavelength of a photon of maximum energy change?

55. (a) How many electrons per second strike the target in an X-ray tube operating at a current of 50 mA?
    (b) If the potential difference between the anode and cathode is 100 kV, how much power is expended?

56. If the binding energies for electrons in the K, L, and M shells of an element are, respectively, 8979 eV, 951 eV, and 74 eV, what are the energies of the $K_\alpha$ and $K_\beta$ characteristic X rays? (These values are representative of Cu without the fine structure.)

57. Given that the $K_{\alpha 1}$ characteristic X ray of copper has an energy of 8.05 keV, estimate the energy of the $K_{\alpha 1}$ X ray of tin.

58. The oxygen atom has a K-shell binding energy of 532 eV and L-shell binding energies of 23.7 eV and 7.1 eV. What are the possible energies of its Auger electrons?

**2.14**

**Answers**

1. $1.61 \times 10^{23}$
2. 1530 L
7. $1.81 \times 10^{17}$ kg m$^{-3}$
9. 23.6 MeV
11. $8.23 \times 10^{-8}$ N
12. $5.27 \times 10^{-34}$ J s
14. 122 eV
16. 0.705 Å
18. 2
19. 0.661 eV; 0.967 eV
20. 4030, 4100, 4200,
    4340, 4540, 4690,
    4860, 5420, and
    6560 Å
21. 7
22. 1.05 mA
26. (a) 6
    (b) $4\hbar$
27. $0.99986m$
29. (a) $2.84 \times 10^{-13}$ m

   (b) 2.53 keV
33. $6.63 \times 10^{-27}$ kg m s$^{-1}$
35. $5.19 \times 10^{-5}$ Å
36. 533 eV
38. 0.0222 Å
40. 0.294 MeV;
    0.621 MeV
41. (a) 946 eV
    (b) 31.0 keV
43. $5.27 \times 10^{-25}$ kg m s$^{-1}$;
    same
45. 0.2 Å
47. (a) 3
    (b) 4
51. (a) $\pm 2$
    (b) 14
53. 0.155 Å
55. (a) $3.12 \times 10^{17}$ s$^{-1}$
    (b) 5000 W
57. 24 keV

# 3
# The Nucleus and Nuclear Radiation

## 3.1
## Nuclear Structure

The nucleus of an atom of atomic number $Z$ and mass number $A$ consists of $Z$ protons and $N = A - Z$ neutrons. The atomic masses of all individual atoms are nearly integers, and $A$ gives the total number of *nucleons* (i.e., protons and neutrons) in the nucleus. A species of atom, characterized by its nuclear constitution—its values of $Z$ and $A$ (or $N$)—is called a *nuclide*. It is conveniently designated by writing the appropriate chemical symbol with a subscript giving $Z$ and superscript giving $A$. For example, $^1_1\text{H}$, and $^2_1\text{H}$, and $^{238}_{92}\text{U}$ are nuclides. Nuclides of an element that have different $A$ (or $N$) are called *isotopes*; nuclides having the same number of neutrons are called *isotones*; for example, $^{206}_{82}\text{Pb}$ and $^{204}_{80}\text{Hg}$ are isotones with $N = 124$. Hydrogen has three isotopes, $^1_1\text{H}$, $^2_1\text{H}$, and $^3_1\text{H}$, all of which occur naturally. Deuterium, $^2_1\text{H}$, is stable; tritium, $^3_1\text{H}$, is radioactive. Fluorine has only a single naturally occurring isotope, $^{19}_9\text{F}$; all of its other isotopes are man-made, radioactive, and short lived. The measured atomic weights of the elements reflect the relative abundances of the isotopes found in nature, as the next example illustrates.

*Example*

Chlorine is found to have two naturally occurring isotopes: $^{35}_{17}\text{Cl}$, which is 76% abundant, and $^{37}_{17}\text{Cl}$, which is 24% abundant. The atomic weights of the two isotopes are 34.97 and 36.97. Show that this isotopic composition accounts for the observed atomic weight of the element.

*Solution*

Taking the weighted average of the atomic weights of the two isotopes, we find for the atomic weight of Cl, $0.76 \times 34.97 + 0.24 \times 36.97 = 35.45$, as observed. (See periodic table in back of book.)

Since the electron configuration of the different isotopes of an element is the same, isotopes cannot be separated chemically. The existence of isotopes does cause a very slight perturbation in atomic energy levels, leading to an observed "isotope shift" in some spectral lines. In addition, the different nuclear spins of

*Atoms, Radiation, and Radiation Protection.* James E. Turner
Copyright © 2007 WILEY-VCH Verlag GmbH & Co. KGaA, Weinheim
ISBN: 978-3-527-40606-7

different isotopes of the same element are responsible for hyperfine structure in the spectra of elements. As we mentioned at the end of Section 2.8, the existence of isotopes has a big effect on the vibration–rotation spectra of molecules.

Nucleons are bound together in a nucleus by the action of the strong, or nuclear, force. The range of this force is only of the order of nuclear dimensions, $\sim 10^{-15}$ m, and it is powerful enough to overcome the Coulomb repulsion of the protons in the nucleus. Figure 3.1(a) schematically shows the potential energy of a proton as a function of the distance $r$ separating its center and the center of a nucleus. The potential energy is zero at large separations. As the proton comes closer, its potential energy increases, due to the work done against the repulsive Coulomb force that

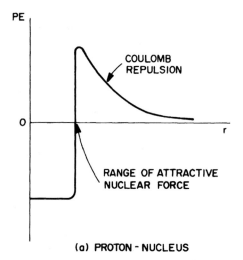

(a) PROTON – NUCLEUS

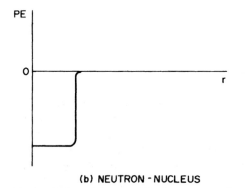

(b) NEUTRON – NUCLEUS

**Fig. 3.1** (a) Potential energy (PE) of a proton as a function of its separation $r$ from the center of a nucleus, (b) Potential energy of a neutron and a nucleus as a function of $r$. The uncharged neutron has no repulsive Coulomb barrier to overcome when approaching a nucleus.

acts between the two positive charges. Once the proton comes within range of the attractive nuclear force, though, its potential energy abruptly goes negative and it can react with the nucleus. If conditions are right, the proton's total energy can also become negative, and the proton will then occupy a bound state in the nucleus. As we learned in the Rutherford experiment in Section 2.2, a positively charged particle requires considerable energy in order to approach a nucleus closely. In contrast, the nucleus is accessible to a neutron of any energy. Because the neutron is uncharged, there is no Coulomb barrier for it to overcome. Figure 3.1(b) shows the potential-energy curve for a neutron and a nucleus.

*Example*

Estimate the minimum energy that a proton would have to have in order to react with the nucleus of a stationary Cl atom.

*Solution*

In terms of Fig. 3.1(a), the proton would have to have enough energy to overcome the repulsive Coulomb barrier in a head-on collision. This would allow it to just reach the target nucleus. We can use Eq. (2.2) to estimate how far apart the centers of the proton and nucleus would then be, when they "just touch." With $A = 1$ and $A = 35$ in Eq. (2.2), we obtain for the radii of the proton ($r_p$) and the chlorine nucleus ($r_{Cl}$)

$$r_p = 1.3 \times 1^{1/3} \times 10^{-15} = 1.3 \times 10^{-15} \text{ m},\qquad(3.1)$$

$$r_{Cl} = 1.3 \times 35^{1/3} \times 10^{-15} = 4.3 \times 10^{-15} \text{ m}.\qquad(3.2)$$

The proton has unit positive charge, $e = 1.60 \times 10^{-19}$ C, and the chlorine ($Z = 17$) nucleus has a charge $17e$. The potential energy of the two charges separated by the distance $r_p + r_{Cl} = 5.6 \times 10^{-15}$ m is therefore (Appendix C)

$$\text{PE} = \frac{8.99 \times 10^9 \times 17 \times (1.60 \times 10^{-19})^2}{5.6 \times 10^{-15}}$$

$$= 7.0 \times 10^{-13} \text{ J} = 4.4 \text{ MeV}.\qquad(3.3)$$

(Problem 9 in Chapter 2 is worked like this example.)

Like an atom, a nucleus is itself a quantum-mechanical system of bound particles. However, the nuclear force, acting between nucleons, is considerably more complicated and more uncertain than the electromagnetic force that governs the structure and properties of atoms and molecules. In addition, wave equations describing nuclei cannot be solved with the same degree of numerical precision that atomic wave equations can. Nevertheless, many detailed properties of nuclei have been worked out and verified experimentally. Both the proton and the neutron are "spin-$\frac{1}{2}$" particles and hence obey the Pauli exclusion principle. Just as excited electron states exist in atoms, excited states can exist in nuclei. Whereas an atom has an infinite number of bound excited states, however, a nucleus has only a finite number, if any. This difference in atomic and nuclear structure is attributable to

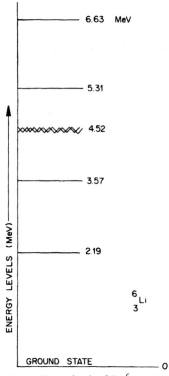

6.63 MeV

5.31

4.52

3.57

2.19

$^{6}_{3}$Li

GROUND STATE 0

**Fig. 3.2** Energy levels of the $^{6}_{3}$Li nucleus, relative to the ground state of zero energy.

the infinite range of the Coulomb force as opposed to the short range and limited, though large, strength of the nuclear force. The energy-level diagram of the $^{6}_{3}$Li nucleus in Fig. 3.2 shows that it has a number of bound excited states.[1] The deuteron and alpha particle (nuclei of $^{2}_{1}$H and $^{4}_{2}$He) are examples of nuclei that have no bound excited states.

## 3.2
## Nuclear Binding Energies

Changes can occur in atomic nuclei in a number of ways, as we shall see throughout this book. Nuclear reactions can be either exothermic (releasing energy) or endothermic (requiring energy in order to take place). The energies associated with nuclear changes are usually in the MeV range. They are thus $\sim 10^{6}$ times

---

1 The "level" at 4.52 MeV is very short lived and therefore does not have a sharp energy. All quantum-mechanical energy levels have a natural width, a manifestation of the uncertainty relation for energy and time, $\Delta E \, \Delta t \gtrsim \hbar$ [Eq. (2.31)]. The lifetimes of atomic states ($\sim 10^{-8}$ s) are long and permit precise knowledge of their energies ($\Delta E \sim 10^{-7}$ eV). For many excited nuclear states, the lifetime $\Delta t$ is so short that the uncertainty in their energy, $\Delta E$, is large, as is the case here.

greater than the energies associated with the valence electrons that are involved in chemical reactions. This factor characterizes the enormous difference in the energy released when an atom undergoes a nuclear transformation as compared with a chemical reaction.

The energy associated with exothermic nuclear reactions comes from the conversion of mass into energy. If the mass loss is $\Delta M$, then the energy released, $Q$, is given by Einstein's relation, $Q = (\Delta M)c^2$, where $c$ is the velocity of light. In this section we discuss the energetics of nuclear transformations.

We first establish the quantitative relationship between atomic mass units (AMU) and energy (MeV). By definition, the $^{12}$C atom has a mass of exactly 12 AMU. Since its gram atomic weight is 12 g, it follows that

$$1 \text{ AMU} = 1/(6.02 \times 10^{23}) = 1.66 \times 10^{-24} \text{ g} = 1.66 \times 10^{-27} \text{ kg}. \tag{3.4}$$

Using the Einstein relation and $c = 3 \times 10^8 \text{ m s}^{-1}$, we obtain

$$1 \text{ AMU} = (1.66 \times 10^{-27})(3 \times 10^8)^2$$

$$= 1.49 \times 10^{-10} \text{ J} \tag{3.5}$$

$$= \frac{1.49 \times 10^{-10} \text{ J}}{1.6 \times 10^{-13} \text{ J MeV}^{-1}} = 931 \text{ MeV}. \tag{3.6}$$

More precisely, 1 AMU = 931.49 MeV.

We now consider one of the simplest nuclear reactions, the absorption of a thermal neutron by a hydrogen atom, accompanied by emission of a gamma ray. This reaction, which is very important for understanding the thermal-neutron dose to the body, can be represented by writing

$$^1_0 n + {}^1_1 H \rightarrow {}^2_1 H + {}^0_0 \gamma, \tag{3.7}$$

the photon having zero charge and mass. The reaction can also be designated $^1_1 H (n, \gamma) {}^2_1 H$. To find the energy released, we compare the total masses on both sides of the arrow. Appendix D contains data on nuclides which we shall frequently use. The atomic weight $M$ of a nuclide of mass number $A$ can be found from the mass difference, $\Delta$, given in column 3. The quantity $\Delta = M - A$ gives the difference between the nuclide's atomic weight and its atomic mass number, expressed in MeV. (By definition, $\Delta = 0$ for the $^{12}$C atom.) Since we are interested only in energy differences in the reaction (3.7), we obtain the energy released, $Q$, directly from the values of $\Delta$, without having to calculate the actual masses of the neutron and individual atoms. Adding the $\Delta$ values for $^1_0 n$ and $^1_1 H$ and subtracting that for $^2_1 H$, we find

$$Q = 8.0714 + 7.2890 - 13.1359 = 2.2245 \text{ MeV}. \tag{3.8}$$

This energy appears as a gamma photon emitted when the capture takes place (the thermal neutron has negligible kinetic energy).

The process (3.7) is an example of energy release by the fusion of light nuclei. The binding energy of the deuteron is 2.2245 MeV, which is the energy required to separate the neutron and proton. As the next example shows, the binding energy of any nuclide can be calculated from a knowledge of its atomic weight (obtainable from $\Delta$) together with the known individual masses of the proton, neutron, and electron.

*Example*

Find the binding energy of the nuclide $^{24}_{11}\text{Na}$.

*Solution*

One can work in terms of either AMU or MeV. The atom consists of 11 protons, 13 neutrons, and 11 electrons. The total mass in AMU of these separate constituents is, with the help of the data in Appendix A,

$$11(1.0073) + 13(1.0087) + 11(0.00055) = 24.199 \text{ AMU}. \tag{3.9}$$

From Appendix D, $\Delta = -8.418$ MeV gives the difference $M - A$. Thus, the mass of the $^{24}_{11}\text{Na}$ nuclide is less than 24 by the amount 8.418 MeV/(931.49 MeV AMU$^{-1}$) = 0.0090371 AMU. Therefore, the nuclide mass is $M = 23.991$ AMU. Comparison with (3.9) gives for the binding energy

$$\text{BE} = 24.199 - 23.991 = 0.208 \text{ AMU} = 194 \text{ MeV}. \tag{3.10}$$

This figure represents the total binding energy of the atom—nucleons plus electrons. However, the electron binding energies are small compared with nuclear binding, which accounts for essentially all of the 194 MeV. Thus the binding energy per nucleon in $^{24}_{11}\text{Na}$ is $194/24 = 8.08$ MeV. [Had we worked in MeV, rather than AMU, the data from Appendix A give, in place of (3.9), $2.2541 \times 10^4$ MeV. Expressed in MeV, $A = 24 \times 931.49 = 2.2356 \times 10^4$ MeV. With $\Delta = -8.418$ MeV we have $M = A + \Delta = 2.2348 \times 10^4$ MeV. Thus the binding energy of the atom is $(2.2541 - 2.2348) \times 10^4 = 193$ MeV.]

The average binding energy per nucleon is plotted as a function of atomic mass number in Fig. 3.3. The curve has a broad maximum at about 8.5 MeV from $A = 40$ to 120.[2] It then drops off as one goes either to lower or higher $A$. The implication from this curve is that the *fusion* of light elements releases energy, as does *the fission* of heavy elements. Both transformations are made exothermic through the increased average nucleon binding energy that results. The $^1_1\text{H}(n,\gamma)^2_1\text{H}$ reaction considered earlier is an example of the release of energy through fusion. With a few exceptions, the average binding energies for all nuclides fall very nearly on the single curve shown. The nuclides $^4_2\text{He}$, $^{12}_6\text{C}$, and $^{16}_8\text{O}$ show considerably tighter

---

2   The fact that the average nucleon binding energy is nearly constant over such a wide range of $A$ is a manifestation of the saturation property of nuclear forces, mentioned at the end of Section 2.2.

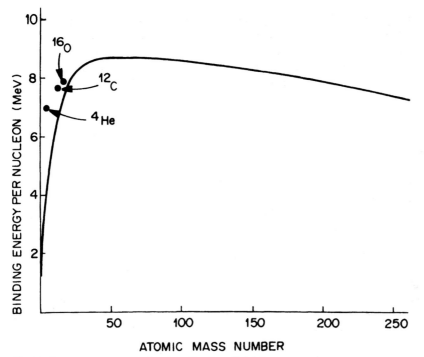

**Fig. 3.3** Average energy per nucleon as a function of atomic mass number.

binding than their immediate neighbors. These nuclei are all "multiples" of the alpha particle, which appears to be a particularly stable nuclear subunit. (No nuclides with $A = 5$ exist for longer than $\sim 10^{-21}$ s.[3])

The loss of mass that accompanies the binding of particles is not a specifically nuclear phenomenon. The mass of the hydrogen atom is smaller than the sum of the proton and electron masses by $1.46 \times 10^{-8}$ AMU. This is equivalent to an energy $1.46 \times 10^{-8}$ AMU $\times$ 931 MeV/AMU$^{-1} = 1.36 \times 10^{-5}$ MeV $= 13.6$ eV, the binding energy of the H atom.

We turn now to the subject of radioactivity, the property that some atomic species, called radionuclides, have of undergoing spontaneous nuclear transformation. All of the heaviest elements are radioactive; $^{209}_{83}$Bi is the only stable nuclide with $Z > 82$. All elements have radioactive isotopes, the majority being man-made. The various kinds of radioactive decay and their associated nuclear energetics are described in the following sections.

3  Various forms of shell models have been studied for nuclei, analogous to an atomic shell model. The alpha particle consists of two spin-$\frac{1}{2}$ protons and two spin-$\frac{1}{2}$ neutrons in s states, forming the most tightly bound, "inner" nuclear shell. Generally, nuclei with even numbers of protons and neutrons ("even–even" nuclei) have the largest binding energies per nucleon.

## 3.3
## Alpha Decay

Almost all naturally occurring alpha emitters are heavy elements with $Z \geq 83$. The principal features of alpha decay can be learned from the example of $^{226}$Ra:

$$^{226}_{88}\text{Ra} \rightarrow {}^{222}_{86}\text{Rn} + {}^{4}_{2}\text{He}. \tag{3.11}$$

The energy $Q$ released in the decay arises from a net loss in the masses $M_{\text{Ra,N}}$, $M_{\text{Rn,N}}$, and $M_{\text{He,N}}$, of the radium, radon, and helium nuclei:

$$Q = M_{\text{Ra,N}} - M_{\text{Rn,N}} - M_{\text{He,N}}. \tag{3.12}$$

This nuclear mass difference is very nearly equal the atomic mass difference, which, in turn, is equal to the difference in $\Delta$ values.[4] Letting $\Delta_\text{P}$, $\Delta_\text{D}$, and $\Delta_\text{He}$ denote the values of the parent, daughter, and helium atoms, we can write a general equation for obtaining the energy release in alpha decay:

$$Q_\alpha = \Delta_\text{P} - \Delta_\text{D} - \Delta_\text{He}. \tag{3.13}$$

Using the values in Appendix D for the decay of $^{226}_{88}$Ra to the ground state of $^{222}_{86}$Rn, we obtain

$$Q = 23.69 - 16.39 - 2.42 = 4.88 \text{ MeV}. \tag{3.14}$$

The $Q$ value (3.14) is shared by the alpha particle and the recoil radon nucleus, and we can calculate the portion that each acquires. Since the radium nucleus was at rest, the momenta of the two decay products must be equal and opposite. Letting $m$ and $v$ represent the mass and initial velocity of the alpha particle and $M$ and $V$ those of the recoil nucleus, we write

$$mv = MV. \tag{3.15}$$

Since the initial kinetic energies of the products must be equal to the energy released in the decay, we have

$$\tfrac{1}{2}mv^2 + \tfrac{1}{2}MV^2 = Q. \tag{3.16}$$

Substituting $V = mv/M$ from Eq. (3.16) into (3.16) and solving for $v^2$, one finds

$$v^2 = \frac{2MQ}{m(m+M)}. \tag{3.17}$$

---

4 Specifically, the relatively slight difference in the binding energies of the 88 electrons on either side of the arrow in (3.11) is neglected when atomic mass loss is equated to nuclear mass loss. In principle, nuclear masses are needed; however, atomic masses are much better known. These small differences are negligible for most purposes.

One thus obtains for the alpha-particle energy

$$E_\alpha = \tfrac{1}{2}mv^2 = \frac{MQ}{m+M}.$$ 

(3.18)

With the roles of the two masses interchanged, it follows that the recoil energy of the nucleus is

$$E_N = \tfrac{1}{2}MV^2 = \frac{mQ}{m+M}.$$ 

(3.19)

As a check, we see that $E_\alpha + E_N = Q$. Because of its much smaller mass, the alpha particle, having the same momentum as the nucleus, has much more energy. For $^{226}$Ra, it follows from (3.14) and (3.18) that

$$E_\alpha = \frac{222 \times 4.88}{4+222} = 4.79 \text{ MeV}.$$ 

(3.20)

The radon nucleus recoils with an energy of only 0.09 MeV.

The conservation of momentum and energy, Eqs. (3.15) and (3.16), fixes the energy of an alpha particle uniquely for given values of $Q$ and $M$. Alpha particles therefore occur with discrete values of energy.

Appendix D gives the principal radiations emitted by various nuclides. We consider each of those listed for $^{226}$Ra. Two alpha-particle energies are shown: 4.785 MeV, occurring with a frequency of 94.4% of all decays, and 4.602 MeV, occurring 5.5% of the time. The $Q$ value for the less frequent alpha particle can be found from Eq. (3.18):

$$Q = \frac{(m+M)E_\alpha}{M} = \frac{226 \times 4.60}{222} = 4.68 \text{ MeV}.$$ 

(3.21)

The decay in this case goes to an excited state of the $^{222}$Rn nucleus. Like excited atomic states, excited nuclear states can decay by photon emission. Photons from the nucleus are called gamma rays, and their energies are generally in the range from tens of keV to several MeV. Under the gamma rays listed in Appendix D for $^{226}$Ra we find a 0.186-MeV photon emitted in 3.3% of the decays, in addition to another that occurs very infrequently (following alpha decay to still another excited level of higher energy in the daughter nucleus). We conclude that emission of the higher energy alpha particle ($E_\alpha = 4.79$ MeV) leaves the daughter $^{222}$Rn nucleus in its ground state. Emission of the 4.60 MeV alpha particle leaves the nucleus in an excited state with energy $4.79 - 4.60 = 0.19$ MeV above the ground state. A photon of this energy can then be emitted from the nucleus, and, indeed, one of energy 0.186 MeV is listed for 3.3% of the decays. As an alternative to photon emission, under certain circumstances an excited nuclear state can decay by ejecting an atomic electron, usually from the K or L shell. This process, which produces the electrons listed (e⁻), is called internal conversion, and will be discussed in Section 3.6.[5] For

5   In atoms, an Auger electron can be ejected from a shell in place of a photon, accompanying an electronic transition (Sect. 2.11).

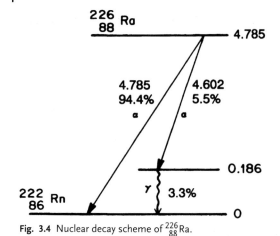

**Fig. 3.4** Nuclear decay scheme of $^{226}_{88}$Ra.

$^{226}$Ra, since the excited state occurs in 5.5% of the total disintegrations and the 0.186 MeV photon is emitted only 3.3% of the time, it follows that internal conversion occurs in about 2.2% of the total decays. As we show in more detail in Section 3.6, the energy of the conversion electron is equal to the excited-state energy (in this case 0.186 MeV) minus the atomic-shell binding energy. The listing in Appendix D shows one of the e$^-$ energies to be 0.170 MeV. In addition, since internal conversion leaves a K- or L-shell vacancy in the daughter atom, one also finds among the photons emitted the characteristic X rays of Rn. Finally, as noted in the radiations listed in Appendix D for $^{226}$Ra, various kinds of radiation are emitted from the radioactive daughters, in this case $^{222}$Rn, $^{218}$Po, $^{214}$Pb, $^{214}$Bi, and $^{214}$Po.

Decay-scheme diagrams, such as that shown in Fig. 3.4 for $^{226}$Ra, conveniently summarize the nuclear transformations. The two arrows slanting downward to the left[6] show the two modes of alpha decay along with the alpha-particle energies and frequencies. Either changes the nucleus from that of $^{226}$Ra to that of $^{222}$Rn. When the lower energy particle is emitted, the radon nucleus is left in an excited state with energy 0.186 MeV above the ground state. (The vertical distances in Fig. 3.4 are not to scale.) The subsequent gamma ray of this energy, which is emitted almost immediately, is shown by the vertical wavy line. The frequency 3.3% associated with this photon emission implies that an internal-conversion electron is emitted in the other 2.2% of the total number of disintegrations. Radiations not emitted directly from the nucleus (i.e., the Rn X rays and the internal-conversion electron) are not shown on such a diagram, which represents the nuclear changes. Relatively infrequent modes of decay could also be shown, but are not included in Fig. 3.4 (see Fig. 3.7). (A small round-off error occurs in the energies.)

The most energetic alpha particles are found to come from radionuclides having relatively short half-lives. An early empirical finding, known as the Geiger–Nuttall law, implies that there is a linear relationship between the logarithm of the range $R$

6    By convention, going left represents a            Photon emission is represented by a vertical
      decrease in Z and right, an increase in Z.          wavy line.

of an alpha particle in air and the logarithm of the emitter's half-life $T$. The relation can be expressed in the form

$$-\ln T = a + b \ln R, \tag{3.22}$$

where $a$ and $b$ are empirical constants.

To conclude this section, we briefly consider the possible radiation-protection problems that alpha emitters can present. As we shall see in Chapter 5, alpha particles have very short ranges and cannot even penetrate the outer, dead layer of skin. Therefore, they generally pose no direct external hazard to the body. Inhaled, ingested, or entering through a wound, however, an alpha source can present a hazard as an internal emitter. Depending upon the element, internal emitters tend to seek various organs and irradiate them. Radium, for example, seeks bone, where it can become lodged and irradiate an individual over his or her lifetime. In addition to the internal hazard, one can generally expect gamma rays to occur with an alpha source, as is the case with radium. Also, many alpha emitters have radioactive daughters that present radiation-protection problems.

## 3.4

## Beta Decay ($\beta^-$)

In beta decay, a nucleus simultaneously emits an electron, or negative beta particle, $_{-1}^{0}\beta$, and an antineutrino, $_{0}^{0}\bar{\nu}$. Both of these particles are created at the moment of nuclear decay. The antineutrino, like its antiparticle[7] the neutrino, $_{0}^{0}\nu$, has no charge and little or no mass;[8] they have been detected only in rather elaborate experiments.

As an example of beta decay, we consider $^{60}$Co:

$$_{27}^{60}\text{Co} \rightarrow {}_{28}^{60}\text{Ni} + {}_{-1}^{0}\beta + {}_{0}^{0}\bar{\nu}. \tag{3.23}$$

In this case, the value of $Q$ is equal to the difference between the mass of the $^{60}$Co nucleus, $M_{\text{Co,N}}$, and that of the $^{60}$Ni nucleus, $M_{\text{Ni,N}}$, plus one electron ($m$):

$$Q = M_{\text{Co,N}} - (M_{\text{Ni,N}} + m). \tag{3.24}$$

The nickel atom has one more electron than the cobalt atom. Therefore, if we neglect differences in atomic-electron binding energies, Eq. (3.24) implies that $Q$ is

---

7   The Dirac equation predicts the existence of an antiparticle for every spin-$\frac{1}{2}$ particle and describes its relationship to the particle. Other examples include the positron, $_{+1}^{0}\beta$, antiparticle to the electron; the antiproton; and the antineutron. Creation of a spin-$\frac{1}{2}$ particle is always accompanied by creation of a related particle, which can be the antiparticle, such as happens in the creation of an electron–positron pair. Particle–antiparticle pairs can annihilate, as electrons and positrons do. A bar over a symbol is used to denote an antiparticle: for example, $\nu, \bar{\nu}$. Several kinds of neutrinos have been found—electron, muon, and tau.

8   Experimentally, the neutrino and antineutrino masses cannot be larger than about 30 eV.

simply equal to the difference in the masses of the $^{60}$Co and $^{60}$Ni *atoms*.[9] There-fore, it follows that one can compute the energy released in beta decay from the difference in the values $\Delta_P$ and $\Delta_D$, of the parent and daughter atoms:

$$Q_{\beta^-} = \Delta_P - \Delta_D. \tag{3.25}$$

Using the data from Appendix D, we find for the energy released in a $\beta^-$ transformation of $^{60}$Co to the ground state of $^{60}$Ni

$$Q = -61.651 - (-64.471) = 2.820 \text{ MeV}. \tag{3.26}$$

In accordance with (3.23), this energy is shared by the beta particle, antineutrino, and recoil $^{60}$Ni nucleus. The latter, because of its relatively large mass, receives negligible energy, and so

$$E_{\beta^-} + E_{\bar{\nu}} = Q, \tag{3.27}$$

where $E_{\beta^-}$ and $E_{\bar{\nu}}$ are the initial kinetic energies of the electron and antineutrino. Depending on the relative directions of the momenta of the three decay products ($\beta^-$, $\bar{\nu}$, and recoil nucleus), $E_{\beta^-}$ and $E_{\bar{\nu}}$ can each have any value between zero and $Q$, subject to the condition (3.27) on their sum. Thus the spectrum of beta-particle energies $E_{\beta^-}$ is continuous, with $0 \le E_{\beta^-} \le Q$, in contrast to the discrete spectra of alpha particles, as required by Eq. (3.18). Alpha particles are emitted in a decay into two bodies, which must share energy and momentum in a unique way, giving rise to discrete alpha spectra. Beta particles are emitted in a decay into three bodies, which can share energy and momentum in a continuum of ways, resulting in continuous beta spectra. The shape of a typical spectrum is shown in Fig. 3.5.[10] The maximum beta-particle energy is always equal to the $Q$ value for the nuclear transition. As a rule of thumb, the average beta energy is about one-third of $Q$: $\bar{E}_{\beta^-} \sim Q/3$.

To construct the decay scheme for $^{60}$Co we consult Appendix D. We see that 99 + % of the decays occur with $Q = 0.318$ MeV and that both of the gamma photons occur with almost every disintegration. Therefore, almost every decay must go through an excited state of the daughter $^{60}$Ni nucleus with an energy at least $1.173 + 1.332 = 2.505$ MeV above the ground state. Adding the maximum beta energy to this gives $2.505 + 0.318 = 2.823$ MeV, the value [Eq. (3.26), except for round-off] calculated for a transition all the way to the ground state of the $^{60}$Ni nucleus. Therefore, we conclude that the $^{60}$Co nucleus first emits a beta particle, with $Q = 0.318$ MeV, which is followed successively by the two gamma rays. It remains

---

9   We can think of adding and subtracting 27 electron masses in Eq. (3.24), giving $Q = (M_{Co,N} + 27m) - (M_{Ni,N} + 28m)$. Neglecting the difference in electron binding energies, then, we have $Q = M_{Co,A} - M_{Ni,A}$, where the subscript $A$ denotes the atomic masses. It follows that $Q$ is equal to the difference in $\Delta$ values for the two atoms.

10   Beta-ray spectra exhibit a variety of shapes. The spectra of some 100 nuclides of importance for radiation protection and biomedical applications are presented by W. G. Gross, H. Ing, and N. Freedman, "A Short Atlas of Beta-Ray Spectra," Phys. Med. Biol. **28**, 1251–1260 (1983).

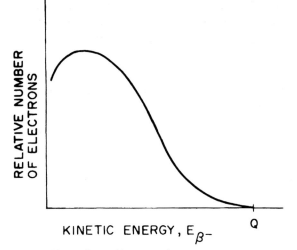

**Fig. 3.5** Shape of typical beta-particle energy spectrum.

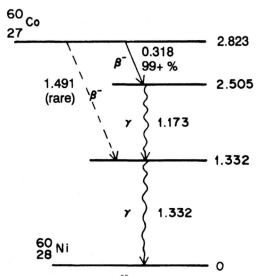

**Fig. 3.6** Decay scheme of $^{60}_{27}$Co.

to determine the energy of the nuclear excited state from which the second photon is emitted: 1.173 MeV or 1.332 MeV? Appendix D lists a rare beta particle with
$Q = 1.491$ MeV. This decay must go to a level in the daughter nucleus having an energy $2.823 - 1.491 = 1.332$ MeV above the ground state. Thus we can conclude that
the 1.332 MeV photon is emitted last in the transition to the ground state. The decay
scheme is shown in Fig. 3.6. The arrows drawn slanting toward the right indicate
the increase in atomic number that results from $\beta^-$ decay. The rare mode is shown
with a dashed line. No significant internal conversion occurs with this radionuclide.

A number of beta emitters have no accompanying gamma rays. Examples of such pure beta emitters are $^3$H, $^{14}$C, $^{32}$P, $^{90}$Sr, and $^{90}$Y. Mixed beta–gamma emitters include $^{60}$Co, $^{137}$Cs, and many others. A number of radionuclides emit beta particles in decaying to several levels of the daughter nucleus, thus giving rise to complex beta spectra. A few radioisotopes can decay by emission of either an alpha or a beta particle. For example, $^{212}_{83}$Bi decays by alpha emission 36% of the time and by beta emission 64% of the time.

Beta rays can have sufficient energy to penetrate the skin and thus be an external radiation hazard. Internal beta emitters are also a hazard. As is the case with $^{60}$Co, many beta radionuclides also emit gamma rays. High-energy beta particles (i.e., in the MeV range) can emit bremsstrahlung, particularly in heavy-metal shielding. The bremsstrahlung from a beta source may be the only radiation that escapes the containment.

## 3.5
### Gamma-Ray Emission

As we have seen, one or more gamma photons can be emitted from the excited states of daughter nuclei following radioactive decay. Transitions that result in gamma emission leave $Z$ and $A$ unchanged and are called *isomeric*; nuclides in the initial and final states are called *isomers*.

As the examples in the last two sections illustrate, the gamma-ray spectrum from a radionuclide is discrete. Furthermore, just as optical spectra are characteristic of the chemical elements, a gamma-ray spectrum is characteristic of the particular radionuclides that are present. By techniques of gamma-ray spectroscopy (Chapter 10), the intensities of photons at various energies can be measured to determine the distribution of radionuclides in a sample. When $^{60}$Co is present, for example, photons of energy 1.173 MeV and 1.332 MeV are observed with equal frequency. (Although these are called "$^{60}$Co gamma rays," we note from Fig. 3.6 that they are actually emitted by the daughter $^{60}$Ni nucleus.) Radium can also be easily detected by its gamma-ray spectrum, which is more complex than indicated by Fig. 3.4. Since individual photons are registered in a spectrometer, gamma rays from infrequent modes of radioactive decay can often be readily measured. Figure 3.7 shows a more detailed decay scheme for $^{226}$Ra, which involves three excited states of the daughter $^{222}$Rn nucleus and the emission of photons of four different energies. Transitions from the highest excited level (0.601 MeV) to the next (0.448 MeV) and from there to ground are "forbidden" by selection rules.

*Example*

Like $^{60}_{27}$Co, another important gamma-ray source is the radioisotope $^{137}_{55}$Cs. Consult Appendix D and work out its decay scheme.

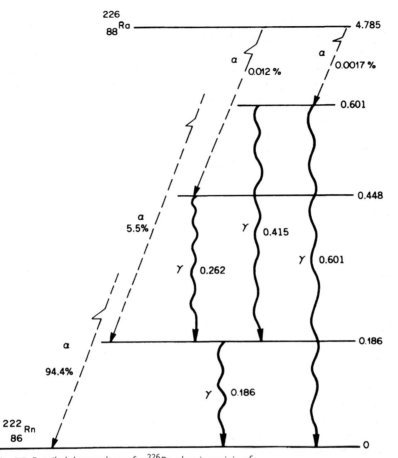

**Fig. 3.7** Detailed decay scheme for $^{226}_{88}$Ra, showing origin of photons found in its gamma spectrum (position of initial $^{226}_{88}$Ra energy level not to scale).

## Solution

Also like $^{60}_{27}$Co, $^{137}_{55}$Cs is a $\beta^-$ emitter that leaves its daughter, stable $^{137}_{56}$Ba, in an excited state that results in gamma emission. The decay is represented by

$$^{137}_{55}\text{Cs} \rightarrow {}^{137}_{56}\text{Ba} + {}^{0}_{-1}\beta + {}^{0}_{0}\bar{\nu}. \tag{3.28}$$

From the $\Delta$ values in Appendix D, we obtain for decay to the daughter ground state $Q = -86.9 + 88.0 = 1.1$ MeV. Comparison with the radiations listed in the Appendix indicates that decay by this mode takes place 5% of the time, releasing 1.174 MeV. Otherwise, the decay in 95% of the cases leaves the daughter nucleus in an excited state with energy $1.174 - 0.512 = 0.662$ MeV. A photon of this energy is shown with 85% frequency. Therefore, internal conversion occurs in $95 - 85 = 10\%$ of the disintegrations, giving rise to the conversion electrons, $e^-$, with the energies shown. Characteristic Ba X rays are emitted following the inner-shell vacancies created in

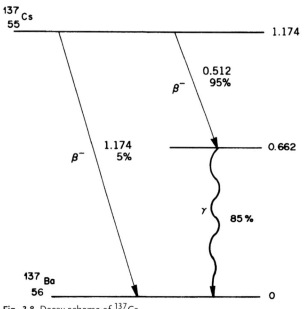

**Fig. 3.8** Decay scheme of $^{137}_{55}$Cs.

the atom by internal conversion. The decay scheme of $^{137}_{55}$Cs is drawn in Fig. 3.8; the spectrum of electrons emitted by the source is shown schematically in Fig. 3.9.

The lifetimes of nuclear excited states vary, but $\sim 10^{-10}$ s can be regarded as typical. Thus, gamma rays are usually emitted quickly after radioactive decay to an excited daughter state. In some cases, however, selection rules prevent photon emission for an extended period of time. The excited state of $^{137}_{56}$Ba following the decay of $^{137}_{55}$Cs has a half-life of 2.55 min. Such a long-lived nuclear state is termed *metastable* and is designated by the symbol m: $^{137m}_{56}$Ba.

Another example of a metastable nuclide is $^{99m}_{43}$Tc, which results from the beta decay of the molybdenum isotope $^{99}_{42}$Mo. $^{99m}_{43}$Tc has a half-life of 6.02 h in making an isomeric transition (IT) to the ground state:

$$^{99m}_{43}\text{Tc} \rightarrow {}^{99}_{43}\text{Tc} + {}^{0}_{0}\gamma. \tag{3.29}$$

The energy released in an isomeric transition is simply equal to the difference in $\Delta$ values of the parent and daughter atoms:

$$Q_{IT} = \Delta_P - \Delta_D. \tag{3.30}$$

*Example*

Work out the decay scheme of $^{99m}_{43}$Tc with the help of the data given in Appendix D.

*Solution*

Using Eq. (3.30) and the given values of $\Delta$, we obtain for the energy released in going to the ground state in the transition (3.29), $Q = 87.33 - 87.18 = 0.15$ MeV. This

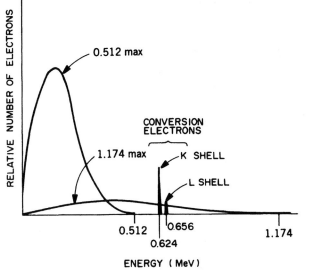

**Fig. 3.9** Sources of electrons from $^{137}_{55}$Cs and their energy spectra. There are two modes of $\beta^-$ decay, with maximum energies of 0.512 MeV (95%) and 1.174 MeV (5%). Internal-conversion electrons also occur, with discrete energies of 0.624 MeV (from the K shell) and 0.656 MeV (L shell) with a total frequency of 10%. See decay scheme in Fig. 3.8. The total spectrum of emitted electrons is the sum of the curves shown here.

transition is responsible for the gamma photon listed in Appendix D, 0.140 (89%). By implication, internal conversion must occur the other 11% of the time, and one finds two electron energies (e⁻), one a little less than the photon energy (by an amount that equals the L-shell electron binding energy in the Tc atom). Because internal conversion leaves inner-shell vacancies, a $^{99m}_{43}$Tc source also emits characteristic Tc X rays, as listed. Since $^{99}_{43}$Tc decays by $\beta^-$ emission into stable $^{99}_{44}$Ru, this daughter radiation also occurs.

The way in which gamma rays penetrate matter is fundamentally different from that of alpha and beta particles. Because of their charge, the latter lose energy almost continually as a result of electromagnetic forces that the electrons in matter exert on them. A shield of sufficient thickness can be used to absorb a beam of charged particles completely. Photons, on the other hand, are electrically neutral. They can therefore travel some depth in matter without being affected. As discussed in Section 8.5, monoenergetic photons, entering a uniform medium, have an exponential distribution of flight distances before they experience their first interaction. Although the intensity of a beam of gamma rays is steadily attenuated by passage through matter, some photons can traverse even thick shields with no interaction. Protection from gamma and X radiation is the subject of Chapter 15.

## 3.6
## Internal Conversion

Internal conversion is the process in which the energy of an excited nuclear state is transferred to an atomic electron, most likely a K- or L-shell electron, ejecting it from the atom. It is an alternative to emission of a gamma photon from the nucleus.[11] We had examples of internal conversion in discussing the decay of $^{137}$Cs and $^{99m}$Tc. In the case of the latter, however, the situation is somewhat more involved than described in the last example. The dominant mode of the isomeric transition to the ground state takes place in two steps. The first, internal conversion with release of nuclear energy by ejection of a 2-keV orbital electron, consumes virtually all of the 6.02-h half-life of $^{99m}$Tc. The second step, nuclear emission of the 140-keV gamma photon, then follows almost immediately, in $<10^{-9}$ s. The relatively soft gamma ray and the 6.02-h half-life, together with the ease of production (neutron bombardment of $^{98}$Mo in a reactor) as well as the special chemical properties of the element, endow $^{99m}$Tc with extremely useful properties for medical imaging. An example is shown in Fig. 3.10.

The internal conversion coefficient $\alpha$ for a nuclear transition is defined as the ratio of the number of conversion electrons $N_e$ and the number of competing gamma photons $N_\gamma$ for that transition:

$$\alpha = \frac{N_e}{N_\gamma}. \tag{3.31}$$

The kinetic energy $E_e$ of the ejected atomic electron is very nearly equal to the excitation energy $E^*$ of the nucleus minus the binding energy $E_B$ of the electron in its atomic shell:

$$E_e = E^* - E_B. \tag{3.32}$$

The conversion coefficients $\alpha$ increase as $Z^3$, the cube of the atomic number, and decrease with $E^*$. Internal conversion is thus prevalent in heavy nuclei, especially in the decay of low-lying excited states (small $E^*$). Gamma decay predominates in light nuclei.

## 3.7
## Orbital Electron Capture

Some nuclei undergo a radioactive transformation by capturing an atomic electron, usually from the K shell, and emitting a neutrino. An isotope of palladium under-

---

11 Internal conversion does not occur as a two-step process in which a photon is emitted by the nucleus and then absorbed by the atomic electron. The mechanism is entirely different. A similar observation was made in regard to Auger electrons in Section 2.11.

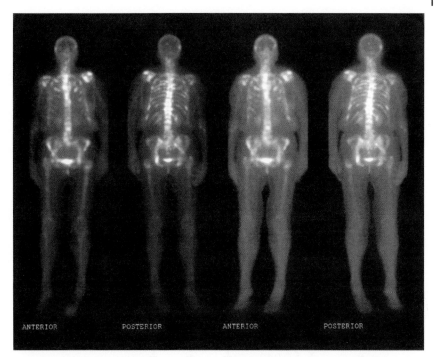

**Fig. 3.10** Images of a woman with a history of breast carcinoma, now with extensive metastatic disease throughout the spine, ribs, and pelvis. Additional lesions can be seen in the skull, shoulders, and proximal femurs. Patient was administered 25 mCi $^{99m}$Tc-hydroxymethylene diphosphonate ($^{99m}$Tc-HDP) and imaged 2–3 hours post injection. Anterior and posterior whole-body images were acquired simultaneously in approximately 15 min with a large field of view, dual head gamma camera. Images on left are displayed with a linear gray scale; those on right use a logarithmic gray scale, enhancing the soft-tissue activity. (Courtesy Glenn J. Hathaway, School of Nuclear Medicine, University of Tennessee Medical Center, Knoxville, TN.)

goes this process of electron capture (EC), going to a metastable state of the nucleus of the daughter rhodium:

$$^{103}_{46}\text{Pd} + {}^{0}_{-1}e \rightarrow {}^{103m}_{45}\text{Rh} + {}^{0}_{0}\nu. \tag{3.33}$$

The neutrino acquires the entire energy $Q$ released by the reaction.

To find $Q$, we note that the captured electron releases its total mass, $m - E_B$, to the nucleus when it is absorbed there, $E_B$ being the mass equivalent of the binding energy of the electron in the atomic shell. Therefore, in terms of the masses $M_{\text{Pd,N}}$ and $M_{m_{\text{Rh}},\text{N}}$ of the parent and daughter nuclei, the energy released by the reaction (3.33) is given by

$$Q = M_{\text{Pd,N}} + m - E_B - M_{m_{\text{Rh}},\text{N}}. \tag{3.34}$$

Since the palladium atom has one more electron than the rhodium atom, it follows that (neglecting the small difference in the electron binding energies) $Q$ is equal to

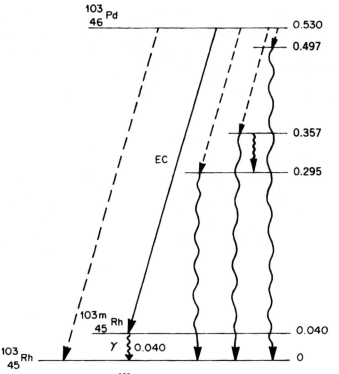

**Fig. 3.11** Decay scheme of $^{103}_{46}$Pd.

the difference in the two *atomic* masses, less the energy $E_B$.[12] Since the difference in the atomic masses is equal to the difference in the parent and daughter $\Delta$ values, we can write the general expression for the energy release by electron capture:

$$Q_{EC} = \Delta_P - \Delta_D - E_B. \tag{3.35}$$

Orbital electron capture thus cannot take place unless $\Delta_P - \Delta_D > E_B$. For the K shell of palladium, $E_B = 0.024$ MeV. Using the $\Delta$ values from Appendix D in Eq. (3.35), we find that, for the decay to $^{103m}$Rh,

$$Q = -87.46 - (-87.974) - 0.024 = 0.490 \text{ MeV}. \tag{3.36}$$

In subsequently decaying to the ground state, $^{103m}$Rh releases an energy $-87.974 + 88.014 = 0.040$ MeV, as found from the values in Appendix D.

A decay scheme for $^{103}$Pd is given in Fig. 3.11. Since electron capture decreases the atomic number of the nucleus, it is symbolized by an arrow pointing downward toward the left. The solid arrow represents the transition to $^{103m}$Rh that we just analyzed. The presence of the gamma rays listed for $^{103}$Pd in Appendix D implies that

12 A similar argument was given in the footnote
after Eq. (3.24), except that $E_B$ was not
involved there.

EC sometimes leaves the nucleus in other excited levels, as shown. These transitions, as well as one directly to the ground state, are indicated by the dashed arrows. It is not possible from the information given in Appendix D to specify the frequency of these transitions relative to that represented by the solid arrow. Since electron capture necessarily leaves an inner-atomic-shell vacancy, characteristic X rays of the daughter are always emitted. (Electron capture is detected through the observation of characteristic X rays and Auger electrons as well as the recoil of the daughter nucleus.)

The radiations listed for $^{103\text{m}}$Rh in Appendix D can also be explained. The photon with energy 0.040 MeV is shown in Fig. 3.11. Its 0.07% frequency implies that 99.03% of the time the metastable nucleus decays to the ground state by internal conversion, resulting in the ejection of atomic electrons ($e^-$) with the energies shown. (The present instance affords an example of internal conversion being favored over gamma emission in the decay of low-lying excited states in heavy nuclei, mentioned at the end of the last section.) Internal conversion also leaves a vacancy in an atomic shell, and hence the characteristic X rays of Rh are also found with a $^{103\text{m}}$Rh source.

The neutrino emitted in electron capture has a negligible interaction with matter and offers no radiation hazard, as far as is known. Characteristic X rays of the daughter will always be present. In addition, if the capture does not leave the daughter in its ground state, gamma rays will occur.

## 3.8
## Positron Decay ($\beta^+$)

Some nuclei, such as $^{22}_{11}$Na, disintegrate by emitting a positively charged electron (positron, $\beta^+$) and a neutrino:

$$^{22}_{11}\text{Na} \rightarrow {}^{22}_{10}\text{Ne} + {}^{0}_{1}\beta + {}^{0}_{0}\nu. \tag{3.37}$$

Positron decay has the same net effect as electron capture, reducing $Z$ by one unit and leaving $A$ unchanged. The energy released is given in terms of the masses $M_{\text{Na,N}}$ and $M_{\text{Ne,N}}$ of the sodium and neon nuclei by

$$Q = M_{\text{Na,N}} - M_{\text{Ne,N}} - m. \tag{3.38}$$

Thus the mass of the parent nucleus must be greater than that of the daughter nucleus by at least the mass $m$ of the positron it creates. As before, we need to express $Q$ in terms of atomic masses, $M_{\text{Na,A}}$ and $M_{\text{Ne,A}}$. Since Na has 11 electrons and Ne 10, we write

$$Q = M_{\text{Na,N}} + 11m - (M_{\text{Ne,N}} + 10m) - 2m \tag{3.39}$$

$$= M_{\text{Na,A}} - M_{\text{Ne,A}} - 2m, \tag{3.40}$$

where the difference in the atomic binding of the electrons has been neglected in writing the last equality. In terms of the values $\Delta_P$ and $\Delta_D$ of the parent and daughter, the energy released in positron decay is given by

$$Q_{\beta^+} = \Delta_P - \Delta_D - 2mc^2. \tag{3.41}$$

Therefore, for positron emission to be possible, the mass of the parent atom must be greater than that of the daughter by at least $2mc^2 = 1.022$ MeV. Using the information from Appendix D, we find for the energy released via positron emission in the decay (3.37) to the ground state of $_{10}^{22}$Ne

$$Q_{\beta^+} = -5.182 - (-8.025) - 1.022 = 1.821 \text{ MeV}. \tag{3.42}$$

Electron capture, which results in the same net change as positron decay, can compete with (3.37):

$$_{-1}^{0}e + {}_{11}^{22}\text{Na} \rightarrow {}_{10}^{22}\text{Ne} + {}_{0}^{0}\nu. \tag{3.43}$$

Neglecting the electron binding energy in the $_{11}^{22}$Na atom, we obtain from Eq. (3.35) for the energy released by electron capture

$$Q_{EC} = -5.182 + 8.025 = 2.843 \text{ MeV}. \tag{3.44}$$

[Comparison of Eqs. (3.35) and (3.41) shows that the $Q$ value for EC is greater than that for $\beta^+$ decay by 1.022 MeV when $E_B$ is neglected.]

We next develop the decay scheme for $_{11}^{22}$Na. Appendix D indicates that $\beta^+$ emission occurs 89.8% of the time and EC 10.2%. A gamma ray with energy 1.275 MeV occurs with 100% frequency, indicating that either $\beta^+$ emission or EC leaves the daughter nucleus in an excited state with this energy. The positron decay scheme is shown in Fig. 3.12(a) and that for electron capture in (b). The two are combined in (c) to show the complete decay scheme for $_{11}^{22}$Na. The energy levels are drawn relative to the ground state of $_{10}^{22}$Ne as having zero energy. The starting EC level is $2mc^2$ higher than the starting level for $\beta^+$ decay.

Additional radiations are given in Appendix D for $_{11}^{22}$Na. Gamma rays of energy 0.511 MeV are shown with 180% frequency. These are annihilation photons that are present with all positron emitters. A positron slows down in matter and then annihilates with an atomic electron, giving rise to two photons, each having energy $mc^2 = 0.511$ MeV and traveling in opposite directions. Since a positron is emitted in about 90% of the decay processes, the frequency of an annihilation photon is 1.8 per disintegration of a $_{11}^{22}$Na atom. The remaining radiation shown, Ne X rays, comes as the result of the atomic-shell vacancy following electron capture.

As this example shows, electron capture and positron decay are competitive processes. However, whereas positron emission cannot take place when the parent–daughter atomic mass difference is less than $2mc^2$, electron capture can, the only restriction being $\Delta_P - \Delta_D > E_B$, as implied by Eq. (3.35). The nuclide $_{53}^{126}$I can decay by three routes: EC (60.2%), $\beta^-$ (36.5%), or $\beta^+$ (3.3%).[13]

---

13 In general, the various possible decay modes for a nuclide are those for which $Q > 0$ for a transition to the daughter ground state.

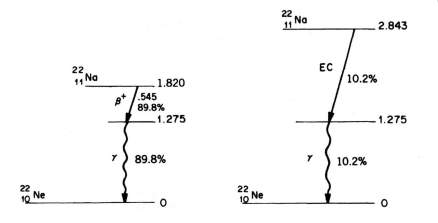

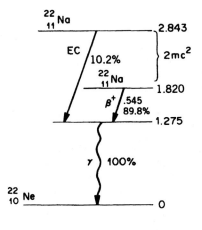

Fig. 3.12 Decay scheme of $^{22}_{11}$Na.

The radiation-protection problems associated with positron emitters include all those of $\beta^-$ emitters (direct radiation and possible bremsstrahlung) and then some. As already mentioned, the 0.511-MeV annihilation photons are always present. In addition, because of the competing process of electron capture, characteristic X rays can be expected.

*Example*

Refer to Appendix D and deduce the decay scheme of $^{26}_{13}$Al.

*Solution*

This nuclide decays by $\beta^+$ emission (81.8%) and EC (18.2%) into $^{26}_{12}$Mg. The energy release for EC with a transition to the daughter ground state is, from the $\Delta$ values,

$$Q_{EC} = -12.211 + 16.214 = 4.003 \text{ MeV.} \qquad (3.45)$$

Here we have neglected the small binding energy of the atomic electron. The corresponding value for $\beta^+$ decay to the ground state is $Q_{\beta^+} = 4.003 - 1.022 = 2.981$ MeV. A 1.809-MeV gamma photon is emitted with 100% frequency, and so we can assume that both EC and $\beta^+$ decay modes proceed via an excited daughter state of this energy. Adding this to the maximum $\beta^+$ energy, we have $1.809 + 1.174 = 2.983$ MeV $= Q_{\beta^+}$. Therefore, the positron decay occurs as shown in Fig. 3.13(a). Its 81.8% frequency accounts for the annihilation photons listed with 164% frequency in Appendix D. The other 18.2% of the decays via EC also go through the level at 1.809 MeV. An additional photon of energy 1.130 MeV and frequency 2.5% is listed in Appendix D. This can arise if a fraction of the EC transformations go to a level with energy $1.809 + 1.130 = 2.939$ MeV above ground. The complete decay scheme is shown in Fig. 3.13(b). (Some small inconsistencies result from round-off.)

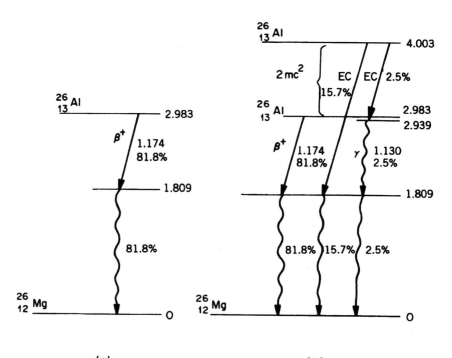

**Fig. 3.13** Decay scheme of $^{26}_{13}$Al (see example in text).

**Table 3.1** Formulas for Energy Release, $Q$, in Terms of Mass Differences, $\Delta_P$ and $\Delta_D$, of Parent and Daughter Atoms

| Type of decay | Formula | Reference |
|---|---|---|
| $\alpha$ | $Q_\alpha = \Delta_P - \Delta_D - \Delta_{He}$ | Eq. (3.13) |
| $\beta^-$ | $Q_{\beta^-} = \Delta_P - \Delta_D$ | Eq. (3.25) |
| $\gamma$ | $Q_{IT} = \Delta_P - \Delta_D$ | Eq. (3.30) |
| EC | $Q_{EC} = \Delta_P - \Delta_D - E_B$ | Eq. (3.35) |
| $\beta^+$ | $Q_{\beta^+} = \Delta_P - \Delta_D - 2mc^2$ | Eq. (3.41) |

This completes the description of the various types of radioactive decay. The formulas for finding the energy release $Q$ from the mass differences $\Delta$ of the parent and daughter atoms are summarized in Table 3.1.

## 3.9
## Suggested Reading

1 Attix, F. H., *Introduction to Radiological Physics and Radiation Dosimetry*, Wiley, New York, NY (1986). [Types of radioactive disintegrations and decay schemes are discussed in Chapter 5.]

2 Cember, H., *Introduction to Health Physics*, 3rd ed., McGraw-Hill, New York, NY (1996). [See especially Chapter 4 on radioactivity.]

3 Evans, R. D., *The Atomic Nucleus*, McGraw-Hill, New York, NY (1955). [A comprehensive and classic text.]

4 Faw, R. E., and Shultis, J. K., *Radiological Assessment*, Prentice-Hall, Englewood Cliffs, NJ (1993). [Appendix B gives detailed decay-scheme data for some 200 radionuclides.]

5 Glasstone, S., *Sourcebook on Atomic Energy*, 3rd ed., D. Van Nostrand, Princeton, NJ (1967). [Describes much of the development of nuclear physics.]

6 Martin, J. E., *Physics for Radiation Protection*, John Wiley, New York, NY (2006). [Treats basic physics to deal with practical problems in radiation protection. Much useful data on radioactivity and radionuclides for the practitioner.]

7 Walker, Philip M. and Carroll, James J., "Ups and Downs of Nuclear Isomers," *Physics Today* 58(6), 39–44 (2005).

A wealth of information and detailed data on radioactivity, decay schemes, isotopes, and related subjects is available through key-word searches on the World Wide Web. One extensive source, for example, with links to a number of related sites, is Lawrence Berkeley National Laboratory's Isotope Project Home Page, http://ie.lbl.gov.

### 3.10
### Problems

1. Gallium occurs with two natural isotopes, $^{69}$Ga (60.2% abundant) and $^{71}$Ga (39.8%), having atomic weights 68.93 and 70.92. What is the atomic weight of the element?

2. The atomic weight of lithium is 6.941. It has two natural isotopes, $^6$Li and $^7$Li, with atomic weights of 6.015 and 7.016. What are the relative abundances of the two isotopes?

3. What minimum energy would an alpha particle need in order to react with a $^{238}$U nucleus?

4. Calculate the energy released when a thermal neutron is absorbed by deuterium.

5. Calculate the total binding energy of the alpha particle.

6. How much energy is released when a $^6$Li atom absorbs a thermal neutron in the reaction $^6_3$Li$(n, \alpha)^3_1$H?

7. What is the mass of a $^6$Li atom in grams?

8. Calculate the average binding energy per nucleon for the nuclide $^{40}_{19}$K.

9. The atomic weight of $^{32}$P is 31.973910. What is the value of $\Delta$ in MeV?

10. Show that 1 AMU $= 1.49 \times 10^{-10}$ J.

11. Calculate the gamma-ray threshold for the reaction $^{12}$C$(\gamma, n)^{11}$C.

12. (a) Calculate the energy released by the alpha decay of $^{222}_{86}$Rn.
    (b) Calculate the energy of the alpha particle.
    (c) What is the energy of the recoil polonium atom?

13. The $^{238}_{92}$U nucleus emits a 4.20-MeV alpha particle. What is the total energy released in this decay?

14. The $^{226}_{88}$Ra nucleus emits a 4.60-MeV alpha particle 5.5% of the time when it decays to $^{222}_{86}$Rn.
    (a) Calculate the $Q$ value for this decay.
    (b) What is the recoil energy of the $^{222}$Rn atom?

15. The $Q$ value for alpha decay of $^{239}_{94}$Pu is 5.25 MeV. Given the masses of the $^{239}$Pu and $^4$He atoms, 239.052175 AMU and 4.002603 AMU, calculate the mass of the $^{235}_{92}$U atom in AMU.

16. Calculate the $Q$ value for the beta decay of the free neutron into a proton, $^1_0$n $\rightarrow$ $^1_1$p $+ ^0_{-1}\beta + ^0_0\bar{\nu}$.

17. (a) Calculate the energy released in the beta decay of $^{32}_{15}$P.
    (b) If a beta particle has 650 keV, how much energy does the antineutrino have?

18. Calculate the $Q$ value for tritium beta decay.

19. Draw the decay scheme for $^{42}_{19}$K.

20. A $^{108}_{49}$In source emits a 633-keV gamma photon and a 606-keV internal-conversion electron from the K shell. What is the binding energy of the electron in the K shell?

21. (a) Draw the decay scheme for $^{198}_{79}$Au.
    (b) Estimate the K-shell electron binding energy from the data given in Appendix D.

22. Draw the decay scheme for $^{59}_{26}$Fe, labeling energies and frequencies (percentages) for each transition.

23. Draw the decay scheme for $^{203}_{80}$Hg.

24. Nuclide A decays into nuclide B by $\beta^+$ emission (24%) or by electron capture (76%). The major radiations, energies (MeV), and frequencies per disintegration are, in the notation of Appendix D:

    $\beta^+$: 1.62 max (16%), 0.98 max (8%)
    $\gamma$: 1.51 (47%), 0.64 (55%), 0.511 (48%, $\gamma^\pm$)
    Daughter X rays
    $e^-$: 0.614

    (a) Draw the nuclear decay scheme, labeling type of decay, percentages, and energies.
    (b) What leads to the emission of the daughter X rays?

25. Draw the decay scheme for $^{84}_{37}$Rb.

26. (a) Calculate the Q values for the decay of $^{57}_{28}$Ni by positron emission and by electron capture.
    (b) Draw the decay scheme.

27. A parent nuclide decays by beta-particle emission into a stable daughter. The major radiations, energies (MeV), and frequencies are, in the notation of Appendix D:

    $\beta^-$: 3.92 max (7%), 3.10 max (5%), 1.60 max (88%)
    $\gamma$: 2.32 (34%), 1.50 (54%), 0.820 (49%)
    $e^-$: 0.818, 0.805

    (a) Draw the decay scheme.
    (b) What is the maximum energy that the antineutrino can receive in this decay?
    (c) What is the value of the internal-conversion coefficient?
    (d) Estimate the L-shell electron binding energy of the daughter nuclide.
    (e) Would daughter X rays be expected also? Why or why not?

28. Calculate the recoil energy of the technetium atom as a result of photon emission in the isomeric transition $^{99m}_{43}$Tc $\rightarrow$ $^{99}_{43}$Tc + $\gamma$.

29. Refer to the decay scheme of $^{137}_{55}$Cs in Fig. 3.8. The binding energies of the K- and L-shell electrons of the daughter $^{137}_{56}$Ba atom are 38 keV and 6 keV.
    (a) What are the energies of the internal-conversion electrons ejected from these shells?

(b) What is the wavelength of the barium $K_\alpha$ X ray emitted when an L-shell electron makes a transition to the K shell?

(c) What is the value of the internal-conversion coefficient?

30. (a) Calculate the Q value for K orbital-electron capture by the $^{37}_{18}\text{Ar}$ nucleus, neglecting the electron binding energy.

(b) Repeat (a), including the binding energy, 3.20 keV, of the K-shell electron in argon.

(c) What becomes of the energy released as a result of this reaction?

31. What is the maximum possible positron energy in the decay of $^{35}_{18}\text{Ar}$?

32. Explain the origins of the radiations listed in Appendix D for $^{85}_{39}\text{Y}$. Draw the decay scheme.

33. The nuclide $^{65}_{30}\text{Zn}$ decays by electron capture (98.5%) and by positron emission (1.5%).

(a) Calculate the Q value for both modes of decay.

(b) Draw the decay scheme for $^{65}\text{Zn}$.

(c) What are the physical processes responsible for each of the major radiations listed in Appendix D?

(d) Estimate the binding energy of a K-shell electron in copper.

34. Does $^{26m}_{13}\text{Al}$ decay to the ground state of its daughter $^{26}_{12}\text{Mg}$?

35. Show that $^{55}_{26}\text{Fe}$, which decays by electron capture, cannot decay by positron emission.

36. The isotope $^{126}_{53}\text{I}$ can decay by EC, $\beta^-$, and $\beta^+$ transitions.

(a) Calculate the Q values for the three modes of decay to the ground states of the daughter nuclei.

(b) Draw the decay scheme.

(c) What kinds of radiation can one expect from a $^{126}\text{I}$ source?

## 3.11
### Answers

2. 7.49% $^6\text{Li}$, 92.51% $^7\text{Li}$
4. 6.26 MeV
9. −24.303 MeV
15. 235.0439 AMU
16. 0.782 MeV
20. 27 keV

28. 0.106 eV
29. (a) 0.624 MeV and 0.656 MeV
 (b) 0.388 Å
 (c) 0.118
31. 4.94 MeV

# 4
# Radioactive Decay

## 4.1
## Activity

The rate of decay, or transformation, of a radionuclide is described by its activity, that is, by the number of atoms that decay per unit time. The unit of activity is the becquerel (Bq), defined as one disintegration per second: $1 \text{ Bq} = 1 \text{ s}^{-1}$. The traditional unit of activity is the curie (Ci), which was originally the activity ascribed to 1 g of $^{226}$Ra. The curie is now defined as $1 \text{ Ci} = 3.7 \times 10^{10}$ Bq, exactly.

## 4.2
## Exponential Decay

The activity of a pure radionuclide decreases exponentially with time, as we now show. If $N$ represents the number of atoms of a radionuclide in a sample at any given time, then the change $dN$ in the number during a short time $dt$ is proportional to $N$ and to $dt$. Letting $\lambda$ be the constant of proportionality, we write

$$dN = -\lambda N \, dt. \tag{4.1}$$

The negative sign is needed because $N$ decreases as the time $t$ increases. The quantity $\lambda$ is called the decay, or transformation, constant; it has the dimensions of inverse time (e.g., $s^{-1}$). The decay rate, or activity, $A$, is given by

$$A = -\frac{dN}{dt} = \lambda N. \tag{4.2}$$

We separate the variables in Eq. (4.1) by writing

$$\frac{dN}{N} = -\lambda \, dt. \tag{4.3}$$

Integration of both sides gives

$$\ln N = -\lambda t + c, \tag{4.4}$$

*Atoms, Radiation, and Radiation Protection.* James E. Turner
Copyright © 2007 WILEY-VCH Verlag GmbH & Co. KGaA, Weinheim
ISBN: 978-3-527-40606-7

where $c$ is an arbitrary constant of integration, fixed by the initial conditions. If we specify that $N_0$ atoms of the radionuclide are present at time $t = 0$, then Eq. (4.4) implies that $c = \ln N_0$. In place of (4.4) we write

$$\ln N = -\lambda t + \ln N_0, \tag{4.5}$$

$$\ln \frac{N}{N_0} = -\lambda t \tag{4.6}$$

or

$$\frac{N}{N_0} = e^{-\lambda t}. \tag{4.7}$$

Equation (4.7) describes the exponential radioactive decay law. Since the activity of a sample and the number of atoms present are proportional, activity follows the same rate of decrease,

$$\frac{A}{A_0} = e^{-\lambda t}, \tag{4.8}$$

where $A_0$ is the activity at time $t = 0$. The dose rate at a given location in the neighborhood of a fixed radionuclide source also falls off at the same exponential rate.

The function (4.8) is plotted in Fig. 4.1. During successive times $T$, called the half-life of the radionuclide, the activity drops by factors of one-half, as shown. To

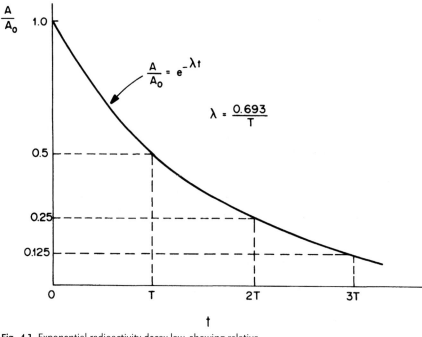

Fig. 4.1 Exponential radioactivity decay law, showing relative activity, $A/A_0$, as a function of time $t$; $\lambda$ is the decay constant and $T$ the half-life.

find $T$ in terms of $\lambda$, we write from Eq. (4.8) at time $t = T$,

$$\tfrac{1}{2} = e^{-\lambda T}. \tag{4.9}$$

Taking the natural logarithm of both sides gives

$$-\lambda T = \ln\left(\tfrac{1}{2}\right) = -\ln 2, \tag{4.10}$$

and therefore

$$T = \frac{\ln 2}{\lambda} = \frac{0.693}{\lambda}. \tag{4.11}$$

Written in terms of the half-life, the exponential decay laws (4.7) and (4.8) become

$$\frac{N}{N_0} = \frac{A}{A_0} = e^{-0.693t/T}. \tag{4.12}$$

The decay law (4.12) can be derived simply on the basis of the half-life. If, for example, the activity decreases to a fraction $A/A_0$ of its original value after passage of time $t/T$ half-lives, then we can write

$$\frac{A}{A_0} = \left(\frac{1}{2}\right)^{t/T}. \tag{4.13}$$

Taking the logarithm of both sides of Eq. (4.13) gives

$$\ln \frac{A}{A_0} = -\frac{t}{T} \ln 2 = -\frac{0.693t}{T}, \tag{4.14}$$

from which Eq. (4.12) follows.

### Example
Calculate the activity of a 30-MBq source of $^{24}_{11}\mathrm{Na}$ after 2.5 d. What is the decay constant of this radionuclide?

### Solution
The problem can be worked in several ways. We first find $\lambda$ from Eq. (4.11) and then the activity from Eq. (4.8). The half-life $T = 15.0$ h of the nuclide is given in Appendix D. From (4.11),

$$\lambda = \frac{0.693}{T} = \frac{0.693}{15.0\,\mathrm{h}} = 0.0462\,\mathrm{h}^{-1}. \tag{4.15}$$

With $A_0 = 30$ MBq and $t = 2.5\,\mathrm{d} \times 24\,\mathrm{h\,d}^{-1} = 60.0$ h,

$$A = 30\,e^{-(0.0462\ h^{-1} \times 60\ h)} = 1.88\ \mathrm{MBq}. \tag{4.16}$$

Note that the time units employed for $\lambda$ and $t$ must be the same in order that the exponential be dimensionless.

*Example*

A solution contains 0.10 $\mu$Ci of $^{198}$Au and 0.04 $\mu$Ci of $^{131}$I at time $t = 0$. What is the total beta activity in the solution at $t = 21$ d? At what time will the total activity decay to one-half its original value?

*Solution*

Both isotopes decay to stable daughters, and so the total beta activity is due to these isotopes alone. (A small fraction of $^{131}$I decays into $^{131m}$Xe, which does not contribute to the beta activity.) From Appendix D, the half-lives of $^{198}$Au and $^{131}$I are, respectively, 2.70 days and 8.05 days. At the end of 21 days, the activities $A_{Au}$ and $A_I$ of the nuclides are, from Eq. (4.12),

$$A_{Au} = 0.10e^{-0.693 \times 21/2.70} = 4.56 \times 10^{-4} \ \mu\text{Ci} \tag{4.17}$$

and

$$A_I = 0.04e^{-0.693 \times 21/8.05} = 6.56 \times 10^{-3} \ \mu\text{Ci.} \tag{4.18}$$

The total activity at $t = 21$ days is the sum of these two activities, $7.02 \times 10^{-3}$ $\mu$Ci. To find the time $t$ in days at which the activity has decayed to one-half its original value of $0.10 + 0.04 = 0.14$ Ci, we write

$$0.07 = 0.1e^{-0.693t/2.70} + 0.04e^{-0.693t/8.05}. \tag{4.19}$$

This is a transcendental equation, which cannot be solved in closed form for $t$. The solution can be found either graphically or by trial and error, focusing in between two values of $t$ that make the right-hand side of (4.19) >0.07 and <0.07. We present a combination of both methods. The decay constants of the two nuclides are, for Au, $0.693/2.70 = 0.257$ d$^{-1}$ and, for I, $0.693/8.05 = 0.0861$ d$^{-1}$. The activities in $\mu$Ci, as functions of time $t$, are

$$A_{Au}(t) = 0.10e^{-0.257t} \tag{4.20}$$

and

$$A_I(t) = 0.04e^{-0.0861t}. \tag{4.21}$$

Figure 4.2 shows a plot of these two activities and the total activity, $A(t) = A_{Au} + A_I$, calculated as functions of $t$ from these two equations. Plotted to scale, the total activity $A(t)$ is found to reach the value 0.07 $\mu$Ci near $t = 3.50$ d. We can improve on this approximate graphical solution. Direct calculation from Eqs. (4.20) and (4.21) shows that $A(3.50) = 0.0703$ and $A(3.60) = 0.0689$. Linear interpolation suggests the solution $t = 3.52$ d; indeed, one can verify that $A(3.52) = 0.0700$ $\mu$Ci.

The average, or mean, life $\tau$ of a radionuclide is defined as the average of all of the individual lifetimes that the atoms in a sample of the radionuclide experience. It is equal to the mean value of $t$ under the exponential curve in Fig. 4.3. Therefore,

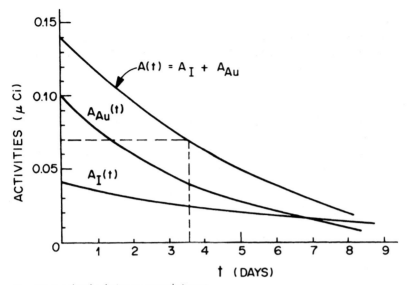

Fig. 4.2 Graphical solution to example in text.

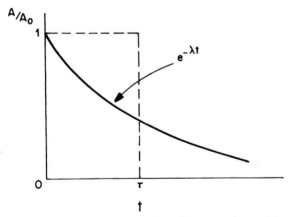

Fig. 4.3 The average life $\tau$ of a radionuclide is given by $\tau = 1/\lambda$.

$\tau$ defines a rectangle, as shown, with area equal to the area under the exponential curve:

$$1 \times \tau = \int_0^\infty e^{-\lambda t}\, dt = -\frac{1}{\lambda} e^{-\lambda t}\Big|_0^\infty = \frac{1}{\lambda}. \tag{4.22}$$

Thus the mean life is the reciprocal of the decay constant. In terms of the half-life, we have

$$\tau = \frac{1}{\lambda} = \frac{T}{0.693}, \tag{4.23}$$

showing that $\tau > T$.

## 4.3
## Specific Activity

The specific activity of a sample is defined as its activity per unit mass, for example, $Bq\,g^{-1}$ or $Ci\,g^{-1}$. If the sample is a pure radionuclide, then its specific activity SA is determined by its decay constant $\lambda$, or half-life $T$, and by its atomic weight $M$ as follows. Since the number of atoms per gram of the nuclide is $N = 6.02 \times 10^{23}/M$, Eq. (4.2) gives for the specific activity

$$SA = \frac{6.02 \times 10^{23}\,\lambda}{M} = \frac{4.17 \times 10^{23}}{MT}. \tag{4.24}$$

If $T$ is in seconds, then this formula gives the specific activity in $Bq\,g^{-1}$. In practice, using the atomic mass number $A$ in place of $M$ usually gives sufficient accuracy.

*Example*
Calculate the specific activity of $^{226}Ra$ in $Bq\,g^{-1}$.

*Solution*
From Appendix D, $T = 1600$ y and $M = A = 226$. Converting $T$ to seconds, we have

$$SA = \frac{4.17 \times 10^{23}}{226 \times 1600 \times 365 \times 24 \times 3600} \tag{4.25}$$

$$= 3.66 \times 10^{10}\,s^{-1}\,g^{-1} = 3.7 \times 10^{10}\,Bq\,g^{-1}. \tag{4.26}$$

This, by definition, is an activity of 1 Ci.

The fact that $^{226}Ra$ has unit specific activity in terms of $Ci\,g^{-1}$ can be used in place of Eq. (4.24) to find SA for other radionuclides. Compared with $^{226}Ra$, a nuclide of shorter half-life and smaller atomic mass number $A$ will have, in direct proportion, a higher specific activity than $^{226}Ra$. The specific activity of a nuclide of half-life $T$ and atomic mass number $A$ is therefore given by

$$SA = \frac{1600}{T} \times \frac{226}{A} Ci\,g^{-1}, \tag{4.27}$$

where $T$ is expressed in years. (The equation gives $SA = 1\,Ci\,g^{-1}$ for $^{226}Ra$.)

*Example*
What is the specific activity of $^{14}C$?

*Solution*
With $T = 5730$ y and $A = 14$, Eq. (4.27) gives

$$SA = \frac{1600}{5730} \times \frac{226}{14} = 4.51\,Ci\,g^{-1}. \tag{4.28}$$

Alternatively, we can use Eq. (4.24) with $T = 5730 \times 365 \times 24 \times 3600 = 1.81 \times 10^{11}$ s, obtaining

$$SA = \frac{4.17 \times 10^{23}}{14 \times 1.81 \times 10^{11}} = 1.65 \times 10^{11} \, \text{Bq} \, \text{g}^{-1} \tag{4.29}$$

$$= \frac{1.65 \times 10^{11} \, \text{Bq} \, \text{g}^{-1}}{3.7 \times 10^{10} \, \text{Bq} \, \text{Ci}^{-1}} = 4.46 \, \text{Ci} \, \text{g}^{-1}, \tag{4.30}$$

in agreement with (4.28).

Specific activity need not apply to a pure radionuclide. For example, $^{14}$C produced by the $^{14}$N(n,p)$^{14}$C reaction can be extracted chemically as a "carrier-free" radionuclide, that is, without the presence of nonradioactive carbon isotopes. Its specific activity would be that calculated in the previous example. A different example is afforded by $^{60}$Co, which is produced by neutron absorption in a sample of $^{59}$Co (100% abundant), the reaction being $^{59}$Co(n,$\gamma$)$^{60}$Co. The specific activity of the sample depends on its radiation history, which determines the fraction of cobalt atoms that are made radioactive. Specific activity is also used to express the concentration of activity in solution; for example, $\mu$Ci mL$^{-1}$ or Bq L$^{-1}$.

## 4.4
## Serial Radioactive Decay

In this section we describe the activity of a sample in which one radionuclide produces one or more radioactive offspring in a chain. Several important cases will be discussed.

### Secular Equilibrium ($T_1 \gg T_2$)

First, we calculate the total activity present at any time when a long-lived parent (1) decays into a relatively short-lived daughter (2), which, in turn, decays into a stable nuclide. The half-lives of the two radionuclides are such that $T_1 \gg T_2$; and we consider intervals of time that are short compared with $T_1$, so that the activity $A_1$ of the parent can be treated as constant. The total activity at any time is $A_1$ plus the activity $A_2$ of the daughter, on which we now focus. The rate of change, $dN_2/dt$, in the number of daughter atoms $N_2$ per unit time is equal to the rate at which they are produced, $A_1$, minus their rate of decay, $\lambda_2 N_2$:

$$\frac{dN_2}{dt} = A_1 - \lambda_2 N_2. \tag{4.31}$$

To solve for $N_2$, we first separate variables by writing

$$\frac{dN_2}{A_1 - \lambda_2 N_2} = dt, \tag{4.32}$$

where $A_1$ can be regarded as constant. Introducing the variable $u = A_1 - \lambda_2 N_2$, we have $du = -\lambda_2 dN_2$ and, in place of Eq. (4.32),

$$\frac{du}{u} = -\lambda_2 \, dt. \tag{4.33}$$

Integration gives

$$\ln(A_1 - \lambda_2 N_2) = -\lambda_2 t + c, \tag{4.34}$$

where $c$ is an arbitrary constant. If $N_{20}$ represents the number of atoms of nuclide (2) present at $t = 0$, then we have $c = \ln(A_1 - \lambda_2 N_{20})$. Equation (4.34) becomes

$$\ln \frac{A_1 - \lambda_2 N_2}{A_1 - \lambda_2 N_{20}} = -\lambda_2 t, \tag{4.35}$$

or

$$A_1 - \lambda_2 N_2 = (A_1 - \lambda_2 N_{20})e^{-\lambda_2 t}. \tag{4.36}$$

Since $\lambda_2 N_2 = A_2$, the activity of nuclide (2), and $\lambda_2 N_{20} = A_{20}$ is its initial activity, Eq. (4.36) implies that

$$A_2 = A_1(1 - e^{-\lambda_2 t}) + A_{20}e^{-\lambda_2 t}. \tag{4.37}$$

In many practical instances one starts with a pure sample of nuclide (1) at $t = 0$, so that $A_{20} = 0$, which we now assume. The activity $A_2$ then builds up as shown in Fig. 4.4. After about seven daughter half-lives ($t \gtrsim 7T_2$), $e^{-\lambda_2 t} \ll 1$ and Eq. (4.37) reduces to the condition $A_1 = A_2$, at which time the daughter activity is equal to

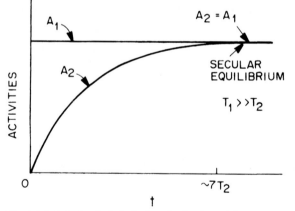

**Fig. 4.4** Activity $A_2$ of relatively short-lived radionuclide daughter ($T_2 \ll T_1$) as a function of time $t$ with initial condition $A_{20} = 0$. Activity of daughter builds up to that of the parent in about seven half-lives ($\sim 7T_2$). Thereafter, daughter decays at the same rate it is produced ($A_2 = A_1$), and secular equilibrium is said to exist.

that of the parent. This condition is called secular equilibrium. The total activity is $2A_1$. In terms of the numbers of atoms, $N_1$ and $N_2$, of the parent and daughter, secular equilibrium can be also expressed by writing

$$\lambda_1 N_1 = \lambda_2 N_2. \tag{4.38}$$

A chain of $n$ short-lived radionuclides can all be in secular equilibrium with a long-lived parent. Then the activity of each member of the chain is equal to that of the parent and the total activity is $n + 1$ times the activity of the original parent.

### General Case

When there is no restriction on the relative magnitudes of $T_1$ and $T_2$, we write in place of Eq. (4.31)

$$\frac{dN_2}{dt} = \lambda_1 N_1 - \lambda_2 N_2. \tag{4.39}$$

With the initial condition $N_{20} = 0$, the solution to this equation is

$$N_2 = \frac{\lambda_1 N_{10}}{\lambda_2 - \lambda_1} (e^{-\lambda_1 t} - e^{-\lambda_2 t}), \tag{4.40}$$

as can be verified by direct substitution into (4.39). This general formula yields Eq. (4.38) when $\lambda_2 \gg \lambda_1$ and $A_{20} = 0$, and hence also describes secular equilibrium.

### Transient Equilibrium ($T_1 \gtrsim T_2$)

Another practical situation arises when $N_{20} = 0$ and the half-life of the parent is greater than that of the daughter, but not greatly so. According to Eq. (4.40), $N_2$ and hence the activity $A_2 = \lambda_2 N_2$ of the daughter initially build up steadily. With the continued passage of time, $e^{-\lambda_2 t}$ eventually becomes negligible with respect to $e^{-\lambda_1 t}$, since $\lambda_2 > \lambda_1$. Then Eq. (4.40) implies, after multiplication of both sides by $\lambda_2$, that

$$\lambda_2 N_2 = \frac{\lambda_2 \lambda_1 N_{10} e^{-\lambda_1 t}}{\lambda_2 - \lambda_1}. \tag{4.41}$$

Since $A_1 = \lambda_1 N_1 = \lambda_1 N_{10} e^{-\lambda_1 t}$ is the activity of the parent as a function of time, this relation says that

$$A_2 = \frac{\lambda_2 A_1}{\lambda_2 - \lambda_1}. \tag{4.42}$$

Thus, after initially increasing, the daughter activity $A_2$ goes through a maximum and then decreases at the same rate as the parent activity. Under this condition, illustrated in Fig. 4.5, transient equilibrium is said to exist. The total activity also

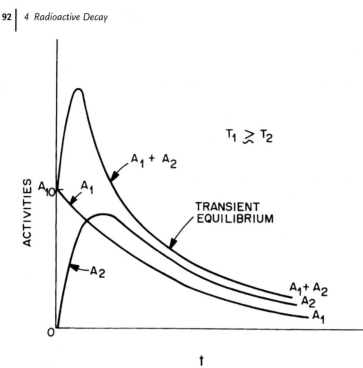

**Fig. 4.5** Activities as functions of time when $T_1$ is somewhat larger than $T_2$ ($T_1 \gtrsim T_2$) and $N_{20} = 0$. Transient equilibrium is eventually reached, in which all activities decay with the half-life $T_1$ of the parent.

reaches a maximum, as shown in the figure, at a time earlier than that of the maximum daughter activity. Equation (4.42) can be differentiated to find the time at which the daughter activity is largest. The result is (Problem 25)

$$t = \frac{1}{\lambda_2 - \lambda_1} \ln \frac{\lambda_2}{\lambda_1}, \quad \text{for maximum } A_2. \tag{4.43}$$

The total activity is largest at the earlier time (Problem 26)

$$t = \frac{1}{\lambda_2 - \lambda_1} \ln \frac{\lambda_2^2}{2\lambda_1\lambda_2 - \lambda_1^2}, \quad \text{for maximum } A_1 + A_2. \tag{4.44}$$

The time at which transient equilibrium is established depends on the individual magnitudes of $T_1$ and $T_2$. Secular equilibrium can be viewed as a special case of transient equilibrium in which $\lambda_2 \gg \lambda_1$ and the time of observation is so short that the decay of the activity $A_1$ is negligible. Under these conditions, the curve for $A_1$ in Fig. 4.5 would be flat, $A_2$ would approach $A_1$, and the figure would resemble Fig. 4.4.

## No Equilibrium ($T_1 < T_2$)

When the daughter, initially absent ($N_{20} = 0$), has a longer half-life than the parent, its activity builds up to a maximum and then declines. Because of its shorter half-life, the parent eventually decays away and only the daughter is left. No equilibrium occurs. The activities in this case exhibit the patterns shown in Fig. 4.6.

*Example*

Starting with a 10.0-GBq ($= 10^{10}$ Bq) sample of pure $^{90}$Sr at time $t = 0$, how long will it take for the total activity ($^{90}$Sr + $^{90}$Y) to build up to 17.5 GBq?

*Solution*

Appendix D shows that $^{90}_{38}$Sr $\beta^-$ decays with a half-life of 29.12 y into $^{90}_{39}$Y, which $\beta^-$ decays into stable $^{90}_{40}$Zr with a half-life of 64.0 h. These two isotopes illustrate a long-lived parent ($T_1 = 29.12$ y) decaying into a short-lived daughter ($T_2 = 64.0$ h). Secular equilibrium is reached in about seven daughter half-lives, that is, in $7 \times 64 = 448$ h. At the end of this time, the $^{90}$Sr activity $A_1$ has not diminished appreciably, the $^{90}$Y activity $A_2$ has increased to the level $A_2 = A_1 = 10.0$ GBq, and the total activity is 20.0 GBq. In the present problem we are asked, in effect, to find the time at which the $^{90}$Y activity reaches 7.5 GBq. The answer will be less than 448 h. Equation (4.37) with $A_{20} = 0$ applies here.[1] The decay constant for $^{90}$Y is $\lambda_2 = 0.693/T_2 = 0.693/64.0 = 0.0108$ h$^{-1}$. With $A_1 = 10.0$ GBq, $A_2 = 7.5$ GBq, and $A_{20} = 0$, Eq. (4.37) gives

$$7.5 = 10.0(1 - e^{-0.0108t}), \tag{4.45}$$

where $t$ is in hours. Rearranging, we have

$$e^{-0.0108t} = \tfrac{1}{4}, \tag{4.46}$$

giving $t = 128$ h. (In this example note that the $^{90}$Y activity increases in an inverse fashion to the way a pure sample of $^{90}$Y would decay. It takes two half-lives, $2T_2 = 128$ h, for the activity to build up to three-fourths its final value at secular equilibrium.)

*Example*

How many grams of $^{90}$Y are in secular equilibrium with 1 mg of $^{90}$Sr?

*Solution*

The amount of $^{90}$Y will be that having the same activity as 1 mg of $^{90}$Sr. The specific activity, SA, of $^{90}$Sr ($T_1 = 29.12$ y) is [from Eq. (4.27)]

$$SA_1 = \frac{1600}{29.12} \times \frac{226}{90} = 138 \text{ Ci g}^{-1}. \tag{4.47}$$

---

1   Equation (4.40), describing the general case without restriction on the relative magnitudes of $T_1$ and $T_2$, can always be applied. To the degree of accuracy with which we are working, one will obtain the same numerical answer from the simplified Eq. (4.37), which already contains the appropriate approximations.

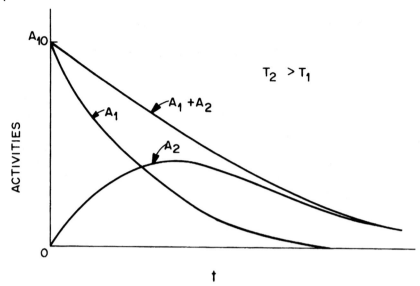

**Fig. 4.6** Activities as functions of time when $T_2 > T_1$ and $N_{20} = 0$. No equilibrium conditions occur. Eventually, only the daughter activity remains.

Therefore, the activity of the 1 mg sample of $^{90}$Sr is

$$A_1 = 10^{-3} \text{ g} \times 138 \text{ Ci g}^{-1} = 0.138 \text{ Ci}, \tag{4.48}$$

which is also equal to the activity $A_2$ of the $^{90}$Y. The latter has a specific activity

$$SA_2 = \frac{1600 \text{ y}}{64.0 \text{ h} \times \dfrac{1 \text{ d}}{24 \text{ h}} \times \dfrac{1 \text{ y}}{365 \text{ d}}} \times \frac{226}{90} \tag{4.49}$$

$$= 5.50 \times 10^5 \text{ Ci g}^{-1}. \tag{4.50}$$

Therefore, the mass of $^{90}$Y in secular equilibrium with 1 mg of $^{90}$Sr is

$$\frac{0.138 \text{ Ci}}{5.50 \times 10^5 \text{ Ci g}^{-1}} = 2.51 \times 10^{-7} \text{ g} = 0.251 \text{ } \mu\text{g}. \tag{4.51}$$

*Example*

A sample contains 1 mCi of $^{191}$Os at time $t = 0$. The isotope decays by $\beta^-$ emission into metastable $^{191m}$Ir, which then decays by $\gamma$ emission into $^{191}$Ir. The decay and half-lives can be represented by writing

$$^{191}_{76}\text{Os} \xrightarrow[15.4 \text{ d}]{\beta^-} {}^{191m}_{77}\text{Ir} \xrightarrow[4.94 \text{ s}]{\gamma} {}^{191}_{77}\text{Ir}. \tag{4.52}$$

(a) How many grams of $^{191}$Os are present at $t = 0$?
(b) How many millicuries of $^{191m}$Ir are present at $t = 25$ d?

(c) How many atoms of $^{191m}$Ir decay between $t = 100$ s and $t = 102$ s?

(d) How many atoms of $^{191m}$Ir decay between $t = 30$ d and $t = 40$ d?

*Solution*

As in the last two examples, the parent half-life is large compared with that of the daughter. Secular equilibrium is reached in about $7 \times 4.9 = 34$ s. Thereafter, the activities $A_1$ and $A_2$ of the $^{191}$Os and $^{191m}$Ir remain equal, as they are in secular equilibrium. During the periods of time considered in (b) and in (d), however, the osmium will have decayed appreciably; and so one deals with an example of transient equilibrium. The problem can be solved as follows.

(a) The specific activity of $^{191}$Os is, from Eq. (4.27),

$$SA_1 = \frac{1600 \times 365}{15.4} \times \frac{226}{191} = 4.49 \times 10^4 \ Ci\,g^{-1}. \tag{4.53}$$

The mass of the sample, therefore, is

$$\frac{10^{-3} \ Ci}{4.49 \times 10^4 \ Ci\,g^{-1}} = 2.23 \times 10^{-8} \ g. \tag{4.54}$$

(b) At $t = 25$ d,

$$A_2 = A_1 = 1 \times e^{-0.693 \times 25/15.4} = 0.325 \ mCi. \tag{4.55}$$

(c) Between $t = 100$ s and $102$ s secular equilibrium exists with the osmium source essentially still at its original activity. Thus the $^{191m}$Ir decay rate at $t = 100$ s is $A_2 = 1$ mCi $= 3.7 \times 10^7$ s$^{-1}$. During the next 2 s the number of $^{191m}$Ir atoms that decay is $2 \times 3.7 \times 10^7 = 7.4 \times 10^7$.

(d) This part is like (c), except that the activities $A_1$ and $A_2$ do not stay constant during the time between 30 and 40 d. Since transient equilibrium exists, the numbers of atoms of $^{191m}$Ir and $^{191}$Os that decay are equal. The number of $^{191m}$Ir atoms that decay, therefore, is equal to the integral of the $^{191}$Os activity during the specified time ($t$ in days):

$$3.7 \times 10^7 \int_{30}^{40} e^{-0.693t/15.4} \ dt = \frac{3.7 \times 10^7}{-0.0450} e^{-0.0450t} \Big|_{30}^{40} \tag{4.56}$$

$$= -8.22 \times 10^8 (0.165 - 0.259)$$

$$= 7.73 \times 10^7. \tag{4.57}$$

## 4.5
**Natural Radioactivity**

All of the heavy elements ($Z > 83$) found in nature are radioactive and decay by alpha or beta emission. The nuclide $^{209}_{83}$Bi is the only one with atomic number greater than that of lead (82) that is stable. The heaviest elements decay into successive radioactive daughters, forming series of radionuclides that end when a stable species is produced. It is found that all of the naturally occurring heavy radionuclides belong to one of three series. Since the atomic mass number can change by only four units (viz., when alpha emission occurs), a given nuclide can be easily identified as belonging to one series or another by noting the remainder obtained when its mass number is divided by four. The uranium series, for example, begins with $^{238}_{92}$U and ends with stable $^{206}_{82}$Pb. When divided by four the mass number of every member of the uranium series has remainder two. The thorium series, starting with $^{232}_{90}$Th and ending with $^{208}_{82}$Pb, has remainder zero. The third group, the actinium series, which begins with $^{235}_{92}$U and ends with $^{207}_{82}$Pb, has remainder three. A fourth series, with remainder one, is the neptunium series. However, its longest-lived member, $^{237}_{93}$Np, has a half-life of $2.2 \times 10^6$ years, which is short on a geological time scale. Neptunium is not found in nature, but has been produced artificially, starting with $^{241}_{94}$Pu and ending with $^{209}_{82}$Pb. All four series contain one gaseous member (an isotope of Rn) and end in a stable isotope of Pb.

Primordial $^{238}$U would be found in secular equilibrium with its much shorter-lived daughters, if undisturbed by physical or chemical processes in nature. It is more likely, however, that secular equilibrium will be found only among certain subsets of nuclides in the series. In this regard, a significant change occurs when $^{226}$Ra decays into $^{222}$Rn. The daughter, radon, is a noble gas, not bound chemically in the material where its parents resided. The half-life of $^{222}$Rn is long enough for much of the gas to work its way out into the atmosphere. As seen from Table 1.1, radon (more precisely, its short-lived daughters) contributes an average of about one-half the effective dose to persons from natural background radiation. This important source of human exposure is discussed in the next section.

Several lighter elements have naturally occurring, primordial radioactive isotopes. One of the most important from the standpoint of human exposure is $^{40}$K, which has an isotopic abundance of 0.0118% and a half-life of $1.28 \times 10^9$ years. The nuclide decays by $\beta^-$ emission (89%) or EC (11%). The maximum $\beta^-$ energy is 1.312 MeV. This isotope is an important source of human internal and external radiation exposure, because potassium is a natural constituent of plants and animals. In addition to the beta particle, $^{40}$K emits a penetrating gamma ray (1.461 MeV) following electron capture (11%).

Other naturally occurring radionuclides are of cosmogenic origin. Only those produced as a result of cosmic-ray interactions with constituents of the atmosphere result in any mentionable exposure to man: $^3$H, $^7$Be, $^{14}$C, and $^{22}$Na. The reaction $^{14}$N$(n, p)^{14}$C with atmospheric nitrogen produces radioactive $^{14}$C, which has a half-life of 5730 y. The radioisotope, existing as $CO_2$ in the atmosphere, is utilized by plants and becomes fixed in their structure through photosynthesis. The time at which $^{14}$C was assimilated in a previously living specimen, used to make furniture or paper, for example, can be inferred from the relative amount of the isotope remaining in it today. Thus the age of such objects can be determined by radiocarbon dating. In modern times, the equilibrium of natural $^{14}$C and $^{3}$H in the atmosphere has been upset by the widespread burning of fossil fuels and by the testing of nuclear weapons in the atmosphere.

*Example*

How many alpha and beta particles are emitted by a nucleus of an atom of the uranium series, which starts as $^{238}_{92}$U and ends as stable $^{206}_{82}$Pb?

*Solution*

Nuclides of the four heavy-element radioactive series decay either by alpha or beta emission. A single disintegration, therefore, either (1) reduces the atomic number by 2 and the mass number by 4 or (2) increases the atomic number by 1 and leaves the mass number unchanged. Since the atomic mass numbers of $^{238}_{92}$U and $^{206}_{82}$Pb differ by 32, it follows that 8 alpha particles are emitted in the series. Since this alone would reduce the atomic number by 16, as compared with the actual reduction of 10, a total of 6 beta particles must also be emitted.

## 4.6
## Radon and Radon Daughters

As mentioned in the last section, the noble gas $^{222}$Rn produced in the uranium series can become airborne before decaying. Soil and rocks under houses are ordinarily the principal contributors to indoor radon, which is typically four or five times more concentrated than radon outdoors, where greater air dilution occurs. Additional contributions to indoor radon come from outside air, building materials, and the use of water and natural gas. Circumstances vary widely in time and place, and so exceptions to generalizations are frequent.

Airborne radon itself poses little health hazard. As an inert gas, inhaled radon is not retained in significant amounts by the body. The potential health hazard arises when radon in the air decays, producing nongaseous radioactive daughters. When inhaled, the airborne daughters can be trapped in the respiratory system, where they are likely to decay before being removed by normal lung-clearing mechanisms of the body. Some of the daughter atoms in air are adsorbed onto micron- or submicron-sized aerosols or dust particles (the "attached fraction"). Others (the "unattached fraction") remain in the air as essentially free ions or in small molecular agglomerates (e.g., with several water molecules). Still other decay products plate out on various surfaces. In assessing hazard, it is common to characterize the radon daughters in an atmosphere by specifying the unattached fraction of each. When inhaled, this fraction is trapped efficiently, especially in the upper respiratory tract.

As shown in Fig. 4.7, $^{222}$Rn decays into a series of short-lived daughters, two of which, $^{218}$Po and $^{214}$Po, are alpha emitters.[2] When an alpha particle is emitted in the lung, it deposits all of its energy locally within a small thickness of adjacent tissue. An alpha particle from $^{214}$Po, for example, deposits its 7.69 MeV of energy within about 70 $\mu$m. A 1-MeV beta particle from $^{214}$Bi, on the other hand, deposits its energy over a much larger distance of about 4000 $\mu$m. The dose to the cells of the lung from the beta (and gamma) radiation from radon daughters is very small compared with that from the alpha particles. The "radon problem," technically, is that of alpha-particle irradiation of sensitive lung tissue by the short-lived daughters of radon and the associated risk of lung cancer.

The health hazard from radon is thus closely related to the air concentration of the potential alpha-particle energy of the short-lived daughters. Depending on local conditions, the daughters will be in various degrees of secular equilibrium with one another and with the parent radon in an atmosphere. Rather than using individual concentrations of the various progeny, one can characterize an atmosphere radio-logically by means of a collective quantity: the potential alpha-energy concentration (PAEC). The PAEC is defined as the amount of alpha energy per unit volume of

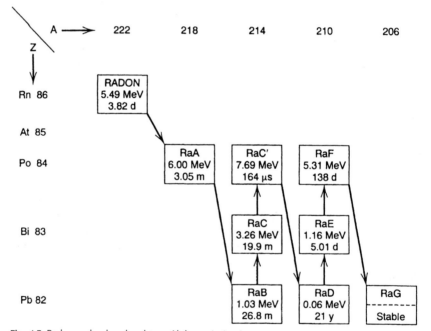

**Fig. 4.7** Radon and radon daughters. Alpha emission is represented by an arrow slanting downward toward the right; beta emission, by a vertical arrow. Alpha-particle and average beta-particle energies and half-lives are shown in the boxes.

2   In earlier terminology, the successive short-lived daughters, $^{218}$Po through $^{210}$Pb, were called RaA, RaB, RaC, RaC', and RaD, respectively.

undisturbed air that would ultimately be released from the particular mixture of short-lived daughters in their decay to $^{210}$Pb. The PAEC can be expressed in J m$^{-3}$ or MeV m$^{-3}$. For a given PAEC, the equilibrium-equivalent decay-product concentration (EEDC) is defined as the concentration of each decay product that would be present if secular equilibrium existed. The ratio of the EEDC and the concentration of radon is called the *equilibrium factor*. By definition, this factor is equal to unity if the radon and all of its short-lived daughters are in secular equilibrium. Equilibrium factors for most indoor atmospheres are in the range of 0.2 to 0.6, a factor of 0.5 often being assumed as a rule of thumb. A limitation of the quantities described in this paragraph is that they do not distinguish between the attached and unattached fractions.

Until now, we have discussed only $^{222}$Rn, which is a member of the uranium series. Radon is also generated in the other two series of naturally occurring radionuclides. However, these isotopes of radon are of lesser radiological importance. The thorium series generates $^{220}$Rn, which is also called thoron. The parent nuclide, $^{232}$Th, is somewhat more abundant than $^{238}$U, but has a longer half-life. As a result, the average rate of production of $^{220}$Rn in the ground is about the same as that of $^{222}$Rn. However, the shorter half-life of $^{220}$Rn, 56 s, as compared with 3.82 d for $^{222}$Rn, gives it a much greater chance to decay before becoming airborne. The contributions of the daughters of $^{220}$Rn to lung dose are usually negligible compared with $^{222}$Rn. The third (actinium) series produces $^{219}$Rn, also called actinon, after several transformations from the relatively rare original nuclide $^{235}$U. Its half-life is only 4 s, and its contribution to airborne radon is insignificant.

*Example*

Measurements of room air show the nuclide activity concentrations given in Table 4.1. Calculate the PAEC for this case.

*Solution*

The PAEC (and EEDC) pertain to the short-lived decay products and do not involve the radon itself, which is not retained by the lungs. To obtain the PAEC, we need to calculate the number of daughter atoms of each type per unit volume of air; multiply these numbers by the potential alpha-particle energy associated with each type of

Table 4.1

| Nuclide | Activity Concentration (Bq m$^{-3}$) |
|---|---|
| $^{222}$Rn | 120 |
| $^{218}$Po | 93 |
| $^{214}$Pb | 90 |
| $^{214}$Bi | 76 |
| $^{214}$Po | 76 |

Table 4.2

| Nuclide | $A$ $(Bq\,m^{-3})$ | $\lambda$ $(s^{-1})$ | $N$ $(m^{-3})$ | $E$ $(MeV)$ | $NE$ $(MeV\,m^{-3})$ |
|---|---|---|---|---|---|
| $^{218}Po$ | 93 | $3.79 \times 10^{-3}$ | $2.45 \times 10^{4}$ | 13.69 | $3.35 \times 10^{5}$ |
| $^{214}Pb$ | 90 | $4.31 \times 10^{-4}$ | $2.09 \times 10^{5}$ | 7.69 | $1.61 \times 10^{6}$ |
| $^{214}Bi$ | 76 | $5.83 \times 10^{-4}$ | $1.30 \times 10^{5}$ | 7.69 | $1.00 \times 10^{6}$ |
| $^{214}Po$ | 76 | $4.23 \times 10^{3}$ | $1.80 \times 10^{-2}$ | 7.69 | $1.38 \times 10^{-1}$ |

atom; and then sum. The number of atoms $N$ of a radionuclide associated with an activity $A$ is given by Eq. (4.2), $N = A/\lambda$, where $\lambda$ is the decay constant. For the first daughter, $^{218}Po$, for example, we find from the half-life $T = 183$ s given in Fig. 4.7 (or Appendix D) that $\lambda = 0.693/T = 3.79 \times 10^{-3}$ s$^{-1}$. From the activity density $A = 93$ Bq m$^{-3}$ given in Table 4.1, it follows that the number density of $^{218}Po$ atoms is

$$N = \frac{A}{\lambda} = \frac{93 \text{ Bq m}^{-3}}{3.79 \times 10^{-3} \text{ s}^{-1}} = 2.45 \times 10^{4} \text{ m}^{-3}, \tag{4.58}$$

where the units Bq and s$^{-1}$ cancel. Each atom of $^{218}Po$ will emit a 6.00-MeV alpha particle. Each will also lead to the emission later of a 7.69-MeV alpha particle with the decay of its daughter $^{214}Po$ into $^{210}Pb$. Thus, the presence of one $^{218}Po$ atom represents a potential alpha-particle energy $E = 6.00 + 7.69 = 13.69$ MeV from the short-lived radon daughters. Using $N$ from Eq. (4.58), we find for the potential alpha-particle energy per unit volume contributed by the atoms of $^{218}Po$, $NE = 3.35 \times 10^{5}$ MeV m$^{-3}$. Similar calculations can be made for the contributions of the other three daughters in Table 4.1 to the PAEC. The only modification for the others is that the potential alpha energy associated with each atom is 7.69 MeV (Fig. 4.7).

The complete calculation is summarized in Table 4.2. The individual nuclide contributions in the last column can be added to give the final answer, PAEC = $2.95 \times 10^{6}$ MeV m$^{-3}$. Note that the half-life of $^{214}Po$, which is in secular equilibrium with $^{214}Bi$ (equal activity densities), is so short that very few atoms are present. Its contribution to the PAEC is negligible.

*Example*
Calculate the EEDC in the last example. What is the equilibrium factor?

*Solution*
By definition, the EEDC is the activity concentration of the short-lived radon daughters that would give a specified value of the PAEC under the condition of secular equilibrium. For the last example, the EEDC is the (equal) concentration that would appear in the second column of Table 4.2 for each nuclide that would result in the given value, PAEC = $2.95 \times 10^{6}$ MeV m$^{-3}$. The solution can be set up in more than one way. We compute the PAEC for secular equilibrium at unit activity density

Table 4.3

| Nuclide | $NE/A$ <br> (MeV Bq$^{-1}$) |
|---|---|
| $^{218}$Po | $3.60 \times 10^3$ |
| $^{214}$Pb | $1.79 \times 10^4$ |
| $^{214}$Bi | $1.32 \times 10^4$ |
| $^{214}$Po | $1.82 \times 10^{-3}$ |

(1 Bq m$^{-3}$), from which the answer follows immediately. The contribution per unit activity from $^{218}$Po, for example, is obtained from Table 4.2:

$$\frac{NE}{A} = \frac{3.35 \times 10^5 \text{ MeV m}^{-3}}{93 \text{ Bq m}^{-3}} = 3.60 \times 10^3 \text{ MeV Bq}^{-1}. \tag{4.59}$$

Values for the four nuclides are shown in Table 4.3. Adding the numbers in the second column gives a total of $3.47 \times 10^4$ MeV Bq$^{-1}$. This is the PAEC (MeV m$^{-3}$) per unit activity concentration (Bq m$^{-3}$) of each daughter in secular equilibrium. Therefore, for the last example we have

$$\begin{aligned} \text{EEDC} &= \frac{\text{PAEC}}{3.47 \times 10^4 \text{ MeV Bq}^{-1}} \\ &= \frac{2.95 \times 10^6 \text{ MeV m}^{-3}}{3.47 \times 10^4 \text{ MeV Bq}^{-1}} = 85.0 \text{ Bq m}^{-3}. \end{aligned} \tag{4.60}$$

The equilibrium factor is the ratio of this activity concentration and that of the radon (Table 4.1): $85.0/120 = 0.708$.

This example shows the useful relationship between an equilibrium concentration of 1 Bq m$^{-3}$ of the short-lived radon daughters and the associated potential alpha-energy concentration:

$$\frac{3.47 \times 10^4 \text{ MeV m}^{-3}}{1 \text{ Bq m}^{-3}} = 3.47 \times 10^4 \text{ MeV Bq}^{-1}. \tag{4.61}$$

Note that the potential alpha energy per unit activity of the daughters in secular equilibrium is independent of the actual concentration of the daughters in air.

An older unit for the PAEC is the working level (WL), defined as a potential alpha-particle energy concentration of $1.3 \times 10^5$ MeV L$^{-1}$ of air for the short-lived radon daughters. This value corresponds to the presence of 100 pCi L$^{-1}$ = 3.7 Bq L$^{-1}$ of the daughters in secular equilibrium, that is, to an EEDC of 3.7 Bq L$^{-1}$. The exposure of persons to radon daughters is often expressed in working-level months (WLM), with a working month defined as 170 h. The WLM represents the integrated exposure of an individual over a specified time period. Concentrations of radon itself are sometimes reported, rather than PAECs or WLs, which pertain to the daughters. As a rule of thumb, 1 WL of radon daughters is often associated with a radon concentration of 200 pCi L$^{-1}$, corresponding to an equilibrium factor of 0.5.

*Example*

A person spends an average of 14 hours per day at home, where the average concentration of radon is 1.5 pCi L$^{-1}$ (a representative value for many residences). What is his exposure in WLM over a six-month period?

*Solution*

Using the rule of thumb just given, we estimate the daughter concentration to be (1.5/200) (1 WL) = 0.0075 WL. The exposure time in hours for the six months (= 183 d) is 183 × 14 = 2560 h. Since there are 170 h in a working month, the exposure time is 2560/170 = 15.1 working months. The person's exposure to radon daughters over the six-month period is therefore 0.0075 × 15.1 = 0.11 WLM.

## 4.7
## Suggested Reading

1 Attix, F. H., *Introduction to Radiological Physics and Radiation Dosimetry*, Wiley, New York, NY (1986). [Chapter 6 discusses radioactive and serial decay.]

2 Bodansky, D., Robkin, M. A., and Stadler, D. R., eds., *Indoor Radon and Its Hazards*, University of Washington Press, Seattle, WA (1987).

3 Cember, H., *Introduction to Health Physics*, 3rd ed., McGraw-Hill, New York, NY (1996). [Chapter 4 treats radioactive transformations, serial decay, and naturally occurring radioactivity.]

4 Evans, R. D., *The Atomic Nucleus*, McGraw-Hill, New York, NY (1955). [Chapter 15 describes serial decay, decay schemes, equilibrium, and other topics.]

5 Faw, R. E., and Shultis, J. K., *Radiological Assessment*, Prentice-Hall, Englewood Cliffs, NJ (1993). [Chapter 4 gives and excellent coverage of exposure to natural sources of radiation, including radon.]

6 Magill, Joseph, and Galy, Jean, *Radioactivity-Radionuclides-Radiation*, Springer Verlag, New York, NY (2005).

[This book discusses a number of environmental and human radiological health issues. It contains extensive nuclear data. A CD-ROM is included.]

7 National Research Council, *Health Risks of Radon and Other Internally Deposited Alpha-Emitters: BEIR IV*, National Academy Press, Washington, DC (1988). [A comprehensive report (602 pp.) from the Committee on the Biological Effects of Ionizing Radiation.]

8 Nazaroff, W. W., and Nero, A. V., Jr., eds., *Radon and its Decay Products in Indoor Air*, Wiley, New York, NY (1988).

9 NCRP Report No. 97, *Measurement of Radon and Radon Daughters in Air*, National Council on Radiation Protection and Measurements, Bethesda, MD (1988). [Contains basic data on the three natural decay series and characteristics of radon daughters discussed here. Main emphasis of the Report is on measurements.]

10 NCRP Report No. 103, *Control of Radon in Houses*, National Council on Radiation Protection and Measurements, Bethesda, MD (1989).

# 4.8
## Problems

1. What is the value of the decay constant of $^{40}$K?
2. What is the decay constant of tritium?
3. The activity of a radioisotope is found to decrease by 30% in 1 wk. What are the values of its
   (a) decay constant
   (b) half-life
   (c) mean life?
4. What percentage of the original activity of a radionuclide remains after
   (a) 5 half-lives
   (b) 10 half-lives?
5. The isotope $^{132}$I decays by $\beta^-$ emission into stable $^{132}$Xe with a half-life of 2.3 h.
   (a) How long will it take for $\frac{7}{8}$ of the original $^{132}$I atoms to decay?
   (b) How long it will take for a sample of $^{132}$I to lose 95% of its activity?
6. A very old specimen of wood contained $10^{12}$ atoms of $^{14}$C in 1986.
   (a) How many $^{14}$C atoms did it contain in the year 9474 B.C.?
   (b) How many $^{14}$C atoms did it contain in 1986 B.C.?
7. A radioactive sample consists of a mixture of $^{35}$S and $^{32}$P. Initially, 5% of the activity is due to the $^{35}$S and 95% to the $^{32}$P. At what subsequent time will the activities of the two nuclides in the sample be equal?
8. The gamma exposure rate at the surface of a shielded $^{198}$Au source is 10 R h$^{-1}$ (roentgen/hour, Sec. 12.2). What will be the exposure rate in this position after 2 wk?
9. Compute the specific activity of
   (a) $^{238}$U
   (b) $^{90}$Sr
   (c) $^{3}$H.
10. How many grams of $^{32}$P are there in a 5 mCi source?
11. How many atoms are there in a 1.16-MBq source of
    (a) $^{24}$Na?
    (b) $^{238}$U?
12. An encapsulated $^{210}$Po radioisotope was used as a heat source, in which an implanted thermocouple junction converts heat into electricity with an efficiency of 15% to power a small transmitter for an early space probe.

(a) How many curies of $^{210}$Po are needed at launch time if the transmitter is to be supplied with 100 W of electricity 1 y after launch?

(b) Calculate the number of grams of $^{210}$Po needed.

(c) If the transmitter shuts off when the electrical power to it falls below 1 W, how long can it be expected to operate after launch?

(d) What health physics precautions would you recommend during fabrication, encapsulation, and handling of the device?

13. The Cassini spacecraft went into orbit about the planet, Saturn, in July 2004, after a nearly seven-year journey from Earth. On-board electrical systems were powered by heat from three radioisotope thermoelectric generators, which together utilized a total of 32.7 kg of $^{238}$Pu, encapsulated as $PuO_2$. The isotope has a half-life of 86.4 y and emits an alpha particle with an average energy of 5.49 MeV. The daughter $^{234}$U has a half-life of $2.47 \times 10^5$ y.

(a) Calculate the specific thermal-power generation rate of $^{238}$Pu in $W\,g^{-1}$.

(b) How much total thermal power is generated in the spacecraft?

14. A 0.2-g sample of $^{85}_{36}$Kr gas, which decays into stable $^{85}_{37}$Rb, is accidentally broken and escapes inside a sealed warehouse measuring $40 \times 30 \times 20$ m. What is the specific activity of the air inside?

15. A 6.2-mg sample of $^{90}$Sr is in secular equilibrium with its daughter $^{90}$Y.

(a) How many Bq of $^{90}$Sr are present?

(b) How many Bq of $^{90}$Y are present?

(c) What is the mass of $^{90}$Y present?

(d) What will the activity of the $^{90}$Y be after 100 y?

16. A sample contains 1.0 GBq of $^{90}$Sr and 0.62 GBq of $^{90}$Y.

(a) What will be the total activity of the sample 10 days later?

(b) What will be the total activity of the sample 29.12 years later?

17. Consider the following $\beta^-$ nuclide decay chain with the half-lives indicated:

$$^{210}_{82}\text{Pb} \xrightarrow[22\ y]{\beta^-} {}^{210}_{83}\text{Bi} \xrightarrow[5.0\ d]{\beta^-} {}^{210}_{84}\text{Po}.$$

A sample contains 30 MBq of $^{210}$Pb and 15 MBq of $^{210}$Bi at time $t = 0$.

(a) Calculate the activity of $^{210}$Bi at time $t = 10$ d.

(b) If the sample was originally pure $^{210}$Pb, then how old is it at time $t = 0$?

18. What is the mean life of a $^{226}$Ra atom?

19. $^{59}$Fe has a half-life of 45.53 d.
    (a) What is the mean life of a $^{59}$Fe atom?
    (b) Calculate the specific activity of $^{59}$Fe.
    (c) How many atoms are there in a 10-mCi source of $^{59}$Fe?

20. At time $t = 0$ a sample consists of 2 Ci of $^{90}$Sr and 8 Ci of $^{90}$Y.
    (a) What will the activity of $^{90}$Y be in the sample after 100 h?
    (b) At what time will the $^{90}$Y activity be equal to 3 Ci?

21. $^{136}$Cs (half-life $= 13.7$ d) decays ($\beta^-$) into $^{136m}$Ba (half-life $= 0.4$ s), which decays ($\gamma$) into stable $^{136}$Ba:

$$^{136}\text{Cs} \xrightarrow[13.7\ \text{d}]{\beta^-} {}^{136m}\text{Ba} \xrightarrow[0.4\ \text{s}]{\gamma} {}^{136}\text{Ba}.$$

    (a) Calculate the decay constant of $^{136}$Cs.
    (b) Calculate the specific activity of $^{136}$Cs.
    (c) Starting with a pure $10^{10}$-Bq sample of $^{136}$Cs at time $t = 0$, how many atoms of $^{136m}$Ba decay between time $t_1 = 13.7$ d (exactly) and time $t_2 = 13.7$ d $+ 5$ s (exactly)?

22. Show that Eq. (4.40) leads to secular equilibrium, $A_1 = A_2$, under the appropriate conditions.

23. Show by direct substitution that the solution given by Eq. (4.40) satisfies Eq. (4.39).

24. A 40-mg sample of pure $^{226}$Ra is encapsulated.
    (a) How long will it take for the activity of $^{222}$Rn to build up to 10 mCi?
    (b) What will be the activity of $^{222}$Rn after 2 years?
    (c) What will be the activity of $^{222}$Rn after 1000 years?
    (d) What is the ratio of the specific activity of $^{222}$Rn to that of $^{226}$Ra?

25. Verify Eq. (4.43).

26. (a) Verify Eq. (4.44).
    (b) Show that the time of maximum total activity occurs earlier than the time of maximum daughter activity in Fig. 4.5.
    (c) Does Eq. (4.43) apply to $A_2$ when there is no equilibrium (Fig. 4.6)?

27. To which of the natural series do the following heavy radionuclides belong: $^{213}_{83}$Bi, $^{215}_{84}$Po, $^{230}_{90}$Th, $^{233}_{92}$U, and $^{224}_{88}$Ra?

28. The average mass of potassium in the human body is about 140 g. From the abundance and half-life given in Appendix D, estimate the average activity of $^{40}$K in the body.

29. An atmosphere contains radon and its short-lived daughters in secular equilibrium at a concentration of 52 Bq m$^{-3}$.

(a) What is the PAEC?

(b) The equilibrium factor?

30. An air sample taken from a room shows the nuclide activity concentrations given in Table 4.4. Calculate

(a) the potential alpha-energy concentration

(b) the EEDC

(c) the equilibrium factor.

**Table 4.4** Problem 30

| Nuclide | Concentration $(Bq\,L^{-1})$ |
|---------|------------------------------|
| $^{222}Rn$ | 9.2 |
| $^{218}Po$ | 4.6 |
| $^{214}Pb$ | 2.7 |
| $^{214}Bi$ | 2.0 |
| $^{214}Po$ | 2.0 |

31. A 5-L sample of air contains the activities (disintegrations per minute, dpm) of radon daughters shown in Table 4.5. Calculate

(a) the potential alpha-energy concentration

(b) the equilibrium-equivalent decay-product concentration.

**Table 4.5** Problem 31

| Nuclide | dpm |
|---------|-----|
| $^{218}Po$ | 1690 |
| $^{214}Pb$ | 1500 |
| $^{214}Bi$ | 1320 |
| $^{214}Po$ | 1320 |

32. A room contains $^{222}Rn$ at a concentration of 370 Bq m$^{-3}$. The PAEC is $7.8 \times 10^6$ MeV m$^{-3}$. What is the equilibrium factor?

33. (a) Show that the EEDC for the short-lived $^{222}Rn$ daughters is given by

$$EEDC = 0.104C(^{218}Po) + 0.516C(^{214}Pb) + 0.380C(^{214}Bi),$$

where $C(^{218}Po)$, $C(^{214}Pb)$, and $C(^{214}Bi)$ are the concentrations of the daughters indicated.

(b) What units are implied in this expression?

(c) Why is the expression independent of $C(^{214}Po)$?

34. If the radon concentration ($^{222}$Rn) inside a building is 0.85 pCi L$^{-1}$ and the equilibrium factor is 0.6, what is the rate of release of alpha-particle energy in MeV L$^{-1}$ h$^{-1}$?

35. Show that
    (a) 1 WL $= 2.1 \times 10^{-5}$ J m$^{-3}$
    (b) 1 WLM $= 0.0036$ J h m$^{-3}$.

36. A 3-L air sample contains the following radon-daughter activities: $^{218}$Po, 16.2 Bq; $^{214}$Pb, 15.0 Bq; $^{214}$Bi, 12.2 Bq; and $^{214}$Po, 12.2 Bq. Calculate the WL concentration.

37. A person spends an average of 10 hours per day, 5 days per week, in an atmosphere where the average radon-daughter concentration is 0.68 WL. What is his exposure in WLM after one year of this activity?

38. A basement measures 12 m $\times$ 10 m $\times$ 2.5 m. The air inside contains the nuclide inventory shown in Table 4.6.
    (a) Calculate the WL concentration.
    (b) If the given activities are average and a person occupies the basement 10 hours per day, 7 days per week, for 12 months, what will be his exposure in WLM?

**Table 4.6** Problem 38

| Nuclide | Activity ($\mu$Ci) |
|---|---|
| $^{222}$Rn | 0.81 |
| $^{218}$Po | 0.69 |
| $^{214}$Pb | 0.44 |
| $^{214}$Bi | 0.25 |
| $^{214}$Po | 0.25 |

39. A room measures 10 m $\times$ 8 m $\times$ 3 m. It contains 80 pCi L$^{-1}$ of $^{218}$Po, 60 pCi L$^{-1}$ of $^{214}$Pb, and 25 pCi L$^{-1}$ each of $^{214}$Bi and $^{214}$Po.
    (a) Calculate the WL concentration in the room.
    (b) Calculate the total potential alpha-particle energy in the room.
    (c) What is the concentration of $^{214}$Po atoms in the air?
    (d) If secular equilibrium existed at this working-level concentration, what would be the activity concentration of $^{214}$Pb atoms?
    (e) What would be the exposure in WLM of an individual who occupied the room 12 hours per day, 6 days per week, for one year?

**40.** An atmosphere contains the numbers of atoms per liter shown in Table 4.7.
   **(a)** Calculate the PAEC in $J\,m^{-3}$.
   **(b)** Calculate the EEDC.
   **(c)** Calculate the equilibrium factor.

**Table 4.7** Problem 40

| Nuclide | Atoms $L^{-1}$ |
|---------|----------------|
| $^{222}Rn$ | $2.34 \times 10^5$ |
| $^{218}Po$ | 52 |
| $^{214}Pb$ | 407 |
| $^{214}Bi$ | 214 |
| $^{214}Po$ | 2 |

## 4.9
## Answers

**2.** $0.0564\,y^{-1}$

**5.** **(a)** 6.90 h
   **(b)** 9.94 h

**6.** **(a)** $4.00 \times 10^{12}$
   **(b)** $1.62 \times 10^{12}$

**7.** 72.5 d

**10.** $1.75 \times 10^{-8}$ g

**12.** **(a)** $1.30 \times 10^5$ Ci
   **(b)** 28.9 g
   **(c)** 1280 d

**13.** **(a)** $0.575\,W\,g^{-1}$
   **(b)** 18.8 kW

**14.** 120 MBq $m^{-3}$

**15.** **(a)** $3.13 \times 10^{10}$ Bq
   **(b)** $3.13 \times 10^{10}$ Bq
   **(c)** $1.56\,\mu g$
   **(d)** $2.90 \times 10^9$ Bq

**16.** **(a)** 1.97 GBq
   **(b)** 1.00 GBq

**21.** **(a)** $0.0506\,d^{-1}$
   **(b)** $2.59 \times 10^{18}$ Bq $kg^{-1}$
   **(c)** $2.5 \times 10^{10}$

**30.** **(a)** $9.15 \times 10^7$ MeV $m^{-3}$
   **(b)** 2630 Bq $m^{-3}$
   **(c)** 0.29

**31.** **(a)** $1.68 \times 10^8$ MeV $m^{-3}$
   **(b)** 4830 Bq $m^{-3}$

**36.** 1.25 WL

**37.** 10.4 WLM

**39.** **(a)** 0.481 WL
   **(b)** $1.50 \times 10^{10}$ MeV
   **(c)** $2.19 \times 10^{-4}\,L^{-1}$
   **(d)** 48.1 pCi $L^{-1}$
   **(e)** 10.6 WLM

# 5

# Interaction of Heavy Charged Particles with Matter

This chapter and the next four deal with the mechanisms by which different types of ionizing radiation interact with matter. Knowledge of the basic physics of radiation interaction and energy transfer is fundamental to radiation detection, measurement, and control, as well as to understanding the biological effects of radiation on living tissue. We consider "heavy" charged particles first, that is charged particles other than the electron and positron.

## 5.1
## Energy-Loss Mechanisms

A heavy charged particle traversing matter loses energy primarily through the ionization and excitation of atoms. (Except at low velocities, a heavy charged particle loses a negligible amount of energy in nuclear collisions.) The moving charged particle exerts electromagnetic forces on atomic electrons and imparts energy to them. The energy transferred may be sufficient to knock an electron out of an atom and thus ionize it, or it may leave the atom in an excited, nonionized state. As we show in the next section, a heavy charged particle can transfer only a small fraction of its energy in a single electronic collision. Its deflection in the collision is negligible. Thus, a heavy charged particle travels an almost straight path through matter, losing energy almost continuously in small amounts through collisions with atomic electrons, leaving ionized and excited atoms in its wake. Occasionally, however, as observed in Rutherford's experiments with alpha-particle scattering from a gold foil, a heavy charged particle will undergo a substantial deflection due to elastic scattering from an atomic nucleus.

Electrons and positrons also lose energy almost continuously as they slow down in matter. However, they can lose a large fraction of their energy in a single collision with an atomic electron (having equal mass), thereby suffering relatively large deflections. Because of their small mass, electrons are frequently scattered through large angles by nuclei. In contrast to heavy charged particles, electrons and positrons do not generally travel through matter in straight lines. An electron can also be sharply deflected by an atomic nucleus, causing it to emit photons in the process called bremsstrahlung (braking radiation). Figure 5.1 shows the con-

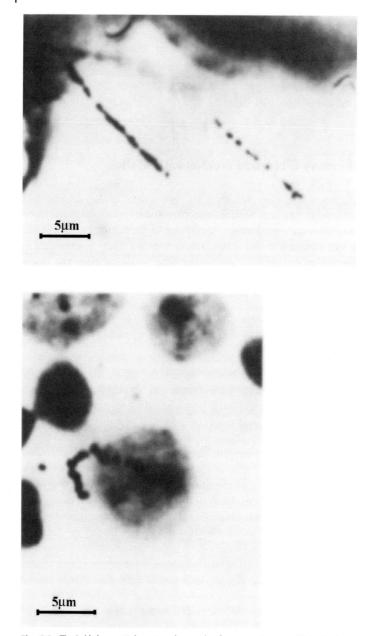

**Fig. 5.1** (Top) Alpha-particle autoradiograph of rat bone after inhalation of $^{241}$Am. Biological preparation by R. Masse and N. Parmentier. (Bottom) Beta-particle autoradiograph of isolated rat-brain nucleus. The $^{14}$C-thymidine incorporated in the nucleolus is located at the track origin of the electron emitted by the tracer element. Biological preparation by M. Wintzerith and P. Mandel. (Courtesy R. Rechenmann and E. Wittendorp-Rechenmann, Laboratoire de Biophysique des Rayonnements et de Methodologie INSERM U.220, Strasbourg, France.)

trast between the straight tracks of two alpha particles and the tortuous track of a beta particle in photographic emulsion.

## 5.2
## Maximum Energy Transfer in a Single Collision

In this section we calculate the maximum energy that a charged particle can lose in colliding with an atomic electron. We assume that the particle moves rapidly compared with the electron and that the energy transferred is large compared with the binding energy of the electron in the atom. Under these conditions the electron can be considered to be initially free and at rest, and the collision is elastic. We treat the problem classically and then give the relativistic results.

Figure 5.2(a) shows schematically a charged particle (mass $M$ and velocity $V$) approaching an electron (mass $m$, at rest). After the collision, which for maximum energy transfer is head-on, the particles in (b) move with speeds $V_1$ and $v_1$ along the initial line of travel of the incident particle. Since the total kinetic energy and momentum are conserved in the collision, we have the two relationships

$$\frac{1}{2}MV^2 = \frac{1}{2}MV_1^2 + \frac{1}{2}mv_1^2 \tag{5.1}$$

and

$$MV = MV_1 + mv_1. \tag{5.2}$$

If we solve Eq. (5.2) for $v_1$ and substitute the result into (5.1), we obtain

$$V_1 = \frac{(M-m)V}{M+m}. \tag{5.3}$$

Using this expression for $V_1$, we find for the maximum energy transfer

$$Q_{max} = \frac{1}{2}MV^2 - \frac{1}{2}MV_1^2 = \frac{4mME}{(M+m)^2}, \tag{5.4}$$

where $E = MV^2/2$ is the initial kinetic energy of the incident particle.

**M,V**          m

— — — ●———————▶    ● — — — — — — — — —

(a) **Before collision**

— — — — — — — — —●———▶    ●——▶ — — — — — —
                   **M,V$_1$**      **m, v$_1$**

(b) **After collision**

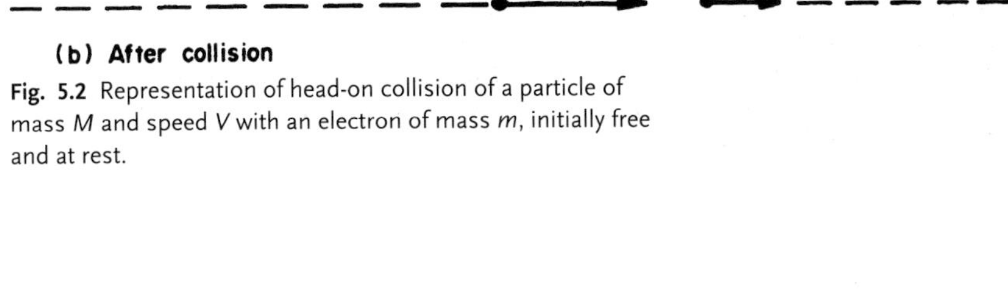

**Fig. 5.2** Representation of head-on collision of a particle of mass $M$ and speed $V$ with an electron of mass $m$, initially free and at rest.

When the incident particle is an electron or positron, the special circumstance arises in which its mass is the same as that of the struck particle: $M = m$. Equation (5.4) then implies that $Q_{max} = E$, and so its entire energy can be transferred in a single, billiard-ball-type collision. As already mentioned, electrons and positrons can thus experience relatively large energy losses and deflections, which contribute to their having tortuous paths in matter. The next particle more massive than the electron is the muon, having a mass $M = 207m$.[1] The maximum fraction of energy that a muon can transfer in a single collision is, from Eq. (5.4),

$$\frac{Q_{max}}{E} = \frac{4m(207m)}{(208m)^2} \cong \frac{4}{208} = 0.0192. \tag{5.5}$$

Thus, the muon (and all heavy charged particles) travel essentially straight paths in matter, except for occasional large-angle deflections by atomic nuclei.

The exact relativistic expression for the maximum energy transfer, with $m$ and $M$ denoting the rest masses of the electron and the heavy particle, is

$$Q_{max} = \frac{2\gamma^2 m V^2}{1 + 2\gamma m/M + m^2/M^2}, \tag{5.6}$$

where $\gamma = 1/\sqrt{1 - \beta^2}$, $\beta = V/c$, and $c$ is the speed of light (Appendix C). Except at extreme relativistic energies, $\gamma m/M \ll 1$, in which case (5.6) reduces to

$$Q_{max} = 2\gamma^2 m V^2 = 2\gamma^2 mc^2 \beta^2, \tag{5.7}$$

which is the usual relativistic result.

*Example*

Calculate the maximum energy that a 10-MeV proton can lose in a single electronic collision.

*Solution*

For a proton of this energy the nonrelativistic formula (5.4) is accurate. Neglecting $m$ compared with $M$, we have $Q_{max} = 4mE/M = 4 \times 1 \times 10/1836 = 2.18 \times 10^{-2}$ MeV = 21.8 keV, which is only 0.22% of the proton's energy.

*Example*

Use the relativistic formula (5.7) to calculate the maximum possible energy loss in a single collision of the 10-MeV proton in the last example.

*Solution*

We first find $\gamma$. Since the proton rest energy is $Mc^2 = 938$ MeV (Appendix A), we can use the formula in Appendix C for the relativistic kinetic energy, $T = 10$ MeV, to write

1   The muon ($M = 207m$), pion ($270m$), and kaon ($967m$) are unstable particles with masses intermediate between those of the electron ($m$) and proton ($1836m$). They occur with cosmic radiation and can also be generated in particle accelerators.

**Table 5.1** Maximum Possible Energy Transfer, $Q_{max}$, in Proton Collision with Electron

| Proton Kinetic Energy $E$ (MeV) | $Q_{max}$ (MeV) | Maximum Percentage Energy Transfer $100 Q_{max}/E$ |
|---|---|---|
| 0.1 | 0.00022 | 0.22 |
| 1 | 0.0022 | 0.22 |
| 10 | 0.0219 | 0.22 |
| 100 | 0.229 | 0.23 |
| $10^3$ | 3.33 | 0.33 |
| $10^4$ | 136. | 1.4 |
| $10^5$ | $1.06 \times 10^4$ | 10.6 |
| $10^6$ | $5.38 \times 10^5$ | 53.8 |
| $10^7$ | $9.21 \times 10^6$ | 92.1 |

$10 = 938(\gamma - 1)$. It follows that $\gamma = 1.01066$ and $\beta^2 = 0.02099$. Since the electron rest energy is $mc^2 = 0.511$ MeV (Appendix A), Eq. (5.7) yields $Q_{max} = 21.9$ keV.

Table 5.1 gives numerical results for a range of proton energies. Except at extreme relativistic energies, where Eq. (5.6) must be used, the maximum fractional energy loss for a heavy charged particle is small. At these extreme energies, the rest energy of the colliding particle contributes little to its total energy (Appendix C). The difference in rest mass between it and the struck electron then has little effect on the collision. One sees that $Q_{max}/E$ approaches 100% in Table 5.1. Encounters in which an amount of energy comparable to $Q_{max}$ is transferred are very rare, though, particularly at high energies. The probabilities for losing different amounts of energy are described by single-collision energy-loss spectra, discussed in the next section.

Equations (5.4), (5.6), and (5.7) for maximum energy loss are kinematic in nature. That is, they follow from the simultaneous conservation of momentum and kinetic energy, independently of the kinds of forces that act. Under the conditions stated at the beginning of this section, it is a good approximation to calculate $Q_{max}$ as though the struck electron were not bound, the collision then being elastic. Charged-particle energy losses to atomic electrons are, in fact, inelastic; the stated conditions do not apply when a small amount of energy is transferred. In Chapter 9 we shall apply Eq. (5.4) to the elastic scattering of neutrons by atomic nuclei, such collisions being truly elastic. The special case $M = m$ and $Q_{max} = E$ arises then for neutron scattering by hydrogen.

## 5.3
## Single-Collision Energy-Loss Spectra

The foregoing analysis of $Q_{max}$ enables one to understand part of the physical basis for the different kinds of trajectories seen for electrons and for heavy charged particles in matter. Further details about charged-particle penetration are embodied in the *spectra* of single-collision energy losses to atomic electrons. These spectra include the effects of the binding of electrons in atoms. As pointed out in the last paragraph, the electronic collisions by which charged particles transfer energy to matter are inelastic. Kinetic energy is lost in overcoming the binding energies of the struck electrons.

Figure 5.3 shows single-collision energy-loss spectra calculated for 50-eV and 150-eV electrons and 1-MeV protons in liquid water. The ordinate gives the probability density $W(Q)$ per eV, such that $W(Q)\,dQ$ is the probability that a given collision will result in an energy loss between $Q$ and $Q + dQ$, with $Q$ expressed in eV. (The curves are normalized to unit area.) Similar spectra calculated for more energetic electrons and protons lie almost on top of the curves for the 150-eV electrons and 1-MeV protons. For other energetic particles, the principal difference in the curves is their association with different values of $Q_{max}$, where the functions $W(Q)$ are very small.

The most striking feature of energy-loss spectra for fast charged particles (i.e., those with speeds greater than the orbital speeds of the atomic electrons) is their remarkable similarity in the region from about 10 eV to 70 eV, where energy losses are most probable. As discussed in more detail in Section 5.5, this universality is attributable to other factors being generally of secondary importance when a quantum transition in an atom is induced by the very swift passage of charge. Such an encounter is referred to as a "sudden" collision.

The energy-loss spectra of slow charged particles differ from one another. The time of interaction is longer than for fast particles, the binding of the electrons is more important, and the most probable energy losses are closer to $Q_{max}$. In addition, slow particles have a greater tendency to excite atoms rather than ionize them. Figure 5.3 shows by way of example the different shape of the energy-loss spectrum for 50-eV electrons.

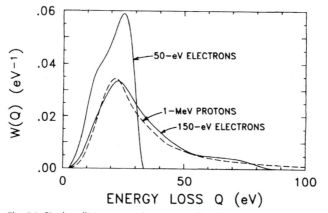

**Fig. 5.3** Single-collision energy-loss spectra for 50-eV and 150-eV electrons and 1-MeV protons in liquid water. (Courtesy Oak Ridge National Laboratory, operated by Martin Marietta Energy Systems, Inc., for the Department of Energy.)

One also sees in Fig. 5.3 that energy is not lost by a charged particle in arbitrarily small amounts. The binding of the electrons in the discrete energy states of the medium requires that a minimum, or threshold, energy, $Q_{min} > 0$, be transferred for excitation or ionization of an atom.

*Example*

Estimate the probability that a 50-MeV proton will lose between 30 eV and 40 eV in a collision with an atomic electron in penetrating the soft tissue of the body.

*Solution*

Soft tissue is similar in atomic composition to liquid water (Table 12.3), and so we use Fig. 5.3 to make the estimate. As implied in the text, the energy-loss spectrum for 50-MeV protons is close to that for 1-MeV protons, except that it extends out to a different value of $Q_{max}$. We see that the area under the curve between 30 eV and 40 eV in Fig. 5.3 is approximately

$$W(Q)\Delta Q = (0.019 \text{ eV}^{-1})(40 - 30) \text{ eV} = 0.19. \tag{5.8}$$

Thus, a 50-MeV proton has about a 20% chance of losing between 30 eV and 40 eV in a single electronic collision in soft tissue.

## 5.4
## Stopping Power

The average linear rate of energy loss of a heavy charged particle in a medium (expressed, for example, in MeV cm$^{-1}$) is of fundamental importance in radiation physics and dosimetry. This quantity, designated $-dE/dx$, is called the stopping power of the medium for the particle. It is also referred to as the linear energy transfer (LET) of the particle, usually expressed as keV $\mu$m$^{-1}$ in water. Stopping power and LET are closely associated with the dose delivered by charged particles (or charged recoil particles produced by incident photons or neutrons) and with the biological effectiveness of different kinds of radiation (Section 7.3).

Stopping powers can be calculated from energy-loss spectra like those discussed in the last section. For a given type of charged particle at a given energy, the stopping power is given by the product of (1) the probability $\mu$ per unit distance of travel that an electronic collision occurs and (2) the average energy loss per collision, $Q_{avg}$. The former is called the macroscopic cross section, or attenuation coefficient, and has the dimensions of inverse length. The latter is given by

$$Q_{avg} = \int_{Q_{min}}^{Q_{max}} Q W(Q) \, dQ, \tag{5.9}$$

where $Q_{min}$ was introduced at the end of the last section. Thus, the stopping power is given by

$$-\frac{dE}{dx} = \mu Q_{avg} = \mu \int_{Q_{min}}^{Q_{max}} Q W(Q) \, dQ. \tag{5.10}$$

If $\mu$ is expressed in cm$^{-1}$ and $Q$ in MeV, then Eq. (5.10) gives the stopping power in MeV cm$^{-1}$. The quantities $\mu$ and $Q_{avg}$, and hence $-dE/dx$, depend upon the type of particle, its energy, and the medium traversed.

*Example*

The macroscopic cross section for a 1-MeV proton in water is 410 $\mu$m$^{-1}$, and the average energy lost in an electronic collision is 72 eV. What is the stopping power in MeV cm$^{-1}$ and in J m$^{-1}$?

*Solution*

With $\mu = 410$ $\mu$m$^{-1}$ and $Q_{avg} = 72$ eV, Eq. (5.10) gives

$$-\frac{dE}{dx} = \mu Q_{avg} = 410 \times 72 = 2.95 \times 10^4 \text{ eV} \, \mu\text{m}^{-1}.$$

Since 1 eV = 10$^{-6}$ MeV and 1 $\mu$m = 10$^{-4}$ cm, we obtain $-dE/dx = 295$ MeV cm$^{-1}$. Converting units further, we have $-dE/dx = 295$ MeV cm$^{-1} \times 1.60 \times 10^{-13}$ J MeV$^{-1} \times 100$ cm m$^{-1} = 4.72 \times 10^{-9}$ J m$^{-1}$.

In this example, note that the average energy loss of 72 eV for a 1-MeV proton is considerably larger than the range of the most probable energy losses in Fig. 5.3. Table 5.1 shows that the maximum energy loss is 2200 eV, which lies beyond the horizontal scale in the figure by a factor of 22. The energy-loss distribution $W(Q)$ for heavy charged particles is thus very skewed for large losses out to $Q_{max}$. Although $W(Q)$ is small when $Q$ is large, the stopping power reflects an energy-loss-weighted average, $QW(Q)$, and hence substantial contributions from the tail of the energy-loss distribution. In traveling short distances in matter, when the total number of collisions is relatively small, the average energy lost by a charged particle (as implied by the stopping power) and the most probable energy lost can differ substantially. This phenomenon of energy straggling is discussed in Section 7.5.

To understand what "short distances" mean in the present context, we recall that the macroscopic cross section $\mu$ is the probability per unit distance of travel that an electronic collision takes place. Its role in charged-particle penetration is analogous to that of the decay constant $\lambda$, which is the probability of disintegration per unit time in radioactive decay. Equation (4.22) showed that the reciprocal of the decay constant is equal to the mean life. In the same way, the reciprocal of $\mu$ is the mean distance of travel, or mean free path, of a charged particle between collisions. In the last example, the mean free path of the 1-MeV proton is $1/\mu = 1/(410 \, \mu\text{m}^{-1}) = 0.0024$ $\mu$m = 24 Å. Atomic diameters are of the order of 1 Å to 2 Å.

## 5.5
### Semiclassical Calculation of Stopping Power

Quantum mechanically, stopping power is the mean, or expectation, value of the linear rate of energy loss. In 1913 Bohr derived an explicit formula giving the stopping power for heavy charged particles. Because quantum mechanics had not yet been discovered, Bohr relied on intuition and insight to obtain the proper semiclassical representation of atomic collisions. He calculated the energy loss of a heavy particle in a collision with an electron at a given distance of passing and then averaged over all possible distances and energy losses. The nonrelativistic formula that Bohr obtained gave the correct physical features of stopping power as borne out by experiment and by the later quantum mechanical theory of Bethe. We present here a derivation along the lines of Bohr.

In Fig. 5.4 we consider a heavy particle (charge $ze$ and velocity $V$) that travels swiftly past an electron (charge $-e$ and mass $m$) in a straight line at a distance $b$, called the *impact parameter*. We assume that the electron is initially free and at rest at the origin of the $XY$ coordinate system shown. We assume, further, that the collision is sudden: it takes place rapidly and is over before the electron moves appreciably. Perpendicular components $F_x$ and $F_y$ of the Coulomb force $F = k_0 ze^2/r^2$ that the particle exerts on the electron at a given instant are shown in Fig. 5.4. With the approximation that the electron remains stationary, the component $F_x$ transfers no net momentum to it over the duration of the collision. (This component acts symmetrically, first toward the left and then toward the right, its net effect being zero). The charged particle transfers momentum to the electron through the action of the other, perpendicular, force component $F_y$. The total momentum imparted to the electron in the collision is

$$p = \int_{-\infty}^{\infty} F_y \, dt = \int_{-\infty}^{\infty} F \cos\theta \, dt = k_0 ze^2 \int_{-\infty}^{\infty} \frac{\cos\theta}{r^2} \, dt. \tag{5.11}$$

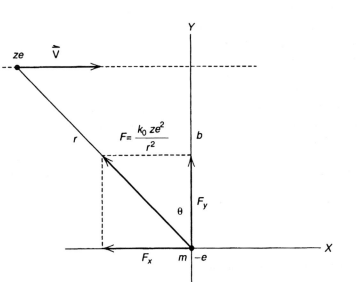

**Fig. 5.4** Representation of the sudden collision of a heavy charged particle with an electron, located at the origin of $XY$ coordinate axes shown. See text.

To carry out the integration, we let $t = 0$ represent the time at which the heavy charged particle crosses the Y-axis in Fig. 5.4. Since $\cos\theta = b/r$ and the integral is symmetric in time, we write

$$\int_{-\infty}^{\infty} \frac{\cos\theta}{r^2}\, dt = 2\int_0^{\infty} \frac{b}{r^3}\, dt = 2b\int_0^{\infty} \frac{dt}{(b^2 + V^2 t^2)^{3/2}}$$

$$= 2b\left[\frac{t}{b^2(b^2 + V^2 t^2)^{1/2}}\right]_0^{\infty} = \frac{2}{Vb}. \tag{5.12}$$

Combining this result with (5.11) gives, for the momentum transferred to the electron in the collision,[2]

$$p = \frac{2k_0 z e^2}{Vb}. \tag{5.13}$$

The energy transferred is

$$Q = \frac{p^2}{2m} = \frac{2k_0^2 z^2 e^4}{m V^2 b^2}. \tag{5.14}$$

In traversing a distance $dx$ in a medium having a uniform density of $n$ electrons per unit volume, the heavy particle encounters $2\pi n b\, db\, dx$ electrons at impact parameters between $b$ and $b + db$, as indicated in Fig. 5.5. The energy lost to these electrons per unit distance traveled is therefore $2\pi n Q b\, db$. The total linear rate of energy loss is found by integration over all possible energy loses. Using Eq. (5.14), we find that

$$-\frac{dE}{dx} = 2\pi n \int_{Q_{min}}^{Q_{max}} Q b\, db = \frac{4\pi k_0^2 z^2 e^4 n}{m V^2} \int_{b_{min}}^{b_{max}} \frac{db}{b} = \frac{4\pi k_0^2 z^2 e^4 n}{m V^2} \ln\frac{b_{max}}{b_{min}}. \tag{5.15}$$

Here the energy limits of integration have been replaced by maximum and minimum values of the impact parameter. It remains to evaluate these quantities explicitly.

The maximum value of the impact parameter can be estimated from the physical principle that a quantum transition is likely only when the passage of the charged particle is rapid compared with the period of motion of the atomic electron. We denote the latter time by $1/f$, where $f$ is the orbital frequency. The duration of the collision is of the order of $b/V$. Thus, the important impact parameters are restricted to values approximately given by

$$\frac{b}{V} < \frac{1}{f} \quad \text{or} \quad b_{max} \sim \frac{V}{f}. \tag{5.16}$$

2   If one assumes that a constant force
$F \sim k_0 z e^2 / b^2$ (equal to that at the distance of
closest approach) acts on the electron for a
time $t \sim b/V$, then it follows that the
momentum transferred is $p = Ft \sim k_0 z e^2 / Vb$.
This simple estimate differs by a factor of 2
from (5.13), which is exact within the
conditions specified.

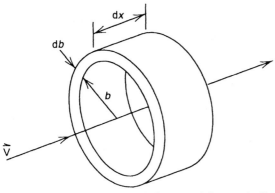

**Fig. 5.5** Annular cylinder of length dx centered about path of heavy charged particle. See text.

For the minimum impact parameter, the analysis implies that the particles' positions remain separated by a distance $b_{min}$ at least as large as their de Broglie wavelengths during the collision. This condition is more restrictive for the less massive electron than for the heavy particle. In the rest frame of the latter, the electron has a de Broglie wavelength $\lambda = h/mV$, since it moves approximately with speed $V$ relative to the heavy particle. Accordingly, we choose

$$b_{min} \sim \frac{h}{mV}. \tag{5.17}$$

Combining the relations (5.15), (5.16), and (5.17) gives the semiclassical formula for stopping power,

$$-\frac{dE}{dx} = \frac{4\pi k_0^2 z^2 e^4 n}{mV^2} \ln \frac{mV^2}{hf}. \tag{5.18}$$

We see that only the charge $ze$ and velocity $V$ of the heavy charged particle enter the expression for stopping power. This fact is consistent with the universality of charged-particle energy-loss spectra in sudden collisions, mentioned in Section 5.3. For the medium, only the electron density $n$ (appearing merely as a multiplicative factor) and the orbital frequency $f$ appear in (5.18). The quantity $hf$ in the denominator of the logarithmic term is to be interpreted as an average energy associated with the electronic quantum states of the medium. This energy is well defined in the quantum-mechanical derivation. The essential correctness of much of the physics in Bohr's derivation was vindicated by the later quantum stopping-power formula, to which we turn in the next section.

*Example*

Calculate the maximum and minimum impact parameters for electronic collisions for an 8-MeV proton. To estimate the orbital frequency $f$, assume that it is about the same as that of the electron in the ground state of the $He^+$ ion.

*Solution*

The proton velocity is given by $V = (2T/M)^{1/2}$, where $T$ is the kinetic energy and $M$ is the mass:

$$V = \left[\frac{2 \times 8 \text{ MeV} \times 1.60 \times 10^{-13} \text{J MeV}^{-1}}{1.67 \times 10^{-27} \text{ kg}}\right]^{1/2} = 3.92 \times 10^7 \text{ m s}^{-1}. \tag{5.19}$$

The orbital frequency of the electron in the ground state of $He^+$ can be found by using Eqs. (2.8) and (2.9) with $Z = 2$ and $n = 1$:

$$f = \frac{v_1}{2\pi r_1} = \frac{4.38 \times 10^6 \text{ m s}^{-1}}{2\pi \times 2.65 \times 10^{-11} \text{ m}} = 2.63 \times 10^{16} \text{ s}^{-1}. \tag{5.20}$$

Equation (5.16) gives for the maximum impact parameter

$$b_{max} \sim \frac{V}{f} = \frac{3.92 \times 10^7 \text{ m s}^{-1}}{2.63 \times 10^{16} \text{ s}^{-1}} = 1.49 \times 10^{-9} \text{ m} = 15 \text{ Å}. \tag{5.21}$$

The minimum impact parameter is, from Eq. (5.17),

$$b_{min} \sim \frac{h}{mV} = \frac{6.63 \times 10^{-34} \text{ J s}}{9.11 \times 10^{-31} \text{ kg} \times 3.92 \times 10^7 \text{ m s}^{-1}}$$

$$= 1.86 \times 10^{-11} \text{ m} = 0.19 \text{ Å}. \tag{5.22}$$

This example quantitatively illustrates some of the concepts that entered the semi-classical theory of stopping power. Additional numerical analysis will be carried out next with the quantum theory.

## 5.6
## The Bethe Formula for Stopping Power

Using relativistic quantum mechanics, Bethe derived the following expression for the stopping power of a uniform medium for a heavy charged particle:

$$-\frac{dE}{dx} = \frac{4\pi k_0^2 z^2 e^4 n}{mc^2 \beta^2} \left[\ln \frac{2mc^2 \beta^2}{I(1-\beta^2)} - \beta^2\right]. \tag{5.23}$$

In this relation
$k_0 = 8.99 \times 10^9 \text{ N m}^2 \text{ C}^{-2}$ (Appendix C),
$z$ = atomic number of the heavy particle,
$e$ = magnitude of the electron charge,
$n$ = number of electrons per unit volume in the medium,
$m$ = electron rest mass,
$c$ = speed of light in vacuum,
$\beta = V/c$ = speed of the particle relative to $c$,
$I$ = mean excitation energy of the medium.

One sees that the (nonrelativistic) result (5.18) is identical with the Bethe formula when $\beta \ll 1$, and we write $V = \beta c$ and $hf = I/2$. Whereas the energy $hf$ in the semiclassical theory has only a rather vague meaning, the mean excitation energy $I$ is explicitly defined in the quantum theory in terms of the properties of the target atoms. This quantity is discussed in the next section.

Figure 5.6 shows the stopping power of liquid water in MeV cm$^{-1}$ for a number of charged particles as a function of their energy. The logarithmic term in Eq. (5.23) leads to an increase in stopping power at very high energies (as $\beta \to 1$), just discernable for muons in the figure. At low energies, the factor in front of the bracket in (5.23) increases as $\beta \to 0$. However, the logarithm term then decreases, causing a peak (called the Bragg peak) to occur. The linear rate of energy loss is a maximum there.

The mass stopping power of a material is obtained by dividing the stopping power by the density $\rho$. Common units for mass stopping power, $-dE/\rho\,dx$, are MeV cm$^2$ g$^{-1}$. The mass stopping power is a useful quantity because it expresses the rate of energy loss of the charged particle per g cm$^{-2}$ of the medium traversed. In a gas, for example. $-dE/dx$ depends on pressure, but $-dE/\rho dx$ does not, because dividing by the density exactly compensates for the pressure. In addition, the mass stopping power does not differ greatly for materials with similar atomic composition. For example, for 10-MeV protons the mass stopping power of $H_2O$ is 45.9 MeV cm$^2$ g$^{-1}$ and that of anthracene ($C_{14}H_{10}$) is 44.2 MeV cm$^2$ g$^{-1}$. The curves in Fig. 5.6 for water can be scaled by density and used for tissue, plastics, hydrocarbons, and other materials that consist primarily of light elements. For Pb ($Z = 82$), on the other hand, $-dE/\rho dx = 17.5$ MeV cm$^2$ g$^{-1}$ for 10-MeV protons. Generally, heavy atoms are less efficient on a g cm$^{-2}$ basis for slowing down heavy charged particles, because many of their electrons are too tightly bound in the inner shells to participate effectively in the absorption of energy.

## 5.7
### Mean Excitation Energies

Mean excitation energies $I$ for a number of elements have been calculated from the quantum-mechanical definition obtained in the derivation of Eq. (5.23). They can also be measured in experiments in which all of the quantities in Eq. (5.23) except $I$ are known. The following approximate empirical formulas can be used to estimate the $I$ value in eV for an element with atomic number $Z$:

$$I \cong \begin{cases} 19.0 \text{ eV}, \ Z = 1 \text{ (hydrogen)} & (5.24) \\ 11.2 + 11.7 \ Z \text{ eV}, \ 2 \le Z \le 13 & (5.25) \\ 52.8 + 8.71 \ Z \text{ eV}, \ Z > 13. & (5.26) \end{cases}$$

Since only the logarithm of $I$ enters the stopping-power formula, values obtained by using these formulas are accurate enough for most applications. The value of $I$ for an element depends only to a slight extent on the chemical compound in which

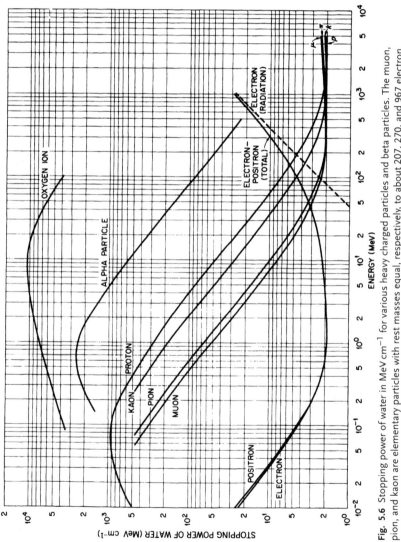

**Fig. 5.6** Stopping power of water in MeV cm$^{-1}$ for various heavy charged particles and beta particles. The muon, pion, and kaon are elementary particles with rest masses equal, respectively, to about 207, 270, and 967 electron rest masses. (Courtesy Oak Ridge National Laboratory, operated by Martin Marietta Energy Systems, Inc., for the Department of Energy.)

the element is found and on the state of condensation of the material, solid, liquid, or gas (Bragg additivity rule).

When the material is a compound or mixture, the stopping power can be calculated by simply adding the separate contributions from the individual constituent elements. If there are $N_i$ atoms $cm^{-3}$ of an element with atomic number $Z_i$ and mean excitation energy $I_i$, then in formula (5.23) one makes the replacement

$$n \ln I = \sum_i N_i Z_i \ln I_i, \tag{5.27}$$

where $n$ is the total number of electrons $cm^{-3}$ in the material ($n = \sum_i N_i Z_i$). In this way the composite $\ln I$ value for the material is obtained from the individual elemental $\ln I_i$ values weighted by the electron densities $N_i Z_i$ of the various elements. When the material is a pure compound, the electron *densities* $n$ and $N_i Z_i$ in Eq. (5.27) can be replaced by the electron *numbers* in a single molecule, as shown in the next example.

*Example*
Calculate the mean excitation energy of $H_2O$.

*Solution*
We obtain the $I$ values for H and O from Eqs. (5.24) and (5.25), and then apply (5.27). For H, $I_H = 19.0$ eV, and for O, $I_O = 11.2 + 11.7 \times 8 = 105$ eV. The electronic densities $N_i Z_i$ and $n$ can be computed in a straightforward way. However, only the ratios $N_i Z_i / n$ are needed to find $I$, and these are much simpler to use. Since the $H_2O$ molecule has 10 electrons, 2 of which belong to H ($Z = 1$) and 8 to O ($Z = 8$), we may write from Eq. (5.27)

$$\ln I = \frac{2 \times 1}{10} \ln 19.0 + \frac{1 \times 8}{10} \ln 105 = 4.312, \tag{5.28}$$

giving $I = 74.6$ eV.

## 5.8
## Table for Computation of Stopping Powers

In this section we develop a numerical table to facilitate the computation of stopping power for a heavy charged particle in any material. In the next section we use the table to calculate the proton stopping power of $H_2O$ as a function of energy.

The multiplicative factor in Eq. (5.23) can be written with the help of the constants in Appendix A and Appendix C as

$$\frac{4\pi k_0^2 z^2 e^4 n}{mc^2 \beta^2} = \frac{4\pi (8.99 \times 10^9)^2 z^2 (1.60 \times 10^{-19})^4 n}{9.11 \times 10^{-31} (3.00 \times 10^8)^2 \beta^2}$$

$$= 8.12 \times 10^{-42} \frac{z^2 n}{\beta^2} \, J\,m^{-1}. \tag{5.29}$$

The units are those of $k_0^2 e^4 n / mc^2$:

$$\frac{(\mathrm{N\,m^2\,C^{-2}})^2 \mathrm{C^4\,m^{-3}}}{\mathrm{J}} = \frac{\mathrm{N^2\,m}}{\mathrm{N\,m}} = \mathrm{J\,m^{-1}}. \tag{5.30}$$

Converting to the more common units, MeV cm$^{-1}$, we have

$$\frac{8.12 \times 10^{-42} z^2 n}{\beta^2} \frac{\mathrm{J}}{\mathrm{m}} \times \frac{1}{1.60 \times 10^{-13}} \frac{\mathrm{MeV}}{\mathrm{J}} \times \frac{1}{100} \frac{\mathrm{m}}{\mathrm{cm}}$$

$$= \frac{5.08 \times 10^{-31} z^2 n}{\beta^2} \; \mathrm{MeV\,cm^{-1}}. \tag{5.31}$$

In the dimensionless logarithmic term in (5.23) we express the energies conveniently in eV. The stopping power is then

$$-\frac{\mathrm{d}E}{\mathrm{d}x} = \frac{5.08 \times 10^{-31} z^2 n}{\beta^2} \left[ \ln \frac{1.02 \times 10^6 \beta^2}{I_{\mathrm{eV}}(1-\beta^2)} - \beta^2 \right] \mathrm{MeV\,cm^{-1}}. \tag{5.32}$$

This general formula for any heavy charged particle in any medium can be written

$$-\frac{\mathrm{d}E}{\mathrm{d}x} = \frac{5.08 \times 10^{-31} z^2 n}{\beta^2} [F(\beta) - \ln I_{\mathrm{eV}}] \; \mathrm{MeV\,cm^{-1}}, \tag{5.33}$$

where

$$F(\beta) = \ln \frac{1.02 \times 10^6 \beta^2}{1-\beta^2} - \beta^2. \tag{5.34}$$

*Example*

Compute $F(\beta)$ for a proton with kinetic energy $T = 10$ MeV.

*Solution*

In the second example in Section 5.2 we found that $\beta^2 = 0.02099$. Substitution of this value into Eq. (5.34) gives $F(\beta) = 9.972$.

The quantities $\beta^2$ and $F(\beta)$ are given for protons of various energies in Table 5.2. Since, for a given value of $\beta$, the kinetic energy of a particle is proportional to its rest mass, the table can also be used for other heavy particles as well. For example, the ratio of the kinetic energies $T_{\mathrm{d}}$ and $T_{\mathrm{p}}$ of a deuteron and a proton traveling at the same speed is

$$\frac{T_{\mathrm{d}}}{T_{\mathrm{p}}} = \frac{M_{\mathrm{d}}}{M_{\mathrm{p}}} = 2. \tag{5.35}$$

The value of $F(\beta) = 9.973$ that we just computed for a 10-MeV proton applies, therefore, to a 20-MeV deuteron. Linear interpolation can be used where needed in the table.

**Table 5.2** Data for Computation of Stopping Power for Heavy Charged Particles

| Proton Kinetic Energy (MeV) | $\beta^2$ | $F(\beta)$ Eq. (5.34) |
|---|---|---|
| 0.01 | 0.000021 | 2.179 |
| 0.02 | 0.000043 | 3.775 |
| 0.04 | 0.000085 | 4.468 |
| 0.06 | 0.000128 | 4.873 |
| 0.08 | 0.000171 | 5.161 |
| 0.10 | 0.000213 | 5.384 |
| 0.20 | 0.000426 | 6.077 |
| 0.40 | 0.000852 | 6.771 |
| 0.60 | 0.001278 | 7.175 |
| 0.80 | 0.001703 | 7.462 |
| 1.00 | 0.002129 | 7.685 |
| 2.00 | 0.004252 | 8.376 |
| 4.00 | 0.008476 | 9.066 |
| 6.00 | 0.01267 | 9.469 |
| 8.00 | 0.01685 | 9.753 |
| 10.00 | 0.02099 | 9.972 |
| 20.00 | 0.04133 | 10.65 |
| 40.00 | 0.08014 | 11.32 |
| 60.00 | 0.1166 | 11.70 |
| 80.00 | 0.1510 | 11.96 |
| 100.0 | 0.1834 | 12.16 |
| 200.0 | 0.3205 | 12.77 |
| 400.0 | 0.5086 | 13.36 |
| 600.0 | 0.6281 | 13.73 |
| 800.0 | 0.7088 | 14.02 |
| 1000. | 0.7658 | 14.26 |

## 5.9
## Stopping Power of Water for Protons

For protons, $z = 1$ in Eq. (5.33). The gram molecular weight of water is 18.0 g, and the number of electrons per molecule is 10. Since 1 m$^3$ of water has a mass of $10^6$ g, the density of electrons is

$$n = 6.02 \times 10^{23} \times \frac{10^6 \text{ g m}^{-3}}{18.0 \text{ g}} \times 10 = 3.34 \times 10^{29} \text{ m}^{-3}. \tag{5.36}$$

Also, as found at the end of Section 5.7, $\ln I_{eV} = 4.312$. From Eq. (5.33) it follows that the stopping power of water for a proton of speed $\beta$ is given by

$$-\frac{dE}{dx} = \frac{0.170}{\beta^2}[F(\beta) - 4.31] \text{ MeV cm}^{-1}. \tag{5.37}$$

At 1 MeV, for example, we find in Table 5.2 that $\beta^2 = 0.00213$ and $F(\beta) = 7.69$; therefore Eq. (5.37) gives

$$-\frac{dE}{dx} = \frac{0.170}{0.00213}(7.69 - 4.31) = 270 \text{ MeV cm}^{-1}. \tag{5.38}$$

This is numerically equal to the value of the stopping power plotted in Fig. 5.6 for water. The curves in the figure were obtained by such calculations.

## 5.10
## Range

The range of a charged particle is the distance it travels before coming to rest. The reciprocal of the stopping power gives the distance traveled per unit energy loss. Therefore, the range $R(T)$ of a particle of kinetic energy $T$ is the integral of this quantity down to zero energy:

$$R(T) = \int_0^T \left(-\frac{dE}{dx}\right)^{-1} dE. \tag{5.39}$$

Table 5.3 gives the mass stopping power and range of protons in water. The latter is expressed in $g \, cm^{-2}$; that is, the range in cm multiplied by the density of water ($\rho = 1 \, g \, cm^3$). Like mass stopping power, the range in $g \, cm^{-2}$ applies to all materials of similar atomic composition.

Although the integral in (5.39) cannot be evaluated in closed form, the explicit functional form of (5.33) enables one to scale the proton ranges in Table 5.3 to obtain the ranges of other heavy charged particles in water. Inspection of Eqs. (5.33) and (5.39) shows that the range of a heavy particle is given by an equation of the form

$$R(T) = \frac{1}{z^2} \int_0^T \frac{dE'}{G(\beta')}, \tag{5.40}$$

in which $z$ is the charge and the function $G(\beta')$ depends only on the velocity $\beta'$. Since $E' = Mc^2/\sqrt{1-\beta'^2}$, where $M$ is the particle's rest mass, the variable of integration in (5.40) can be expressed as $dE' = Mg(\beta') \, d\beta'$, where $g$ is another function of velocity alone. It follows that Eq. (5.40) has the form

$$R(\beta) = \frac{M}{z^2} \int_0^\beta \frac{g(\beta')}{G(\beta')} d\beta' = \frac{M}{z^2} f(\beta), \tag{5.41}$$

where the function $f(\beta)$ depends only on the initial velocity of the heavy charged particle. The structure of Eq. (5.41) enables one to scale ranges for different particles in the following manner. Since $f(\beta)$ is the same for two heavy charged particles at the same initial speed $\beta$, the ratio of their ranges is simply

$$\frac{R_1(\beta)}{R_2(\beta)} = \frac{z_2^2 M_1}{z_1^2 M_2}, \tag{5.42}$$

**Table 5.3** Mass Stopping Power $-dE/\rho dx$ and Range $R_p$ for Protons in Water

| Kinetic Energy (MeV) | $\beta^2$ | $-dE/\rho dx$ (MeV cm$^2$ g$^{-1}$) | $R_p$ (g cm$^{-2}$) |
|---|---|---|---|
| 0.01 | .000021 | 500. | $3 \times 10^{-5}$ |
| 0.04 | .000085 | 860. | $6 \times 10^{-5}$ |
| 0.05 | .000107 | 910. | $7 \times 10^{-5}$ |
| 0.08 | .000171 | 920. | $9 \times 10^{-5}$ |
| 0.10 | .000213 | 910. | $1 \times 10^{-4}$ |
| 0.50 | .001065 | 428. | $8 \times 10^{-4}$ |
| 1.00 | .002129 | 270. | 0.002 |
| 2.00 | .004252 | 162. | 0.007 |
| 4.00 | .008476 | 95.4 | 0.023 |
| 6.00 | .01267 | 69.3 | 0.047 |
| 8.00 | .01685 | 55.0 | 0.079 |
| 10.0 | .02099 | 45.9 | 0.118 |
| 12.0 | .02511 | 39.5 | 0.168 |
| 14.0 | .02920 | 34.9 | 0.217 |
| 16.0 | .03327 | 31.3 | 0.280 |
| 18.0 | .03731 | 28.5 | 0.342 |
| 20.0 | .04133 | 26.1 | 0.418 |
| 25.0 | .05126 | 21.8 | 0.623 |
| 30.0 | .06104 | 18.7 | 0.864 |
| 35.0 | .07066 | 16.5 | 1.14 |
| 40.0 | .08014 | 14.9 | 1.46 |
| 45.0 | .08948 | 13.5 | 1.80 |
| 50.0 | .09867 | 12.4 | 2.18 |
| 60.0 | .1166 | 10.8 | 3.03 |
| 70.0 | .1341 | 9.55 | 4.00 |
| 80.0 | .1510 | 8.62 | 5.08 |
| 90.0 | .1675 | 7.88 | 6.27 |
| 100. | .1834 | 7.28 | 7.57 |
| 150. | .2568 | 5.44 | 15.5 |
| 200. | .3207 | 4.49 | 25.5 |
| 300. | .4260 | 3.52 | 50.6 |
| 400. | .5086 | 3.02 | 80.9 |
| 500. | .5746 | 2.74 | 115. |
| 600. | .6281 | 2.55 | 152. |
| 700. | .6721 | 2.42 | 192. |
| 800. | .7088 | 2.33 | 234. |
| 900. | .7396 | 2.26 | 277. |
| 1000. | .7658 | 2.21 | 321. |
| 2000. | .8981 | 2.05 | 795. |
| 4000. | .9639 | 2.09 | 1780. |

where $M_1$ and $M_2$ are the rest masses and $z_1$ and $z_2$ are the charges. If particle number 2 is a proton ($M_2 = 1$ and $z_2 = 1$), then we can write for the range $R$ of the other particle (mass $M_1 = M$ proton masses and charge $z_1 = z$)

$$R(\beta) = \frac{M}{z^2} R_p(\beta), \tag{5.43}$$

where $R_p(\beta)$ is the proton range.

*Example*

Use Table 5.3 to find the range of an 80-MeV $^3\text{He}^{2+}$ ion in soft tissue.

*Solution*

Applying (5.43), we have $z^2 = 4$, $M = 3$, and $R(\beta) = 3R_p(\beta)/4$. Thus the desired range is three-quarters that of a proton traveling with the speed of an 80-MeV $^3\text{He}^{2+}$ ion. At this speed, the proton has an energy of $80/3 = 26.7$ MeV, that is, an energy smaller than that of the helium ion by the ratio of the masses. Interpolation in Table 5.3 gives for the proton range at this energy $R_p = 0.705$ g cm$^{-2}$. It follows that the range of the 80-MeV $^3\text{He}^{2+}$ particle is $(\frac{3}{4})(0.705) = 0.529$ g cm$^{-2}$, or 0.529 cm in unit-density soft tissue.

Figure 5.7 shows the ranges in g cm$^{-2}$ of protons, alpha particles, and electrons in water or muscle (virtually the same), bone, and lead. For a given proton energy, the range in g cm$^{-2}$ is greater in Pb than in $H_2O$, consistent with the smaller mass stopping power of Pb, as mentioned at the end of Section 5.6. The same comparison is true for electrons in Pb and water at the lower energies in Fig. 5.7 ($\lesssim 20$ MeV). At higher energies, bremsstrahlung greatly increases the rate of energy loss for electrons in Pb, reducing the range in g cm$^{-2}$ below that in $H_2O$.

Figure 5.8 gives the range in cm of protons, alpha particles, and electrons in air at standard temperature and pressure. For alpha particles in air at 15°C and 1-atm pressure, the following approximate empirical relations[3] fit the observed range $R$ in cm as a function of energy $E$ in MeV:

$$R = 0.56E, \qquad E < 4; \tag{5.44}$$

$$R = 1.24E - 2.62, \quad 4 < E < 8. \tag{5.45}$$

Alpha rays from sources external to the body present little danger because their range is less than the minimum thickness of the outermost, dead layer of cells of the skin (epidermis, minimum thickness $\sim$7 mg cm$^{-2}$). The next example illustrates the nature of the potential hazard from one important alpha emitter when inhaled and trapped in the lung.

---

3  U.S. Public Health Service, *Radiological Health Handbook*, Publ. No. 2016, Bureau of Radiological Health, Rockville, MD (1970).

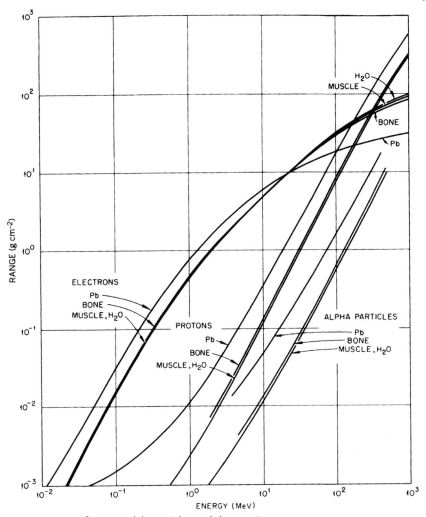

**Fig. 5.7** Ranges of protons, alpha particles, and electrons in water, muscle, bone, and lead, expressed in g cm⁻². (Courtesy Oak Ridge National Laboratory, operated by Martin Marietta Energy Systems, Inc., for the Department of Energy.)

*Example*

The radon daughter $^{214}_{84}$Po (Section 4.6), which emits a 7.69-MeV alpha particle, is present in the atmosphere of uranium mines. What is the range of this particle in soft tissue? Describe briefly the nature of the radiological hazard from inhalation of this nuclide.

*Solution*

We use the proton range in Table 5.3 to find the alpha-particle range in tissue. Applied to alpha rays, Eq. (5.43) gives ($z^2 = 4$ and $M = 4$) $R_\alpha(\beta) = R_p(\beta)$. Thus,

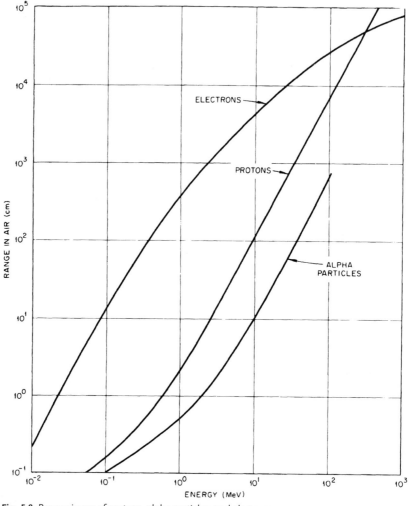

**Fig. 5.8** Ranges in cm of protons, alpha particles, and electrons in air at STP. (Courtesy Oak Ridge National Laboratory, operated by Martin Marietta Energy Systems, Inc., for the Department of Energy.)

the ranges of an alpha particle and a proton with the same velocity are the same. The ratio of the kinetic energies at the same speed is $T_\alpha/T_p = M_\alpha/M_p = 4$, and so $T_p = T_\alpha/4 = 7.69/4 = 1.92$ MeV. The alpha-particle range is, therefore, equal to the range of a 1.92-MeV proton. Interpolation in Table 5.3 gives $R_p = R_\alpha = 0.0066$ cm in tissue of unit density. The $^{214}$Po alpha particles thus cannot penetrate the 0.007 cm minimum epidermal thickness from outside the body to reach living cells. On the other hand, inhaled particulate matter containing $^{214}$Po can be deposited in the lung. There the range of the alpha particles is sufficient to reach the basal cells of the bronchial epithelium. The increase in lung-cancer incidence among uranium miners over that normally expected has been linked to the alpha-particle dose from inhaled radon daughters.

Because of the statistical nature of energy losses by atomic collisions, all particles of a given type and initial energy do not travel exactly the same distance before coming to rest in a medium. This phenomenon, called range straggling, is discussed in Section 7.6. The quantity defined by Eq. (5.39) provides the range in what is called the *continuous-slowing-down approximation*, or *csda*. It ignores fluctuations of energy loss in collisions and assumes that a charged particle loses energy continuously along its path at the linear rate given by the instantaneous stopping power. Unless otherwise indicated, we shall use the term "range" to mean the csda range. For all practical purposes, the csda range at a fixed initial energy is the same as the average pathlength that a charged particle travels in coming to rest. Heavy charged particles of a given type with the same initial energy travel almost straight ahead in a medium to about the same depth, distributed narrowly about the csda range. As we shall see in the next chapter, electrons travel tortuous paths, with the result that there is no simple relationship between their range and the depth to which a given electron will penetrate.[4] Electron transport is discussed more fully in the next two chapters.

## 5.11
## Slowing-Down Time

We can use the stopping-power formula to calculate the mean rate at which a heavy charged particle slows down. The time rate of energy loss, $-dE/dt$, can be expressed in terms of the stopping power by using the chain rule of differentiation: $-dE/dt = -(dE/dx)/(dt/dx) = V(-dE/dx)$, where $V = dx/dt$ is the velocity of the particle. For a proton with kinetic energy $T = 0.5$ MeV in water, for example, the rate of energy loss is $-dE/dt = 4.19 \times 10^{11}$ MeV s$^{-1}$.

A rough estimate can be made of the time it takes a heavy charged particle to stop in matter, if one assumes that the slowing-down rate is constant. For a particle with kinetic energy $T$, this time is approximately,

$$\tau \sim \frac{T}{-dE/dt} = \frac{T}{V(-dE/dx)}. \tag{5.46}$$

For a 0.5-MeV proton in water, $\tau \sim (0.5 \text{ MeV})/(4.19 \times 10^{11} \text{ MeV s}^{-1}) = 1.2 \times 10^{-12}$ s. Slowing-down rates and estimated stopping times for protons of other energies are given in Table 5.4. Because, as seen from Fig. 5.6, the stopping power increases as a proton slows down, actual stopping times are shorter than the estimates.

---

4  "Range" is sometimes used in the literature to mean the depth of penetration for electrons.

**Table 5.4** Calculated Slowing-Down Rates, $-dE/dt$, and Estimated Stopping Times $\tau$ for Protons in Water

| Proton Energy T (MeV) | Slowing-Down Rate $-dE/dt$ (MeV s$^{-1}$) | Estimated Stopping Time $\tau$ (s) |
|---|---|---|
| 0.5 | $4.19 \times 10^{11}$ | $1.2 \times 10^{-12}$ |
| 1.0 | $3.74 \times 10^{11}$ | $2.7 \times 10^{-12}$ |
| 10.0 | $2.00 \times 10^{11}$ | $5.0 \times 10^{-11}$ |
| 100.0 | $9.35 \times 10^{10}$ | $1.1 \times 10^{-9}$ |
| 1000.0 | $5.81 \times 10^{10}$ | $1.7 \times 10^{-8}$ |

## 5.12
### Limitations of Bethe's Stopping-Power Formula

The stopping-power formula (5.23) is valid at high energies as long as the inequality $\gamma m/M \ll 1$, mentioned before Eq. (5.7), holds (e.g., up to $\sim 10^6$ MeV for protons). Other physical factors, not included in Bethe's theory, come into play at higher energies. These include forces on the atomic electrons due to the particle's spin and magnetic moment as well as its internal electric and magnetic structures (particle form factors). Bethe's formula is also based on the assumption that the particle moves much faster than atomic electrons. At low energies the formula (5.23) fails because the term $\ln 2mc^2\beta^2/I$ eventually becomes negative, giving a negative value for the stopping power.

In the low-energy region, also, a positively charged particle captures and loses electrons as it moves, thus reducing its net charge and stopping power. Electron capture becomes important when the speed $V$ of the heavy particle is comparable to or less than the speed that an electron needs in order to orbit about the particle as a nucleus. Based on Eq. (2.9) of the discussion of Bohr's theory, the orbital speed of an electron in the ground state about a nucleus of charge $ze$ is $k_0 ze^2/\hbar$. Thus, as a condition for electron capture and loss one has $k_0 ze^2/\hbar V \gtrsim 1$. For electron capture by protons ($z = 1$), we see from Eq. (2.9) that $V = 2.2 \times 10^6$ m s$^{-1}$, corresponding to a kinetic energy of $\sim 25$ keV.

The dependence of the Bethe formula on $z^2$, the square of the charge of the heavy particle, implies that pairs of particles with the same mass and energy but opposite charge, such as pions, $\pi^{\pm}$, and muons, $\mu^{\pm}$, have the same stopping power and range. Departures from this prediction have been measured and theoretically explained by the inclusion of $z^3$ and higher powers of the charge in the stopping-power formula. Bethe's formula is obtained by calculating the stopping power in the first Born approximation in quantum mechanics. Successive Born approximations yield terms proportional to the higher powers $z^3$, $z^4$, and so on, of the incident particle's charge.

The validity of the Born-approximation formalism rests on the assumption that the speed $V$ of the incident particle is large compared with the speeds of the atomic electrons. (This assumption is akin to use of the impulse approximation in Section 5.5 for the semiclassical computation of stopping power.) Since speeds cannot exceed that of light, $V$ often is not large compared with orbital electron speeds, especially for the inner shells of heavy elements. To compensate for this deficiency, successive shell-correction terms $C_K$, $C_L$, and so on, can be calculated and added to the terms in the square brackets in Eq. (5.23).

Derivation of the Bethe formula is based on the additive interaction (called Bragg additivity) of energy loss by the incident particle to individual atoms. In condensed media, other effects can come into play, such as the coherent (plasmon) oscillations of many electrons collectively. The electric field of the incident charged particle can also polarize the condensed medium, thus reducing the rate of energy loss. Polarization, also called the density effect, depends on the dielectric constant of the medium and becomes important at relativistic speeds.

## 5.13
## Suggested Reading

1 Attix, F. H., *Introduction to Radiological Physics and Radiation Dosimetry*, Wiley, New York, NY (1986). [Chapter 8 has material on stopping powers.]

2 Fano, U., "Penetration of Protons, Alpha Particles and Mesons", *Ann. Rev. Nucl. Sci.* **13**, 1–66 (1963). [A comprehensive review of the modern theory of stopping power and various associated aspects of heavy charged-particle penetration in matter.]

3 ICRU Report 49, *Stopping Powers and Ranges for Protons and Alpha Particles*, International Commission on Radiation Units and Measurements, Bethesda, MD (1993). [Available with data disk as Report 49D. Tables give data for protons with energies from 1 keV to 10 GeV and helium ions from 1 keV to 1 GeV in 74 materials of particular importance for radiation dosimetry. Tabulations include collision (electronic), nuclear, and total mass stopping powers as well as csda ranges. Available on line, http://physics.nist.gov/PhysRefData/. A companion volume (ICRU Report 37) for electrons and positrons is listed in Section 6.8.]

4 ICRU Report 73, *Stopping of Ions Heavier than Helium*, Journal of the ICRU, Vol. 5, No. 1 (2005), International Commission on Radiation Units and Measurements, Oxford Univ. Press, Oxford, UK. [Tables of electronic stopping powers for ions from lithium through argon and for iron in 25 elements and compound materials (essentially the materials in Reports 37 and 49) with energies from 0.025 to 1000 MeV per atomic mass unit.]

5 Knoll, G. F., *Radiation Detection and Measurement*, 3rd ed., Wiley, New York (2000). [Chapter 2 discusses energy loss, stopping powers, and scaling laws for heavy charges particles and electrons.]

6 Turner, J. E., "Interaction of Ionizing Radiation with Matter," *Health Phys.* **86**, 228–252 (2004). [This review article describes basic physical mechanisms by which heavy charged particles, electrons, photons, and neutrons interact with matter. Aspects related to practices in health physics are considered.]

The Physical Reference Data Website, http://physics.nist.gov/PhysRefData/, of the National Institute of Standards and Technology is a valuable resource for information and numerical data in many areas of physics. Included are data bases for the following subjects: physical constants, atomic and molecular data, X and gamma rays, radiation dosimetry, nuclear physics, and others. The stopping-power and range information for protons, helium ions, and electrons comprise all of the tables for 26 elements and 48 compounds and mixtures in ICRU Reports 37 and 49 and much more. Calculations can be performed on line. Also, FORTRAN-77 source codes for the three charged particles can be downloaded. Total, electronic, and nuclear mass stopping powers are calculated as well as csda and projected ranges, and detour factors. Radiative mass stopping powers are included for electrons. The user can specify energies desired, and the results can be obtained in graphical, as well as numerical, form. Default energy ranges are from 1 keV to 10 GeV for protons, 1 keV to 1 GeV for helium ions, and 1 keV to 10 GeV for electrons. For electrons, computations can also be performed for any user-specified material.

## 5.14
## Problems

1. Derive Eq. (5.4).
2. Derive Eq. (5.6).
3. Show that Eq. (5.7) follows from Eq. (5.6).
4. (a) Calculate the maximum energy that a 3-MeV alpha particle can transfer to an electron in a single collision.
   (b) Repeat for a 100-MeV pion.
5. According to Eq. (5.7), what would be the relationship between the kinetic energies $T_p$ and $T_d$ of a proton and a deuteron that could transfer the same maximum energy to an atomic electron?
6. Which can transfer more energy to an electron in a single collision—a proton or an alpha particle? Explain.
7. Calculate the maximum energy that a 10-MeV muon can lose in a single collision with an electron.
8. (a) Estimate the probability that a 1-MeV proton will lose between 70 eV and 80 eV in a collision with an atomic electron in water.
   (b) Is the collision elastic?
9. (a) What is the magnitude, approximately, of the most probable energy loss by a fast charged particle in a single collision with an atomic electron?
   (b) Are the most probable and the average energy losses comparable in magnitude? Explain.

10. If the macroscopic cross section for a charged particle is $62 \; \mu m^{-1}$, what is the average distance of travel before having a collision?

11. Use the semiclassical theory in Section 5.5 to calculate the momentum transferred by a 15-MeV proton to an electron at an impact parameter of 10 Å.

12. Calculate the energy transferred in the last problem.

13. Compute the mean excitation energy of (a) Be, (b) Al, (c) Cu, (d) Pb.

14. Calculate the mean excitation energy of $C_6H_6$.

15. Compute the mean excitation energy of $SiO_2$.

16. What is the $I$ value of air? Assume a composition of 4 parts $N_2$ to 1 part $O_2$ by volume.

17. Show that the stopping-power formula (5.23) gives $-dE/dx$ in the dimensions of energy/length.

18. (a) Calculate $F(\beta)$ directly from Eq. (5.34) for a 52-MeV proton.
    (b) Use Table 5.2 to obtain $F(\beta)$ by interpolation.

19. Find $F(\beta)$ from Table 5.2 for a 500-MeV alpha particle.

20. (a) Use Table 5.2 to determine $F(\beta)$ for a 5-MeV deuteron.
    (b) What is the stopping power of water for a 5-MeV deuteron?

21. (a) What is $F(\beta)$ for a 100-MeV muon?
    (b) Calculate the stopping power of copper for a 100-MeV muon.

22. Using Eq. (5.37), calculate the stopping power of water for
    (a) a 7-MeV proton,
    (b) a 7-MeV pion,
    (c) a 7-MeV alpha particle.
    Compare answers with Fig. 5.6.

23. Using Table 5.3 for the proton mass stopping power of water, estimate the stopping power of Lucite (density $= 1.19 \; g \, cm^{-3}$) for a 35-MeV proton.

24. Refer to Fig. 5.6.
    (a) By what factor can the stopping power of water for alpha particles exceed that for protons?
    (b) By what factor does the maximum alpha-particle stopping power exceed the maximum proton stopping power?
    (c) Why is the answer to (b) not 4, the ratio of the square of their charges?
    (d) What is the value of the maximum stopping power for pions?

25. From Table 5.3 determine the minimum energy that a proton must have to penetrate 30 cm of tissue, the approximate thickness of the human body.

26. What is the range of a 15-MeV $^3He^{2+}$ particle in water?

27. Write a formula that gives the range of a $\pi^+$ at a given velocity in terms of the range of a proton at that velocity.

28. How much energy does an alpha particle need to penetrate the minimal protective epidermal layer of skin (thickness $\sim 7$ mg cm$^{-2}$)?

29. Use Table 5.3 to determine the range in cm of an 11-MeV proton in air at STP.

30. A proton and an alpha particle with the same velocity are incident on a soft-tissue target. Which will penetrate to a greater depth?

31. What is the range of a 5-MeV deuteron in soft tissue?

32. **(a)** What is the range of a 4-MeV alpha particle in tissue?
    **(b)** Using the answer for (a), estimate the range in cm in air at STP. Compare with Fig. 5.8.

33. $^{239}$Pu emits a 5.16-MeV alpha particle. What is its range in cm in
    **(a)** muscle,
    **(b)** bone of density 1.9 g cm$^{-3}$,
    **(c)** air at 22°C and 750 mm Hg?

34. Convert the formulas (5.44) and (5.45), which apply to alpha particles in air at 15°C and 1 atm, to air at STP.

35. Calculate the slowing-down rate of a 10-MeV proton in water.

36. Calculate the slowing-down rate of a 6-MeV alpha particle in water.

37. Estimate the time it takes for a 6-MeV proton to stop in water.

38. **(a)** How does the stopping time for a 6-MeV proton in water (Problem 37) compare with the stopping time in lead?
    **(b)** How does the stopping time in water compare with that in air?

39. Estimate the time required for a 2.5-MeV alpha particle to stop in tissue.

40. **(a)** Estimate the slowing-down time for a 2-MeV pion in water.
    **(b)** Repeat for a 2-MeV muon.
    **(c)** Give a physical reason for the difference in the times.

41. Estimate the energy at which an alpha particle begins to capture and lose electrons when slowing down.

42. Estimate the energy at which electron capture and loss become important when a positive pion slows down.

**5.15**

**Answers**

4. (a) 1.63 keV
   (b) 2.02 MeV
5. $T_d = 2T_p$
13. (a) 58.0 eV
   (b) 163 eV
   (c) 305 eV
   (d) 767 eV
16. 95.4 eV
19. 12.31
20. (a) 8.549
   (b) 136 MeV cm$^{-1}$
22. (a) 61.1 MeV cm$^{-1}$

   (b) 12.9 MeV cm$^{-1}$
   (c) 711 MeV cm$^{-1}$
26. 0.0263 cm
27. $R_\pi(\beta) = 0.147 R_p(\beta)$
28. ~8 MeV
32. (a) 0.002 cm
   (b) 1.6 cm
35. $2.00 \times 10^{11}$ MeV s$^{-1}$
39. $1.5 \times 10^{-12}$ s
40. (a) $1.1 \times 10^{-11}$ s
   (b) $1.2 \times 10^{-11}$ s
41. 0.40 MeV

# 6
# Interaction of Electrons with Matter

## 6.1
## Energy-Loss Mechanisms

We treat electron and positron energy-loss processes together, referring to both simply as "electrons" or "beta particles." Their stopping powers and ranges are virtually the same, except at low energies, as can be seen from Fig. 5.6. Energetic gamma photons produced by the annihilation of positrons with atomic electrons (Sect. 8.5) present a radiation problem with $\beta^+$ sources that does not occur with $\beta^-$ emitters.

Like heavy charged particles, beta particles can excite and ionize atoms. In addition, they can also radiate energy by bremsstrahlung. As seen from Fig. 5.6, the radiative contribution to the stopping power (shown by the dashed line) becomes important only at high energies. At 100 MeV, for example, radiation accounts for about half the total rate of energy loss in water. We consider separately the collisional stopping power $(-dE/dx)_{col}$ and the radiative stopping power $(-dE/dx)_{rad}$ for beta particles. Beta particles can also be scattered elastically by atomic electrons, a process that has a significant effect on beta-particle penetration and diffusion in matter at low energies.

## 6.2
## Collisional Stopping Power

The collisional stopping power for beta particles is different from that of heavy charged particles because of two physical factors. First, as mentioned in Section 5.1, a beta particle can lose a large fraction of its energy in a single collision with an atomic electron, which has equal mass. Second, a $\beta^-$ particle is identical to the atomic electron with which it collides and a $\beta^+$ is the electron's antiparticle. In quantum mechanics, the identity of the particles implies that one cannot distinguish experimentally between the incident and struck electron after a collision. Energy loss is defined in such a way that the electron of lower energy after collision is treated as the struck particle. Unlike heavy charged particles, the identity of $\beta^-$

*Atoms, Radiation, and Radiation Protection*. James E. Turner
Copyright © 2007 WILEY-VCH Verlag GmbH & Co. KGaA, Weinheim
ISBN: 978-3-527-40606-7

and the relation of $\beta^+$ to atomic electrons imposes certain symmetry requirements on the equations that describe their collisions with atoms.

The collisional stopping-power formulas for electrons and positrons can be written

$$\left(-\frac{dE}{dx}\right)^{\pm}_{col} = \frac{4\pi k_0^2 e^4 n}{mc^2 \beta^2}\left[\ln\frac{mc^2 \tau \sqrt{\tau + 2}}{\sqrt{2}I} + F^{\pm}(\beta)\right], \tag{6.1}$$

where

$$F^-(\beta) = \frac{1-\beta^2}{2}\left[1 + \frac{\tau^2}{8} - (2\tau + 1)\ln 2\right] \tag{6.2}$$

is used for electrons and

$$F^+(\beta) = \ln 2 - \frac{\beta^2}{24}\left[23 + \frac{14}{\tau + 2} + \frac{10}{(\tau + 2)^2} + \frac{4}{(\tau + 2)^3}\right] \tag{6.3}$$

for positrons. Here $\tau = T/mc^2$ is the kinetic energy $T$ of the $\beta^-$ or $\beta^+$ particle expressed in multiples of the electron rest energy $mc^2$. The other symbols in these equations, including $I$, are the same as in Eq. (5.23). Similar to Eq. (5.33), we have from (6.1)

$$\left(-\frac{dE}{dx}\right)^{\pm}_{col} = \frac{5.08 \times 10^{-31}n}{\beta^2}\left[\ln\frac{3.61 \times 10^5 \tau \sqrt{\tau + 2}}{I_{eV}} + F^{\pm}(\beta)\right]\text{ MeV cm}^{-1}. \tag{6.4}$$

As with heavy charged particles, this can be put into a general form:

$$\left(-\frac{dE}{dx}\right)^{\pm}_{col} = \frac{5.08 \times 10^{-31}n}{\beta^2}[G^{\pm}(\beta) - \ln I_{eV}]\text{ MeV cm}^{-1}, \tag{6.5}$$

where

$$G^{\pm}(\beta) = \ln(3.61 \times 10^5 \tau \sqrt{\tau + 2}) + F^{\pm}(\beta). \tag{6.6}$$

*Example*

Calculate the collisional stopping power of water for 1-MeV electrons.

*Solution*

This quantity is $(-dE/dx)^-_{col}$, given by Eq. (6.5). We need to compute $\beta^2$, $\tau$, $F^-(\beta)$ and then $G^-(\beta)$. As in Section 5.9, we have $n = 3.34 \times 10^{29}$ m$^{-3}$ and $\ln I_{eV} = 4.31$. Using the relativistic formula for kinetic energy with $T = 1$ MeV and $mc^2 = 0.511$ MeV, we write

$$1 = 0.511\left(\frac{1}{\sqrt{1-\beta^2}} - 1\right), \tag{6.7}$$

giving $\beta^2 = 0.886$. Also, $\tau = T/mc^2 = 1/0.511 = 1.96$. From Eq. (6.2),

$$F^-(\beta) = \frac{1-0.886}{2}\left[1 + \frac{(1.96)^2}{8} - (2 \times 1.96 + 1)\ln 2\right], \tag{6.8}$$

giving $F^-(\beta) = -0.110$. From Eq. (6.6),

$$G^-(\beta) = \ln(3.61 \times 10^5 \times 1.96\sqrt{1.96 + 2}) - 0.110 = 14.0. \qquad (6.9)$$

Finally, applying Eq. (6.5), we find

$$\left(-\frac{dE}{dx}\right)^-_{col} = \frac{5.08 \times 10^{-31} \times 3.34 \times 10^{29}}{0.886}[14.0 - 4.31]$$

$$= 1.86 \text{ MeV cm}^{-1}. \qquad (6.10)$$

It is of interest to compare this result with that for a 1-MeV positron. The quantities $\beta^2$ and $\tau$ are the same. Calculation gives $F^+(\beta) = -0.312$, which is a little larger in magnitude than $F^-(\beta)$. In place of (6.9) and (6.10) one finds $G^+(\beta) = 13.8$ and $(-dE/dx)^+_{col} = 1.82$ MeV cm$^{-1}$. The $\beta^+$ collisional stopping power is practically equal to that for $\beta^-$ at 1 MeV in water.

The collisional, radiative, and total mass stopping powers of water as well as the radiation yield and range for electrons are given in Table 6.1. The total stopping power for $\beta^-$ or $\beta^+$ particles is the sum of the collisional and radiative contributions:

$$\left(-\frac{dE}{dx}\right)^\pm_{tot} = \left(-\frac{dE}{dx}\right)^\pm_{col} + \left(-\frac{dE}{dx}\right)^\pm_{rad}, \qquad (6.11)$$

with a similar relation holding for the mass stopping powers. Radiative stopping power, radiation yield, and range are treated in the next three sections. Table 6.1 can also be used for positrons with energies above about 10 keV.

The calculated mass stopping power of liquid water for electrons at low energies is shown in Fig. 6.1. (Measurement of this important quantity does not appear to be technically feasible.) The radiative stopping power is negligible at these energies, and so no subscript is needed to distinguish between the total and collisional stopping powers. The threshold energy ($Q_{min}$) required for excitation to the lowest lying electronic quantum state is estimated to be 7.4 eV. The curve in Fig. 6.1 joins smoothly onto the electron stopping power at $10^{-2}$ MeV in Fig. 5.6. The electron-transport computer code, NOREC, was used to calculate the stopping power in Fig. 6.1.[1]

The relative importance of ionization, excitation, and elastic scattering at energies up to 1 MeV can be seen from the plot of the respective attenuation coefficients, $\mu$, in Fig. 6.2, also from the NOREC code. The ordinate gives the values of $\mu$ in units of reciprocal micrometers. Recall from the discussion at the end of Section 5.4 that $\mu$ represents the probability of interaction per unit distance traveled, which is also the inverse of the mean free path. Elastic scattering is the dominant process at the lowest energies. Slow electrons undergo almost a random diffusion, changing direction through frequent elastic collisions without energy loss. Eventually, the occasional competing inelastic excitation and ionization collisions bring

1   See Semenenko, V. A., Turner, J. E., and
    Borak, T. B. in Section 6.8.

**Table 6.1** Electron Collisonal, Radiative, and Total Mass Stopping Powers; Radiation Yield; and Range in Water

| Kinetic Energy | $\beta^2$ | $-\frac{1}{\rho}\left(\frac{dE}{dx}\right)^-_{col}$ (MeV cm$^2$ g$^{-1}$) | $-\frac{1}{\rho}\left(\frac{dE}{dx}\right)^-_{rad}$ (MeV cm$^2$ g$^{-1}$) | $-\frac{1}{\rho}\left(\frac{dE}{dx}\right)^-_{tot}$ (MeV cm$^2$ g$^{-1}$) | Radiation Yield | Range (g cm$^{-2}$) |
|---|---|---|---|---|---|---|
| 10 eV | 0.00004 | 4.0 | — | 4.0 | — | $4 \times 10^{-8}$ |
| 30 | 0.00012 | 44. | — | 44. | — | $2 \times 10^{-7}$ |
| 50 | 0.00020 | 170. | — | 170. | — | $3 \times 10^{-7}$ |
| 75 | 0.00029 | 272. | — | 272. | — | $4 \times 10^{-7}$ |
| 100 | 0.00039 | 314. | — | 314. | — | $5 \times 10^{-7}$ |
| 200 | 0.00078 | 298. | — | 298. | — | $8 \times 10^{-7}$ |
| 500 eV | 0.00195 | 194. | — | 194. | — | $2 \times 10^{-6}$ |
| 1 keV | 0.00390 | 126. | — | 126. | — | $5 \times 10^{-6}$ |
| 2 | 0.00778 | 77.5 | — | 77.5 | — | $2 \times 10^{-5}$ |
| 5 | 0.0193 | 42.6 | — | 42.6 | — | $8 \times 10^{-5}$ |
| 10 | 0.0380 | 23.2 | — | 23.2 | 0.0001 | 0.0002 |
| 25 | 0.0911 | 11.4 | — | 11.4 | 0.0002 | 0.0012 |
| 50 | 0.170 | 6.75 | — | 6.75 | 0.0004 | 0.0042 |
| 75 | 0.239 | 5.08 | — | 5.08 | 0.0006 | 0.0086 |
| 100 | 0.301 | 4.20 | — | 4.20 | 0.0007 | 0.0140 |
| 200 | 0.483 | 2.84 | 0.006 | 2.85 | 0.0012 | 0.0440 |
| 500 | 0.745 | 2.06 | 0.010 | 2.07 | 0.0026 | 0.174 |
| 700 keV | 0.822 | 1.94 | 0.013 | 1.95 | 0.0036 | 0.275 |
| 1 MeV | 0.886 | 1.87 | 0.017 | 1.89 | 0.0049 | 0.430 |
| 4 | 0.987 | 1.91 | 0.065 | 1.98 | 0.0168 | 2.00 |
| 7 | 0.991 | 1.93 | 0.084 | 2.02 | 0.0208 | 2.50 |
| 10 | 0.998 | 2.00 | 0.183 | 2.18 | 0.0416 | 4.88 |
| 100 | 0.999+ | 2.20 | 2.40 | 4.60 | 0.317 | 32.5 |
| 1000 MeV | 0.999+ | 2.40 | 26.3 | 28.7 | 0.774 | 101. |

the energies of the electrons down into the region where they can react chemically with water molecules or became hydrated. The excitation and ionization probabilities rise steeply from the lowest energies. At about 200 eV, ionization and elastic scattering are comparable and considerably more probable than excitation. Thereafter, the attenuation-coefficient curves do not cross. At the highest energies, elastic scattering occurs increasingly in the forward direction. While elastic scattering affects electron transport through redirection of the electron paths, it does not contribute to the stopping power, because there is no associated energy loss.

Figures 6.1 and 6.2 are basic to radiation physics and the subsequent chemistry that takes place in irradiated living systems. The end result of the absorption of any kind of ionizing radiation in tissue is the production of large numbers of secondary electrons. These, in turn, cause additional ionizations as they lose energy and slow down, until all liberated electrons reach energies in the eV range. Since it takes an average of only about 22 eV to produce a secondary electron in liquid water, radiation produces low-energy electrons in abundance. A 10-keV electron,

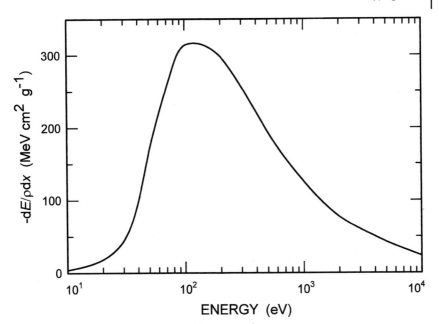

**Fig. 6.1** Mass stopping power of water for low-energy electrons.

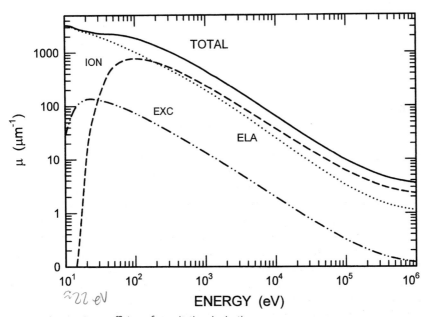

≈22 eV

**Fig. 6.2** Attenuation coefficients for excitation, ionization, elastic scattering, and total interaction for electrons in liquid water as functions of energy.

for example, ultimately produces a total of about 450 secondary electrons, a large fraction of which occur with initial energies of less than 100 eV. The details of electron transport and charged-particle track structure and their relation to chemical and biological effects will be considered in Chapter 13.

## 6.3
## Radiative Stopping Power

The acceleration of a heavy charged particle in an atomic collision is usually small, and except under extreme conditions negligible radiation occurs. A beta particle, on the other hand, having little mass can be accelerated strongly by the same electromagnetic force within an atom and thereby emit radiation, called bremsstrahlung. Bremsstrahlung occurs when a beta particle is deflected in the electric field of a nucleus and, to a lesser extent, in the field of an atomic electron. At high beta-particle energies, the radiation is emitted mostly in the forward direction, that is, in the direction of travel of the beta particle. As indicated in Fig. 6.3, this circumstance is observed in a betatron or synchrotron, a device that accelerates electrons to high energies in circular orbits. Most of the synchrotron radiation, as it is called, is emitted in a narrow sweeping beam nearly in the direction of travel of the electrons that produce it.

Energy loss by an electron in radiative collisions was studied quantum mechanically by Bethe and Heitler. If the electron passes near a nucleus, the field in which it is accelerated is essentially the bare Coulomb field of the nucleus. If it passes at a greater distance, the partial screening of the nuclear charge by the atomic electrons becomes important, and the field is no longer coulombic. Thus, depending on how close the electron comes to the nucleus, the effect of atomic-electron screening will be different. The screening and subsequent energy loss also depend on the energy of the incident beta particle. The maximum energy that a bremsstrahlung photon can have is equal to the kinetic energy of the beta particle. The photon energy spectrum is approximately flat out to this maximum.

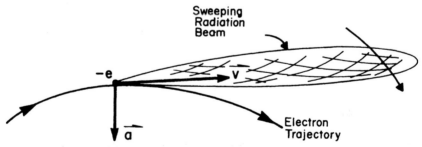

**Fig. 6.3** Synchrotron radiation. At high energies, photons are emitted by electrons (charge −e) in circular orbits in the direction (crosshatched area) of their instantaneous velocity **v**. The direction of the electrons' acceleration **a** is also shown.

Unlike collisional energy losses, no single analytic formula exists for calculating the radiative stopping power $(-dE/dx)^{\pm}_{rad}$. Instead, numerical procedures are used to obtain values, such as those in Table 6.1. Details of the analysis show that energy loss by radiation behaves quite differently from that by ionization and excitation. The efficiency of bremsstrahlung in elements of different atomic number $Z$ varies nearly as $Z^2$. Thus, for beta particles of a given energy, bremsstrahlung losses are considerably greater in high-$Z$ materials, such as lead, than in low-$Z$ materials, such as water. As seen from Eq. (6.1), the collisional energy-loss rate in an element is proportional to $n$ and hence to $Z$. In addition, the radiative energy-loss rate increases nearly linearly with beta-particle energy, whereas the collisional rate increases only logarithmically. At high energies, therefore, bremsstrahlung becomes the predominant mechanism of energy loss for beta particles, as can be seen from Table 6.1.

The following approximate formula gives the ratio of radiative and collisional stopping powers for an electron of total energy $E$, expressed in MeV, in an element of atomic number $Z$:

$$\frac{(-dE/dx)^-_{rad}}{(-dE/dx)^-_{col}} \cong \frac{ZE}{800}. \tag{6.12}$$

This formula shows that in lead ($Z = 82$), for example, the two rates of energy loss are approximately equal at a total energy given by

$$\frac{82E}{800} \cong 1. \tag{6.13}$$

Thus $E \cong 9.8$ MeV, and the electron's kinetic energy is $T = E - mc^2 \cong 9.3$ MeV. In oxygen ($Z = 8$), the two rates are equal when $E \cong 100$ MeV $\cong T$, an order-of-magnitude higher energy than in lead. The radiative stopping power $(-dE/dx)^-_{rad}$ for electrons is shown by the dashed curve in Fig. 5.6.

At very high energies the dominance of radiative over collisional energy losses gives rise to electron-photon cascade showers. Since the bremsstrahlung photon spectrum is approximately flat out to its maximum (equal to the electron's kinetic energy), high-energy beta particles emit high-energy photons. These, in turn, produce Compton electrons and electron-positron pairs, which then produce additional bremsstrahlung photons, and so on. These repeated interactions result in an electron-photon cascade shower, which can be initiated by either a high-energy beta particle or a photon.

## 6.4
### Radiation Yield

We have discussed the relative rates of energy loss by collision and by radiation. Radiation yield is defined as the average fraction of its energy that a beta particle radiates as bremsstrahlung in slowing down completely. Radiation yields are given in Table 6.1 for electrons of various energies in water. At 100 MeV, for example, the rates of energy loss by collision and by radiation are approximately equal. As the electron slows down, however, the relative amount lost by radiation decreases steadily. In slowing down completely, a 100-MeV electron loses an average of 0.317 of its initial energy (i.e., 31.7 MeV) by radiation. Radiation yield increases with electron energy. A 1000-MeV electron stopping in water will lose an average of 0.774 of its energy by bremsstrahlung. For an electron of given energy, radiation yield also increases with atomic number.

An estimate of radiation yield can give an indication of the potential bremsstrahlung hazard of a beta-particle source. If electrons of initial kinetic energy $T$ in MeV are stopped in an absorber of atomic number $Z$, then the radiation yield is given approximately by the formula[1]

$$Y \cong \frac{6 \times 10^{-4} ZT}{1 + 6 \times 10^{-4} ZT}. \tag{6.14}$$

To keep bremsstrahlung to a minimum, low-$Z$ materials can be used as a shield to stop beta particles. Such a shield can, in turn, be surrounded by a material of high $Z$ to efficiently absorb the bremsstrahlung photons.

*Example*

Estimate the fraction of the energy of a 2-MeV beta ray that is converted into bremsstrahlung when the particle is absorbed in aluminum and in lead.

*Solution*

For Al, $ZT = 13 \times 2 = 26$, and $Y \cong 0.016/1.016 = 0.016$. For Pb ($Z = 82$), $Y \cong 0.090$. Thus, about 1.6% of the electron kinetic energy is converted into photons in Al, while the corresponding figure for Pb is about 9.0%.

For radiation-protection purposes, conservative assumptions can be made in order to apply Eq. (6.14) to the absorption of beta particles from a radioactive source. To this end, the *maximum* beta-particle energy is used for $T$. This assumption overestimates the energy converted into radiation because bremsstrahlung efficiency is less at the lower electron energies. Furthermore, assuming that all bremsstrahlung photons have the energy $T$ will also give a conservative estimate of the actual photon hazard.

*Example*

A small $3.7 \times 10^8$ Bq $^{90}$Y source is enclosed in a lead shield just thick enough to absorb the beta particles, which have a maximum energy of 2.28 MeV and an average energy of 0.94 MeV. Estimate the rate at which energy is radiated as bremsstrahlung. For protection purposes, estimate the photon fluence rate at a distance of 1 m from the source.

1   H. W. Koch and J. W. Motz, "Bremsstrahlung Cross Section Formulas and Related Data," *Rev. Mod. Phys.* **31**, 920–955 (1959).

*Solution*

Setting $T = 2.28$ and $Z = 82$ in Eq. (6.14) gives for the fraction of the beta-particle energy converted into photons $Y \cong 0.10$. The total beta-particle energy released per second from the source is $(3.7 \times 10^8 \text{ s}^{-1}) (0.94 \text{ MeV}) = 3.48 \times 10^8 \text{ MeV s}^{-1}$. Multiplication by $Y$ gives, for the rate of energy emission by bremsstrahlung, $\sim 3.48 \times 10^7$ MeV. The energy fluence rate at a distance of 1 m is therefore $\sim (3.48 \times 10^7 \text{ MeV s}^{-1})/(4\pi \times 100^2 \text{ cm}^2) = 277 \text{ MeV cm}^{-2} \text{ s}^{-1}$. For assessing the radiation hazard, we assume that the photons have an energy of 2.28 MeV. Therefore, the photon fluence rate at this distance is $\sim 227/2.28 = 121$ photons cm$^{-2}$ s$^{-1}$. For comparison, we note that use of an aluminum $(Z = 13)$ shield to stop the beta particles would give $Y \cong 0.017$, reducing the bremsstrahlung by a factor of 5.9.

## 6.5

## Range

The range of a beta particle is defined like that of heavy particles by Eq. (5.39) in which the total stopping power $(-dE/dx)^{\pm}_{\text{tot}}$ is used. Unlike a heavy particle, however, its range is only a poor indicator of the depth to which a given electron is likely to go into a target. Nevertheless, we shall employ electron csda ranges and the assumption of straight-ahead travel in order to make at least rough estimates in working problems. One must always bear in mind that this procedure overestimates electron penetration in matter. A more quantitative relationship between the pathlength and the maximum penetration depth is considered in Section 6.7.

Table 6.1 gives electron ranges in water down to 10 eV. As with heavy charged particles, the ranges expressed in g cm$^{-2}$ are approximately the same in different materials of similar atomic composition. Electron ranges in $H_2O$, muscle, bone, Pb, and air are included in Figs. 5.7 and 5.8. For the same reasons as with heavy charged particles (discussed at the end of Sect. 5.6), the collisional mass stopping power for beta particles is smaller in high-$Z$ materials, such as lead, than in water. In Fig. 5.7, this fact accounts for the greater range of electrons in Pb compared with $H_2O$ at energies below about 20 MeV. At higher energies, the radiative energy-loss rate in Pb more than compensates for the difference in the collisional rate, and the electron range in Pb is less than in $H_2O$.

The following empirical equations for electrons in low-$Z$ materials relate the range $R$ in g cm$^{-2}$ to the kinetic energy $T$ in MeV:
For $0.01 \leq T \leq 2.5$ MeV,

$$R = 0.412 T^{1.27-0.0954 \ln T} \tag{6.15}$$

or

$$\ln T = 6.63 - 3.24(3.29 - \ln R)^{1/2}; \tag{6.16}$$

for $T > 2.5$ MeV,

$$R = 0.530 T - 0.106, \tag{6.17}$$

or

$$T = 1.89R + 0.200. \tag{6.18}$$

These relations fit the curve plotted in Fig. 6.4.[1]

*Example*

How much energy does a 2.2-MeV electron lose on the average in passing through 5 mm of Lucite (density $\rho = 1.19 \text{ g cm}^{-3}$)?

*Solution*

Lucite is a low-$Z$ material, and so we may apply Eqs. (6.15)–(6.18) or use Fig. 6.4 directly. We shall employ the equations and check the results against the figure. First, we find how far the 2.2-MeV electron can travel in Lucite. From Eq. (6.15) with $T = 2.2$ MeV,

$$R = 0.412(2.2)^{1.27 - 0.0954 \ln 2.2} = 1.06 \text{ g cm}^{-2}, \tag{6.19}$$

in agreement with Fig. 6.4. This range gives a distance

$$d = \frac{R}{\rho} = \frac{1.06 \text{ g cm}^{-2}}{1.19 \text{ g cm}^{-3}} = 0.891 \text{ cm.} \tag{6.20}$$

Since the Lucite is only 0.5 cm thick, the electron emerges with enough energy $T'$ to carry it another 0.391 cm, or 0.465 g cm$^{-2}$. The energy $T'$ can be found from Eq. (6.16):

$$\ln T' = 6.63 - 3.24(3.29 - \ln 0.465)^{1/2} = 0.105, \tag{6.21}$$

and so $T' = 1.11$ MeV, again in agreement with Fig. 6.4. It follows that the energy lost by the electron is $T - T' = 2.20 - 1.11 = 1.09$ MeV. The analysis and numerical values found here are the same also for a 2.2-MeV positron.

Unlike alpha particles, beta rays from many radionuclides have a range greater than the thickness of the epidermis. As seen from Fig. 6.4, a 70-keV electron has a range equal to the minimum thickness of 7 mg cm$^{-2}$ of the epidermal layer. $^{90}$Y, for example, emits a beta particle with a maximum energy of 2.28 MeV, which has a range of over 1 g cm$^{-2}$ in tissue. In addition to being an internal radiation hazard, beta emitters can potentially damage the skin and eyes.

## 6.6
## Slowing-Down Time

The rate of slowing down and the stopping time for electrons and positrons can be estimated by the methods used for heavy charged particles in Section 5.11, the *total*

---

1   Figure 6.4 and the relations (6.15)–(6.18) are taken from the U.S. Public Health Service, *Radiological Health Handbook*, Publ. No. 2016, Bureau of Radiological Health, Rockville, MD (1970).

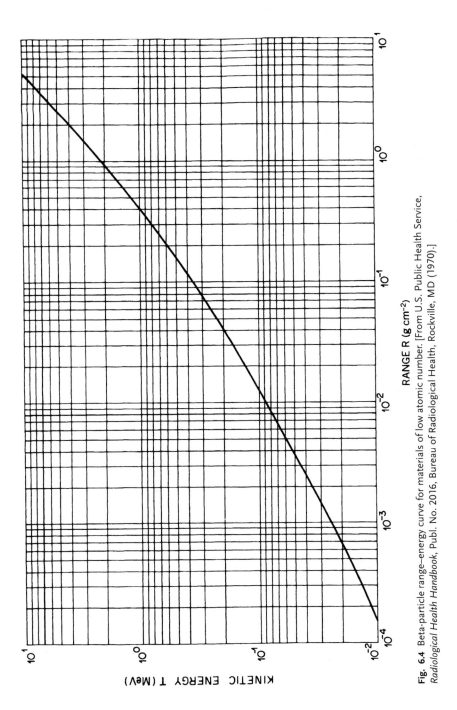

**Fig. 6.4** Beta-particle range–energy curve for materials of low atomic number. [From U.S. Public Health Service, *Radiological Health Handbook*, Publ. No. 2016, Bureau of Radiological Health, Rockville, MD (1970).]

stopping power being employed for beta rays. Estimating the stopping time as the ratio of the initial energy and the total slowing-down rate is not grossly in error. For a 1-MeV electron, for example, this ratio is $\tau = 1.9 \times 10^{-11}$ s; numerical integration over the actual total stopping rate as a function of energy gives $1.3 \times 10^{-11}$s.

*Example*

Calculate the slowing-down rate of an 800-keV electron in water and estimate the stopping time.

*Solution*

The slowing-down rate for a beta particle of speed $v$ is given by $-dE/dt = v(-dE/dx)^-_{tot}$. For an 800-keV beta particle, we find by interpolating in Table 6.1, $\beta^2 = 0.843$. The velocity of the electron is $v = \beta c = \sqrt{0.843} \times 3 \times 10^{10} = 2.75 \times 10^{10}$ cm s$^{-1}$. The interpolated total stopping power at 800 keV from Table 6.1 is $(-dE/dx)^-_{tot} = 1.93$ MeV cm$^{-1}$. The slowing-down rate is

$$-\frac{dE}{dt} = v\left(-\frac{dE}{dx}\right)^-_{tot} = 2.75 \times 10^{10} \frac{cm}{s} \times 1.93 \frac{MeV}{cm}$$

$$= 5.31 \times 10^{10} \text{ MeV s}^{-1}. \tag{6.22}$$

With $T = 0.800$ MeV, the stopping time is

$$\tau \cong \frac{T}{-dE/dt} = \frac{0.800 \text{ MeV}}{5.31 \times 10^{10} \text{ MeV s}^{-1}} = 1.5 \times 10^{-11} \text{ s.} \tag{6.23}$$

## 6.7
## Examples of Electron Tracks in Water

Figure 6.5 shows a three-dimensional representation of three electron "tracks" calculated by a Monte Carlo computer code to simulate electron transport in water. Each primary electron starts with an energy of 5 keV from the origin and moves initially toward the right along the horizontal axis. Each dot represents the location, at $10^{-11}$ s, of a chemically active species produced by the action of the primary electron or one of its secondaries. The Monte Carlo code randomly selects collision events from specified distributions of flight distance, energy loss, and angle of scatter in order to calculate the fate of individual electrons, simulating as nearly as is known what actually happens in nature. All electrons are transported in the calculation until their energies fall below the threshold of 7.4 eV for electronic excitation. Ticks along the axes are $0.1 \ \mu m = 1000$ Å apart.

These examples illustrate a number of characteristics of the tracks of electrons that stop in matter. As we have mentioned already, the tracks tend to wander, due to large-angle deflections that electrons can experience in single collisions. The wandering is augmented at low energies, near the end of a track, by the greatly increased and almost isotropic elastic scattering that occurs there. In addition, energy-loss events are more sparsely distributed at the beginning of the track,

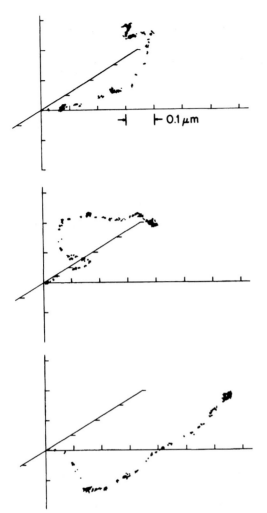

⊢ 0.1 μm

**Fig. 6.5** Three calculated tracks of 5-keV electrons in liquid water. Each electron starts from the origin and initially travels along the horizontal axis toward the right. Each dot gives the position of a chemically active species at $10^{-11}$ s. [From J. E. Turner, J. L. Magee, H. A. Wright, A. Chatterjee, R. N. Hamm, and R. H. Ritchie, "Physical and Chemical Development of Electron Tracks in Liquid Water," *Rad. Res.* **96**, 437–449 (1983). Courtesy Oak Ridge National Laboratory, operated by Martin Marietta Energy Systems, Inc., for the Department of Energy.]

where the primary electron is moving faster. This is generally true of charged-particle tracks, since the stopping power is smaller at high energies than near the end of the range. Note also the clustering of events, particularly in the first part of the tracks. Such groupings, called "spurs," are due to the production of a secondary electron with just enough energy to produce several additional ionizations and excitations. The range of the original secondary electron is usually not great enough to take it appreciably away from the region through which the primary electron passes. Clustering occurs as a result of the broad shape of the single-collision spectra in Fig. 5.3 with most of the area covering energy losses $\lesssim 70$ eV.

Figure 6.6(a) gives a stereoscopic representation of another 5-keV electron track, traveling out of the page toward the reader, calculated in water. The same track is shown from the side in (b), except that the primary electron was forced to move straight ahead in the calculation.

As another example, Fig. 6.7 displays ten tracks randomly calculated for 740-keV electrons in liquid water. They are normally incident from the left at the origin of the $X$–$Y$ axes. The dots show the coordinates of every one-hundredth inelastic event in the track projected onto the $X$–$Y$ plane. Of the ten electrons comprising the figure, nine slow down and stop in the phantom, and one is backscattered into the space $x < 0$. The diverse, tortuous paths of the electrons are in stark contrast to those of heavy charged particles. Under identical initial conditions, the latter travel in almost straight lines to about the same depth.

Whereas the application of range-energy tables and graphs for shielding and dosimetry with heavy charged particles is relatively straightforward, their use with beta particles warrants a closed look. As related at the beginning of Section 6.5, the range of an electron of given energy is the average pathlength that it travels in coming to rest. Figure 6.7 illustrates that there is a considerable difference between electron range and the depth of penetration in matter. The calculated distributions

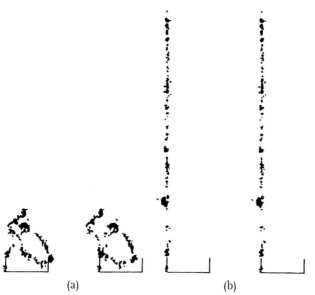

(a)                                   (b)

**Fig. 6.6** (a) Stereoscopic view of a 5-keV electron track in water. (b) Lateral view of same track in which the primary electron is forced to always move straight ahead. (Courtesy R. N. Hamm, Oak Ridge National Laboratory, operated by Martin Marietta Energy Systems, Inc., for the Department of Energy.)

of pathlengths and of maximum depths of penetration for 740-keV electrons in water are shown in Fig. 6.8. They were compilied from the histories of $1.5 \times 10^6$ tracks randomly generated on the computer. The skewed shape of the single-collision energy-loss spectrum with a high-energy tail for energetic electrons (Fig. 5.3), is reflected in the skewed shape for the pathlength curve in Fig. 6.8. The computed mean pathlength, or range, was 3,000 $\mu$m and the average maximum depth of penetration was 1,300 $\mu$m. It is seen that the maximum depth reached by relatively few electrons is even close to the range. As a rule of thumb, the average deepest penetration for beta particles is roughly one-half the range. The particular range chosen for discussion here is the nominal tissue depth of 0.3 cm for the lens of the eye, specified in regulatory documents.

This chapter covers much of the basic physics underlying our understanding of some effects of ionizing radiation in matter. All ionizing radiation produces

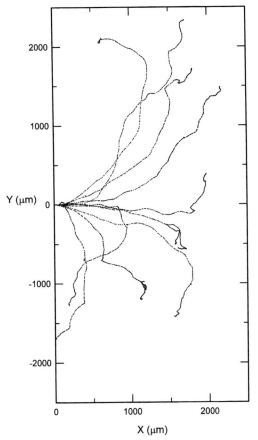

**Fig. 6.7** Calculated tracks (projected into the X–Y plane of the figure) of ten 740-keV electrons entering a water slab normally from the left at the origin. One electron is seen to be scattered back out of the slab.

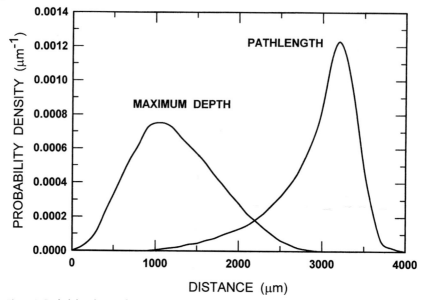

**Fig. 6.8** Probability density functions for maximum depth of
penetration and pathlength for 740-keV electrons normally
incident on a water slab.

low-energy electrons in great abundance. Studies with water, a main constituent
of living systems, can be partially checked by radiochemical measurements. They
shed considerable light on the physical and chemical changes induced by radiation
that must ultimately lead to biological effects. The subsequent chemical evolution
that follows energy deposition within the tracks of charged particles in water is
described in Chapter 13.

## 6.8
## Suggested Reading

1  Attix, F. H., *Introduction to Radiological Physics and Radiation Dosimetry,*
   Wiley, New York (1986). [See Chapter 8.]

2  ICRU Report 37, *Stopping Powers
   for Electrons and Positrons,* International Commission on Radiation
   Units and Measurements, Bethesda,
   MD (1984). [Tables give collisional,
   radiative, and total mass stopping
   powers; ranges; radiation yields;
   and other data for electrons and
   positrons (10 keV–1 GeV) for a number of elements, compounds, and
   mixtures. These tabulations as well
   as other data are available on-line,
   http://physics.nist.gov/PhysRefData/.
   See note at end of Section 5.13.]

3  Semenenko, V. A., Turner, J. E., and
   Borak, T. B., "NOREC," a Monte Carlo
   Code for Simulating Electron Tracks
   in Liquid Water," *Radiat. Env. Biophys.*
   **42**, 213–217 (2003). [This paper documents revisions made to the electron-
   transport code OREC, developed at
   the Oak Ridge National Laboratory.
   References to original documents are
   cited.]

4  Turner, J. E., "Interaction of Ionizing Radiation with Matter," *Health Phys.* 86, 228–252 (2004). [Review article includes electron interactions with matter and some related health-physics applications.]

## 6.9
## Problems

1. Calculate $F^-(\beta)$ for a 600-keV electron.
2. Calculate $F^+(\beta)$ for a 600-keV positron.
3. Derive Eq. (6.4) from Eq. (6.1).
4. Calculate the collisional stopping power of water for 600-keV electrons.
5. Calculate the collisional stopping power of water for 600-keV positrons.
6. Show that, when $\beta^2 \ll 1$, the collisional stopping-power formula for an electron with kinetic energy $T$ can be written

$$\left(-\frac{dE}{dx}\right)^-_{col} = \frac{\pi k_0^2 e^4 n}{T}\left(\ln\frac{T^2}{2I^2} + 1\right).$$

7. Use the formula from the last problem to calculate the stopping power of $CO_2$ at STP for 9.5-keV electrons.
8. (a) From Fig. 6.2, estimate for a 100-eV electron the probability that a given energy-loss event will result in excitation, rather than ionization, in water.
   (b) What fraction of the collisions at 100 eV are due to elastic scattering?
9. Use Fig. 6.2 to estimate the fraction of the collisions of a 100-keV electron in water that are due to
   (a) ionization
   (b) excitation
   (c) elastic scattering.
10. Estimate the kinetic energy at which the collisional and radiative stopping powers are equal for electrons in
    (a) Be
    (b) Cu
    (c) Pb.
11. What is the ratio of the collisional and radiative stopping powers of Al for electrons of energy
    (a) 10 keV
    (b) 1 MeV
    (c) 100 MeV ?
12. Estimate the radiation yield for 10-MeV electrons in
    (a) Al

(b) Fe

(c) Au.

13. A small 1.85-GBq $^{198}$Au source (maximum beta-particle energy 0.961 MeV, average 0.315 MeV) is enclosed in a lead shield just thick enough to absorb all the beta particles.

    (a) Estimate the energy fluence rate due to bremsstrahlung at a point 50 cm away.

    (b) What is the estimated bremsstrahlung photon fluence rate at that point for the purpose of assessing the potential radiation hazard?

14. A 12-mA electron beam is accelerated through a potential difference of $2 \times 10^5$ V in an X-ray tube with a tungsten target. X rays are generated as bremsstrahlung from the electrons stopping in the target. Neglect absorption in the tube.

    (a) Estimate the fraction of the beam power that is emitted as radiation.

    (b) How much power is radiated as bremsstrahlung?

15. Estimate the range in cm of the maximum-energy beta ray (2.28 MeV) from $^{90}$Y in bone of density 1.9 g cm$^{-3}$.

16. Use Eq. (6.15) to estimate the range of a 400-keV beta particle in water. How does the answer compare with Table 6.1?

17. Derive Eq. (6.16) from Eq. (6.15).

18. Use Table 6.1 to estimate the range in cm in air at STP for electrons of energy

    (a) 50 keV

    (b) 830 keV

    (c) 100 MeV.

19. A positron emerges normally from a 4-mm-thick plastic slab (density 1.14 g cm$^{-3}$) with an energy of 1.62 MeV. What was its energy when it entered the slab?

20. Use the range curve in Fig. 5.7 and make up a formula or formulas like Eqs. (6.15) and (6.17), giving the range of electrons in lead as a function of energy.

21. A cell culture (Fig. 6.9) is covered with a 1-cm sheet of Lucite (density $= 1.19$ g cm$^{-3}$). What thickness of lead (in cm) is needed on top of the Lucite to prevent 10-MeV beta rays from reaching the culture? Use the approximate empirical formulas relating range $R$ in g cm$^{-2}$ to electron kinetic energy $T$ in MeV:

    Lucite:  $R = 0.334T^{1.48}$,  $0 \leq T \leq 4$

    lead:   $R = 0.426T^{1.14}$,  $0.1 \leq T \leq 10$.

22. To protect the cell culture in the last problem from radiation, what advantage would be gained if the positions of the Lucite and the lead were swapped, so that the Lucite was on top?

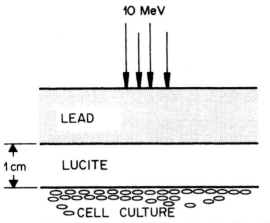

10 MeV

LEAD

1 cm    LUCITE

CELL CULTURE

Fig. **6.9** Cell culture covered with Lucite and lead (Problems 21 and 22).

23. Estimate the slowing-down time in water for positrons of energy
    (a) 100 keV
    (b) 1 MeV.
24. (a) Use the information from Problems 6 and 7 to calculate the slowing-down rate of a 9.5-keV electron in $CO_2$ at STP.
    (b) Estimate the stopping time for the electron.
25. (a) Estimate the stopping time of a 9.5-keV electron in soft tissue.
    (b) Why is this time considerably shorter than the time in Problem 24(b)?
26. (a) Calculate the ratio of the slowing-down times of a 1-MeV proton and a 1-MeV electron in water.
    (b) Calculate the ratio for a 250-MeV proton and a 0.136-MeV electron ($\beta = 0.614$ for both).
    (c) Discuss physical reasons for the time difference in (a) and (b).
27. (a) Approximately how many secondary electrons are produced when a 5-MeV electron stops in water?
    (b) What is the average number of ions per cm (specific ionization) along its track?
28. For a 150-eV electron, the ordinate in Fig. 5.3 has an average value of about 0.03 in the energy-loss interval between 19 eV and 28 eV. What fraction of the collisions of 150-eV electrons in water result in energy losses between 19 eV and 28 eV?
29. (a) Use Fig. 6.8 to estimate the probability that a normally incident, 740-keV electron will penetrate a water phantom to a maximum depth between 1,500 $\mu$m and 2,000 $\mu$m.

(b) What is the probability that the pathlength will be between these two distances?

30. In Problem 21, the range $R$ represents the mean value of the pathlength in each medium. In view of Fig. 6.8, make a rough, but more realistic, estimate of the lead thickness needed.

## 6.10
## answers

1. −0.122
4. 1.97 MeV cm$^{-1}$
7. 0.0383 MeV cm$^{-1}$
9. (a) 120
   (b) 41
   (c) 0.61
12. (a) 0.072
    (b) 0.13
    (c) 0.32
13. (a) 840 MeV cm$^{-2}$ s$^{-1}$
    (b) 870 cm$^{-2}$ s$^{-1}$
14. (a) 0.0088

(b) 21 W
18. (a) 3.3 cm
    (b) 270 cm
    (c) 250 m
19. 2.2 MeV
21. 0.42 cm
24. (a) $2.19 \times 10^8$ MeV s$^{-1}$
    (b) $4.3 \times 10^{-11}$ s
25. (a) $6.6 \times 10^{-14}$ s
27. (a) ~227,000
    (b) ~$10^5$
28. 0.27

# 7
# Phenomena Associated with Charged-Particle Tracks

## 7.1
### Delta Rays

A heavy charged particle or an electron traversing matter sometimes produces a
secondary electron with enough energy to leave the immediate vicinity of the primary particle's path and produce a noticeable track of its own. Such a secondary
electron is called a delta ray. Figure 7.1 shows a number of examples of delta rays
along calculated tracks of protons and alpha particles at several energies with the
same speeds. The 20-MeV alpha particle produces a very-high-energy delta ray,
which itself produces another delta ray. There is no sharp distinction in how one
designates one secondary electron along a track as a delta ray and another not, except that its track be noticeable or distinct from that of the primary charged particle.
Delta rays can also be seen along the electron tracks in Fig. 6.7. (Note the difference
in scale compared with Fig. 7.1.)

## 7.2
### Restricted Stopping Power

As will be discussed in Chapter 12, radiation dose is defined as the energy absorbed
per unit mass in an irradiated material. Absorbed energy thus plays a preeminent
role in dosimetry and in radiation protection.

Stopping power gives the energy *lost* by a charged particle in a medium. This is
not always equal to the energy *absorbed* in a target, especially if the target is small
compared with the ranges of secondary electrons produced. On the biological scale,
many living cells have diameters of the order of microns ($10^{-4}$ cm). Subcellular
structures can be many times smaller; the DNA double helix, for example, has a
diameter of about 20 Å. Delta rays and other secondary electrons can effectively
transport energy out of the original site in which it is lost by a primary particle.

The concept of restricted stopping power has been introduced to associate energy loss in a target more closely with the energy that is actually absorbed there.
The restricted stopping power, written $(-dE/dx)_\Delta$, is defined as the linear rate of
energy loss due only to collisions in which the energy transfer does not exceed a

*Atoms, Radiation, and Radiation Protection*. James E. Turner
Copyright © 2007 WILEY-VCH Verlag GmbH & Co. KGaA, Weinheim
ISBN: 978-3-527-40606-7

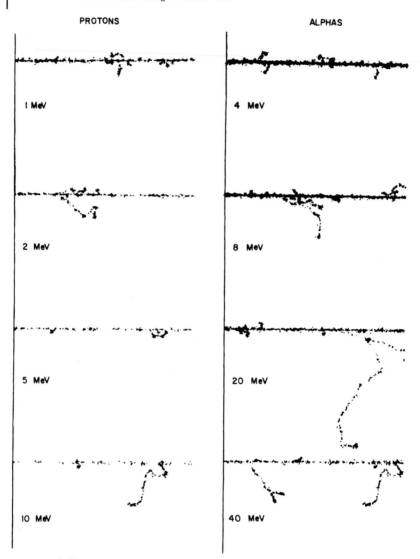

**Fig. 7.1** Calculated track segments (0.7 $\mu$m) of protons and alpha particles having the same velocities in water. (Courtesy Oak Ridge National Laboratory, operated by Martin Marietta Energy Systems, Inc., for the Department of Energy.)

specified value $\Delta$. To calculate this quantity, one integrates the weighted energy-loss spectrum only up to $\Delta$, rather than $Q_{\max}$. In place of Eq. (5.10) one defines the restricted stopping power as

$$\left(-\frac{\mathrm{d}E}{\mathrm{d}x}\right)_{\Delta} = \mu \int_{Q_{\min}}^{\Delta} Q W(Q)\,\mathrm{d}Q. \tag{7.1}$$

Depending on the application, different values of $\Delta$ can be selected, for example, $(-dE/dx)_{100 \text{ eV}}$, $(-dE/dx)_{1 \text{ keV}}$, and so forth. In the context of restricted stopping power, the subscript $Q_{max}$ or $\infty$ is used to designate the usual stopping power:

$$-\frac{dE}{dx} = \left(-\frac{dE}{dx}\right)_{Q_{max}} = \left(-\frac{dE}{dx}\right)_{\infty}. \tag{7.2}$$

We see from Table 6.1 that restricting single-collision energy losses by electrons in water to 100 eV or less, for example, limits the range of secondary electrons to $\sim 5 \times 10^{-7}$ cm, or about 50 Å. With $\Delta = 1$ keV, the maximum range of the secondary electrons contributing to the restricted stopping power is $5 \times 10^{-6}$ cm, or about 500 Å.

*Example*
A sample of viruses, assumed to be in the shape of spheres of diameter 300 Å, is to be irradiated by a charged-particle beam in an experiment. Estimate a cutoff value that would be appropriate for determining a restricted stopping power that would be indicative of the actual energy *absorbed* in the individual virus particles.

*Solution*
As an approximation, one can specify that the range of the most energetic delta ray should not exceed 300 Å $= 3 \times 10^{-6}$ cm. We assume that the virus sample has unit density. Table 6.1 shows that this distance is approximately the range of a 700-eV secondary electron. Therefore, we choose $\Delta = 700$ eV and use the restricted stopping power $(-dE/dx)_{700 \text{ eV}}$ as a measure of the average energy absorbed in an individual virus particle from a charged particle traversing it.

Restricted mass stopping powers of water for protons are given in Table 7.1. At energies of 0.05 MeV and below, collisions that transfer more than 100 eV do not contribute significantly to the total stopping power, and so $(-dE/\rho\, dx)_{100 \text{ eV}} = (-dE/\rho\, dx)_{\infty}$. In fact, at 0.05 MeV, $Q_{max} = 109$ eV. At 0.10 MeV, on the other hand, $Q_{max} = 220$ eV; and so the restricted mass stopping power with $\Delta = 1$ keV is significantly larger than that with $\Delta = 100$ eV. At 1 MeV, a negligible number of collisions result in energy transfers of more than 10 keV; at 10 MeV, about 8% of the stopping

**Table 7.1** Restricted Mass Stopping Power of Water, $(-dE/\rho dx)_{\Delta}$ in MeV cm$^2$ g$^{-1}$, for Protons

| Energy (MeV) | $\left(-\frac{dE}{\rho dx}\right)_{100 \text{ eV}}$ | $\left(-\frac{dE}{\rho dx}\right)_{1 \text{ keV}}$ | $\left(-\frac{dE}{\rho dx}\right)_{10 \text{ keV}}$ | $\left(-\frac{dE}{\rho dx}\right)_{\infty}$ |
|---|---|---|---|---|
| 0.05 | 910. | 910. | 910. | 910. |
| 0.10 | 711. | 910. | 910. | 910. |
| 0.50 | 249. | 424. | 428. | 428. |
| 1.00 | 146. | 238. | 270. | 270. |
| 10.0 | 24.8 | 33.5 | 42.2 | 45.9 |
| 100. | 3.92 | 4.94 | 5.97 | 7.28 |

**Table 7.2** Restricted Collisional Mass Stopping Power of Water, $(-dE/\rho dx)_\Delta$ in MeV cm$^2$ g$^{-1}$, for Electrons

| Energy (MeV) | $\left(-\frac{dE}{\rho dx}\right)_{100 \text{ eV}}$ | $\left(-\frac{dE}{\rho dx}\right)_{1 \text{ keV}}$ | $\left(-\frac{dE}{\rho dx}\right)_{10 \text{ keV}}$ | $\left(-\frac{dE}{\rho dx}\right)_{\infty}$ |
|---|---|---|---|---|
| 0.0002 | 298. | 298. | 298. | 298. |
| 0.0005 | 183. | 194. | 194. | 194. |
| 0.001 | 109. | 126. | 126. | 126. |
| 0.003 | 40.6 | 54.4 | 60.1 | 60.1 |
| 0.005 | 24.9 | 34.0 | 42.6 | 42.6 |
| 0.01 | 15.1 | 20.2 | 23.2 | 23.2 |
| 0.05 | 4.12 | 5.26 | 6.35 | 6.75 |
| 0.10 | 2.52 | 3.15 | 3.78 | 4.20 |
| 1.00 | 1.05 | 1.28 | 1.48 | 1.89 |

power is due to collisions that transfer more than 10 keV. Corresponding data for the restricted collisional mass stopping power for electrons are presented in Table 7.2. Here, the restricted stopping powers are different at much lower energies than in Table 7.1.

## 7.3
## Linear Energy Transfer (LET)

The concept of linear energy transfer, or LET, was introduced in the early 1950s to characterize the rate of energy transfer per unit distance along a charged-particle track. As such, LET and stopping power were synonymous. In studying radiation effects in terms of LET, the distinction was made between the energy transferred from a charged particle in a target and the energy actually absorbed there. In 1962 the International Commission on Radiation Units and Measurements (ICRU) defined LET as the quotient $-dE_L/dx$, where $dE_L$ is the "average energy locally imparted" to a medium by a charged particle in traversing a distance $dx$. The words "locally imparted," however, were not precisely specified, and LET was not always used with exactly the same meaning. In 1980, the ICRU defined LET$_\Delta$ as the restricted stopping power for energy losses not exceeding $\Delta$:

$$\text{LET}_\Delta = \left(-\frac{dE}{dx}\right)_\Delta, \tag{7.3}$$

with the symbol LET$_\infty$ denoting the usual (unrestricted) stopping power.

LET is often found in the literature with no subscript or other clarification. It can generally be assumed then that the unrestricted stopping power is implied.

*Example*

Use Table 7.1 to determine LET$_{1 \text{ keV}}$ and LET$_{5 \text{ keV}}$ for 1-MeV protons in water.

*Solution*

Note that Eq. (7.3) for LET involves stopping power rather than mass stopping power. Since $\rho = 1$ for water, the numbers in Table 7.1 also give $(-dE/dx)_\Delta$ in MeV cm$^{-1}$. We find LET$_{1\ keV}$ = 238 MeV cm$^{-1}$ given directly in the table. Linear interpolation gives LET$_{5\ keV}$ = 252 MeV cm$^{-1}$.

LET is often expressed in units of keV $\mu$m$^{-1}$ of water. Conversion of units shows that 1 keV $\mu$m$^{-1}$ = 10 MeV cm$^{-1}$ (Problem 10).

In 1998, the ICRU introduced the following new definition, also called "linear energy transfer, or restricted linear electronic stopping power, $L_\Delta$":

$$L_\Delta = -\frac{dE_\Delta}{dx}. \tag{7.4}$$

Here $E_\Delta$ is the total energy lost by the charged particle due to electronic collisions in traversing a distance $dx$, minus the sum of the kinetic energies of all electrons released with energies in excess of $\Delta$. Compared with the 1980 definition, Eq. (7.3), there are two important differences. First, the binding energies for all collisions are included in (7.4). Second, the threshold kinetic energy of the secondary electrons for a collision is now $\Delta$, rather than $\Delta$ minus the binding energy. Equation (7.4) can be written in the alternate form,

$$L_\Delta = -\frac{dE}{dx} - \frac{dE_{ke,\Delta}}{dx}, \tag{7.5}$$

where $dE_{ke,\Delta}$ is the sum of the kinetic energies greater than $\Delta$ of the secondary electrons. We shall not deal further with the newer quantity. The reader is referred to the 1980 and 1998 ICRU Reports 33 and 60 listed in Section 7.8.

## 7.4
## Specific Ionization

The average number of ion pairs that a particle produces per unit distance traveled is called the specific ionization. This quantity, which expresses the density of ionizations along a track, is often considered in studying the response of materials to radiation and in interpreting some biological effects. The specific ionization of a particle at a given energy is equal to the stopping power divided by the average energy required to produce an ion pair at that particle energy. The stopping power of air for a 5-MeV alpha particle is 1.23 MeV cm$^{-1}$, and an average of about 36 eV is needed to produce an ion pair. Thus, the specific ionization of a 5-MeV alpha particle in air is $(1.23 \times 10^6$ eV cm$^{-1})/(36$ eV$) = 34{,}200$ cm$^{-1}$. For a 5-MeV alpha particle in water or soft tissue, $-dE/dx = 950$ MeV cm$^{-1}$ (Fig. 5.6). Since about 22 eV is required to produce an ion pair, the specific ionization is $4.32 \times 10^7$ cm$^{-1}$.

## 7.5
## Energy Straggling

As a charged particle penetrates matter, statistical fluctuations occur in the number of collisions along its path and in the amount of energy lost in each collision. As a result, a number of identical particles starting out under identical conditions will show (1) a distribution of energies as they pass a given depth and (2) a distribution of pathlengths traversed before they stop. The phenomenon of unequal energy losses under identical conditions is called *energy straggling* and the existence of different pathlengths is referred to as *range straggling*. We examine these two forms of straggling in this and the next section.

Energy straggling can be observed experimentally by the setup shown schematically in Fig. 7.2. A monoenergetic beam of protons (or other charged particles) is passed through a gas-filled cylindrical proportional counter parallel to its axis. The ends of the cylinder can be thin aluminum or other material that absorbs little energy. Each proton makes a number of electronic collisions and produces a single pulse in the counter, which is operated so that the pulse height is proportional to the total energy that the proton deposits in the counter gas. (Proportional counters are described in Chapter 10.) Thus, by measuring the distribution of pulse heights, called the pulse-height spectrum, in an experiment one obtains the distribution of proton energy losses in the gas. By changing the gas pressure and repeating the experiment, one can study how the energy-loss distribution depends on the amount of matter traversed.

Some data are shown in Fig. 7.3, based on experiments reported by Gooding and Eisberg, using 37-MeV protons and a 10-cm-long counter filled with a mixture of 96% Ar and 4% $CO_2$ at pressures up to 1.2 atm. The average energy needed to produce an ion pair in the gas is about 25 eV. Data are provided for gas pressures of 0.2 atm and 1.2 atm. The ordinate shows the relative number of counts at different

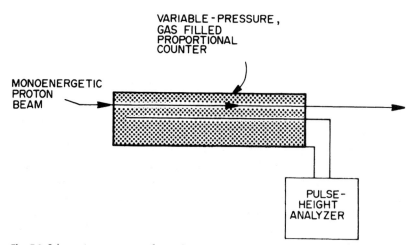

**Fig. 7.2** Schematic arrangement for studying energy straggling experimentally.

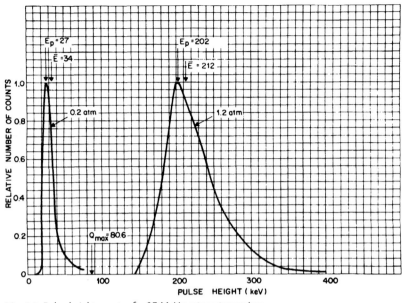

**Fig. 7.3** Pulse-height spectra for 37-MeV protons traversing proportional counter with gas at 0.2-atm and 1.2-atm pressure. See text. [Based on T. J. Gooding and R. M. Eisberg, "Statistical Fluctuations in Energy Losses of 37-MeV Protons," Phys. Rev. **105**, 357–360 (1957).]

pulse heights given by the abscissa. We can examine both curves quantitatively. For reference, the value $Q_{max} = 80.6$ keV for a single collision of a 37-MeV proton with an electron is also shown.

At 0.2 atm, the most probable energy loss measured for a proton traversing the gas is $E_p = 27$ keV and the average loss is $\overline{E} = 34$ keV. Since about 25 eV is needed to produce an ion pair, the average number of secondary electrons is $34{,}000/25 = 1360$ per proton. Some of these electrons are produced directly by the proton and the others are produced by secondary electrons. If we assume that the proton directly ejects secondary electrons with a mean energy of $\sim$60 eV, then the proton makes approximately $34{,}000/60 = 570$ collisions in traversing the gas. At this lower pressure, where the proton mean energy loss is considerably less than $Q_{max}$ and only a few hundred collisions take place, the pulse-height spectrum shows the skewed distribution characteristic of the single-collision spectrum (Fig. 5.3). The relative separation of the peak and mean energies is $(\overline{E} - E_p)/\overline{E} = (34 - 27)/34 = 0.21$.

At 1.2 atm, an average of six times as many proton collisions takes place. The observed mean energy loss $\overline{E} = 212$ keV is some three times $Q_{max}$. The pulse-height spectrum, while still skewed, is somewhat more symmetric. The relative separation of the peak and mean energies is $(212 - 202)/212 = 0.05$, which is considerably less than that at 0.2 atm. At still higher gas pressures, the pulse-height spectrum shifts further to the right and becomes more symmetric, approaching a Gaussian shape

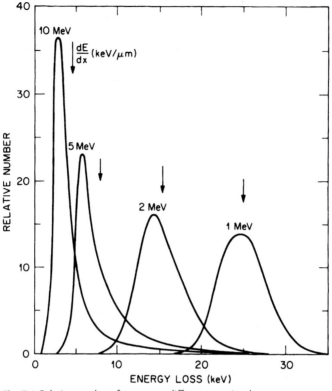

**Fig. 7.4** Relative number of protons at different energies that experience energy losses shown on the abscissa in traversing 1 $\mu$m of water. (Courtesy Oak Ridge National Laboratory, operated by Martin Marietta Energy Systems, Inc., for the Department of Energy.)

with $E_p = \overline{E}$. Statistical aspects of radiation interaction and energy loss in matter will be discussed in Chapter 11.

Figure 7.4 shows the energy straggling calculated for protons traversing a thickness of 1 $\mu$m of water at several energies. The curves are normalized to the same area. The energy loss is given in keV, and so its mean value at each proton energy (indicated by the vertical arrows) is numerically equal to the stopping power in keV $\mu$m$^{-1}$. The stopping power is smallest for the 10-MeV protons; they make the fewest collisions on average and show the most skewed straggling distribution. The most probable energy loss for them is only about 60% of the average. Thus, using the value of the stopping power as an estimate of energy loss for 10-MeV protons in micron-sized volumes of tissue can be misleading. Most protons lose considerably less energy than the mean, while a few experience energy losses several times greater. At the lower proton energies the number of collisions and the stopping power increase progressively. The straggling curves assume more closely a Gaussian shape, with the average and the most probable values being close. In contrast to 10 MeV, at 1 MeV it is relatively unlikely that a given proton will experience an energy loss 20% or more different from the mean (i.e., outside an interval ±5 keV about the mean of 25 keV). The relation between single-collision energy-loss spectra, stopping power, and energy straggling was discussed in Section 5.4.

The physics behind Fig. 7.4 is of fundamental importance to radiation biology. As pointed out in Section 7.2, radiation damages living tissue directly on a scale of microns and below. At this level, energy straggling, as well as (related) delta-ray effects, play important roles in the interaction of radiation with matter. The subject of microdosimetry (Section 12.10) deals with the distribution and fluctuations in energy loss and deposition in small volumes of tissue.

## 7.6
## Range Straggling

Range straggling for heavy charged particles can be measured with the experimental arrangement shown in Fig. 7.5(a). A monoenergetic beam is directed on an absorber whose thickness can be varied by using additional layers of the material. A count-rate meter is used to measure the relative number of beam particles that emerge from the absorber as a function of its thickness. A plot of relative count

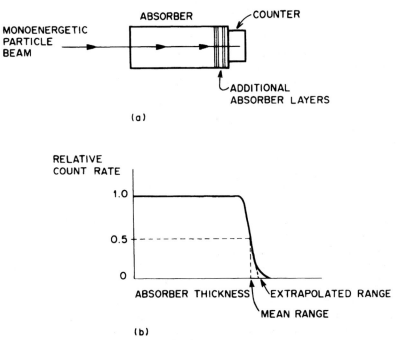

**Fig. 7.5** (a) Experimental arrangement for observing range straggling, (b) Plot of relative count rate vs. absorber thickness, showing the mean and extrapolated ranges.

rate versus thickness for energetic heavy charged particles has the form shown in Fig. 7.5(b). At first the absorber serves only to reduce the energy of the particles traversing it, and therefore the count-rate curve is flat. When additional absorber material is added, the curve remains flat until the thickness approaches the range of the particles. Then the number of particles emerging from the absorber begins to decrease rapidly and almost linearly as more material is added until all of the particles are stopped in the absorber.

The mean range is defined as the absorber thickness at which the relative count rate is 0.50, as shown in Fig. 7.5(b). The extrapolated range is determined by extending the straight portion of the curve to the abscissa. The distribution of stopping depths about the mean range is nearly Gaussian in shape.

Range straggling is not large for heavy charged particles. For 100-MeV protons in tissue, for example, the root-mean-square fluctuation in pathlength is about 0.09 cm. The range is 7.57 cm (Table 5.3), and so the relative spread in stopping distances is $(0.09/7.57) \times 100 = 1.2\%$.

When monoenergetic electrons are used with an experimental setup like that in Fig. 7.5, a different penetration pattern is found. Figure 7.6 shows the relative number of 100-keV electrons that pass through a water absorber of variable thickness. The relative number slightly exceeds unity for the thinnest absorbers because of the

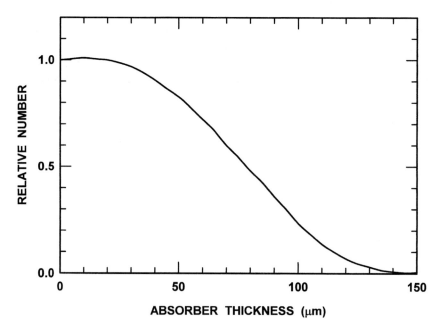

**Fig. 7.6** Relative number of normally incident, 100-keV electrons that get through water slabs of different thicknesses with arrangement like that in Fig. 7.5(a). The reduction in transmission as a function of absorber thickness differs markedly from that in Fig. 7.5(b) for monoenergetic heavy charged particles.

buildup of secondary electrons as the incident beam initially penetrates and ionizes the target. With increasing absorber thickness the relative number that penetrate falls off steadily out to about 110 $\mu$m, and then tails off. Mean and extrapolated ranges can be similarly defined as in Fig. 7.5. The csda range for 100-keV electrons in water is 140 $\mu$m.

## 7.7
## Multiple Coulomb Scattering

We have seen how straggling affects the penetration of charged particles and introduces some fuzziness into the concept of range. Another phenomenon, elastic scattering from atomic nuclei via the Coulomb force, further complicates the analysis of particle penetration. The path of a charged particle in matter—even a fast electron or a heavy charged particle—deviates from a straight line because it undergoes frequent small-angle nuclear scattering events.

Figure 7.7 illustrates how a heavy particle, starting out along the $X$-axis at the origin O in an absorber might be deviated repeatedly by multiple Coulomb scattering until coming to rest at a depth $X_0$. The total pathlength traveled, $R$, which is the quantity calculated from Eq. (5.39) and given in the tables, is greater than the depth of penetration $X_0$. The latter is sometimes called the projected range. The difference between $R$ and $X_0$ for heavy charged particles is typically $\lesssim 1\%$, and so $R$ is usually considered to be the same as $X_0$.

Another effect of multiple Coulomb scattering is to spread a pencil beam of charged particles into a diverging beam as it penetrates a target, as illustrated in Fig. 7.8. The magnitude of the spreading increases with the atomic number of the material. When a pencil beam of 120-MeV protons penetrates 1 cm of water, for example, about 4% of the particles emerge outside an angle $\varphi = 1.5°$ in Fig. 7.8.

The arrangement in Fig. 7.8 is basically the same as that used by Rutherford when he investigated alpha-particle scattering through thin metal foils (Section 2.2). The occasional, unexpected, large-angle scattering of a particle led him to the discovery of the nucleus.

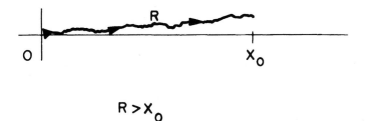

$$R > X_0$$

**Fig. 7.7** Schematic representation of the effect of multiple Coulomb scattering on the path of a heavy charged particle that starts moving from the origin O toward the right along the $X$-axis. The displacement lateral to the $X$-axis is exaggerated for illustrative purposes.

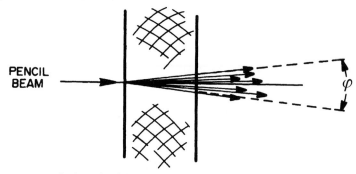

**Fig. 7.8** Multiple Coulomb scattering causes spread in a pencil beam of charged particles as they penetrate matter.

In radiotherapy with charged-particle beams, multiple Coulomb scattering can significantly diminish the dose that can be concentrated in a tumor, particularly when it is located at some depth in the body.

## 7.8
## Suggested Reading

1 Ahlen, S. P., "Theoretical and Experimental Aspects of the Energy Loss of Relativistic Heavily Ionizing Particles," *Rev. Mod. Phys.* **52**, 121–173 (1980). [Review of stopping-power theory, including track effects discussed here, and also the response of charged-particle detectors.]

2 Attix, F. H., *Introduction to Radiological Physics and Radiation Dosimetry*, Wiley, New York (1986). [See Chapter 8.]

3 Fano, U., "Penetration of Protons, Alpha Particles, and Mesons," *Ann. Rev. Nucl. Sci.* **13**, 1–66 (1963). [Track phenomena discussed in considerable detail.]

4 ICRU Report 33, *Radiation Quantities and Units*, International Commission on Radiation Units and Measurements, Washington, DC (1980). [Reference on LET and restricted stopping power.]

5 ICRU Report 36, *Microdosimetry*, International Commission on Radiation Units and Measurements, Bethesda, MD (1983).

6 ICRU Report 60, *Fundamental Quantities and Units for Ionizing Radiation*, International Commission on Radiation Units and Measurements, Bethesda, MD (1998). [Reference on LET and restricted stopping power.]

7 Turner, J. E., "An Introduction to Microdosimetry," *Rad. Prot. Management* **9**(3), 25–58 (1992). [This review goes deeper into some topics introduced here. Microdosimetric implications for the biological effects of radiation and for biological-effects modeling are discussed.]

## 7.9
## Problems

1. A 4-MeV proton in water produces a delta ray with energy $0.1 Q_{max}$. How large is the range of this delta ray compared with the range of the proton?

2. How does the maximum energy of a delta ray that can be produced by a 3-MeV alpha particle compare with that from a 3-MeV proton?

3. What is the energy of a proton that can produce a delta ray with enough energy to traverse a cell having a diameter of 2.5 $\mu$m?

4. Find $(-dE/dx)_{500\ eV}$ for a 500-keV proton in water.

5. (a) For 0.05-MeV protons in water, what is the smallest value of $\Delta$ for which $(-dE/dx)_\Delta = (-dE/dx)_\infty$?
   (b) Repeat for 0.10-MeV protons.
   (c) Are your answers consistent with Table 7.1?

6. Use Table 7.1 to estimate the restricted mass stopping power $(-dE/\rho dx)_{2\ keV}$ of water for 10-MeV protons.

7. Use Table 7.2 to estimate the ratio of the restricted stopping power of water $(-dE/dx)_{500\ eV}$ to the total stopping power for 50-keV electrons.

8. Find $LET_{1\ keV}$ for 10-MeV protons in
   (a) soft tissue,
   (b) bone of density 1.93 g cm$^{-3}$

9. From Fig. 5.6 find $LET_\infty$ for a
   (a) 2-MeV alpha particle,
   (b) 100-MeV muon,
   (c) 100-keV positron.

10. Show that the common LET unit, keV $\mu$m$^{-1}$, is equal to 10 MeV cm$^{-1}$.

11. What is the specific ionization of a 12-MeV proton in tissue if an average of 22 eV is needed to produce an ion pair?

12. In Section 7.4 we calculated the specific ionization of a 5-MeV alpha particle in air (3.42 × 10$^4$ cm$^{-1}$) and in water (4.32 × 10$^7$ cm$^{-1}$). Why is the specific ionization so much greater in water?

13. How does the maximum specific ionization of an alpha particle in tissue compare with that of a proton?

14. (a) Define energy straggling.
    (b) Does energy straggling cause range straggling?

15. (a) From the numerical analysis of Fig. 7.3 given in Section 7.5, how many total ionizations per proton would be expected at a pressure of 0.7 atm?

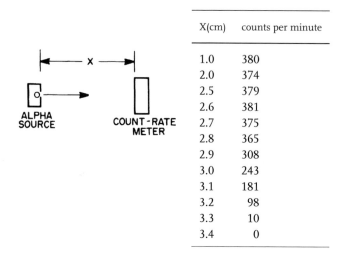

| X(cm) | counts per minute |
|-------|-------------------|
| 1.0   | 380               |
| 2.0   | 374               |
| 2.5   | 379               |
| 2.6   | 381               |
| 2.7   | 375               |
| 2.8   | 365               |
| 2.9   | 308               |
| 3.0   | 243               |
| 3.1   | 181               |
| 3.2   | 98                |
| 3.3   | 10                |
| 3.4   | 0                 |

**Fig. 7.9** Count rate from a collimated, monoenergetic alpha-particle beam (Problem 16).

(b) How many of these ionizations would be produced by secondary electrons?

16. Figure 7.9 shows an experimental arrangement in which the count rate from a collimated, monoenergetic alpha-particle beam is measured at different separations $x$ in air. From the given data, determine
   (a) the mean range,
   (b) the extrapolated range.

## 7.10
## Answers

1. $\sim 1.8 \times 10^{-4}$
3. 5.0 MeV
4. 330 MeV cm$^{-1}$
6. 34 MeV cm$^2$ g$^{-1}$

7. 0.69
13. $\sim 2.7$
16. (a) 3.1 cm
    (b) 3.3 cm

# 8
# Interaction of Photons with Matter

## 8.1
### Interaction Mechanisms

Unlike charged particles, photons are electrically neutral and do not steadily lose energy as they penetrate matter. Instead, they can travel some distance before interacting with an atom. How far a given photon will penetrate is governed statistically by a probability of interaction per unit distance traveled, which depends on the specific medium traversed and on the photon energy. When the photon interacts, it might be absorbed and disappear or it might be scattered, changing its direction of travel, with or without loss of energy.

Thomson and Raleigh scattering are two processes by which photons interact with matter without appreciable transfer of energy. In Thomson scattering an electron, assumed to be free, oscillates classically in response to the electric vector of a passing electromagnetic wave. The oscillating electron promptly emits radiation (photons) of the same frequency as the incident wave. The net effect of Thomson scattering, which is elastic, is the redirection of some incident photons with no transfer of energy to the medium. In the modern, quantum-mechanical theory of photon–electron interactions, Thomson scattering represents the low-energy limit of Compton scattering, as the incident photon energy approaches zero.

Raleigh scattering of a photon results from the combined, coherent action of an atom as a whole. The scattering angle is usually very small. There is no appreciable loss of energy by the photon to the atom, which, however, does "recoil" enough to conserve momentum. We shall not consider Thomson or Raleigh scattering further.

The principal mechanisms of energy deposition by photons in matter are photoelectric absorption, Compton scattering, pair production, and photonuclear reactions. We treat these processes in some detail.

*Atoms, Radiation, and Radiation Protection*. James E. Turner
Copyright © 2007 WILEY-VCH Verlag GmbH & Co. KGaA, Weinheim
ISBN: 978-3-527-40606-7

## 8.2
### Photoelectric Effect

The ejection of electrons from a surface as a result of light absorption is called the photoelectric effect. The arrangement in Fig. 8.1 can be used to study this process experimentally. Monochromatic light passes into an evacuated glass tube through a quartz window (which allows ultraviolet light to be used) and strikes an electrode 1 causing photoelectrons to be ejected. Electrode 1 can be made of a metal to be studied or have its surface covered with such a metal. The current $I$ that flows during illumination can be measured as a function of the variable potential difference $V_{21}$ applied between the two electrodes, 1 and 2, of the tube. Curves (a) and (b) represent data obtained at two different intensities of the incident light. With the surface illuminated, there will be some current even with $V_{21} = 0$. When $V_{21}$ is made positive and increased, the efficiency of collecting photoelectrons at electrode 2 increases; the current rises to a plateau when all of the electrons are be-

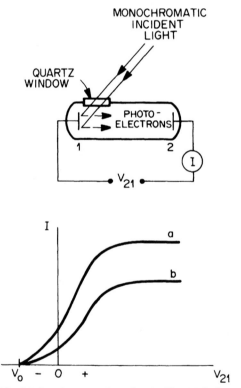

**Fig. 8.1** Experiment on photoelectric effect. With electrode 1 illuminated with monochromatic light of constant intensity, the current $I$ is measured as a function of the potential difference $V_{21}$ between electrodes 2 and 1. Curves (a) and (b) represent data at two different intensities of the incident light.

ing collected. The ratio of the plateau currents is equal to the relative light intensity used for curves (a) and (b).

When the polarity of the potential difference is reversed ($V_{21} < 0$), photoelectrons ejected from the illuminated electrode 1 now experience an attractive force back toward it. Making $V_{21}$ more negative allows only the most energetic photoelectrons to reach electrode 2, thus causing the current $I$ to decrease. Independently of the light intensity, the photoelectric current drops to zero when the reversed potential difference reaches a magnitude $V_0$, called the stopping potential. The potential energy $eV_0$, where $e$ is the magnitude of the electronic charge, is equal to the maximum kinetic energy, $T_{max}$, of the photoelectrons:

$$T_{max} = eV_0. \tag{8.1}$$

The stopping potential $V_0$ varies linearly with the frequency $\nu$ of the monochromatic light used. A threshold frequency $\nu_0$ is found below which no photoelectrons are emitted, even with intense light. The value of $\nu_0$ depends on the metal used for electrode 1.

The photoelectric effect is of special historical significance. The experimental findings are incompatible with the classical wave theory of light, which had been so successful in the latter part of the nineteenth century. Based on a wave concept, one would expect the maximum kinetic energy of photoelectrons, $T_{max}$ in Eq. (8.1), to increase as the intensity of the incident light is increased. Yet the value of $V_0$ for a metal was found to be independent of the intensity (Fig. 8.1). Furthermore, one would also expect some photoelectrons to be emitted by light of any frequency, simply by making it intense enough. However, a threshold frequency $\nu_0$ exists for every metal.

To explain the photoelectric effect, Einstein in 1905 proposed that the incident light arrives in discrete quanta (photons), having an energy given by $E = h\nu$, where $h$ is Planck's constant. He further assumed that a photoelectron is produced when a single electron in the metal completely absorbs a single photon. The kinetic energy $T$ with which the photoelectron is emitted from the metal is equal to the photon energy minus an energy $\varphi$ that the electron expends in escaping the surface:

$$T = h\nu - \varphi. \tag{8.2}$$

The energy $\varphi$ may result from collisional losses (stopping power) and from work done against the net attractive forces that normally keep the electron in the metal. A minimum energy, $\varphi_0$, called the work function of the metal, is required to remove the most loosely bound electron from the surface. The maximum kinetic energy that a photoelectron can have is given by

$$T_{max} = h\nu - \varphi_0. \tag{8.3}$$

Einstein received the Nobel Prize in 1921 "for his contributions to mathematical physics, and especially for his discovery of the law of the photoelectric effect."

The probability of producing a photoelectron when light strikes an atom is strongly dependent on the atomic number $Z$ and the energy $h\nu$ of the photons.

It is largest for high-$Z$ materials and low-energy photons with frequencies above the threshold value $\nu_0$. The probability varies as $Z^4/(h\nu)^3$.

*Example*

(a) What threshold energy must a photon have to produce a photoelectron from Al, which has a work function of 4.20 eV? (b) Calculate the maximum energy of a photoelectron ejected from Al by UV light with a wavelength of 1500 Å. (c) How does the maximum photoelectron energy vary with the intensity of the UV light?

*Solution*

(a) The work function $\varphi_0 = 4.20$ eV represents the minimum energy that a photon must have to produce a photoelectron. (b) The energy of the incident photons in eV is given in terms of the wavelength $\lambda$ in angstroms by Eq. (2.26):

$$E = h\nu = \frac{12400}{\lambda} = \frac{12400}{1500} = 8.27 \text{ eV.} \tag{8.4}$$

From Eq. (8.3),

$$T_{max} = 8.27 - 4.20 = 4.07 \text{ eV.} \tag{8.5}$$

(c) $T_{max}$ is independent of the light intensity.

## 8.3
### Energy–Momentum Requirements for Photon Absorption by an Electron

As with charged particles, when the energy transferred by a photon to an atomic electron is large compared with its binding energy, then the electron can be treated as initially free and at rest. We now show that the conservation of energy and momentum prevents the *absorption* of a photon by an electron under these conditions. Thus, the binding of an electron and its interaction with the rest of the atom are essential for the photoelectric effect to occur. However, a photon can be *scattered* from a free electron, either with a reduction in its energy (Compton effect, next section) or with no change in energy (Thomson scattering).

If an electron, initially free and at rest (rest energy $mc^2$), absorbs a photon of energy $h\nu$ and momentum $h\nu/c$ (Appendix C), then the conservation of energy and momentum requires, respectively, that

$$mc^2 + h\nu = \gamma mc^2 \tag{8.6}$$

and

$$h\nu/c = \gamma mc\beta. \tag{8.7}$$

Here $\gamma = (1-\beta^2)^{-1/2}$ is the relativistic factor and $\beta = v/c$ is the ratio of the speed of the electron after absorbing the photon and the speed of light $c$. Multiplying both sides of Eq. (8.7) by $c$ and subtracting from (8.6) gives

$$mc^2 = \gamma mc^2(1-\beta). \tag{8.8}$$

This equation has only the trivial solution $\beta = 0$ and $\gamma = 1$, which, by Eq. (8.6), leads to the condition $h\nu = 0$. We conclude that the photoelectric effect occurs because the absorbing electron interacts with the nucleus and the other electrons in the atom to conserve the total energy and momentum of all interacting partners.

We now turn to the scattering of photons by electrons.

## 8.4
## Compton Effect

Figure 8.2 illustrates the experimental arrangement used by Compton in 1922. Molybdenum $K_\alpha$ X rays (photon energy 17.4 keV, wavelength $\lambda = 0.714$ Å) were directed at a graphite target and the wavelengths $\lambda'$ of scattered photons were measured at various angles $\theta$ with respect to the incident photon direction. The intensities of the scattered radiation versus $\lambda'$ for three values of $\theta$ are sketched in Fig. 8.3. Each plot shows peaks at two values of $\lambda'$: one at the wavelength $\lambda$ of the incident photons and another at a longer wavelength, $\lambda' > \lambda$. The appearance of scattered radiation at a longer wavelength is called the Compton effect. The Compton shift in wavelength, $\Delta\lambda = \lambda' - \lambda$, was found to depend only on $\theta$; it is independent of the incident-photon wavelength $\lambda$. In the crucial new experiment in 1922, Compton measured the shift $\Delta\lambda = 0.024$ Å at $\theta = 90°$.

The occurrence of scattered radiation at the same wavelength as that of the incident radiation can be explained by classical electromagnetic wave theory. The electric field of an incident wave accelerates atomic electrons back and forth at the same frequency $\nu = c/\lambda$ with which it oscillates. The electrons therefore emit radiation with the same wavelength. This Thomson scattering of radiation from atoms with no change in wavelength was known before Compton's work. The occurrence of the scattered radiation at longer wavelengths contradicted classical expectations.

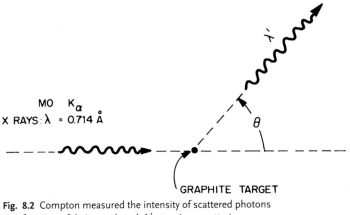

MO $K_\alpha$
X RAYS: $\lambda = 0.714$ Å

$\theta$

GRAPHITE TARGET

**Fig. 8.2** Compton measured the intensity of scattered photons as a function of their wavelength $\lambda'$ at various scattering angles $\theta$. Incident radiation was molybdenum $K_\alpha$ X rays, having a wavelength $\lambda = 0.714$ Å.

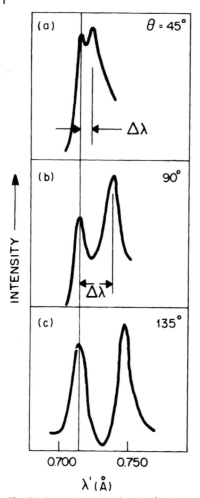

**Fig. 8.3** Intensity vs. wavelength $\lambda'$ of photons scattered at
angles (a) $\theta = 45°$, (b) 90°, and (c) 135°.

To account for his findings, Compton proposed the following quantum model. In Fig. 8.4(a), a photon of energy $h\nu$ and momentum $h\nu/c$ (wavy line) is incident on a stationary, free electron. After the collision, the photon in (b) is scattered at an angle $\theta$ with energy $h\nu'$ and momentum $h\nu'/c$. The struck electron recoils at an angle $\varphi$ with total energy $E'$ and momentum $P'$. Conservation of total energy in the collision requires that

$$h\nu + mc^2 = h\nu' + E'. \tag{8.9}$$

Conservation of the components of momentum in the horizontal and vertical directions gives the two equations

$$\frac{h\nu}{c} = \frac{h\nu'}{c} \cos\theta + P' \cos\varphi \tag{8.10}$$

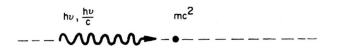

$h\nu, \dfrac{h\nu}{c}$           $mc^2$

(a) BEFORE COLLISION

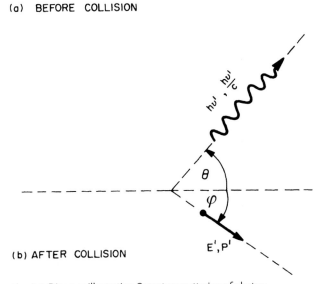

(b) AFTER COLLISION

**Fig. 8.4** Diagram illustrating Compton scattering of photon (energy $h\nu$, momentum $h\nu/c$) from electron, initially free and at rest with total energy $mc^2$. As a result of the collision, the photon is scattered at an angle $\theta$ with reduced energy $h\nu'$ and momentum $h\nu'/c$, and the electron recoils at angle $\varphi$ with total energy $E'$ and momentum $P'$.

and

$$\frac{h\nu'}{c}\sin\theta = P'\sin\varphi. \tag{8.11}$$

Eliminating $P'$ and $\varphi$ from these three equations and solving for $\nu'$, one finds that (Problem 12)

$$h\nu' = \frac{h\nu}{1 + (h\nu/mc^2)(1-\cos\theta)}. \tag{8.12}$$

With this result, the Compton shift is given by

$$\Delta\lambda = \lambda' - \lambda = c\left(\frac{1}{\nu'} - \frac{1}{\nu}\right) = \frac{h}{mc}(1-\cos\theta). \tag{8.13}$$

Thus, as Compton found experimentally, the shift does not depend on the incident photon frequency $\nu$. The magnitude of the shift at $\theta = 90°$ is

$$\Delta\lambda = \frac{h}{mc} = \frac{6.63 \times 10^{-34}}{9.11 \times 10^{-31} \times 3.00 \times 10^8} = 2.43 \times 10^{-12}\ \text{m}, \tag{8.14}$$

which agrees with the measured value. The quantity $h/mc = 0.0243$ Å is called the Compton wavelength.

*Example*

A 1.332-MeV gamma photon from [60]Co is Compton scattered at an angle of 140°. Calculate the energy of the scattered photon and the Compton shift in wavelength. What is the momentum of the scattered photon?

*Solution*

The energy of the scattered photons is given by Eq. (8.12):

$$h\nu' = \frac{1.332 \text{ MeV}}{1 + (1.332/0.511)[1 - (-0.766)]} = 0.238 \text{ MeV}. \tag{8.15}$$

The Compton shift is given by Eq. (8.13) with $h/mc = 0.0243$ Å:

$$\Delta\lambda = (0.0243 \text{ Å})[1 - (-0.766)] = 0.0429 \text{ Å}. \tag{8.16}$$

The momentum of the scattered photon is

$$\frac{h\nu'}{c} = \frac{0.238 \text{ MeV} \times 1.6 \times 10^{-13} \text{ J MeV}^{-1}}{3 \times 10^{8} \text{ m s}^{-1}} = 1.27 \times 10^{-22} \text{ kg m s}^{-1}, \tag{8.17}$$

where we have used the fact that $1 \text{ J} = 1 \text{ kg m}^2 \text{ s}^{-2}$ (Appendix B).

We next examine some of the details of energy transfer in Compton scattering. The kinetic energy acquired by the secondary electron is given by

$$T = h\nu - h\nu'. \tag{8.18}$$

Substituting (8.12) for $h\nu'$ and carrying out a few algebraic manipulations, one obtains (Problem 13)

$$T = h\nu \frac{1 - \cos\theta}{mc^2/h\nu + 1 - \cos\theta}. \tag{8.19}$$

The maximum kinetic energy, $T_{max}$, that a secondary electron can acquire occurs when $\theta = 180°$. In this case, Eq. (8.19) gives

$$T_{max} = \frac{2h\nu}{2 + mc^2/h\nu}. \tag{8.20}$$

When the photon energy becomes very large compared with $mc^2$, $T_{max}$ approaches $h\nu$.

The recoil angle of the electron, $\varphi$ in Fig. 8.4, is related to $h\nu$ and $\theta$. Using Eqs. (8.10) and (8.11) together with the trigonometric identity $\sin\varphi/\cos\varphi = \tan\varphi$, one obtains

$$\tan\varphi = \frac{h\nu' \sin\theta}{h\nu - h\nu' \cos\theta}. \tag{8.21}$$

Substituting from Eq. (8.12) for $hv'$ gives

$$\tan \varphi = \frac{\sin \theta}{(1 + hv/mc^2)(1 - \cos \theta)}. \tag{8.22}$$

The trigonometric term in $\theta$ can be conveniently expressed in terms of the half-angle. Since $\sin \theta = 2 \sin(\theta/2) \cos(\theta/2)$ and $1 - \cos \theta = 2 \sin^2(\theta/2)$, it reduces to

$$\frac{\sin \theta}{1 - \cos \theta} = \frac{2 \sin(\theta/2) \cos(\theta/2)}{2 \sin^2(\theta/2)} = \cot \frac{\theta}{2}. \tag{8.23}$$

Equation (8.22) can thus be written in the compact form

$$\cot \frac{\theta}{2} = \left(1 + \frac{hv}{mc^2}\right) \tan \varphi. \tag{8.24}$$

When $\theta$ is small, $\cot \theta/2$ is large and $\varphi$ is near 90°. In this case, the photon travels in the forward direction, imparting relatively little energy to the electron, which moves off nearly at right angles to the direction of the incident photon. As $\theta$ increases from 0° to 180°, $\cot \theta/2$ decreases from $\infty$ to 0. Therefore, $\varphi$ decreases from 90° to 0°. The electron recoil angle $\varphi$ in Fig. 8.4 is thus always confined to the forward direction ($0 \leq \varphi \leq 90°$), whereas the photon can be scattered in any direction.

*Example*

In the previous example a 1.332-MeV photon from $^{60}$Co was scattered by an electron at an angle of 140°. Calculate the energy acquired by the recoil electron. What is the recoil angle of the electron? What is the maximum fraction of its energy that this photon could lose in a single Compton scattering?

*Solution*

Substitution into Eq. (8.19) gives the electron recoil energy,

$$T = 1.332 \frac{1 - (-0.766)}{0.511/1.332 + 1 - (-0.766)} = 1.094 \text{ MeV}. \tag{8.25}$$

Note from Eq. (8.15) that $T + hv' = 1.332 \text{ MeV} = hv$, as it should. The angle of recoil of the electron can be found from Eq. (8.24). We have

$$\tan \varphi = \frac{\cot (140°/2)}{1 + 1.332/0.511} = 0.101, \tag{8.26}$$

from which it follows that $\varphi = 5.76°$. This is a relatively hard collision in which the photon is backscattered, retaining only the fraction $0.238/1.332 = 0.179$ of its energy and knocking the electron in the forward direction. From Eq. (8.20),

$$T_{max} = \frac{2 \times 1.332}{2 + 0.511/1.332} = 1.118 \text{ MeV}. \tag{8.27}$$

The maximum fractional energy loss is $T_{max}/hv = 1.118/1.332 = 0.839$.

All of the details that we have worked out thus far for Compton scattering follow kinematically from the energy and momentum conservation requirements expressed by Eqs. (8.9)–(8.11). We have said nothing about how the photon and electron interact or about the *probability* that the photon will be scattered in the direction $\theta$. The quantum-mechanical theory of Compton scattering, based on the specific photon–electron interaction, gives for the angular distribution of scattered photons the Klein–Nishina formula

$$\frac{d_e\sigma}{d\Omega} = \frac{k_0^2 e^4}{2m^2c^4}\left(\frac{\nu'}{\nu}\right)^2\left(\frac{\nu}{\nu'} + \frac{\nu'}{\nu} - \sin^2\theta\right) \text{ m}^2 \text{ sr}^{-1}.$$

(8.28)

Here $d_e\sigma/d\Omega$, called the *differential scattering cross section*, is the probability per unit solid angle in steradians (sr) that a photon, passing normally through a layer of material containing one electron m$^{-2}$, will be scattered into a solid angle $d\Omega$ at angle $\theta$. The integral of the differential cross section over all solid angles, $d\Omega = 2\pi \sin\theta d\theta$, is called the *Compton collision cross section*. It gives the probability $_e\sigma$ that the photon will have a Compton interaction per electron m$^{-2}$:

$$_e\sigma = 2\pi \int \frac{d_e\sigma}{d\Omega} \sin\theta d\theta \text{ m}^2.$$

(8.29)

The Compton cross section, $_e\sigma$, can be thought of as the cross-sectional area, like that of a target, presented to a photon for interaction by one electron m$^{-2}$. It is thus rigorously the stated interaction *probability*. However, the area $_e\sigma$, which depends on the energy of the photon, is not the physical size of the electron.

The dependence of the differential cross section (8.28) on $\theta$ can be written explicitly with the help of Eq. (8.12) (Problem 24). Equation (8.28) can then be used with the kinematic equations we derived to calculate various quantities of interest. For example, the energy spectrum of Compton recoil electrons produced by 1-MeV photons is shown in Fig. 8.5. The relative number of recoil electrons decreases from $T = 0$ until it begins to rise rapidly as $T$ approaches $T_{max} = 0.796$ MeV, where the spectrum has its maximum value (called the *Compton edge* in gamma-ray spectroscopy). The most probable collisions are those that transfer relatively large amounts of energy. The Compton electron energy spectra are similar in shape for photons of other energies.

Of special importance for dosimetry is the average recoil energy, $T_{avg}$, of Compton electrons. For photons of a given energy $h\nu$, one can write the differential Klein–Nishina energy-transfer cross section (per electron m$^{-2}$)

$$\frac{d_e\sigma_{tr}}{d\Omega} = \frac{T}{h\nu}\frac{d_e\sigma}{d\Omega}.$$

(8.30)

The average recoil energy is then given by

$$T_{avg} = h\nu \frac{_e\sigma_{tr}}{_e\sigma},$$

(8.31)

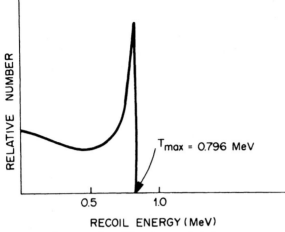

Fig. 8.5 Relative number of Compton recoil electrons as a function of their energy for 1-MeV photons.

where $_e\sigma$ is given by Eq. (8.29) and $_e\sigma_{tr}$ is the Compton energy-transfer cross section (per electron $m^{-2}$). The latter is found by integrating the differential cross section (8.30) over all solid angles:

$$_e\sigma_{tr} = 2\pi \int \frac{d_e\sigma_{tr}}{d\Omega} \sin\theta \, d\theta. \tag{8.32}$$

This cross section gives the average fraction of the incident photon energy that is transferred to Compton electrons per electron $m^{-2}$ in the material traversed. Table 8.1 shows values of $T_{avg}$ for a range of photon energies. Also shown is the fraction of the incident photon energy that is converted into the kinetic energy of the Compton electrons. This fraction increases steadily with photon energy.

Similarly, the differential cross section for energy scattering (i.e., the energy carried by the scattered photons) is defined by writing

$$\frac{d_e\sigma_s}{d\Omega} = \frac{\nu'}{\nu} \frac{d_e\sigma}{d\Omega}. \tag{8.33}$$

The average energy of the scattered photons for incident photons of energy $h\nu$ is

$$(h\nu')_{avg} = h\nu \frac{_e\sigma_s}{_e\sigma}. \tag{8.34}$$

Here the ratio of $_e\sigma_s$, obtained by integration of both sides of Eq. (8.33) over all solid angles, and $_e\sigma$ gives the fraction of the incident photon energy that is Compton scattered per electron $m^{-2}$. Since

$$\frac{T_{avg}}{h\nu} + \frac{(h\nu')_{avg}}{h\nu} = 1, \tag{8.35}$$

**Table 8.1** Average Kinetic Energy, $T_{avg}$, of Compton Recoil Electrons and Fraction of Incident Photon Energy, $h\nu$

| Photon Energy $h\nu$ (MeV) | Average Recoil Electron Energy $T_{avg}$ (MeV) | Average Fraction of Incident Energy $T_{avg}/h\nu$ |
|---|---|---|
| 0.01 | 0.0002 | 0.0187 |
| 0.02 | 0.0007 | 0.0361 |
| 0.04 | 0.0027 | 0.0667 |
| 0.06 | 0.0056 | 0.0938 |
| 0.08 | 0.0094 | 0.117 |
| 0.10 | 0.0138 | 0.138 |
| 0.20 | 0.0432 | 0.216 |
| 0.40 | 0.124 | 0.310 |
| 0.60 | 0.221 | 0.368 |
| 0.80 | 0.327 | 0.409 |
| 1.00 | 0.440 | 0.440 |
| 2.00 | 1.06 | 0.531 |
| 4.00 | 2.43 | 0.607 |
| 6.00 | 3.86 | 0.644 |
| 8.00 | 5.34 | 0.667 |
| 10.0 | 6.84 | 0.684 |
| 20.0 | 14.5 | 0.727 |
| 40.0 | 30.4 | 0.760 |
| 60.0 | 46.6 | 0.776 |
| 80.0 | 62.9 | 0.787 |
| 100.0 | 79.4 | 0.794 |

the Compton collision cross section is the sum of the energy-transfer and energy-scattering cross sections. Thus, it follows from Eqs. (8.31), (8.34), and (8.35) that

$$_e\sigma = {}_e\sigma\left[\frac{T_{avg}}{h\nu} + \frac{(h\nu')_{avg}}{h\nu}\right] = {}_e\sigma\left[\frac{_e\sigma_{tr}}{_e\sigma} + \frac{_e\sigma_s}{_e\sigma}\right] = {}_e\sigma_{tr} + {}_e\sigma_s. \tag{8.36}$$

If the material traversed consists of $N$ atoms m$^{-3}$ of an element of atomic number $Z$, then the number of electrons m$^{-3}$ is $n = NZ$. The Compton interaction probability per unit distance of travel for the photon in the material is then

$$\sigma = NZ_e\sigma = n_e\sigma. \tag{8.37}$$

The quantity $Z_e\sigma$ is the Compton collision cross section per atom, and $\sigma$ is the Compton macroscopic cross section, or attenuation coefficient, having the dimensions of inverse length. If the material is a compound or mixture, then the cross sections of the individual elements contribute additively to $\sigma$ as $NZ_e\sigma$.

The (Compton) linear attenuation coefficients for energy transfer and energy scattering can be obtained from Eqs. (8.31) and (8.34) after multiplication by the density of electrons. They are

$$\sigma_{tr} = \sigma\frac{T_{avg}}{h\nu} \tag{8.38}$$

and

$$\sigma_s = \sigma \frac{(h\nu')_{avg}}{h\nu}. \tag{8.39}$$

From Eq. (8.36), the (total) Compton attenuation coefficient can be written

$$\sigma = \sigma_{tr} + \sigma_s. \tag{8.40}$$

### Example

The Compton collision cross section for the interaction of an 8-MeV photon with an electron is $5.99 \times 10^{-30}$ m$^2$. For water, find the following quantities for Compton scattering: (a) the collision cross section per molecule; (b) the linear attenuation coefficient; (c) the linear attenuation coefficient for energy transfer; and (d) the linear attenuation coefficient for energy scattering.

### Solution

(a) The Compton collision cross section, defined by Eq. (8.29), is $_e\sigma = 5.99 \times 10^{-30}$ m$^2$. Since there are 10 electrons in a water molecule, the Compton cross section per molecule is $10_e\sigma = 5.99 \times 10^{-29}$ m$^2$. [See discussion in connection with Eq. (8.37).]
(b) Using Eq. (8.37) with $n = 3.34 \times 10^{29}$ m$^{-3}$ (calculated in Section 5.9), we find for the linear attenuation coefficient, $\sigma = n_e\sigma = (3.44 \times 10^{29}$ m$^{-3}) \times (5.99 \times 10^{-30}$ m$^2) = 2.06$ m$^{-1}$.
(c) In Table 8.1 we find that $T_{avg}/h\nu = 0.667$ for 8-MeV photons. It follows from Eq. (8.38) that the linear attenuation coefficient for energy transfer is $\sigma_{tr} = 0.667\sigma = 0.667 \times 2.06 = 1.37$ m$^{-1}$.
(d) From Eq. (8.39) we find that $\sigma_s = 0.333\sigma = 0.333 \times 2.06 = 0.686$ m$^{-1}$. Alternatively, one can use Eq. (8.40) and the answer from (c): $\sigma_s = \sigma - \sigma_{tr} = 2.06 - 1.37 = 0.69$ m$^{-1}$.

Like the photoelectric effect, the Compton effect gave confirmation of the corpuscular nature of light. The discovery of quantum mechanics followed Compton's experiments by a few years. Modern quantum electrodynamics accounts very successfully for the dual wave–photon nature of electromagnetic radiation.

## 8.5
## Pair Production

A photon with an energy of at least twice the electron rest energy, $h\nu \geq 2mc^2$, can be converted into an electron–positron pair in the field of an atomic nucleus. Pair production can also occur in the field of an atomic electron, but the probability is considerably smaller and the threshold energy is $4mc^2$. (This process is often referred to as "triplet" production because of the presence of the recoiling atomic electron in addition to the pair.) When pair production occurs in a nuclear field, the massive nucleus recoils with negligible energy. Therefore, the photon energy $h\nu$ is converted into $2mc^2$ plus the kinetic energies $T_+$ and $T_-$ of the partners:

$$h\nu = 2mc^2 + T_+ + T_-. \tag{8.41}$$

The distribution of the excess energy between the electron and positron is continuous; that is, the kinetic energy of either can vary from zero to a maximum of $h\nu - 2mc^2$. Furthermore, the energy spectra are almost the same for the two particles and depend on the atomic number of the nucleus. The threshold photon wavelength for pair production is 0.012 Å (Problem 28). Pair production becomes more likely with increasing photon energy, and the probability increases with atomic number approximately as $Z^2$.

The inverse process also occurs when an electron and positron annihilate to produce photons. A positron can annihilate in flight, although it is more likely first to slow down, attract an electron, and form positronium. Positronium is the bound system, analogous to the hydrogen atom, formed by an electron–positron pair orbiting about their mutual center of mass. Positronium exists for $\sim 10^{-10}$ s before the electron and positron annihilate. Since the total momentum of positronium before decay is zero, at least two photons must be produced in order to conserve momentum. The most likely event is the creation of two 0.511-MeV photons going off in opposite directions. If the positron annihilates in flight, then the total photon energy will be $2mc^2$ plus its kinetic energy. Three photons are occasionally produced. The presence of 0.511-MeV annihilation photons around any positron source is always a potential radiation hazard.

## 8.6
### Photonuclear Reactions

A photon can be absorbed by an atomic nucleus and knock out a nucleon. This process is called photodisintegration. An example is gamma-ray capture by a $^{206}_{82}$Pb nucleus with emission of a neutron: $^{206}_{82}$Pb$(\gamma, n)^{205}_{82}$Pb. The photon must have enough energy to overcome the binding energy of the ejected nucleon, which is generally several MeV. Like the photoelectric effect, photodisintegration can occur only at photon energies above a threshold value. The kinetic energy of the ejected nucleon is equal to the photon energy minus the nucleon's binding energy.

The probability for photonuclear reactions is orders of magnitude smaller than the combined probabilities for the photoelectric effect, Compton effect, and pair production. However, unlike these processes, photonuclear reactions can produce neutrons, which often pose special radiation-protection problems. In addition, residual nuclei following photonuclear reactions are often radioactive. For these reasons, photonuclear reactions can be important around high-energy electron accelerators that produce energetic photons.

The thresholds for $(\gamma, p)$ reactions are often higher than those for $(\gamma, n)$ reactions because of the repulsive Coulomb barrier that a proton must overcome to escape from the nucleus (Fig. 3.1). Although the probability for either reaction is about the same in the lightest elements, the $(\gamma, n)$ reaction is many times more probable than $(\gamma, p)$ in heavy elements.

Other photonuclear reactions also take place. Two-nucleon knock-out reactions such as $(\gamma, 2n)$ and $(\gamma, np)$ occur, as well as $(\gamma, \alpha)$ reactions. Photon absorption can also induce fission in heavy nuclei.

*Example*

Compute the threshold energy for the $(\gamma, n)$ photodisintegration of $^{206}$Pb. What is the energy of a neutron produced by absorption of a 10-MeV photon?

*Solution*

The mass differences, $\Delta$, from Appendix D, are −23.79 MeV for $^{206}$Pb, −23.77 MeV for $^{205}$Pb, and 8.07 MeV for the neutron. The mass difference after the reaction is −23.77 + 8.07 = −15.70 MeV. The threshold energy needed to remove the neutron from $^{206}$Pb is therefore −15.70 − (−23.79) = 8.09 MeV. Absorption of a 10-MeV photon produces a neutron and recoil $^{205}$Pb nucleus with a total kinetic energy of 10 − 8.09 = 1.91 MeV. The absorbed photon contributes negligible momentum. In analogy with Eq. (3.18) for alpha decay, the energy of the neutron is $(1.91 \times 205)/206 = 1.90$ MeV.

## 8.7
## Attenuation Coefficients

As pointed out at the beginning of this chapter, photon penetration in matter is governed statistically by the probability per unit distance traveled that a photon interacts by one physical process or another. This probability, denoted by $\mu$, is called the linear attenuation coefficient (or macroscopic cross section) and has the dimensions of inverse length (e.g., cm$^{-1}$). The coefficient $\mu$ depends on photon energy and on the material being traversed. The mass attenuation coefficient $\mu/\rho$ is obtained by dividing $\mu$ by the density $\rho$ of the material. It is usually expressed in cm$^2$ g$^{-1}$, and represents the probability of an interaction per g cm$^{-2}$ of material traversed.

Monoenergetic photons are attenuated exponentially in a uniform target, as we now show. Figure 8.6 represents a narrow beam of $N_0$ monoenergetic photons incident normally on a slab. As the beam penetrates the absorber, some photons can be scattered and some absorbed. We let $N(x)$ represent the number of photons that reach a depth $x$ without having interacted. The number that interact within the next small distance d$x$ is proportional to $N$ and to d$x$. Thus we may write

$$dN = -\mu N dx, \tag{8.42}$$

where the constant of proportionality $\mu$ is the linear attenuation coefficient. The solution is[1]

$$N(x) = N_0 e^{-\mu x}. \tag{8.43}$$

---

1   Also, in analogy with Eq. (4.22), $1/\mu$ is the average distance traveled by a photon before interacting; hence $\mu$ is also called the inverse mean free path. This relationship was described for charged particles at the end of Section 5.4.

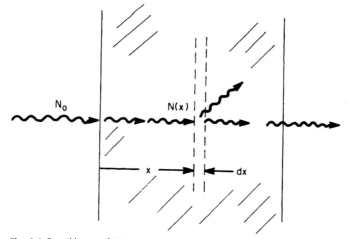

**Fig. 8.6** Pencil beam of $N_0$ monoenergetic photons incident on slab. The number of photons that reach a depth $x$ without having an interaction is given by $N(x) = N_0 e^{-\mu x}$, where $\mu$ is the linear attenuation coefficient.

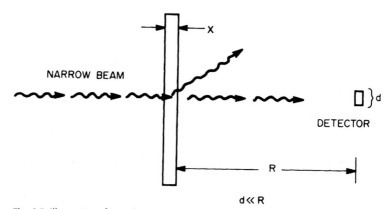

**Fig. 8.7** Illustration of "good" scattering geometry for measuring linear attenuation coefficient $\mu$. Photons from a *narrow* beam that are absorbed or scattered by the absorber do not reach a small detector placed in beam line some distance away.

It follows that $e^{-\mu x}$ is just the probability (i.e., $N/N_0$) that a normally incident photon will traverse a slab of thickness $x$ without interacting. The factor $e^{-\mu x}$ thus generally describes the fraction of "uncollided photons" that go through a shield.

The linear attenuation coefficient can be measured by the experimental arrangement shown in Fig. 8.7. A narrow beam of monoenergetic photons is directed toward an absorbing slab of thickness $x$. A small detector of size $d$ is placed at a distance $R \gg d$ behind the slab directly in the beam line. Under these conditions, referred to as "narrow-beam" or "good" scattering geometry, only photons that traverse the slab without interacting will be detected. One can measure the relative rate at which photons reach the detector as a function of the absorber thickness and then use Eq. (8.43) to obtain the value of $\mu$ (Problem 33).

The linear attenuation coefficient for photons of a given energy in a given material comprises the individual contributions from the various physical processes that can remove photons from the narrow beam in Fig. 8.7. We write

$$\mu = \tau + \sigma + \kappa, \tag{8.44}$$

where $\tau$, $\sigma$, and $\kappa$ denote, respectively, the linear attenuation coefficients for the photoelectric effect, Compton effect [Eq. (8.40)], and pair production. The respective mass attenuation coefficients are $\tau/\rho$, $\sigma/\rho$, and $\kappa/\rho$ for a material of density $\rho$. We could also add the (usually) small contributions to the attenuation due to photonuclear reactions and the Raleigh scattering, but we are neglecting these.

Figures 8.8 and 8.9 give the mass attenuation coefficients for five chemical elements and a number of materials for photons with energies from 0.010 MeV to 100 MeV. The structure of these curves reflects the physical processes we have been discussing. At low photon energies the binding of the atomic electrons is important and the photoelectric effect is the dominant interaction. High-Z materials provide greater attenuation and absorption, which decrease rapidly with increasing photon energy. The coefficients for Pb and U rise abruptly when the photon energy is suffi-

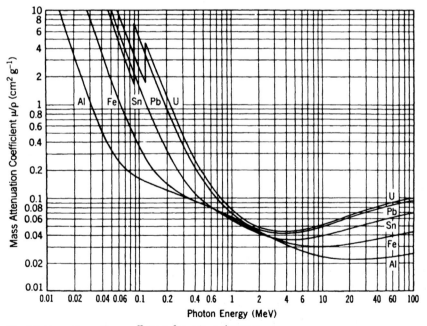

**Fig. 8.8** Mass attenuation coefficients for various elements. [Reprinted with permission from K. Z. Morgan and J. E. Turner, eds., *Principles of Radiation Protection*, Wiley, New York (1967). Copyright 1967 by John Wiley & Sons.]

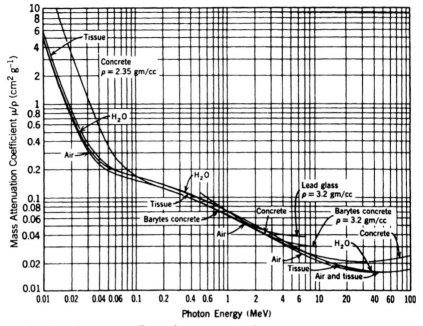

**Fig. 8.9** Mass attenuation coefficients for various materials.
[Reprinted with permission from K. Z. Morgan and J. E. Turner,
eds., *Principles of Radiation Protection*, Wiley, New York (1967).
Copyright 1967 by John Wiley & Sons.]

cient to eject a photoelectron from the K shell of the atom. The curves for the other
elements show the same structure at lower energies. When the photon energy is
several hundred keV or greater, the binding of the atomic electrons becomes relatively unimportant and the dominant interaction is Compton scattering. Since the
elements (except hydrogen) contain about the same number of electrons per unit
mass, there is not a large difference between the values of the mass attenuation
coefficients for the different materials. Compton scattering continues to be important above the 1.022-MeV pair-production threshold until the latter process takes
over as the more probable. Attenuation by pair production is enhanced by a large
nuclear charge of the absorber.

### Example

What thickness of concrete and of lead are needed to reduce the number of 500-
keV photons in a narrow beam to one-fourth the incident number? Compare the
thicknesses in cm and in g cm$^{-2}$. Repeat for 1.5-MeV photons.

### Solution

We use Eq. (8.43) with $N(x)/N_0 = 0.25$. The mass attenuation coefficients $\mu/\rho$ obtained from Figs. 8.8 and 8.9 are shown in Table 8.2. At 500 keV, the linear attenu-

ation coefficient for concrete is $\mu = (0.089 \text{ cm}^2 \text{ g}^{-1}) \ (2.35 \text{ g cm}^{-3}) = 0.209 \text{ cm}^{-1}$; that for lead is $\mu = (0.15)(11.4) = 1.71 \text{ cm}^{-1}$. Using Eq. (8.43), we have for concrete

$$0.25 = e^{-0.209x}, \tag{8.45}$$

giving $x = 6.63$ cm. For lead,

$$0.25 = e^{-1.71x}, \tag{8.46}$$

giving $x = 0.811$ cm. The concrete shield is thicker by a factor of $6.63/0.811 = 8.18$. In $\text{g cm}^{-2}$, the concrete thickness is $6.63 \text{ cm} \times 2.35 \text{ g cm}^{-3} = 15.6 \text{ g cm}^{-2}$, while that for lead is $0.811 \times 11.4 = 9.25 \text{ g cm}^{-2}$. The concrete shield is more massive in thickness by a factor of $15.6/9.25 = 1.69$. Lead is a more efficient attenuator than concrete for 500-keV photons on the basis of mass. Photoelectric absorption is important at this energy, and the higher atomic number of lead is effective. The calculation can be repeated in exactly the same way for 1.5-MeV photons. Instead, we do it a little differently by using the mass attenuation coefficient directly, writing the exponent in Eq. (8.43) as $\mu x = (\mu/\rho)\rho x$. For 1.5-MeV photons incident on concrete,

$$0.25 = e^{-0.052\rho x}, \tag{8.47}$$

giving $\rho x = 26.7 \text{ g cm}^{-2}$ and $x = 11.4$ cm. For lead,

$$0.25 = e^{-0.051\rho x}, \tag{8.48}$$

and so $\rho x = 21.2 \text{ g cm}^{-2}$ and $x = 2.39$ cm. At this energy the Compton effect is the principal interaction that attenuates the beam, and therefore all materials (except hydrogen) give comparable attenuation per g $\text{cm}^{-2}$. Lead is almost universally used when low-energy photon shielding is required. It can be used to line the walls of X-ray rooms, be incorporated into aprons worn by personnel around X-ray equipment, and be fabricated into containers for gamma sources. Lead bricks also afford a convenient and effective way to erect shielding. The design of shielding is described in Chapter 15.

As discussed in connection with Eq. (8.37), the attenuation coefficient for Compton scattering in an elemental medium can be expressed as the product of the number of atoms per unit volume and the Compton cross section per atom. The same kind of relationship holds between the density of atoms and atomic cross sections for the photoelectric effect and pair production. (Unlike the Compton effect, however, the atomic cross sections for the latter two interactions are physically

**Table 8.2** Mass Attenuation Coefficients

| | $\mu/\rho \ (\text{cm}^2 \text{ g}^{-1})$ | |
| --- | --- | --- |
| $h\nu$ | Concrete $\rho = 2.35 \text{ g cm}^{-3}$ | Pb $\rho = 11.4 \text{ g cm}^{-3}$ |
| 500 keV | 0.089 | 0.15 |
| 1.5 MeV | 0.052 | 0.051 |

not the sums of separate electronic cross sections, like $_e\sigma$.) The linear attenuation coefficient $\mu$, given by Eq. (8.44), is also equal to the product of the atomic density $N_A$ and the total atomic cross section $\sigma_A$ for all processes:

$$\mu = N_A \sigma_A. \tag{8.49}$$

The number of atoms cm$^{-3}$ of an element is given by $N_A = (\rho/A)N_0$, where $\rho$ is the density of the material in g cm$^{-3}$, $A$ is the gram atomic weight, and $N_0$ is Avogadro's number. Thus, we can write $\mu = \rho N_0 \sigma_A / A$, or

$$\frac{\mu}{\rho} = \frac{N_0 \sigma_A}{A}, \tag{8.50}$$

giving the relationship between the mass attenuation coefficient and the atomic cross section for photon interaction with any element. For a compound or mixture, one can add the separate contributions from each element to obtain $\mu$.

Cross sections are often expressed in the unit, barn, where 1 barn $= 10^{-24}$ cm$^2$.

*Example*

What is the atomic cross section of lead for 500-keV photons?

*Solution*

From Fig. 8.8, the mass attenuation coefficient is $\mu/\rho = 0.16$ cm$^2$ g$^{-1}$. The gram atomic weight of lead is 207 g. We find from Eq. (8.50) that

$$\sigma_A = \left(\frac{\mu}{\rho}\right)\left(\frac{A}{N_0}\right) = (0.16 \text{ cm}^2 \text{ g}^{-1})\left(\frac{207 \text{ g}}{6.02 \times 10^{23}}\right) = 5.50 \times 10^{-23} \text{ cm}^2. \tag{8.51}$$

Alternatively, $\sigma_A = 55.0$ barn.

## 8.8
### Energy-Transfer and Energy-Absorption Coefficients

In dosimetry we are interested in the energy absorbed in matter exposed to photons. This energy is related to the linear attenuation coefficients given in Eq. (8.44). However, some care is needed in making the connections.

Figure 8.10 shows a uniform, broad, parallel beam of monoenergetic photons normally incident on an absorber of thickness $x$. The incident fluence $\Phi_0$ is the number of photons per unit area that cross a plane perpendicular to the beam.[2] The number that cross per unit area per unit time at any instant is called the *fluence rate*, or flux density: $\dot\Phi_0 = d\Phi_0/dt$ $(= \varphi_0)$. Examples of units for $\Phi_0$ and $\dot\Phi_0$ are, respectively, m$^{-2}$ and m$^{-2}$ s$^{-1}$. The energy that passes per unit area is called the *energy fluence* $\Psi_0$, having the units J m$^{-2}$. The corresponding instantaneous rate of energy flow per unit area per unit time is the energy fluence rate, or energy

---

2   Our notation is consistent with that of ICRU
    Report 60, listed in Section 8.10.

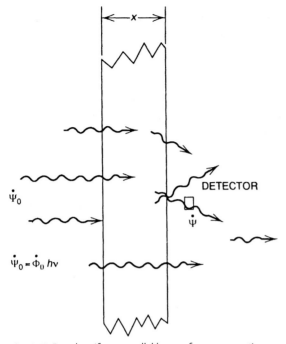

**Fig. 8.10** Broad, uniform, parallel beam of monoenergetic photons normally incident on an absorber of thicknesses $x$. Incident energy fluence rate is $\dot{\Psi}_0$, and transmitted energy fluence rate is $\dot{\Psi}$.

flux density: $\dot{\Psi}_0 = d\Psi_0/dt \ (=\psi_0)$. This quantity is also called the intensity. For the special beam in Fig. 8.10,

$$\Psi_0 = \Phi_0 h\nu \quad \text{and} \quad \dot{\Psi}_0 = \dot{\Phi}_0 h\nu. \tag{8.52}$$

Energy fluence rate (intensity) can be expressed in $J\,m^{-2}\,s^{-1} = W\,m^{-2}$.

To infer the rate of energy absorption in the slab from the uniform beam in Fig. 8.10, one can compare the intensity $\dot{\Psi}$ of the radiation reaching a detector placed right behind the slab to the incident intensity $\dot{\Psi}_0$. Figures 8.7 and 8.10 together indicate that, under the broad-beam conditions, the detector receives uncollided as well as scattered and other (e.g., bremsstrahlung and fluorescence) photons. Thus, not all of the energy of the incoming photons that interact in the slab is necessarily absorbed there. The decrease in beam intensity with increasing $x$ can be expected to be *less* than that described by the linear attenuation coefficient, $e^{-\mu x}$. We consider each of the principal energy-loss mechanisms in turn, discussing first the energy-transfer coefficient and then the energy-absorption coefficient.

In the photoelectric effect, absorption of a photon of energy $h\nu$ by an atom produces a secondary electron with initial kinetic energy $T = h\nu - B$, where $B$ is the binding energy of the ejected electron. Following ejection of the photoelectron, the inner-shell vacancy in the atom is immediately filled by an electron from an upper level. This and subsequent electronic transitions are accompanied by the simulta-neous emission of photons or Auger electrons (Sections 2.10 and 2.11). The frac-tion of the incident intensity transferred to electrons (i.e., the photoelectron and the Auger electrons) can be expressed as $1 - \delta/h\nu$, where $\delta$ is the average energy emitted as fluorescence radiation following photoelectric absorption in the material. Just as the mass attenuation coefficient $\tau/\rho$, defined after Eq. (8.44), describes the fraction of photons that interact by photoelectric absorption per $g\,cm^{-2}$ of matter traversed, the mass energy-transfer coefficient,

$$\frac{\tau_{tr}}{\rho} = \frac{\tau}{\rho}\left(1 - \frac{\delta}{h\nu}\right), \tag{8.53}$$

gives the fraction of the intensity that is transferred to electrons per $g\,cm^{-2}$. To the extent that the photoelectron and Auger electrons subsequently emit photons (as bremsstrahlung), the energy-*transfer* coefficient does not adequately describe energy *absorption* in the slab. We return to this point after defining the mass energy-transfer coefficients for Compton scattering and pair production.

For Compton scattering of monoenergetic photons (Fig. 8.4), the mass energy-transfer coefficient follows directly from Eq. (8.38):

$$\frac{\sigma_{tr}}{\rho} = \frac{\sigma}{\rho}\frac{T_{avg}}{h\nu}. \tag{8.54}$$

The factor $T_{avg}/h\nu$ gives the average fraction of the incident photon energy that is converted into the initial kinetic energy of the Compton electrons. As with the photoelectric effect, the energy-transfer coefficient (8.54) takes no account of sub-sequent bremsstrahlung by the Compton electrons.

A photon of energy $h\nu$ produces an electron–positron pair with a total initial kinetic energy $h\nu - 2mc^2$, where $2mc^2$ is the rest energy of the pair [Eq. (8.41)]. Therefore, the mass energy-transfer coefficient for pair production is related to the mass attenuation coefficient, defined after Eq. (8.44), as follows:

$$\frac{\kappa_{tr}}{\rho} = \frac{\kappa}{\rho}\left(1 - \frac{2mc^2}{h\nu}\right). \tag{8.55}$$

This relationship applies to pair production in the field of an atomic nucleus; we neglect the small contribution from triplet production (Section 8.5).

The total mass energy-transfer coefficient $\mu_{tr}/\rho$ for photons of energy $h\nu$ in a given material is found by combining the last three equations:

$$\frac{\mu_{tr}}{\rho} = \frac{\tau_{tr}}{\rho} + \frac{\sigma_{tr}}{\rho} + \frac{\kappa_{tr}}{\rho} \tag{8.56}$$

$$= \frac{\tau}{\rho}\left(1 - \frac{\delta}{h\nu}\right) + \frac{\sigma}{\rho}\left(\frac{T_{avg}}{h\nu}\right) + \frac{\kappa}{\rho}\left(1 - \frac{2mc^2}{h\nu}\right). \tag{8.57}$$

This coefficient determines the total initial kinetic energy of all electrons produced by the photons, both directly (as in photoelectric absorption, Compton scattering,

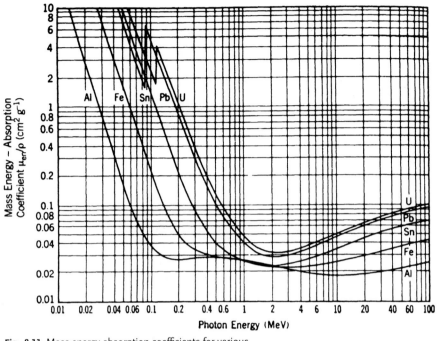

**Fig. 8.11** Mass energy-absorption coefficients for various elements. [Reprinted with permission from K. Z. Morgan and J. E. Turner, eds., *Principles of Radiation Protection*, Wiley, New York (1967). Copyright 1967 by John Wiley & Sons.]

and pair production) and indirectly (as Auger electrons). Except for the subsequent bremsstrahlung that the electrons might emit, the energy absorbed in the immediate vicinity of the interaction site would be the same as the energy transferred there.[3]

Letting g represent the average fraction of the initial kinetic energy transferred to electrons that is subsequently emitted as bremsstrahlung, one defines the mass energy-absorption coefficient as

$$\frac{\mu_{en}}{\rho} = \frac{\mu_{tr}}{\rho}(1-g).$$ (8.58)

Generally, the factor g is largest for materials having high atomic number and for photons of high energy. Figures 8.11 and 8.12 show the mass energy-absorption coefficients for the elements and other materials we considered earlier.

Differences in the various coefficients are illustrated by the data shown for water and lead in Table 8.3. It is seen that bremsstrahlung is relatively unimportant in water for photon energies less than ~10 MeV (i.e., $\mu_{tr}/\rho$ is not much larger than $\mu_{en}/\rho$). In lead, on the other hand, bremsstrahlung accounts for a significant differ-

---

3  A positron that annihilates in flight will also cause the absorbed energy to be less than the energy transferred. We ignore this usually small effect.

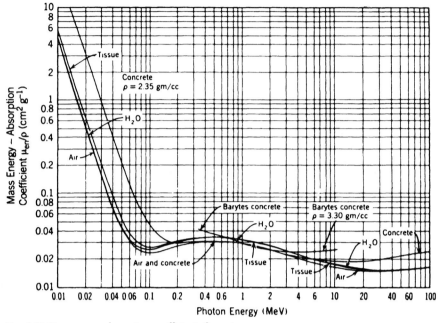

**Fig. 8.12** Mass energy-absorption coefficients for various materials. [Reprinted with permission from K. Z. Morgan and J. E. Turner, eds., *Principles of Radiation Protection*, Wiley, New York (1967). Copyright 1967 by John Wiley & Sons.]

**Table 8.3** Mass Attenuation, Mass Energy-Transfer, and Mass Energy-Absorption Coefficients (cm$^2$ g$^{-1}$) for Photons in Water and Lead

| Photon Energy | Water | | | Lead | | |
|---|---|---|---|---|---|---|
| (MeV) | $\mu/\rho$ | $\mu_{tr}/\rho$ | $\mu_{en}/\rho$ | $\mu/\rho$ | $\mu_{tr}/\rho$ | $\mu_{en}/\rho$ |
| 0.01 | 5.33 | 4.95 | 4.95 | 131. | 126. | 126. |
| 0.10 | 0.171 | 0.0255 | 0.0255 | 5.55 | 2.16 | 2.16 |
| 1.0 | 0.0708 | 0.0311 | 0.0310 | 0.0710 | 0.0389 | 0.0379 |
| 10.0 | 0.0222 | 0.0163 | 0.0157 | 0.0497 | 0.0418 | 0.0325 |
| 100.0 | 0.0173 | 0.0167 | 0.0122 | 0.0931 | 0.0918 | 0.0323 |

*Source*: Based on P. D. Higgins, F. H. Attix, J. H. Hubbell, S. M. Seltzer, M. J. Berger, and C. H. Sibata, *Mass Energy-Transfer and Mass Energy-Absorption Coefficients, Including In-Flight Positron Annihilation for Photon Energies 1 keV to 100 MeV*, NISTIR 4680, National Institute of Standards and Technology, Gaithersburg, MD (1991).

ence in the mass energy-transfer and mass energy-absorption coefficients at much lower energies.

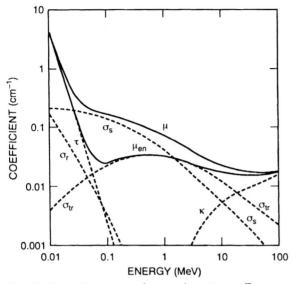

**Fig. 8.13** Linear attenuation and energy-absorption coefficients as functions of energy for photons in water.

It is instructive to see how the individual physical processes contribute to the interaction coefficients as functions of the photon energy. Figure 8.13 for water shows $\tau$, $\sigma_s$, $\sigma_{tr}$, and $\kappa$ as well as the coefficients $\mu$ and $\mu_{en}$. Also shown for comparison is the attenuation coefficient $\sigma_r$ for Raleigh scattering, which we have ignored. At the lowest energies (<15 keV), the photoelectric effect accounts for virtually all of the interaction. As the photon energy increases, $\tau$ drops rapidly and goes below $\sigma_s$. Between about 100 keV and 10 MeV, most of the attenuation in water is due to the Compton effect. Above about 1.5 MeV, $\sigma_{tr} > \sigma_s$. The Compton coefficients then fall off with increasing energy, and pair production becomes the dominant process at high energies.

### 8.9
### Calculation of Energy Absorption and Energy Transfer

It remains to show how the coefficients $\mu_{tr}$ and $\mu_{en}$ are used in computations. We consider again the idealized broad, parallel beam of monoenergetic photons in Fig. 8.10 and ask how one can determine the rate at which energy is absorbed in the slab, given the description of the incident photon field.

We begin by assuming that the slab is thin compared with the mean free paths of the incident and secondary photons, so that (1) multiple scattering of photons in the slab is negligible and (2) virtually all fluorescence and bremsstrahlung photons escape from it. On the other hand, we assume that the secondary electrons pro-

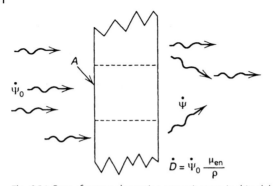

**Fig. 8.14** Rate of energy absorption per unit mass in thin slab (dose rate, $\dot{D}$) is equal to the product of the incident intensity and mass energy-absorption coefficient.

duced by the photons are stopped in the slab. Under these conditions, the transmitted intensity in Fig. 8.10 is given by

$$\dot{\psi} = \dot{\psi}_0 e^{-\mu_{en}x}. \tag{8.59}$$

For $\mu_{en}x \ll 1$, which is consistent with our assumptions, one can write $e^{-\mu_{en}x} \approx 1 - \mu_{en}x$. Equation (8.59) then implies that

$$\dot{\psi}_0 - \dot{\psi} = \dot{\psi}_0 \mu_{en}x. \tag{8.60}$$

With reference to Fig. 8.14, the rate at which energy is absorbed in the slab over an area $A$ is $(\dot{\psi}_0 - \dot{\psi})A = \dot{\psi}_0 \mu_{en}xA$. Since the mass of the slab over this area is $\rho Ax$, where $\rho$ is the density, the rate of energy absorption per unit mass, $\dot{D}$, in the slab is

$$\dot{D} = \frac{\dot{\psi}_0 \mu_{en}xA}{\rho Ax} = \dot{\psi}_0 \frac{\mu_{en}}{\rho}. \tag{8.61}$$

The quantity $\dot{D}$ is, by definition, the average dose rate in the slab. As discussed in Chapter 12, under the condition of electronic equilibrium Eq. (8.61) also implies that the dose rate at a point in a medium is equal to the product of the intensity, or energy fluence rate, at that point and the mass energy-absorption coefficient.

The mass energy-transfer coefficient can be employed in a similar derivation. The quantity thus obtained,

$$\dot{K} = \dot{\psi}_0 \frac{\mu_{tr}}{\rho}, \tag{8.62}$$

is called the average *kerma rate* in the slab. Equation (8.62) also gives the kerma rate at a point in a medium in terms of the energy fluence rate at that point, irrespective of electronic equilibrium. As described in Section 12.10, kerma is defined generally as the total initial kinetic energy of all charged particles liberated by uncharged radiation (photons and/or neutrons) per unit mass of material.

The ideal geometry and other conditions represented by Fig. 8.10 and Eq. (8.59) are approached in practice only to various degrees of approximation. The nonuniformity and finite width of real beams, for example, are two factors that usually deviate significantly from the ideal. The computation of attenuation and energy absorption in thick slabs is treated with the help of buildup factors (Charter 15). Nevertheless, $\mu_{en}$ is frequently useful for estimating absorbed energy in a number of situations, as the following examples illustrate.

*Example*

A parallel beam of 1-MeV photons is normally incident on a 1.2-cm aluminum slab ($\rho = 2.70$ g cm$^{-3}$) at a rate of $10^3$ s$^{-1}$. The mass attenuation and mass energy-absorption coefficients are, respectively, 0.0620 cm$^2$ g$^{-1}$ and 0.0270 cm$^2$ g$^{-1}$. (a) What fraction of the photons is transmitted without interacting? (b) What fraction of the incident photon energy is transmitted by the slab? (c) How much energy is absorbed per second by the slab? (d) What fraction of the transmitted energy is carried by the uncollided photons? (e) If the mass energy-transfer coefficient is 0.0271 cm$^2$ g$^{-1}$, what fraction of the initial kinetic energy transferred to the electrons in the slab is emitted as bremsstrahlung?

*Solution*

(a) The values given for the mass attenuation and mass energy-absorption coefficients can be checked against Figs. 8.8 and 8.11. The attenuation coefficient is $\mu = 0.0620 \times 2.70 = 0.167$ cm$^{-1}$. The fraction of photons that penetrate the 1.2-cm slab without interacting is therefore $e^{-(0.167 \times 1.2)} = 0.818$.

(b) The energy transmitted by the slab can be inferred from Eq. (8.59), irrespective of the beam width and uniformity, provided the slab can be regarded as "thin." This criterion is satisfied to a good approximation, since 0.818 of the incident photons do not interact ($\mu x = 0.200$). The energy absorption coefficient is $\mu_{en} = 0.0270 \times 2.70 = 0.0729$ cm$^{-1}$. It follows from Eq. (8.59) that the fraction of the incident photon energy that is transmitted by the slab is

$$\frac{\dot{E}}{\dot{E}_0} = e^{-\mu_{en}x} = e^{-0.0729 \times 1.2} = 0.916, \tag{8.63}$$

where $\dot{E}$ and $\dot{E}_0$ are the transmitted and incident rates of energy flow. Whereas the result (8.63) is approximate, the answer to part (a) is exact.

(c) The rate of energy absorption by the slab is the difference between the incident and transmitted rates. The result (8.63) implies that the fraction of the incident photon energy that is absorbed by the slab is 0.084. With the incident rate $\dot{E}_0 = (1.0$ MeV$) \times (10^3$ s$^{-1}) = 1.0 \times 10^3$ MeV s$^{-1}$, we find that the rate of energy absorption in the slab is $0.084 \dot{E}_0 = 84.0$ MeV s$^{-1}$.

(d) From part (a) it follows that the rate of energy transmission through the slab by the uncollided photons is (exactly)

$$\dot{E}_0 e^{-\mu x} = 1.0 \times 10^3 \times 0.818 = 818 \text{ MeV s}^{-1}. \tag{8.64}$$

Since 84.0 MeV s$^{-1}$ is absorbed by the slab, $1000 - 84 = 916$ MeV s$^{-1}$ is transmitted. The fraction of the total transmitted energy carried by the uncollided photons is therefore $818/916 = 0.893$.

(e) With $\mu_{en}/\rho = 0.0270$ cm$^2$ g$^{-1}$ and $\mu_{tr}/\rho = 0.0271$ cm$^2$ g$^{-1}$, we have $\mu_{en}/\mu_{tr} = 0.0270/0.0271 = 0.9963$. Equation (8.58) implies that the fraction of the secondary-electron initial kinetic energy that is emitted as bremsstrahlung is

$$g = 1 - \frac{\mu_{en}}{\mu_{tr}} = 1 - 0.9963 = 0.0037. \tag{8.65}$$

*Example*

A $^{137}$Cs source is stored in a laboratory. The photon fluence rate in air at a point in the neighborhood of the source is $5.14 \times 10^7$ m$^{-2}$ s$^{-1}$. Calculate the rate of energy absorption per unit mass (dose rate) in the air at that point.

*Solution*

The desired quantity is given by Eq. (8.61). The mass energy-absorption coefficient of air for the photons emitted by $^{137}$Cs ($h\nu = 0.662$ MeV, Appendix D) is, from Fig. 8.12, $\mu_{en}/\rho = 0.030$ cm$^2$ g$^{-1}$. The incident fluence rate is $\dot{\Phi} = 5.14 \times 10^7$ m$^{-2}$ s$^{-1}$, and so the energy fluence rate is

$$\dot{\Psi} = \dot{\Phi} h\nu = 3.40 \times 10^7 \text{ MeV m}^{-2} \text{ s}^{-1} = 3.40 \times 10^3 \text{ MeV cm}^{-2} \text{ s}^{-1}. \tag{8.66}$$

Thus, Eq. (8.61) gives

$$\dot{D} = \dot{\Psi} \frac{\mu_{en}}{\rho} = 102 \text{ MeV g}^{-1} \text{ s}^{-1}. \tag{8.67}$$

Expressed in SI units,

$$\dot{D} = \frac{102 \text{ MeV}}{\text{g s}} \times 1.60 \times 10^{-13} \frac{\text{J}}{\text{MeV}} \times 10^3 \frac{\text{g}}{\text{kg}} \tag{8.68}$$

$$= 1.63 \times 10^{-8} \text{ J kg}^{-1} \text{ s}^{-1} = 0.0587 \text{ mGy h}^{-1}. \tag{8.69}$$

The unit, J kg$^{-1}$, is called the gray (Gy). Note that the dose rate is independent of the temperature and pressure of the air. An exact balance occurs between the amount of energy absorbed and the amount of mass present, regardless of the other conditions. For a given material, the mass interaction coefficients (Figs. 8.8, 8.9, 8.11, and 8.12) are independent of the density. On the other hand, the value of the linear coefficients does depend on density. For example, Fig. 8.13 applies to water in the liquid phase (unit density), whereas the values given in Figs. 8.9 and 8.11 are the same for both liquid water and water vapor.

## 8.10
## Suggested Reading

1 Attix, F. H., *Introduction to Radiological Physics and Radiation Dosimetry*, Wiley, New York (1986). [Chapter 8 of this excellent text is devoted to X- and gamma-ray interactions with matter. In addition, the book deals thoroughly with virtually all aspects of photon interaction coefficients, covering the fine points. Appendix D gives tables of numerical data for interaction coefficients. One of the best available resources on the subject of this chapter.]

2 Hubbell, J. H., and Seltzer, S. M., *Tables of X-Ray Mass Attenuation Coefficients and Mass Energy-Absorption Coefficients*, National Institute of Standards and Technology, on line, http://physics.nist.gov/PhysRefData/XrayMassCoef. [Tables and graphs of $\mu/\rho$ and $\mu_{en}/\rho$ for elements $Z = 1$ to 92 and 48 compounds and mixtures of radiological interest.]

3 ICRU Report 60, *Fundamental Quantities and Units for Ionizing Radiation*, International Commission on Radiation Units and Measurements, Bethesda, MD (1998). [Gives definitions of radiometric quantities, interaction coefficients, and dosimetric quantities.]

4 Johns, H. E., and Cunningham, J. R., *The Physics of Radiology*, 4th ed., Charles C. Thomas, Springfield, IL (1983). [Chapters 5 and 6 discuss interaction coefficients and the basic physics of photon interactions.]

5 Seltzer, S. M., "Calculation of Photon Mass Energy-Transfer Coefficients and Mass Energy-Absorption Coefficients," *Rad. Res.* **136**, 147–170 (1993). [A critical analysis of all aspects of the subject, with coefficients derived from the cross-section database of the National Institute of Standards and Technology.]

6 Turner, J. E., "Interaction of Ionizing Radiation with Matter," *Health Phys.* **86**, 228–252 (2004). [Review paper describes the physics of photon interactions and various interaction coefficients, with examples in health physics. Reference is made to the photon as the field quantum of the electromagnetic interaction in quantum electrodynamics.]

## 8.11
## Problems

1. The work function for lithium is 2.3 eV.
   (a) Calculate the maximum kinetic energy of photoelectrons produced by photons with energy of 12 eV.
   (b) What is the cutoff frequency?
   (c) What is the threshold wavelength?
2. The threshold wavelength for tungsten is 2700 Å. What is the maximum kinetic energy of photoelectrons produced by photons of wavelength 2200 Å?
3. What stopping potential is needed to turn back a photoelectron having a kinetic energy of $3.84 \times 10^{-19}$ J?
4. The threshold wavelength for producing photoelectrons from sodium is 5650 Å.

(a) What is the work function?

(b) What is the maximum kinetic energy of photoelectrons produced by light with a wavelength of 4000 Å?

5. Light of wavelength 1900 Å is incident on a nickel surface (work function 4.9 eV).

(a) Calculate the stopping potential.

(b) What is the cutoff frequency for nickel?

(c) What is the threshold wavelength?

6. A potential difference of 3.90 V is needed to stop photoelectrons of maximum kinetic energy when ultraviolet light of wavelength 2200 Å is incident on a metal surface. What is the threshold frequency for the metal for the photoelectric effect?

7. In a photoelectric experiment, the stopping potential is found to be 3.11 volts when light, having a wavelength of 1700 Å, is shone on a certain metal.

(a) What is the work function of the metal?

(b) What frequency of light would be needed to double the stopping potential?

8. (a) Make a sketch of the stopping potential versus light frequency for the photoelectric effect in silicon, which has a work function of 4.4 eV.

(b) Write an equation for the stopping potential as a function of frequency.

(c) What stopping potential is needed for light of frequency $1.32 \times 10^{15}$ s$^{-1}$?

9. (a) If the curve in the last problem is extended to values of $v$ below threshold, show that it intersects the vertical axis at a voltage $-\varphi_0/e$, where $\varphi_0$ is the work function and $e$ is the electronic charge.

(b) Are the curves for two metals, having different threshold values, $v_0$, parallel?

10. In an experiment with potassium, a potential difference of 0.80 V is needed to stop the photoelectron current when light of wavelength 4140 Å is used. When the wavelength is changed to 3000 Å, the stopping potential is 1.94 V. Use these data to determine the value of Planck's constant.

11. From the information given in Problem 10, determine

(a) the threshold wavelength,

(b) the cutoff frequency,

(c) the work function of potassium.

12. Derive Eq. (8.12) from Eqs. (8.9)–(8.11).

13. Show that Eq. (8.19) follows from (8.18) and (8.12).

14. A 1-MeV photon is Compton scattered at an angle of 55°. Calculate

(a) the energy of the scattered photon,
(b) the change in wavelength,
(c) the angle of recoil of the electron,
(d) the recoil energy of the electron.

15. If a 1-MeV gamma ray loses 200 keV in a Compton collision, at what angle is it scattered with respect to its original direction?

16. A 662-keV photon is Compton scattered at an angle of 120° with respect to its incident direction.
(a) What is the energy of the scattered electron?
(b) What is the angle between the paths of the scattered electron and photon?

17. A 900-keV photon is Compton scattered at an angle of 115° with respect to its original direction.
(a) How much energy does the recoil electron receive?
(b) At what angle is the electron scattered?

18. (a) What is the maximum recoil energy that an electron can acquire from an 8-MeV photon?
(b) At what angle of scatter will an 8-MeV photon lose 95% of its energy in a Compton scattering?
(c) Sketch a curve showing the fraction of energy lost by an 8-MeV photon as a function of the angle of Compton scattering.

19. At what energy can a photon lose at most one-half of its energy in Compton scattering?

20. In a Compton scattering experiment a photon is observed to be scattered at an angle of 122° while the electron recoils at an angle of 17° with respect to the incident photon direction.
(a) What is the incident photon energy?
(b) What is the frequency of the scattered photon?
(c) How much energy does the electron receive?
(d) What is the recoil momentum of the electron?

21. Monochromatic X rays of wavelength 0.5 Å and 0.1 Å are Compton scattered from a graphite target and the scattered photons are viewed at an angle of 60° with respect to the incident photon direction.
(a) Calculate the Compton shift in each case.
(b) Calculate the photon energy loss in each case.

22. Show that the fractional energy loss of a photon in Compton scattering is given by

$$\frac{T}{h\nu} = \frac{\Delta\lambda}{\lambda + \Delta\lambda}.$$

23. How much energy will $10^4$ scattered photons deposit, on average, in the graphite target in Fig. 8.2? (Use Table 8.1.)

24. Use Eq. (8.12) to write the formula for the Klein–Nishina cross section (8.28) in terms of the photon scattering angle $\theta$:

$$\frac{d_e\sigma}{d\Omega} = \frac{k_0^2 e^4}{2m^2 c^4} \left[\frac{1}{1 + \dfrac{h\nu}{mc^2}(1 - \cos\theta)}\right]^2 (1 + \cos^2\theta)$$

$$\times \left\{1 + \frac{(\frac{h\nu}{mc^2})^2(1 - \cos\theta)^2}{(1 + \cos^2\theta)\left[1 + \dfrac{h\nu}{mc^2}(1 - \cos\theta)\right]}\right\} \text{ m}^2 \text{ sr}^{-1}.$$

25. The Klein–Nishina cross section for the collision of a 1-MeV photon with an electron is $2.11 \times 10^{-25}$ cm$^2$. Calculate, for Compton scattering on aluminum,
    (a) the energy-transfer cross section (per electron cm$^{-2}$)
    (b) the energy-scattering cross section (per electron cm$^{-2}$)
    (c) the atomic cross section
    (d) the linear attenuation coefficient.
26. Which of the quantities (a)–(d) in the last problem are the same for Compton scattering from other chemical elements?
27. Repeat Problem 25 for tin.
28. Show that the threshold photon wavelength for producing an electron–positron pair is 0.012 Å.
29. A 4-MeV photon creates an electron–positron pair in the field of a nucleus. What is the total kinetic energy of the pair?
30. Calculate the threshold energy for the reaction $^{12}_{6}\text{C}(\gamma, n)^{11}_{6}\text{C}$. The mass differences are given in Appendix D.
31. Calculate the energy of the proton ejected in the $^{16}_{8}\text{O}(\gamma, p)^{15}_{7}\text{N}$ reaction with a 20-MeV photon.
32. (a) What is the threshold energy for the $^{206}_{82}\text{Pb}(\gamma, p)^{205}_{81}\text{Tl}$ reaction, which competes with $^{206}\text{Pb}(\gamma, n)^{205}\text{Pb}$?
    (b) Why is the $(\gamma, p)$ threshold lower than that of $(\gamma, n)$?
    (c) Which process is more probable?
33. An experiment is carried out with monoenergetic photons in the "good" geometry shown in Fig. 8.7. The relative count rate of the detector is measured with different thicknesses $x$ of tin used as absorber. The following data are measured:

| $x$ (cm) | 0 | 0.50 | 1.0 | 1.5 | 2.0 | 3.0 | 5.0 |
|---|---|---|---|---|---|---|---|
| Relative count rate | 1.00 | 0.861 | 0.735 | 0.621 | 0.538 | 0.399 | 0.210 |

    (a) What is the value of the linear attenuation coefficient?
    (b) What is the value of the mass attenuation coefficient?
    (c) What is the photon energy?
34. Show that Eq. (8.42) implies that $\mu$ is the probability per unit distance that a photon interacts.

35. A narrow beam of 400-keV photons is incident normally on a 2 mm iron liner.
    (a) What fraction of the photons have an interaction in the liner?
    (b) What thickness of Fe is needed to reduce the fraction of photons that are transmitted without interaction to 10%?
    (c) If Al were used instead of Fe, what thickness would be needed in (b)?
    (d) How do the answers in (b) and (c) compare when expressed in $g\,cm^{-2}$?
    (e) If lead were used in (b), how would its thickness in $g\,cm^{-2}$ compare with those for Al and Fe?

36. The linear attenuation coefficient of copper for 800-keV photons is $0.58\ cm^{-1}$. Calculate the atomic cross section.

37. The mass attenuation coefficient for 1-MeV photons in carbon is $0.0633\ cm\,g^{-1}$.
    (a) Calculate the atomic cross section.
    (b) Estimate the total Compton cross section per electron.

38. The mass attenuation coefficients of Cu and Sn for 200-keV photons are, respectively, $0.15\ cm^2\,g^{-1}$ and $0.31\ cm^2\,g^{-1}$. What are the macroscopic cross sections?

39. Calculate the microscopic cross sections in Problem 38 for Cu and Sn.

40. A bronze absorber (density $= 8.79\ g\,cm^{-3}$), made of 9 parts of Cu and 1 part Sn by weight, is exposed to 200-keV X rays (see Problem 38). Calculate the linear and mass attenuation coefficients of bronze for photons of this energy.

41. The atomic cross sections for 1-MeV photon interactions with carbon and hydrogen are, respectively, 1.27 barns and 0.209 barn.
    (a) Calculate the linear attenuation coefficient for paraffin. (Assume the composition $CH_2$ and density $0.89\ g\,cm^{-3}$.)
    (b) Calculate the mass attenuation coefficient.

42. What is the atomic cross section of Fe for 400-keV photons? What is the atomic energy-absorption cross section?

43. A pencil beam of 200-keV photons is normally incident on a 1.4-cm-thick sheet of aluminum pressed against a 2-mm-thick sheet of lead behind it.
    (a) What fraction of the incident photons penetrate both sheets without interacting?
    (b) What would be the difference if the photons came from the other direction, entering to lead first and then the aluminum?

44. A parallel beam of 1-MeV photons is normally incident on a sheet of uranium, 1.0 mm thick. The incident beam intensity is $10^4$ MeV cm$^{-2}$ s$^{-1}$.
    (a) Calculate the energy fluence rate transmitted by the sheet.
    (b) What fraction of the transmitted energy fluence rate is due to uncollided photons?
    (c) What physical processes are responsible for energy transfer to the sheet?
    (d) What processes are responsible for energy absorption in the sheet?

45. A narrow beam of 500-keV photons is directed normally at a slab of tin, 1.08 cm thick.
    (a) What fraction of the photons interact in the slab?
    (b) If 200 photons per minute are incident on the slab, what is the rate of energy transmission through it?
    (c) What fraction of the transmitted energy is carried by the uncollided photons?

46. A narrow beam of $10^4$ photons s$^{-1}$ is normally incident on a 6-mm aluminum sheet. The beam consists of equal numbers of 200-keV photons and 2-MeV photons.
    (a) Calculate the number of photons s$^{-1}$ of each energy that are transmitted without interaction through the sheet.
    (b) How much energy is removed from the narrow beam per second by the sheet?
    (c) How much energy is absorbed in the sheet per second?

47. (a) Show that a 1.43-mm lead sheet has the same thickness in g cm$^{-2}$ as the aluminum sheet in the last problem.
    (b) Calculate the number of photons s$^{-1}$ of each energy that are transmitted without interaction when the Al sheet in Problem 46 is replaced by 1.43 mm of Pb.
    (c) Give a physical reason for any differences or similarities in the answers to Problems 46(a) and 47(b).

48. From Fig. 8.13 determine the energy at which the photoelectric effect and Compton scattering contribute equally to the attenuation coefficient of water.

49. The mass attenuation coefficient of Pb for 70-keV photons is 3.0 cm$^2$ g$^{-1}$ and the mass energy–absorption coefficient is 2.9 cm$^2$ g$^{-1}$ (Figs. 8.8 and 8.11).
    (a) Why are these two values almost equal?
    (b) How many cm of Pb are needed to reduce the transmitted intensity of a 70-keV X-ray beam to 1% of its original value?
    (c) What percentage of the photons penetrate this thickness without interacting?

50. A broad, uniform, parallel beam of 0.5-MeV gamma rays is normally incident on an aluminum absorber of thickness $8 \, \mathrm{g \, cm^{-2}}$. The beam intensity is $8.24 \, \mathrm{J \, m^{-2} \, s^{-1}}$.
    (a) Calculate the fraction of photons transmitted without interaction.
    (b) What is the total transmitted beam intensity?
    (c) What is the rate of energy absorption per unit mass in the aluminum?

51. What fraction of the energy in a 10-keV X-ray beam is deposited in 5 mm of soft tissue?

52. A broad beam of 2-MeV photons is normally incident on a 20-cm concrete shield.
    (a) What fraction is transmitted without interaction?
    (b) Estimate the fraction of the beam intensity that is transmitted by the shield.

53. A dentist places the window of a 100-kVp (100-kV, peak voltage) X-ray machine near the face of a patient to obtain an X-ray of the teeth. Without filtration, considerable low-energy (assume 20 keV) X rays are incident on the skin.
    (a) If the intervening tissue has a thickness of 5 mm, calculate the fraction of the 20-keV intensity absorbed in it.
    (b) What thickness of aluminum filter would reduce the 20-keV radiation exposure by a factor of 10?
    (c) Calculate the reduction in the intensity of 100-keV X rays transmitted by the filter.
    (d) After adding the filter, the exposure time need not be increased to obtain the same quality of X-ray picture. Why not?

54. A parallel beam of 500-keV photons is normally incident on a sheet of lead 8 mm thick. The rate of energy transmission is $4 \times 10^4 \, \mathrm{MeV \, s^{-1}}$.
    (a) What fraction of the incident photon energy is absorbed in the sheet?
    (b) How many photons per second are incident on the sheet?
    (c) What fraction of the transmitted energy is due to uncollided photons?

## 8.12

## Answers

1. (a) 9.7 eV
   (b) $5.55 \times 10^{14} \, \mathrm{s^{-1}}$
   (c) 5400 Å
3. 2.40 V

5. (a) 1.63 V
   (b) $1.18 \times 10^{15} \, \mathrm{s^{-1}}$
   (c) 2540 Å
6. $4.20 \times 10^{14} \, \mathrm{s^{-1}}$

11. (a) 5636 Å
    (b) $5.32 \times 10^{14}$ s$^{-1}$
    (c) 2.20 eV
14. (a) 0.545 MeV
    (b) 0.0103 Å
    (c) 33.0°
    (d) 0.455 MeV
15. 29.3°
17. (a) 0.643 MeV
    (b) 13.0°
18. (a) 7.75 MeV
    (b) 102°
20. (a) 0.415 MeV
    (b) $4.47 \times 10^{19}$ s$^{-1}$
    (c) 0.230 MeV
    (d) $2.86 \times 10^{-22}$
    kg m s$^{-1}$
23. 5.7 MeV
25. (a) $9.28 \times 10^{-26}$ cm$^2$
    (b) $1.18 \times 10^{-25}$ cm$^2$
    (c) $2.74 \times 10^{-24}$ cm$^2$
    (d) 0.165 cm$^{-1}$
30. 18.72 MeV
33. (a) 0.309 cm$^{-1}$
    (b) 0.0423 cm$^2$ g$^{-1}$
    (c) ~2 MeV

35. (a) 0.135
    (b) 3.18 cm
    (c) 9.28 cm
    (d) both ~25 g cm$^{-2}$
    (e) 11.0 g cm$^{-2}$
36. $6.83 \times 10^{-24}$ cm$^2$
39. 0.158 and 0.611 barn
40. 1.4 cm$^{-1}$;
    0.16 cm$^2$ g$^{-1}$
43. (a) 0.078
    (b) None
44. (a) $9.15 \times 10^3$
    MeV cm$^{-2}$ s$^{-1}$
    (b) 0.946
46. (a) 4120 s$^{-1}$; 4670 s$^{-1}$
    (b) 836 MeV s$^{-1}$
    (c) 394 MeV s$^{-1}$
49. (b) 0.14 cm
    (c) 0.86%
51. 0.93
53. (a) 0.27
    (b) 0.30 cm
    (c) 2.9%
54. (a) 0.598
    (b) $1.99 \times 10^5$ s$^{-1}$
    (c) 0.580

# 9
# Neutrons, Fission, and Criticality

## 9.1
## Introduction

The neutron was discovered by Chadwick in 1932. Nuclear fission, induced by capture of a slow neutron in $^{235}$U, was discovered by Hahn and Strassman in 1939. In principle, the fact that several neutrons are emitted when fission takes place suggested that a self-sustaining chain reaction might be possible. Under Fermi's direction, the world's first man-made nuclear reactor went critical on December 2, 1942.[1] The neutron thus occupies a central position in the modern world of atoms and radiation.

In this chapter we describe the principal sources of neutrons, their interactions with matter, neutron activation, nuclear fission, and criticality. The most important neutron interactions from the standpoint of radiation protection will be stressed.

## 9.2
## Neutron Sources

Nuclear reactors are the most copious sources of neutrons. The energy spectrum of neutrons from the fission of $^{235}$U extends from a few keV to more than 10 MeV. The average energy is about 2 MeV. Research reactors often have ports through which neutron beams emerge into experimental areas outside the main reactor shielding. These neutrons are usually degraded in energy, having passed through parts of the reactor core and coolant as well as structural materials. Figure 9.1 shows an example of a research reactor.

The High Flux Isotope Reactor (HFIR) is one of the most powerful research reactors in the world (Fig. 9.2). The 85-MW unit employs highly enriched uranium fuel,

1   In 1972 the French Atomic Energy Commission found unexpectedly low assays of $^{235}$U/$^{238}$U isotopic ratios in uranium ores from the Oklo deposit in Gabon, Africa. Close examination revealed that several sites in the Oklo mine were natural nuclear reactors in the distant past. As far as is known, the Oklo phenomenon, as it is called, is unique. For more information the reader is referred to the article by M. Maurette, "Fossil Nuclear Reactors," *Ann. Rev. Nucl. Sci.* **26**, 319 (1976). Also see *Physics Today* **57**(12), 9 (2004).

*Atoms, Radiation, and Radiation Protection*. James E. Turner
Copyright © 2007 WILEY-VCH Verlag GmbH & Co. KGaA, Weinheim
ISBN: 978-3-527-40606-7

**Fig. 9.1** Oak Ridge Bulk Shielding Reactor, an early swimming-pool type in which water served as coolant, moderator, and shield. Glow around reactor core is blue light emitted as Cerenkov radiation (Section 10.4) by electrons that travel faster than light in the water. (Photo courtesy Oak Ridge National Laboratory, operated by Martin Marietta Energy Systems, Inc., for the Department of Energy.)

and has a peak thermal-neutron fluence rate of $2.6 \times 10^{15}$ cm$^{-2}$ s$^{-1}$. Its neutrons are used to explore the structure and behavior of irradiated materials such as metals, polymers, high-temperature superconductors, and biological samples. Unique studies of materials after welding and other stresses as well as neutron-activation analyses are conducted. The HFIR produces some 35 primary radioisotopes for medical and industrial purposes. It is the Western World's sole producer of $^{252}$Cf.

Particle accelerators are used to generate neutron beams by means of a number of nuclear reactions. For example, accelerated deuterons that strike a tritium target produce neutrons via the $^3$H(d,n)$^4$He reaction, that is,

$$^2_1\text{H} + ^3_1\text{H} \rightarrow ^4_2\text{He} + ^1_0\text{n}. \tag{9.1}$$

To obtain monoenergetic neutrons with an accelerator, excited states of the product nucleus are undesirable. Therefore, light materials are commonly used as targets for a proton or deuteron beam. Table 9.1 lists some important reactions that are used to obtain monoenergetic neutrons. The first two are exothermic and can be used with ions of a few hundred keV energy in relatively inexpensive accelerators.

(a)

(b)

**Fig. 9.2** (a) Control room of the High Flux Isotope Reactor.
(b) View of workmen over vessel head of the reactor. (Courtesy
William H. Cabage, Oak Ridge National Laboratory, managed
by UT-Battelle, LLC, for the U. S. Department of Energy.)

**Table 9.1** Reactions Used to Produce Monoenergetic Neutrons with Accelerated Protons (p) and Deuterons (d)

| Reaction | $Q$ Value (MeV) |
|---|---|
| $^3$H(d,n)$^4$He | 17.6 |
| $^2$H(d,n)$^3$He | 3.27 |
| $^{12}$C(d,n)$^{13}$N | −0.281 |
| $^3$H(p,n)$^3$He | −0.764 |
| $^7$Li(p,n)$^7$Be | −1.65 |

**Table 9.2** $(\alpha, n)$ Neutron Sources

| Source | Average Neutron Energy (MeV) | Half-life |
|---|---|---|
| $^{210}$PoBe | 4.2 | 138 d |
| $^{210}$PoB | 2.5 | 138 d |
| $^{226}$RaBe | 3.9 | 1600 y |
| $^{226}$RaB | 3.0 | 1600 y |
| $^{239}$PuBe | 4.5 | 24100 y |

For a given ion-beam energy, neutrons leave a thin target with energies that depend on the angle of exit with respect to the incident beam direction.

An alpha source, usually radium, polonium, or plutonium and a light metal, such as beryllium or boron, can be mixed together as powders and encapsulated to make a "radioactive" neutron source. Neutrons are emitted as a result of $(\alpha, n)$ reactions, such as the following:

$$\,_2^4\text{He} + \,_4^9\text{Be} \rightarrow \,_6^{12}\text{C} + \,_0^1\text{n}. \tag{9.2}$$

Light metals are used in order to minimize the Coulomb repulsion between the alpha particle and nucleus. The neutron intensity from such a source dies off with the half-life of the alpha emitter. Neutrons leave the source with a continuous energy spectrum, because the alpha particles slow down by different amounts before striking a nucleus. The neutron and the recoil nucleus [e.g., $^{12}_6$C in (9.2)] share a total energy equal to the sum of the $Q$ value and the kinetic energy that the alpha particle has as it strikes the nucleus. Some common $(\alpha, n)$ sources are shown in Table 9.2.

Similarly, photoneutron sources, making use of $(\gamma, n)$ reactions, are also available. Several examples are listed in Table 9.3. In contrast to $(\alpha, n)$ sources, which emit neutrons with a continuous energy spectrum, monoenergetic photoneutrons can be obtained by selecting a nuclide that emits a gamma ray of a single energy. Photoneutron sources decay in intensity with the half-life of the photon emitter. All the sources in Table 9.3 are monoenergetic except the last; $^{226}$Ra emits gamma rays of several energies. It is important for radiation protection to remember that

(a)

(b)

**Fig. 9.3** (a) Artist's rendering of the Spallation Neutron Source.
(b) Arial photograph of facility in 2006. (Courtesy William H.
Cabage, Oak Ridge National Laboratory, managed by
UT-Battelle, LLC, for the U. S. Department of Energy.)

**Table 9.3** ($\gamma$, n) Neutron Sources

| Source | Neutron Energy (MeV) | Half-life |
|---|---|---|
| $^{24}$NaBe | 0.97 | 15.0 h |
| $^{24}$NaD$_2$O | 0.26 | 15.0 h |
| $^{116}$InBe | 0.38 | 54 min |
| $^{124}$SbBe | 0.024 | 60 d |
| $^{140}$LaBe | 0.75 | 40 h |
| $^{226}$RaBe | 0.7 (maximum) | 1600 y |

all photoneutron sources have gamma-ray backgrounds of >1000 photons per neutron.

Some very heavy nuclei fission spontaneously, emitting neutrons in the process. They can be encapsulated and used as neutron sources. Examples of some important spontaneous-fission sources are $^{254}$Cf, $^{252}$Cf, $^{244}$Cm, $^{242}$Cm, $^{238}$Pu, and $^{232}$U. In most cases the half-life for spontaneous fission is much greater than that for alpha decay. An exception is $^{254}$Cf, which decays almost completely by spontaneous fission with a 60-day half-life.

The Spallation Neutron Source (SNS), shown in Fig. 9.3 at the Oak Ridge National Laboratory, produced its first neutrons in 2006. This cooperative effort among a number of laboratories operates as a user facility and offers an order-of-magnitude improvement in neutron beam intensity compared with other sources. Neutrons are produced by bombarding a target module containing 20 tons of circulating mercury with 1-GeV protons from a linear accelerator, using superconductor technology, and an accumulator-ring system. The SNS will make research possible in a number of heretofore unreachable areas in instrumentation, materials properties and dynamics, high-temperature superconductivity, biological structures, and nanoscience.

## 9.3
### Classification of Neutrons

It is convenient to classify neutrons according to their energies. At the low end of the scale, neutrons can be in approximate thermal equilibrium with their surroundings. Their energies are then distributed according to the Maxwell–Boltzmann formula. The energy of a thermal neutron is sometimes given as 0.025 eV, which is the most probable energy in the distribution at room temperature (20°C). The average energy of thermal neutrons at room temperature is 0.038 eV. Thermal-neutron distributions do not necessarily have to correspond to room temperature. "Cold" neutrons, with lower "temperatures," are produced at some facilities, while others generate neutrons with energy distributions characteristic of temperatures considerably above 20°C. Thermal neutrons gain and lose only small amounts of energy through elastic scattering in matter. They diffuse about until captured by atomic nuclei.

Neutrons of higher energies, up to about 0.01 MeV or 0.1 MeV (the convention is not precise), are known variously as "slow," "intermediate," or "resonance" neutrons. "Fast" neutrons are those in the next-higher-energy classification, up to about 10 MeV or 20 MeV. "Relativistic" neutrons have still higher energies.

## 9.4
## Interactions with Matter

Like photons, neutrons are uncharged and hence can travel appreciable distances in matter without interacting. Under conditions of "good geometry" (cf. Section 8.7) a narrow beam of monoenergetic neutrons is also attenuated exponentially by matter. The interaction of neutrons with electrons, which is electromagnetic in nature,[2] is negligible. In passing through matter a neutron can collide with an atomic nucleus, which can scatter it elastically or inelastically. The scattering is elastic when the total kinetic energy is conserved; that is, when the energy lost by the neutron is equal to the kinetic energy of the recoil nucleus. When the scattering is inelastic, the nucleus absorbs some energy internally and is left in an excited state. The neutron can also be captured, or absorbed, by a nucleus, leading to a reaction, such as (n, p), (n, 2n), (n, $\alpha$), or (n, $\gamma$). The reaction changes the atomic mass number and/or atomic number of the struck nucleus.

Typically, a fast neutron will lose energy in matter by a series of (mostly) elastic scattering events. This slowing-down process is called neutron moderation. As the neutron energy decreases, scattering continues, but the probability of capture by a nucleus generally increases. If a neutron reaches thermal energies, it will move about randomly by elastic scattering until absorbed by a nucleus.

Cross sections for the interactions of neutrons with atomic nuclei vary widely and usually are complicated functions of neutron energy. Figure 9.4 shows the total cross sections for neutron interactions with hydrogen and carbon as functions of energy. Because the hydrogen nucleus (a proton) has no excited states, only elastic scattering and neutron capture are possible. The total hydrogen cross section shown in Fig. 9.4 is the sum of the cross sections for these two processes. The capture cross section for hydrogen is comparatively small, reaching a value of only 0.33 barn (1 barn = $10^{-24}$ cm$^2$) at thermal energies, where it is largest. Thermal-neutron capture is an important interaction in hydrogenous materials.

---

2 Although the neutron is electrically neutral, it has spin, a magnetic moment, and a nonzero *distribution* of electric charge within it. These properties, coupled with the charge and spin of the electron, give rise to electromagnetic forces between neutrons and electrons. These forces, however, are extremely weak. In contrast, the neutron interacts with protons and neutrons at close range by means of the strong, or nuclear, force. The stopping power of matter for neutrons due to their electromagnetic interaction with atomic electrons has been calculated (see Section 9.12).

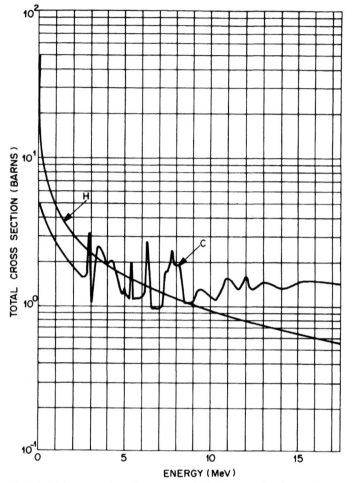

**Fig. 9.4** Total cross sections for neutrons with hydrogen and carbon as functions of energy.

In contrast, the carbon cross section in Fig. 9.4 shows considerable structure, especially in the region 1–10 MeV. The nucleus possesses discrete excited states, which can enhance or depress the elastic and inelastic scattering cross sections at certain values of the neutron energy (cf. Fig. 3.2).

**9.5**
**Elastic Scattering**

As mentioned in the last section, elastic scattering is the most important process for slowing down neutrons; the contribution by inelastic scattering is usually small in comparison. We treat elastic scattering here.

**Table 9.4** Maximum Fraction of Energy Lost, $Q_{max}/E_n$ from Eq. (9.3), by Neutron in Single Elastic Collision with Various Nuclei

| Nucleus | $Q_{max}/E_n$ |
|---------|---------------|
| $^{1}_{1}H$ | 1.000 |
| $^{2}_{1}H$ | 0.889 |
| $^{4}_{2}He$ | 0.640 |
| $^{9}_{4}Be$ | 0.360 |
| $^{12}_{6}C$ | 0.284 |
| $^{16}_{8}O$ | 0.221 |
| $^{56}_{26}Fe$ | 0.069 |
| $^{118}_{50}Sn$ | 0.033 |
| $^{238}_{92}U$ | 0.017 |

The maximum energy that a neutron of mass $M$ and kinetic energy $E_n$ can transfer to a nucleus of mass $m$ in a single (head-on) elastic collision is given by Eq. (5.4):

$$Q_{max} = \frac{4mME_n}{(M+m)^2}. \tag{9.3}$$

Setting $M = 1$, we can calculate the maximum fraction of a neutron's energy that can be lost in a collision with nuclei of different atomic-mass numbers $m$. Some results are shown in Table 9.4 for nuclei that span the periodic system. For ordinary hydrogen, because the proton and neutron masses are equal, the neutron can lose all of its kinetic energy in a head-on, billiard-ball-like collision. As the nuclear mass increases, one can see how the efficiency of a material per collision for moderating neutrons grows progressively worse. As a rule of thumb, the average energy lost per collision is approximately one-half the maximum.

An interesting consequence of the equality of the masses in neutron–proton scattering is that the particles separate at right angles after collision, when the collision is nonrelativistic. Figure 9.5(a) represents a neutron of mass $M$ and momentum $MV$ approaching a stationary nucleus of mass $m$. After collision, in Figure 9.5(b), the nucleus and neutron, respectively, have momenta $mv'$ and $MV'$. The conservation of momentum requires that the sum of the vectors, $mv' + MV'$, be equal to the initial momentum vector $MV$, as shown in Figure 9.5(c). Since kinetic energy is conserved, we have

$$\frac{1}{2}MV^2 = \frac{1}{2}mv'^2 + \frac{1}{2}MV'^2. \tag{9.4}$$

If $M = m$, then $V^2 = v'^2 + V'^2$, which implies the Pythagorean theorem for the triangle in (c). Therefore, $v'$ and $V'$ are at right angles.

The elastic scattering of neutrons plays an important role in neutron energy measurements. As discussed in the next chapter, under suitable conditions the recoil

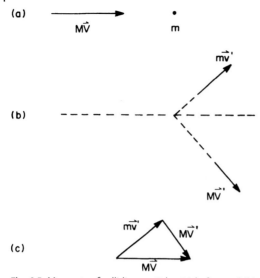

**Fig. 9.5** Momenta of colliding particles (a) before and (b) after collision. (c) Representation of momentum conservation.

energies of nuclei in a proportional-counter gas under neutron bombardment can be measured. The nuclear recoil energy and angle are directly related to the neutron energy. For example, as illustrated in Fig. 9.6, when a neutron of energy $E_n$ strikes a proton, which recoils with energy $Q$ at an angle $\theta$ with respect to the incident-neutron direction, then the conservation of energy and momentum requires that (Problem 9)

$$Q = E_n \cos^2 \theta. \tag{9.5}$$

Thus, if $Q$ and $\theta$ can be measured individually for a number of incident neutrons, one obtains the incident neutron spectrum directly. (The proton-recoil neutron spectrometer is discussed in Section 10.5.) More often, only the energies of the recoil nuclei in the gas (e.g., $^3$He, $^4$He, or $^1$H and $^{12}$C from CH$_4$) are determined, and the neutron energy spectrum must be unfolded from its statistical relationship to the recoil-energy spectra. The unfolding is further complicated by the fact that the recoil tracks do not always lie wholly within the chamber gas (wall effects).

Because of the magnitude of the cross section, the efficiency of energy transfer, and the abundance of hydrogen in soft tissue, neutron–proton (n–p) scattering is usually the dominating mechanism whereby fast neutrons deliver dose to tissue. As we shall see in Section 12.9, over 85% of the "first-collision" dose in soft tissue (composed of H, C, O, and N) arises from n–p scattering for neutron energies below 10 MeV.

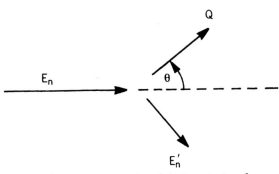

**Fig. 9.6** Schematic representation of elastic scattering of a neutron by a proton. The initial neutron energy $E_n$ is given in terms of the proton recoil energy $Q$ and angle $\theta$ by Eq. (9.5).

## 9.6
## Neutron–Proton Scattering Energy-Loss Spectrum

Like Eq. (8.19) for Compton scattering, Eq. (9.5) for neutron–proton scattering is purely kinematic in nature, reflecting (nonrelativistically) the conservation of kinetic energy and momentum. It provides no information about the probability that the neutron is scattered in the direction $\theta$. Experimentally, for neutron energies up to about 10 MeV, it is observed that neutron–proton scattering is isotropic in the center-of-mass coordinate system. That is, the neutron (as well as the proton) is scattered with equal likelihood in any direction in three dimensions in this coordinate system. One can translate this experimental finding into the probability density for having a proton recoil at an angle $\theta$ as seen in the laboratory system. Equation (9.5) can then be used to compute the probability density $P(Q)$ for $Q$, which is then the neutron energy-loss spectrum. As we now show, isotropic scattering in the center-of-mass system results in a flat energy-loss spectrum for neutron–proton scattering in the laboratory system.

The collision is represented in the laboratory system by Fig. 9.5. In (a), with the two masses equal ($M = m$), the center of mass is located midway between the neutron and proton and moves toward the right with constant speed $\frac{1}{2}V$. The center of mass crosses the collision point at the instant of collision and continues moving toward the right with speed $\frac{1}{2}V$ thereafter. (Its motion is unchanged by interaction of the particles.) Figure 9.7(a) shows the locations of the collision point $O$, the center of mass $C$, the neutron $N$ and proton $P$ at unit time after the collision. At this time, the scattered neutron and proton have displacements equal numerically to $V'$ and $v'$ relative to $O$. (We use the same symbols in this discussion as in Figs. 9.5 and 9.6.) Also, the center of mass $C$ bisects the line $NP$ at a displacement $\frac{1}{2}V$ from $O$. The scattering angle of the proton is $\theta$ in the laboratory system and $\omega$ in the center-of-mass system. Before collision, the neutron and proton approach each other from opposite directions with speeds $\frac{1}{2}V$ in the center-of-mass system. Since momentum is conserved, the particles are scattered "back-to-back" with equal speeds. Because no kinetic energy is lost, the speeds of the neutron and proton in the center-of-mass

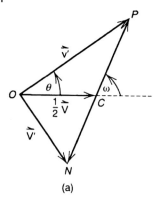

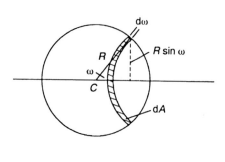

(a)  (b)

**Fig. 9.7** (a) Positions of the proton P and neutron N at unit time following the collision at O. In unit time, their displacements are the same in magnitude and direction as the velocity vectors v' and V'. The center of mass C has displacement V/2 and bisects the line NP. The proton scattering angle is θ in the laboratory system and ω in the center-of-mass system. Drawing is in the laboratory system. (b) Sphere of radius R centered about C in the center-of-mass system. For isotropic scattering, the probability that the proton passes through any area on the surface of the sphere is proportional to the size of that area.

system are not changed by the collision. Therefore, the triangles in Fig. 9.7(a) are isosceles: $OC = CP = CN$. In the upper triangle, since angle $OPC$ and $\theta$ are equal,

$$\omega = 2\theta, \tag{9.6}$$

giving the relationship between the proton recoil angles in the center-of-mass and laboratory systems.

Figure 9.7(b) shows a sphere of radius $R$ centered about the center of mass in that system. Because the scattering is isotropic, the probability that the proton is scattered through any area of the sphere's surface is, by definition, equal to the ratio of that area and the total area $A = 4\pi R^2$ of the sphere. In particular, the probability $P_\omega(\omega)\,d\omega$ that the proton is scattered into the area $dA$ of the band between $\omega$ and $\omega + d\omega$ in Fig. 9.7(b) is

$$P_\omega(\omega)\,d\omega = \frac{dA}{4\pi R^2} = \frac{2\pi R \sin\omega \times R\,d\omega}{4\pi R^2} = \frac{1}{2}\sin\omega\,d\omega. \tag{9.7}$$

The probability $P_\theta(\theta)\,d\theta$ of scattering into the angular interval between $\theta$ and $\theta + d\theta$ in the laboratory system is given by

$$P_\theta(\theta)\,d\theta = P_\omega(\omega)\,d\omega. \tag{9.8}$$

With the help of Eqs. (9.6) and (9.7), this relation gives

$$P_\theta(\theta)\,d\theta = \frac{1}{2}\sin 2\theta\,d(2\theta) = \sin 2\theta\,d\theta = 2\sin\theta\cos\theta\,d\theta. \tag{9.9}$$

The probability that the neutron loses an energy between $Q$ and $Q + dQ$ is

$$P(Q)\,dQ = P_\theta(\theta)\,d\theta = 2\sin\theta\cos\theta\left(\frac{dQ}{d\theta}\right)^{-1}dQ. \tag{9.10}$$

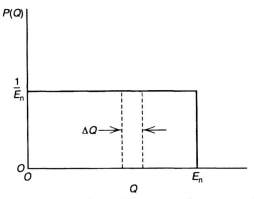

**Fig. 9.8** Normalized energy-loss spectrum for scattering of neutrons of energy $E_n$ from hydrogen. $P(Q)\,dQ$ is the probability that the energy loss is between $Q$ and $Q + dQ$. Isotropic scattering in the center-of-mass system leads to the flat spectrum.

Equation (9.5) implies that $dQ/d\theta = 2E_n \cos\theta \sin\theta$.[3] Therefore, the neutron energy-loss spectrum (9.10) becomes, simply,

$$P(Q)\,dQ = \frac{2\sin\theta \cos\theta \, dQ}{2E_n \cos\theta \sin\theta} = \frac{1}{E_n}\,dQ. \tag{9.11}$$

The (normalized) spectrum for the scattering of neutrons of energy $E_n$ by protons is shown in Fig. 9.8. Because the spectrum is flat and the maximum energy loss is $Q_{max} = E_n$, the probability that a neutron loses an amount of energy $\Delta Q$ is simply the fraction $\Delta Q/E_n$ (independently of where $\Delta Q$ is located in Fig. 9.8). The average energy loss is $Q_{avg} = \frac{1}{2}Q_{max}(= \frac{1}{2}E_n)$. This relationship, mentioned earlier as a rule of thumb for neutron scattering, is exact for isotropic scattering in the center-of-mass system. The energy-loss spectrum is also flat for isotropic center-of-mass scattering when the masses are unequal, and $Q_{avg} = \frac{1}{2}Q_{max}$, with $Q_{max}$ given by Eq. (9.3). The last relation is approximately valid for neutron elastic scattering by C, N, and O in tissue.

*Example*

A 2.6-MeV neutron has a collision with hydrogen. (a) What is the probability that it loses between 0.63 and 0.75 MeV? (b) If the neutron loses 0.75 MeV, at what angle is it scattered? (c) What is the average energy lost by 2.6-MeV neutrons in collisions with carbon? (d) In (b), how much energy does the neutron lose in the center-of-mass system?

*Solution*

(a) In Fig. 9.8, with $E_n = 2.6$ MeV and $\Delta Q = 0.75 - 0.63 = 0.12$ MeV, it follows that the probability of a neutron energy loss in the specified interval is $\Delta Q/E_n = 0.0462$.

---

3   We ignore the negative sign from differentiating $\cos\theta$, which indicates that $Q$ decreases as $\theta$ increases. We write all probabilities as positive definite.

(b) Applying Eq. (9.5) with $Q = 0.75$ MeV and $E_n = 2.6$ MeV, we obtain $\cos^2 \theta = 0.288$, giving $\theta = 57.5°$ for the scattering angle of the proton (Fig. 9.6). Therefore, the neutron scattering angle is $90.0° - 57.5° = 32.5°$.

(c) The average energy loss, $Q_{avg}$, is approximately one-half the maximum possible. With $M = 1$ and $m = 12$ in Eq. (9.3), we find for neutron collisions with carbon

$$Q_{max} = \frac{4 \times 12 \times 1}{(1 + 12)^2} E_n = 0.284 E_n. \tag{9.12}$$

Thus, $Q_{max} = 0.284 \times 2.6 = 0.738$ MeV; and $Q_{avg} = 0.369$ MeV.

(d) As measured by an observer at rest with respect to the center of mass of the colliding neutron and proton, neither particle loses energy in the collision. The two, having equal masses, first approach each other with equal speeds. The energy lost by the neutron in the center-of-mass system is zero.

## Example

(a) In the last problem, what are the energies of the neutron and proton in the center-of-mass system? (b) How much energy is associated with the motion of the center of mass in the laboratory system?

## Solution

(a) Since the speed of both particles in the center-of-mass system is $\frac{1}{2}V$ and the masses are equal, the neutron and proton each have the energy

$$\epsilon = \frac{1}{2} M \left( \frac{1}{2} V \right)^2 = \frac{1}{8} M V^2 = \frac{1}{4} E_n = 0.650 \text{ MeV}. \tag{9.13}$$

(b) To an observer at rest relative to the center of mass, each of the colliding particles has an energy $\epsilon$. The total energy associated with this relative motion is $\epsilon_{rel} = 2\epsilon = \frac{1}{2} E_n = 1.30$ MeV. In the laboratory system, the total energy $E_n$ is the sum of the energy of relative motion, $\epsilon_{rel}$, and the energy of motion of the center of mass, $E_{com}$. Therefore

$$E_{com} = E_n - \epsilon_{rel} = E_n - \frac{1}{2} E_n = \frac{1}{2} E_n = 1.30 \text{ MeV}, \tag{9.14}$$

which is the desired answer. An alternative way of analyzing the problem is the following. As we have seen (Chapter 2, Problem 31), the total energy in the laboratory system is the sum of the energies of the center-of-mass motion and the relative motion. The total mass, $M + M = 2M$, moves with speed $\frac{1}{2}V$. The energy of this motion is

$$E_{com} = \frac{1}{2}(2M)\left( \frac{1}{2} V \right)^2 = \frac{1}{4} M V^2 = \frac{1}{2} E_n, \tag{9.15}$$

in agreement with (9.14). The effective mass for the relative motion is given by Eq. (2.20) for the reduced mass, which with equal masses is $m_r = \frac{1}{2}M$. The relative velocity of the neutron and proton is $V$. Thus, the energy of relative motion is

$$\epsilon_{rel} = \frac{1}{2} m_r V^2 = \frac{1}{4} M V^2 = \frac{1}{2} E_n, \tag{9.16}$$

as found earlier.

The last example shows that the collision of a 2.6-MeV neutron with a stationary proton is the same as that of a 0.650-MeV neutron and a 0.650-MeV proton that approach each other from opposite directions. In high-energy particle accelerators, considerable advantage in effective particle energy accrues from having the laboratory itself be the center-of-mass system. The large electron–positron storage ring (LEP) (Fig. 1.5) is an example of such a colliding-beam facility. The electron and its antiparticle, the positron, are accelerated in opposite directions in the same machine and allowed to collide head-on with equal energies. These collisions occur at relativistic energies, but the same principle applies as found here. The effective collision energy is much greater than that for positrons of the same energy colliding with electrons in a stationary target.

## 9.7
## Reactions

In this section we describe several neutron reactions that are important in various aspects of neutron detection and radiation protection. The way in which the reactions are used for detection will be described in Chapter 10.

### $^1$H$(n, \gamma)^2$H
We have already mentioned the capture of thermal neutrons by hydrogen. This reaction is an example of radiative capture; that is, neutron absorption followed by the immediate emission of a gamma photon. Explicitly,

$$\,_0^1n + \,_1^1H \rightarrow \,_1^2H + \,_0^0\gamma. \tag{9.17}$$

Since the thermal neutron has negligible energy by comparison, the gamma photon has the energy $Q = 2.22$ MeV released by the reaction, which represents the binding energy of the deuteron.[4] When tissue is exposed to thermal neutrons, the reaction (9.17) provides a source of gamma rays that delivers dose to the tissue. The capture cross section for the reaction (9.17) for thermal neutrons is 0.33 barn.

Capture cross sections for low-energy neutrons generally decrease as the reciprocal of the velocity as the neutron energy increases. This phenomenon is often called the "$1/v$ law." Thus, if the capture cross section $\sigma_0$ is known for a given velocity $v_0$ (or energy $E_0$), then the cross section at some other velocity $v$ (or energy $E$) can be estimated from the relations

$$\frac{\sigma}{\sigma_0} = \frac{v_0}{v} = \sqrt{\frac{E_0}{E}}. \tag{9.18}$$

These expressions can generally be used for neutrons of energies up to 100 eV or 1 keV, depending on the absorbing nucleus.

4   The energetics were worked out in Chapter 3
    [Eq. (3.8)].

*Example*

The capture cross section for the reaction (9.17) for thermal neutrons is 0.33 barn. Estimate the cross section for neutrons of energy 10 eV.

*Solution*

The thermal-neutron energy is usually assumed to be the most probable value, $E_0 = 0.025$ eV. Applying Eq. (9.18) with $\sigma_0 = 0.33$ barn and $E = 10$ eV, we find for the capture cross section at 10 eV

$$\sigma = \sigma_0 \sqrt{\frac{E_0}{E}} = 0.33 \sqrt{\frac{0.025}{10}} = 0.0165 \text{ barn.} \tag{9.19}$$

### $^3$He(n, p)$^3$H

The cross section for thermal-neutron capture is 5330 barns and the energy $Q = 765$ keV is released by the reaction following thermal-neutron capture. Some proportional counters used for fast-neutron monitoring contain a little added $^3$He for calibration purposes. A polyethylene sleeve slipped over the tube thermalizes incident neutrons by the time they enter the counter gas. The pulse-height spectrum then shows a peak, which identifies the channel number in which a pulse height of 765 keV is registered. With the energy per channel thus established, the polyethylene sleeve can be removed and the instrument used for fast-neutron monitoring. Other neutron devices use $^3$He as the proportional-counter gas. They measure a continuum of pulse heights due to the recoil $^3$He nuclei from elastic scattering. In addition, when a neutron with kinetic energy $T$ is captured by a $^3$He nucleus, an energy of $T + 765$ keV is released (see Fig. 10.45).

### $^6$Li(n, t)$^4$He

This reaction, which produces a $^3$H nucleus, or triton (t), and has a $Q$ value of 4.78 MeV, is also used for thermal-neutron detection. The cross section is 940 barns, and the isotope $^6$Li is 7.42% abundant. Neutron-sensitive LiI scintillators can be made, and Li can also be added to other scintillators to register neutrons. Lithium enriched in the isotope $^6$Li is available.

### $^{10}$B(n, $\alpha$)$^7$Li

For this reaction, $\sigma = 3840$ barns for thermal neutrons. The isotope $^{10}$B is 19.7% abundant. In 96% of the reactions the $^7$Li nucleus is left in an excited state and emits a 0.48-MeV gamma ray. The total kinetic energy shared by the alpha particle and $^7$Li recoil nucleus is then $Q = 2.31$ MeV. The other 4% of the reactions go to the ground state of the $^7$Li nucleus with $Q = 2.79$ MeV. BF$_3$ is a gas that can be used directly in a neutron counter. Boron is also employed as a liner inside the tubes of proportional counters for neutron detection. It is also used as a neutron shielding material. Boron enriched in the isotope $^{10}$B is available. Additional information is given in Section 10.7.

## $^{14}$N(n, p)$^{14}$C

Since nitrogen is a major constituent of tissue, this reaction, like neutron capture by hydrogen, contributes to neutron dose. The cross section for thermal neutrons is 1.70 barns, and the $Q$ value is 0.626 MeV. Since their ranges in tissue are small, the energies of the proton and $^{14}$C nucleus are deposited locally at the site where the neutron was absorbed. Capture by hydrogen and by nitrogen are the only two processes through which thermal neutrons deliver a significant dose to soft tissue.

## $^{23}$Na(n, $\gamma$)$^{24}$Na

Absorption of a neutron by $^{23}$Na gives rise to the radioactive isotope $^{24}$Na. The latter has a half-life of 15.0 h and emits two gamma rays, having energies of 2.75 MeV and 1.37 MeV, per disintegration. The thermal-neutron capture cross section is 0.534 barn. Since $^{23}$Na is a normal constituent of blood, activation of blood sodium can be used as a dosimetric tool when persons are exposed to relatively high doses of neutrons, for example, in a criticality accident.

## $^{32}$S(n, p)$^{32}$P

For this reaction to occur, the neutron must have an energy of at least 0.957 MeV [Eq. (9.32)]. It is an example of but one of many threshold reactions used for neutron detection. As described in Section 10.7, the simultaneous activation of foils made from a series of nuclides with different thresholds provides a means of estimating neutron spectra. The existence of $^{32}$S in human hair has also been used to help estimate high-energy ($\sim$3.2 MeV) neutron doses to persons exposed in criticality accidents. The product $^{32}$P, a pure beta emitter with a maximum energy of 1.71 MeV and a half-life of 14.3 days, is easily counted.

## $^{113}$Cd(n, $\gamma$)$^{114}$Cd

Because of the large, 21,000-barn, thermal-neutron capture cross section of $^{113}$Cd, cadmium is used as a neutron shield and as a reactor control-rod material. The relative abundance of the $^{113}$Cd isotope is 12.3%. The absorption cross section of $^{113}$Cd for neutrons is large from thermal energies up to $\sim$0.2 eV. It drops off two orders of magnitude between 0.2 eV and 0.6 eV. A method for measuring the ratio of thermal to resonance neutrons consists of comparing the induced activities in two identical foils (e.g., indium), one bare and the other covered with a cadmium shield. The latter absorbs essentially all neutrons with energies below the so-called cadmium cutoff of $\sim$0.4 eV.

## $^{115}$In(n, $\gamma$)$^{116m}$In

The cross section for thermal-neutron capture by $^{115}$In (95.7% abundant) is 157 barns, and the metastable $^{116m}$In decays with a half-life of 54.2 min. The induced activity in indium foils worn by persons suspected of having been exposed to neutrons can be checked as a quick-sort method following a criticality accident. In practical cases the method is sensitive enough to permit detection with an ionization chamber as well as a GM or scintillation survey instrument. The degree of foil activity depends so strongly on the orientation of the exposed person, the neutron-energy spectrum, and other factors that it does not provide a useful basis for even a crude estimate of dose. However, the fact that an exposure occurred (or did not) can thus be established.

### $^{197}$Au(n, $\gamma$)$^{198}$Au

This isotope, which is 100% abundant, has a thermal-neutron capture cross section of 98.8 barns. Although not as sensitive as indium, its longer half-life of 2.70 days permits monitoring at later times after exposure.

### $^{235}$U(n, f)

Fission (f) is discussed in Section 9.10. Because of the large release of energy (~200 MeV), the fission process provides a distinct signature for detecting thermal neutrons, even in high backgrounds of other types of radiation (cf. Section 10.7).

## 9.8
## Energetics of Threshold Reactions

As mentioned in the last section in connection with the reaction $^{32}_{16}$S(n, p)$^{32}_{15}$P, the activation of different nuclides through reactions with different threshold energies provides information on the spectrum of neutrons to which they are exposed. In this section we show how threshold energies can be calculated.

An endothermic reaction, by definition, requires the addition of energy in order to take place. The reaction thus converts energy into mass. (In the notation of Section 3.2, $Q < 0$.) Such a reaction can be brought about by one particle striking another at rest, provided the incident particle has sufficient energy. In Section 8.6 we considered threshold energies for photonuclear reactions. In this instance, the reaction occurs when the photon has an energy $h\nu \geq -Q$ needed to provide the increase in mass. The condition for the threshold energy for a neutron reaction is different. The neutron must have enough energy to supply both the increase in mass, $-Q$, and also the continued motion of the center of mass of the colliding particles after the collision. For photons, the latter is negligible.

To calculate threshold energies we consider a head-on collision. A particle with mass $M_1$ strikes a particle with mass $M_2$, initially at rest. The identity of the particles is changed by the reaction, and so there will generally be different masses, $M_3$ and $M_4$, after the encounter. The collision is shown schematically in Fig. 9.9. The

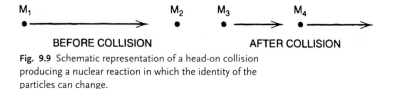

**Fig. 9.9** Schematic representation of a head-on collision producing a nuclear reaction in which the identity of the particles can change.

change in rest energy, $Q = M_1 + M_2 - (M_3 + M_4)$, is negative for the endothermic reaction.[5] The conservation of total energy requires that

$$E_1 = E_3 + E_4 - Q, \tag{9.20}$$

where $E_1$, $E_3$, and $E_4$ are the kinetic energies of the moving particles. Conservation of momentum gives

$$p_1 = p_3 + p_4, \tag{9.21}$$

where $p_1$, $p_3$, and $p_4$ are the magnitudes of the respective momenta. To calculate the threshold energy $E_1$, we eliminate either $E_3$ or $E_4$ from these two equations and solve for the other. This procedure will give the explicit condition that $E_1$ must fulfill.

We eliminate $E_4$. Using the relationship, $E_4 = p_4^2/2M_4$, between energy and momentum, we write with the help of Eq. (9.21)

$$E_4 = \frac{p_4^2}{2M_4} = \frac{1}{2M_4}(p_1 - p_3)^2. \tag{9.22}$$

Substituting $p_1 = (2M_1 E_1)^{1/2}$ and $p_3 = (2M_3 E_3)^{1/2}$ gives

$$E_4 = \frac{1}{M_4}[M_1 E_1 - 2(M_1 M_3)^{1/2}(E_1 E_3)^{1/2} + M_3 E_3]. \tag{9.23}$$

Using this expression in Eq. (9.20) and carrying out some algebraic manipulations, one finds for $E_3$ that

$$E_3 - \frac{2(M_1 M_3 E_1)^{1/2}}{M_3 + M_4}\sqrt{E_3} - \frac{(M_4 - M_1)E_1 + M_4 Q}{M_3 + M_4} = 0. \tag{9.24}$$

This is a quadratic equation in $\sqrt{E_3}$, having the form

$$E_3 - 2A\sqrt{E_3} - B = 0, \tag{9.25}$$

where $A$ and $B$ are the coefficients that appear in Eq. (9.24). The two roots yield

$$E_3 = B + 2A^2\left(1 \pm \frac{1}{A}\sqrt{A^2 + B}\right). \tag{9.26}$$

For $E_3$ to be real, $A^2 + B \geq 0$:

$$\frac{M_1 M_3 E_1}{(M_3 + M_4)^2} + \frac{(M_4 - M_1)E_1 + M_4 Q}{M_3 + M_4} \geq 0, \tag{9.27}$$

or

$$E_1 \geq -Q\left(1 + \frac{M_1}{M_3 + M_4 - M_1}\right). \tag{9.28}$$

---

5   We assume that both particles are in their ground states after the collision. If particle 3 or 4 is a nucleus in an excited state with energy $E^*$ above the ground-state energy, then one must replace $Q$ by $Q - E^*$ in this discussion.

The smallest possible value of $E_1$, satisfying the equality, is the threshold energy for the incident particle:

$$E_{th} = -Q\left(1 + \frac{M_1}{M_3 + M_4 - M_1}\right). \tag{9.29}$$

The smaller the incident-particle mass $M_1$ is, compared with the total mass $M_3 + M_4$, the more nearly the threshold energy is equal to $-Q$. It differs from $-Q$ by the energy needed to keep the center of mass in motion (Fig. 9.9).

*Example*

Calculate the threshold energy for the reaction $^{32}S(n,p)^{32}P$.

*Solution*

For a neutron (mass $M_1$) incident on a sulfur atom at rest in the laboratory, we write

$$^1_0n + ^{32}_{16}S \rightarrow ^1_1H + ^{32}_{15}P. \tag{9.30}$$

Since the number of electrons is the same on both sides of the arrow, we can use the atomic mass differences $\Delta$ given in Appendix D to find $Q$. Taking the values in the order in which they occur in (9.30), we obtain

$$Q = 8.0714 - 26.013 - (7.2890 - 24.303) = -0.9276. \tag{9.31}$$

The threshold energy is, from Eq. (9.29),

$$E_{th} = 0.9276\left(1 + \frac{1}{1 + 32 - 1}\right) = 0.957 \text{ MeV}. \tag{9.32}$$

Nuclear cross sections for neutron reactions with a threshold usually increase steadily from zero at $E_{th}$ to a maximum and then decline at higher energies. The activity induced in a foil is indicative of the relative number of neutrons over a range of energies above the threshold. The energy at which the cross section has approximately its average value is called the *effective threshold energy*, which is larger than $E_{th}$. For $^{32}S$, the effective threshold energy for neutron activation is about 3.2 MeV.

For a nuclear reaction to be induced by a positively charged projectile, the repulsive Coulomb barrier must be overcome. For example, the reaction $^{14}N(\alpha,p)^{17}O$ has the value $Q = -1.19$ MeV. Equation (9.29) implies that the threshold energy for the reaction is $E_{th} = 1.53$ MeV. However, the Coulomb barrier for the alpha particle and the nitrogen nucleus is 3.9 MeV. The effective threshold energy for the reaction to occur at an appreciable rate is about 4.6 MeV.

## 9.9
## Neutron Activation

The time dependence of the activity induced by neutron capture can be described quantitatively. If a sample containing $N_T$ target atoms with cross section $\sigma$ is exposed to a uniform, broad beam of monoenergetic neutrons with a fluence rate $\dot{\Phi}$, then the production rate of daughter atoms from neutron absorption is $\dot{\Phi}\sigma N_T$. If the number of daughter atoms in the sample is $N$ and the decay constant is $\lambda$, then the rate of loss of daughter atoms from the sample is $\lambda N$. Thus, the rate of change $dN/dt$ in the number of daughter atoms present at any time while the sample is being bombarded is given by

$$\frac{dN}{dt} = \dot{\Phi}\sigma N_T - \lambda N. \tag{9.33}$$

To solve this equation, we assume that the fluence rate is constant and that the original number of target atoms is not significantly depleted, so that $N_T$ is also constant. Without the term $\dot{\Phi}\sigma N_T$, which is then constant, Eq. (9.33) would be the same as the linear homogeneous Eq. (4.2). Therefore, we try a solution to (9.33) in the same form as (4.7) for the homogeneous equation plus an added constant. Substituting $N = a + be^{-\lambda t}$ into Eq. (9.33) gives

$$-b\lambda e^{-\lambda t} = \dot{\Phi}\sigma N_T - a\lambda - b\lambda e^{-\lambda t}. \tag{9.34}$$

The exponential terms on both sides cancel, and one finds that $a = \dot{\Phi}\sigma N_T/\lambda$. Thus, the general solution is

$$N = \frac{\dot{\Phi}\sigma N_T}{\lambda} + be^{-\lambda t}. \tag{9.35}$$

The constant $b$ depends on the initial conditions. Specifying that no daughter atoms are present when the neutron irradiation begins (i.e., $N = 0$ when $t = 0$), we find from (9.35) that $b = -\dot{\Phi}\sigma N_T/\lambda$. Thus we obtain the final expression

$$\lambda N = \dot{\Phi}\sigma N_T(1 - e^{-\lambda t}). \tag{9.36}$$

The left-hand side expresses the activity of the daughter as a function of the time $t$. The quantity $\dot{\Phi}\sigma N_T$ is called the saturation activity because it represents the maximum activity obtainable when the sample is irradiated for a long time ($t \to \infty$). A sketch of the function (9.36) is shown in Fig. 9.10.

When the neutrons are not monoenergetic, the terms in Eq. (9.33) can be treated in separate energy groups. Alternatively, Eq. (9.36) can be used as is, provided the proper average cross section is used for $\sigma$.

*Example*

A 3-g sample of $^{32}$S is irradiated with fast neutrons having a constant fluence rate of 155 cm$^{-2}$ s$^{-1}$. The cross section for the reaction $^{32}$S$(n, p)^{32}$P is 0.200 barn, and the half-life of $^{32}$P is $T = 14.3$ d. What is the maximum $^{32}$P activity that can be induced? How many days are needed for the level of the activity to reach three quarters of the maximum?

*Solution*

The total number of target atoms is $N_T = \frac{3}{32} \times 6.02 \times 10^{23} = 5.64 \times 10^{22}$. The maximum (saturation) activity is $\dot{\Phi}\sigma N_T = (155 \text{ cm}^{-2}\text{ s}^{-1})(0.2 \times 10^{-24} \text{ cm}^2)(5.64 \times 10^{22}) =$

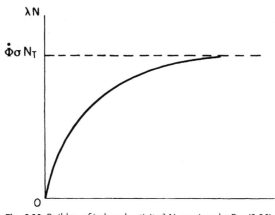

**Fig. 9.10** Buildup of induced activity $\lambda N$, as given by Eq. (9.36), during neutron irradiation at constant fluence rate.

$1.75\ s^{-1} = 1.75$ Bq. [Expressed in curies, the saturation activity is $1.75/(3.7 \times 10^{10}) = 4.73 \times 10^{-11}$ Ci.] The time $t$ needed to reach three-quarters of this value can be found from Eq. (9.36) by writing $\frac{3}{4} = 1 - e^{-\lambda t}$. Then $e^{-\lambda t} = \frac{1}{4}$ and $t = 2T = 28.6$ d. Note that the buildup toward saturation activity is analogous to the approach to secular equilibrium by the daughter of a long-lived parent (Sect. 4.4).

*Example*

Estimate the fraction of the $^{32}$S atoms that would be consumed in the last example in 28.6 days.

*Solution*

The rate at which $^{32}$S atoms are used up is $\dot{\Phi}\sigma N_T = 1.75\ s^{-1}$. Since $t = 28.6\ d = 2.47 \times 10^6$ s, the number of $^{32}$S atoms lost is $1.75 \times 2.47 \times 10^6 = 4.32 \times 10^6$. The fraction of $^{32}$S atoms consumed, therefore, is $4.32 \times 10^6/(5.64 \times 10^{22}) = 7.66 \times 10^{-17}$, a negligible amount. Note that fractional burnup does not depend on the sample size $N_T$. The problem can also be worked by writing for the desired fraction $\dot{\Phi}\sigma N_T t/N_T = \dot{\Phi}\sigma t = 155 \times 0.2 \times 10^{-24} \times 2.47 \times 10^6 = 7.66 \times 10^{-17}$. The assumption of constant $N_T$ for the validity of Eq. (9.36) is therefore warranted. The actual fraction of $^{32}$S atoms consumed during a long time $t$ is, of course, less than $\dot{\Phi}\sigma t$.

## 9.10
## Fission

As described in Section 3.2, the binding energy per nucleon for heavy elements decreases as the atomic mass number increases (Fig. 3.3). Thus, when heavy nuclei are split into smaller pieces, energy is released. Alpha decay is an example of one such process that is spontaneous. With the discovery of nuclear fission, another process was realized, in which the splitting was much more dramatic and the energy release almost two orders of magnitude greater. Nuclear fission can be induced in certain nuclei as a result of absorbing a neutron. With $^{235}$U, $^{239}$Pu, and $^{233}$U, absorption of a thermal neutron can set up vibrations in the nucleus which cause it to become so distended that it splits apart under the mutual electrostatic repulsion of its parts. The thermal-neutron fission cross sections for these isotopes are, respectively, 580, 747, and 525 barns. A greater activation energy is required to cause other nuclei to fission. An example is $^{238}$U, which requires a neutron with a kinetic energy in excess of 1 MeV to fission. Cross sections for such "fast-fission" reactions are much smaller than those for thermal fission. The fast-fission cross section for $^{238}$U, for instance, is 0.29 barn. Also, fission does not always result when a neutron is absorbed by a fissionable nucleus. $^{235}$U fissions only 85% of the time after thermal-neutron absorption.

Nuclei with an odd number of nucleons fission more readily following neutron absorption than do nuclei with an even number of nucleons. This fact is related to the greater binding energy per nucleon found in even–even nuclei, as mentioned in Section 3.2. The $^{235}$U nucleus is even–odd in terms of its proton and neutron numbers. Addition of a neutron transforms it into an even–even nucleus with a larger energy release than that following neutron absorption by $^{238}$U.

Fissionable nuclei break up in a number of different ways. The $^{235}$U nucleus splits in some 40 or so modes following the absorption of a thermal neutron. One typical example is the following:

$$\ {}_{0}^{1}n + {}_{92}^{235}U \rightarrow {}_{57}^{147}La + {}_{35}^{87}Br + 2{}_{0}^{1}n. \tag{9.37}$$

An average energy of about 195 MeV is released in the fission process, distributed as shown in Table 9.5. The major share of the energy (162 MeV) is carried away by the charged fission fragments, such as the La and Br fragments in (9.37). Fission neutrons and gamma rays account for another 12 MeV. Subsequent fission-product decay accounts for 10 MeV and neutrinos carry off 11 MeV. In a new reactor, in which there are no fission products, the energy output is about 175 MeV per fission. In an older reactor, with a significant fission-product inventory, the corresponding figure is around 185 MeV. The energy produced in a reactor is converted mostly into heat from the stopping of charged particles, including the recoil nuclei struck by neutrons and the secondary electrons produced by gamma rays. Neutrinos escape with negligible energy loss.

As exemplified by (9.37), nuclear fission produces asymmetric masses with high probability. The mass distribution of fission fragments from $^{235}$U is thus bimodal.

All fission fragments are radioactive and most decay through several steps to stable daughters. The decay of the collective fission-product activity following the fission of a number of atoms at $t = 0$ is given by

$$A \cong 10^{-16}t^{-1.2} \text{ curies/fission}, \tag{9.38}$$

where $t$ is in days. This expression can be used for estimating residual fission-product activity between about 10 s and 1000 h.

The average number of neutrons produced per fission of $^{235}$U is 2.5. This number must exceed unity in order for a chain reaction to be possible. Some 99.36% of

**Table 9.5** Average Distribution of Energy Among Products
Released by Fission of $^{235}$U

| | |
|---|---|
| Kinetic energy of charged fission fragments | 162 MeV |
| Fission neutrons | 6 |
| Fission gamma rays | 6 |
| Subsequent beta decay | 5 |
| Subsequent gamma decay | 5 |
| Neutrinos | 11 |
| Total | 195 MeV |

the fission neutrons are emitted promptly (in $\sim 10^{-14}$ s) from the fission fragments, while the other (delayed) neutrons are emitted later (up to $\sim 1$ min or more). The delayed neutrons play an important role in the ease of control of a nuclear reactor, as discussed in the next section.

## 9.11
## Criticality

An assembly of fissionable material is said to be critical when, on the average, exactly one of the several neutrons emitted in the fission process causes another nucleus to fission. The power output of the assembly is then constant. The other fission neutrons are either absorbed without fission or else escape from the system. Criticality thus depends upon geometrical factors as well as the distribution and kinds of the material present. If an average of more than one fission neutron produces fission of another nucleus, then the assembly is said to be supercritical, and the power output increases. If less than one fission occurs per fission neutron produced, the unit is subcritical.

Criticality is determined by the extent of neutron multiplication as successive generations are produced. If $N_i$ thermal neutrons are present in a system, their absorption will result in a certain number $N_{i+1}$ of next-generation thermal neutrons. The effective multiplication factor is defined as

$$k_{\text{eff}} = \frac{N_{i+1}}{N_i}. \tag{9.39}$$

The system is critical if $k_{\text{eff}} = 1$, exactly; supercritical if $k_{\text{eff}} > 1$; and subcritical if $k_{\text{eff}} < 1$.

It is useful to discuss $k_{\text{eff}}$ independently of the size and shape of an assembly. Therefore, we introduce the infinite multiplication factor, $k_\infty$, for a system that is infinite in extent. For a finite system one can then write

$$k_{\text{eff}} = L k_\infty, \tag{9.40}$$

where $L$ is the probability that a neutron will not escape. The value of $k_\infty$ will depend on several factors, as we now describe for uranium fuel.

Of $N_i$ total thermal neutrons present in the $i$th generation in an infinite system, generally only a fraction $f$, called the thermal utilization factor, will be absorbed in the fissionable fuel (viz., $^{235}$U and $^{238}$U). The rest will be absorbed by other kinds of atoms (moderator and impurities). If $\eta$ represents the average number of fission neutrons produced per thermal-neutron capture in the uranium fuel, then the disappearance of the $N_i$ thermal neutrons will result in the production of $N_i f \eta$ fission neutrons that belong to the next generation. Some of these neutrons will produce fast fission in $^{238}$U before they have a chance to become thermalized. The fast-fission factor $\epsilon$ is defined as the ratio of the total number of fission neutrons and the number produced by thermal fission. Then the absorption of the original $N_i$ thermal neutrons results in the production of a total of $N_i f \eta \epsilon$ fission neutrons. Not all of these will become thermalized, because they may undergo radiative capture in the fuel ($^{238}$U) and moderator. Many materials have resonances in the (n, $\gamma$) cross section at energies of several hundred eV and downward. Letting $p$ represent the resonance escape probability (i.e., the probability that a fast neutron will slow down to thermal energies without radiative capture), we obtain for the total number $N_{i+1}$ of thermal neutrons in the next generation

$$N_{i+1} = N_i f \eta \epsilon p. \tag{9.41}$$

From the definition (9.39), it follows that the infinite-system multiplication factor is given by

$$k_\infty = f \eta \epsilon p. \tag{9.42}$$

The right-hand side of Eq. (9.42) is called the four-factor formula, describing the multiplication factor for an infinitely large system. The factors $f$, $\epsilon$, and $p$ depend on the composition and enrichment of the fuel and its physical distribution in the moderator. For a pure uranium-metal system, $f = 1$. The thermal utilization factor can be small if competition for thermal absorption by other materials is great. For pure natural uranium, the fast-fission factor has the value $\epsilon = 1.3$; for a homogeneous distribution of fuel and moderator, $\epsilon \sim 1$. Generally, but depending on the particular circumstances, $p$ ranges from $\sim$0.7 to $\sim$1 for enriched systems. For pure $^{235}$U, $p = 1$. For natural uranium metal, $p = 0$; and so such a system—even of infinite extent—will not be critical. The fourth factor, $\eta$, depends only on the fuel. The isotope $^{235}$U emits an average of 2.5 neutrons per fission. However, since thermal capture by $^{235}$U results in fission only 85% of the time, if follows that, for pure $^{235}$U, $\eta = 0.85 \times 2.5 = 2.1$. For other enrichments, $\eta < 2.1$.

*Example*

Given the thermal-neutron fission cross section of $^{235}$U, 580 barns, and the thermal-neutron absorption cross section for $^{238}$U, 2.8 barns, find the value of $\eta$ for natural uranium, which consists of 0.72% $^{235}$U and 99.28% $^{238}$U.

*Solution*

By definition, $\eta$ is the average number of fission neutrons produced per thermal neutron absorbed in the uranium fuel ($^{235}$U plus $^{238}$U), which is the only element present. From the given fission cross section it follows that the thermal-neutron absorption cross section for $^{235}$U, which fissions 85% of the time, is $580/0.85 = 682$ barns. The fraction of thermal-neutron absorption events in the natural uranium fuel that result in fission is therefore $0.0072 \times 580/(0.0072 \times 682 + 0.9928 \times 2.8) = 0.543$. Since, on the average, 2.5 neutrons are emitted when $^{235}$U fissions, it follows that $\eta = 0.543 \times 2.5 = 1.36$.

### Example

Compute $\eta$ for solid uranium metal that is 90% enriched in $^{235}$U.

### Solution

Similar to the last problem, we have here

$$\eta = \frac{0.90 \times 580}{0.90 \times 682 + 0.10 \times 2.8} \times 2.5 = 2.13. \tag{9.43}$$

Note that the largest possible value for $\eta$ (that for pure $^{235}$U) is $\frac{580}{682} \times 2.5 = 0.85 \times 2.5 = 2.13$, as pointed out earlier.

In its basic form, a nuclear reactor is an assembly that consists of fuel (usually enriched uranium); a moderator, preferably of low atomic mass number, for thermalizing neutrons; control rods made of materials with a high thermal-neutron absorption cross section (e.g., cadmium or boron steel); and a coolant to remove the heat generated. With the control rods fully inserted in the reactor, the multiplication constant $k$ is less than unity. As a rod is withdrawn, $k$ increases; and when $k = 1$, the reactor becomes critical and produces power at a constant level. Further withdrawal of control rods makes $k > 1$ and causes a steady increase in the power level. When the desired power level is attained, the rods are partially reinserted to make $k = 1$, and the reactor operates at a steady level.

As mentioned in the last section, 99.36% of the fission neutrons from $^{235}$U are prompt; that is, they are emitted immediately in the fission process, while the other 0.64% are released at times of the order of seconds to over a minute after fission. Since a fission neutron can be thermalized in a fraction of a second, the existence of these delayed neutrons greatly facilitates the control of a uranium reactor. If $1 < k < 1.0064$, the reactor is said to be in a delayed critical condition, since the delayed neutrons are essential to maintaining the chain reaction. The rate of power increase is then sufficiently slow to allow control through mechanical means, such as the physical adjustment of control-rod positions. When $k > 1.0064$, the chain reaction can be maintained by the prompt neutrons alone, and the condition is called prompt critical. The rate of increase in the power level is then much faster than when delayed critical.

Whenever fissionable material is chemically processed, machined, transported, stored, or otherwise handled, care must be taken to prevent accidental criticality. Generally, procedures for avoiding criticality depend on limiting the total mass or concentration of fissionable material present and on the geometry in which it is contained. For example, an infinitely long, water-reflected cylinder of aqueous solution with a concentration of 75 g $^{235}$U per liter is subcritical as long as its diameter is less than 6.3 in. Without water reflection the limiting diameter is 8.7 in. When using such "always safe" geometry, attention must be given to the possibility that two or more subcritical units could become critical in close proximity.

## 9.12
## Suggested Reading

1 Attix, F. H., *Introduction to Radiological Physics and Radiation Dosimetry*, Wiley, New York, NY (1986). [Chapter 16 covers neutron interactions and dosimetry.]

2 Bromley, D. A., "Neutrons in Science and Technology," *Physics Today* **36**(12), 30–39 (Dec. 1983). [This article describes various uses of the neutron as a tool for basic physics research; for industrial applications, such as mining and food preservation; and for art history.]

3 Cember, H., *Introduction to Health Physics*, 3rd ed., McGraw-Hill, New York, NY (1996). [Neutron interactions are described. Chapter 12 is on criticality.]

4 Turner, J. E., Kher, R. K., Arora, D., Bisht, J. S., Neelavathi, V. N., and Vora, R. B., "Quantum Mechanical Calculation of Neutron Stopping Power," *Phys. Rev. B* **8**, 4057–4062 (1973).

## 9.13
## Problems

1. Calculate the neutron energy from a $^{24}$NaD$_2$O source (Table 9.3).

2. (a) What is the maximum energy that a 4-MeV neutron can transfer to a $^{10}$B nucleus in an elastic collision?
   (b) Estimate the average energy transferred per collision.

3. (a) Estimate the average energy that a 2-MeV neutron transfers to a deuteron in a single collision.
   (b) What is the maximum possible energy transfer?

4. Make a rough estimate of the number of collisions that a neutron of 2-MeV initial energy makes with deuterium in order for its energy to be reduced to 1 eV.

5. Repeat Problem 4 for a 4-MeV neutron in carbon.

6. (a) If a neutron starts with an energy of 1 MeV in a graphite moderator, what is the minimum number of collisions it must make with carbon nuclei in order to become thermalized?
   (b) What is the minimum number of collisions if the moderator is hydrogen?

7. A parallel beam of neutrons incident on $H_2O$ produces a 4-MeV recoil proton in the straight-ahead direction.
   (a) What is the maximum energy of the neutrons?
   (b) What is the maximum recoil energy of an oxygen nucleus?
8. (a) Estimate the average recoil energy of a carbon nucleus scattered elastically by 1-MeV neutrons.
   (b) What is the average recoil energy of a hydrogen nucleus?
   (c) Discuss the relative importance of these two reactions as a basis for producing biological effects in soft tissue exposed to 1-MeV neutrons.
9. Derive Eq. (9.5).
10. A parallel beam of 10-MeV neutrons is normally incident on a layer of water, 0.5 cm thick. Ignore neutron collisions with oxygen and multiple collisions with hydrogen.
    (a) How many neutron collisions per second deposit energy at a rate of $10^{-7}$ J s$^{-1}$?
    (b) How many incident neutrons per second are needed to produce this rate of energy deposition?
11. (a) What is the probability that a 5-MeV neutron will lose between 4.0 and 4.2 MeV in a single collision with an atom of hydrogen?
    (b) What is the probability of this energy loss for an 8-MeV neutron?
12. Show that the probability given by Eq. (9.7) is normalized.
13. What is the probability that a fast neutron will be scattered by a proton at an angle between 100° and 120° in the center-of-mass system?
14. (a) What is the probability that a proton will recoil at an angle between 20° and 30° in the laboratory system when struck by a fast neutron?
    (b) What is the probability that a fast neutron will be scattered at an angle between 20° and 30° by a proton in the laboratory system?
    (c) Account for the fact that these probabilities do not depend on the neutron energy.
15. Show that the center of mass in Fig. 9.9 moves in the laboratory system with speed $M_1 V/(M_1 + M_2)$, where $V$ is the speed of the incident particle.
16. Construct a figure like 9.7(a) for the collision of two particles of unequal masses. Show that the relation, $\omega = 2\theta$, also holds in this case for the scattering angles of the struck particle in the two coordinate systems.
17. Why does the photon in the reaction (9.17) get the energy $Q = 2.22$ MeV, while the deuteron gets negligible energy?
18. Estimate the capture cross section for 19-eV neutrons by $^{235}U$.

19. (a) Estimate the capture cross section of $^{10}$B for 100-eV
    neutrons.
    (b) What is the capture probability per cm for a 100-eV
    neutron in pure $^{10}$B (density $= 2.17$ g cm$^{-3}$)?
    (c) Estimate the probability that a 100-eV neutron will
    penetrate a 1 cm $^{10}$B shield and produce a fission in a
    1-mm $^{239}$Pu foil (density 18.5 g cm$^{-3}$) behind it. Neglect
    energy loss of the neutron due to elastic scattering.

20. The thermal-neutron reaction $^{10}_{5}$B$(n, \alpha)^{7}_{3}$Li leaves the $^{7}_{3}$Li
    nucleus in the ground state 4% of the time. Otherwise, the
    reaction leaves the $^{7}_{3}$Li nucleus in an excited state, from which
    it decays to the ground state by emission of a 0.48-MeV gamma
    ray.
    (a) Calculate the $Q$ value of the reaction in both cases.
    (b) Calculate the alpha-particle energy in both cases.

21. How much energy is released per minute by the $^{14}_{7}$N$(n, p)^{14}_{6}$C
    reaction in a 10-g sample of soft tissue bombarded by $2 \times 10^{10}$
    thermal neutrons cm$^{-2}$ s$^{-1}$? Nitrogen atoms constitute 3% of
    the mass of soft tissue.

22. (a) Calculate the $Q$ value for the reaction $^{3}$H$(p, n)^{3}$He.
    (b) What is the threshold energy for this reaction when
    protons are incident on a tritium target?

23. How much kinetic energy is associated with the center-of-mass
    motion in the last problem?

24. Calculate the threshold energy for neutron production by
    protons striking a lithium target via the reaction $^{7}$Li$(p,n)^{7}$Be.

25. Calculate the maximum energy of the neutrons produced when
    10-MeV protons strike a $^{7}$Li target.

26. Verify Eqs. (9.20)–(9.29).

27. A sample containing 127 g of $^{23}$Na (100% abundant) is exposed
    to a beam of thermal neutrons at a constant fluence rate of
    $1.19 \times 10^{4}$ cm$^{-2}$ s$^{-1}$. The thermal-neutron capture cross section
    for the reaction $^{23}$Na$(n, \gamma)^{24}$Na is 0.53 barn.
    (a) Calculate the saturation activity.
    (b) Calculate the $^{24}$Na activity in the sample 24 h after it is
    placed in the beam.
    (c) How many $^{23}$Na atoms are consumed in the first 24 h?

28. What is the saturation activity of $^{24}$Na that can be induced in a
    400-g sample of NaCl with a constant thermal-neutron fluence
    rate of $5 \times 10^{10}$ cm$^{-2}$ s$^{-1}$? The isotope $^{23}$Na is 100% abundant
    and has a thermal-neutron capture cross section of 0.53 barn.

29. A sample containing 62 g of $^{31}$P (100% abundant) is exposed to
    $2 \times 10^{11}$ thermal neutrons cm$^{-2}$ s$^{-1}$. If the thermal-neutron
    capture cross section is 0.19 barn, how much irradiation time
    is required to make a 1-Ci source of $^{32}$P?

30. To make a $^{60}$Co source, a 50-g sample of cobalt metal ($^{59}$Co, 100% abundant) is exposed to thermal neutrons at a constant fluence rate of $10^9$ cm$^{-2}$ s$^{-1}$. The thermal-neutron capture cross section is 37 barns.

    (a) How much exposure time is required to make a 1-mCi source of $^{60}$Co?

    (b) Estimate the number of $^{59}$Co atoms consumed in 1 week.

31. A metal sample to be analyzed for its cobalt content is exposed to thermal neutrons at a constant fluence rate of $7.20 \times 10^{10}$ cm$^{-2}$ s$^{-1}$ for 10 days. The thermal-neutron absorption cross section for $^{59}$Co (100% abundant) to form $^{60}$Co is 37 barns. If, after the irradiation, the sample shows a disintegration rate of 23 min$^{-1}$ from $^{60}$Co, how many grams of cobalt are present?

32. A sample containing an unknown amount of chromium is irradiated at a constant thermal-neutron fluence rate of $10^{12}$ cm$^{-2}$ s$^{-1}$. The isotope $^{50}$Cr (4.31% abundant, by number) absorbs a thermal neutron (cross section 13.5 barns) to form radioactive $^{51}$Cr, which decays with a half-life of 27.8 d. If, after 5 d of irradiation, the induced activity of $^{51}$Cr is 1121 Bq, what is the mass of chromium in the sample?

33. Estimate the fission-product activity 48 h following a criticality accident in which there were $5 \times 10^{15}$ fissions.

34. If the exposure rate at a given location in Problem 33 is 5 R h$^{-1}$ 10 h after the accident, estimate what it will be there exactly 1 wk after the accident.

35. The "seven–ten" rule for early fallout from a nuclear explosion states that, for every sevenfold increase in time after the explosion, the exposure rate decreases by a factor of 10. Using this rule and the exposure rate at 1 h as a reference value, estimate the relative exposure rates at 7, $7 \times 7$, and $7 \times 7 \times 7$ h. Compare with ones obtained with the help of Eq. (9.38).

36. A reactor goes critical for the first time, operates at a power level of 50 W for 3 h, and is then shut down. How many fissions occurred?

37. What is the fission-product inventory in curies for the reactor in the last problem 8 h after shutdown?

38. Calculate $\eta$ for uranium enriched to 3% in $^{235}$U. The thermal-neutron fission cross section of $^{235}$U is 580 barns and the thermal-neutron absorption cross section for $^{238}$U is 2.8 barns.

39. If $k = 1.0012$, by what factor will the neutron population be increased after 10 generations?

40. How many generations are needed in the last problem to increase the power by a factor of 1000?

**9.14**

**Answers**

1. 0.26 MeV
2. (a) 1.32 MeV
   (b) 0.661 MeV
4. 25
6. (a) 53
   (b) 1
8. (a) 0.142 MeV
   (b) 0.500 MeV
10. (a) $1.25 \times 10^5 \ s^{-1}$
    (b) $4.03 \times 10^6 \ s^{-1}$
13. 0.163
18. 24.7 barns
19. (a) 60.7 barns
    (b) $7.93 \ cm^{-1}$
    (c) $1.9 \times 10^{-5}$
21. $1.65 \times 10^{10} \ MeV \, min^{-1}$

22. (a) −0.764 MeV
    (b) 1.02 MeV
27. (a) $2.10 \times 10^4$ Bq
    (b) $1.41 \times 10^4$ Bq
    (c) $1.81 \times 10^9$
28. $1.09 \times 10^{11}$ Bq
30. (a) 5.43 d
    (b) $1.14 \times 10^{16}$
32. $1.42 \times 10^{-6}$ g
33. 0.22 Ci
34. $0.17 \ R \, h^{-1}$
36. $1.9 \times 10^{16}$
37. 6 Ci
38. 1.88
39. 1.0121
40. 5760

# 10
# Methods of Radiation Detection

This chapter describes ways in which ionizing radiation can be detected and measured. Section 10.7 covers special methods applied to neutrons. In Chapter 12 we shall see how these techniques are applied in radiation dosimetry.

## 10.1
### Ionization in Gases

### Ionization Current

Figure 10.1(a) illustrates a uniform, parallel beam of monoenergetic charged particles that steadily enter a gas chamber across an area $A$ with energy $E$ and come to rest in the chamber. A potential difference $V$ applied across the parallel chamber plates $P_1$ and $P_2$ gives rise to a uniform electric field between them. As the particles slow down in the chamber, they ionize gas atoms by ejecting electrons and leaving positive ions behind. The ejected electrons can immediately produce additional ion pairs. If the electric field strength, which is proportional to $V$, is relatively weak, then only a few of the total ion pairs will drift apart under its influence, and a small current $I$ will flow in the circuit. Most of the other ion pairs will recombine to form neutral gas atoms. As shown in Fig. 10.1(b), the current $I$ can be increased by increasing $V$ up to a value $V_0$, at which the field becomes strong enough to collect all of the ion pairs produced by the incident radiation and its secondary electrons. Thereafter, the current remains on a plateau at its saturation value $I_0$ when $V > V_0$.

Since it is readily measurable, it is important to see what information the saturation current gives about the radiation. If the fluence rate is $\dot{\Phi}$ cm$^{-2}$ s$^{-1}$, then the intensity $\dot{\Psi}$ of the radiation (Section 8.8) entering the chamber is given by $\dot{\Psi} = \dot{\Phi}E$. If $W$ denotes the average energy needed to produce an ion pair when a particle of initial energy $E$ stops in the chamber, then the average number $N$ of ion pairs produced by an incident particle and its secondary electrons is $N = E/W$. The average charge (either + or –) produced per particle is $Ne$, where $e$ is the magnitude of the electronic charge. The saturation current $I_0$ in the circuit is equal to the product of $Ne$ and $\dot{\Phi}A$, the total number of particles that enter the chamber per unit time. Therefore, we have

*Atoms, Radiation, and Radiation Protection.* James E. Turner
Copyright © 2007 WILEY-VCH Verlag GmbH & Co. KGaA, Weinheim
ISBN: 978-3-527-40606-7

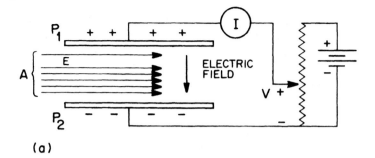

(a)

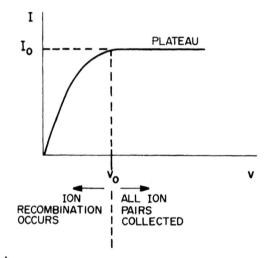

(b)

**Fig. 10.1** (a) Monoenergetic beam of particles stopping in parallel-plate ionization chamber with variable potential difference $V$ applied across plates $P_1$ and $P_2$ (seen edge on). (b) Plot of current $I$ vs. $V$.

$$I_0 = Ne\dot{\Phi}A = \frac{e\dot{\Phi}AE}{W}. \tag{10.1}$$

It follows that

$$\dot{\psi} = \dot{\Phi}E = \frac{I_0 W}{eA}, \tag{10.2}$$

showing that the beam intensity is proportional to the saturation current.

The important relationship (10.2) is of limited use, because it applies to a uniform, parallel beam of radiation. However, since the rate of total energy absorption in the chamber gas, $\dot{E}_{abs}$, is given by $\dot{E}_{abs} = \dot{\psi}A$, we can write in place of Eq. (10.2)

$$\dot{E}_{abs} = \frac{I_0 W}{e}. \tag{10.3}$$

Thus, the saturation current gives a direct measure of the rate of energy absorption in the gas. The relationship (10.3) holds independently of any particular condition on beam geometry, and is therefore of great practical utility. Fortunately, it is the energy absorbed in a biological system that is relevant for dosimetry; the radiation intensity itself is usually of secondary importance.

*Example*

Good electrometers measure currents as small as $10^{-16}$ A. What is the corresponding rate of energy absorption in a parallel-plate ionization chamber containing a gas for which $W = 30$ eV per ion pair (eV ip$^{-1}$)?

*Solution*

From Appendix B, $1\,A = 1\,C\,s^{-1}$. Equation (10.3) gives

$$\dot{E}_{abs} = \frac{(10^{-16}\,C\,s^{-1}) \times 30\,eV}{1.6 \times 10^{-19}\,C} = 1.88 \times 10^4\,eV\,s^{-1}. \tag{10.4}$$

Ionization measurements are very sensitive. This average current would be produced, for example, by a single 18.8-keV beta particle stopping in the chamber per second.

## W Values

Figure 10.2 shows $W$ values for protons (H), alpha particles (He), and carbon and nitrogen ions of various energies in nitrogen gas, $N_2$. The values represent the average energy expended per ion pair when a particle of initial energy $E$ and all of the secondary electrons it produces stop in the gas. The value for electrons, $W_\beta = 34.6$ eV ip$^{-1}$, shown by the horizontal line, is about the same as that for protons at energies $E > 10$ keV. $W$ values for heavy ions, which are constant at high energies, increase with decreasing energy because a larger fraction of energy loss results in excitation rather than ionization of the gas. Elastic scattering of the ions by nuclei also causes a large increase at low energies.

The data in Fig. 10.2 indicate that $W$ values for a given type of charged particle are approximately independent of its initial energy, unless that energy is small. This fact is of great practical significance, since it often enables absorbed energy to be inferred from measurement of the charge collected, independently of the identity or energy spectrum of the incident particles. Alternatively, the rate of energy absorption can be inferred from measurement of the current.

$W$ values for many polyatomic gases are in the range 25–35 eV ip$^{-1}$. Table 10.1 gives some values for alpha and beta particles in a number of gases.

We have defined $W$ as the average energy needed to produce an ion pair and expressed it in eV ip$^{-1}$. Since $1$ eV $= 1.60 \times 10^{-19}$ J and the charge separated per ion pair is $1.60 \times 10^{-19}$ C, it follows that $W$ has the same numerical value when expressed either in eV ip$^{-1}$ or J C$^{-1}$ (Problem 13).

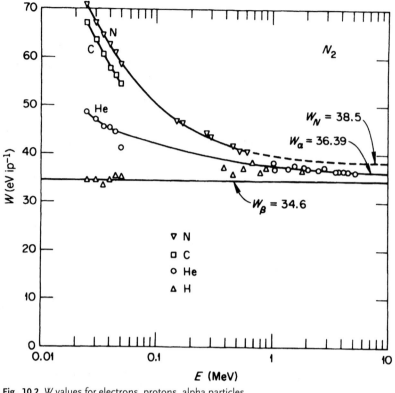

**Fig. 10.2** $W$ values for electrons, protons, alpha particles, carbon ions, and nitrogen ions in nitrogen gas as a function of initial particle energy $E$. The points represent experimental data, through which the curves are drawn. (Courtesy Oak Ridge National Laboratory, operated by Martin Marietta Energy Systems, Inc., for the Department of Energy.)

**Table 10.1** $W$ Values, $W_\alpha$ and $W_\beta$, for Alpha and Beta Particles in Several Gases

| Gas | $W_\alpha$ (eV ip$^{-1}$) | $W_\beta$ (eV ip$^{-1}$) | $W_\alpha/W_\beta$ |
|---|---|---|---|
| He | 43 | 42 | 1.02 |
| $H_2$ | 36 | 36 | 1.00 |
| $O_2$ | 33 | 31 | 1.06 |
| $CO_2$ | 36 | 33 | 1.09 |
| $CH_4$ | 29 | 27 | 1.07 |
| $C_2H_4$ | 28 | 26 | 1.08 |
| Air | 36 | 34 | 1.06 |

**Ionization Pulses**

In addition to measuring absorbed energy, a parallel-plate ionization chamber operated in the plateau region [Fig. 10.1(b)] can also be used to count particles. When a charged particle enters the chamber, the potential difference across the plates momentarily drops slightly while the ions are being collected. After collection, the potential difference returns to its original value. The electrical pulse that occurs during ion collection can be amplified and recorded electronically to register the particle. Furthermore, if the particle stops in the chamber, then, since the number of ion pairs is proportional to its original energy, the size of each pulse can be used to determine the energy spectrum. While such measurements can, in principle, be carried out, pulse ionization chambers are of limited practical use because of the attendant electronic noise.

The noise problem is greatly reduced in a proportional counter. Such a counter utilizes a gas enclosed in a tube often made with a fine wire anode running along the axis of a conducting cylindrical-shell cathode, as shown schematically in Fig. 10.3. The electric field strength at a distance $r$ from the center of the anode in this cylindrical geometry is given by

$$\epsilon(r) = \frac{V}{r\ln(b/a)}, \tag{10.5}$$

where $V$ is the potential difference between the central anode and the cylinder wall, $b$ is the radius of the cylinder, and $a$ is the radius of the anode wire. With this arrangement, very large field strengths are possible when $a$ is small in the region near the anode, where $r$ is also small. This fact is utilized as follows.

Consider the pulse produced by an alpha particle that stops in the counter gas. When the applied voltage is low, the tube operates like an ionization chamber. The number of ion pairs collected, or pulse height, is small if the voltage is low enough

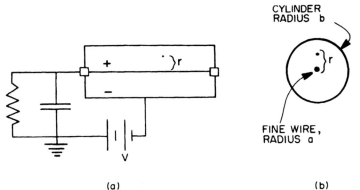

(a)             (b)

**Fig. 10.3** (a) Schematic side view and (b) end view of cylindrical proportional-counter tube. Variation of electric field strength with distance $r$ from center of anode along cylindrical axis is given by Eq. (10.5).

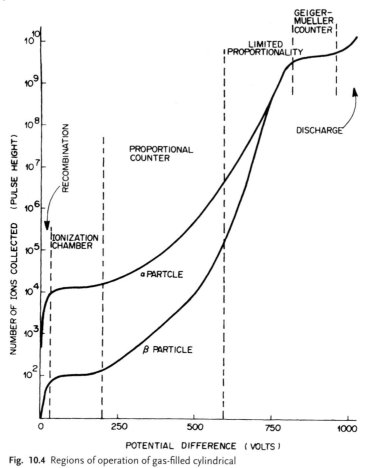

**Fig. 10.4** Regions of operation of gas-filled cylindrical ionization chamber operated in pulse mode.

to allow recombination. As the potential difference is increased, the size of the pulse increases and then levels off over the plateau region, typically up to ∼200 V, as shown in Fig. 10.4. When the potential difference is raised to a few hundred volts, the field strength near the anode increases to the point where electrons produced by the alpha particle and its secondary electrons acquire enough energy there to ionize additional gas atoms. Gas multiplication then occurs, and the number of ions collected in the pulse is proportional to the original number produced by the alpha particle and its secondaries. The tube operates as a proportional counter up to potential differences of ∼700 V and can be used to measure the energy spectrum of individual alpha particles stopping in the gas. Gas multiplication factors of ∼$10^4$ are typical.

When the potential difference is further increased the tube operates with limited proportionality, and then, at still higher voltage, enters the Geiger–Mueller (GM) region. In the latter mode, the field near the anode is so strong that any initial ionization of the gas results in a pulse, the size being independent of the number of initial ion pairs. With still further increase of the voltage, the field eventually becomes so strong that it ionizes gas atoms directly and the tube continually discharges.

Compared with an alpha particle, the pulse-height curve for a beta particle is similar, but lower, as Fig. 10.4 shows. The two curves merge in the GM region.

### Gas-Filled Detectors

Most ionization chambers for radiation monitoring are air-filled and unsealed, although sealed types that employ air or other gases are common. Used principally to monitor beta, gamma, and X radiation, their sensitivity depends on the volume and pressure of the gas and on the associated electronic readout components. The chamber walls are usually air equivalent or tissue equivalent in terms of the secondary-electron spectra they produce in response to the radiation.

Ionization chambers are available both as active and as passive detectors. An active detector, such as that illustrated by Fig. 10.1, gives an immediate reading in a radiation field through direct processing of the ionization current in an external circuit coupled to the chamber. Examples of this type of device include the free-air ionization chamber (Section 12.3) and the traditionally popular cutie pie (Fig. 10.5), a portable beta–gamma survey rate meter still in use today.

Passive pocket ionization chambers were used a great deal in the past. Basically a plastic condenser of known capacitance $C$, the unit is given a charge $Q = CV$ at a fixed potential difference $V$ before use. Exposure to radiation produces ions in the chamber volume. These partially neutralize the charge on the chamber and

**Fig. 10.5** Portable ionization-chamber survey meter (cutie pie). (Courtesy Victoreen, Inc.)

cause a voltage drop, $\Delta V$. The amount of lost charge $\Delta Q$, which depends on the energy absorbed, is directly proportional to the measured voltage change: $\Delta Q = C\Delta V$. Calibrated self-reading pocket ionization chambers, like that in Fig. 10.6, are still in use.

Proportional counters can be used to detect different kinds of radiation and, under suitable conditions, to measure radiation dose (Chapter 12). A variety of gases, pressures, and tube configurations are employed, depending on the intended purposes. The instrument shown in Fig. 10.7 uses a sealed gas proportional-counter

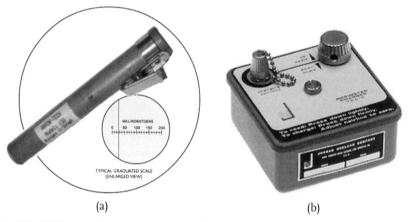

(a)            (b)

**Fig. 10.6** (a) Direct-reading, condenser-type pocket ionization chamber. Amount of exposure to X and gamma radiation can be read on calibrated scale through eyepiece. (b) Charger uses standard D battery. (Courtesy Arrow-Tech, Inc.)

**Fig. 10.7** Proportional-counter monitor for measuring dose and dose rate. See text. (Courtesy Berthold Technologies USA, LLC.)

probe to measure dose and dose rate. It can serve as a portable or stationary monitor. Different probes can be attached for different radiations (alpha, beta, gamma, and neutrons) and for different purposes, such as general surveys, surface contamination monitoring, or air monitoring. The basic control module, which contains extensive software, identifies the attached probe and automatically adjusts for it. Proportional-counter tubes may be either of a sealed or gas-flow type. As illustrated schematically in Fig. 10.8, the latter type of "windowless" counter is useful for counting alpha and soft beta particles, because the sample is in direct contact with the counter gas. Figure 10.9 displays such a system that monitors tritium activity concentration in air. The unit on the left regulates the incoming blend of the counter gas (methane or P-10, a mixture of 90% argon and 10% methane) with air, and also houses the detector and associated electronics. The unit on the right analyzes and displays the data. Single pulses of tritium beta particles (maximum energy 18.6 keV) are differentiated from other events by pulse-shape discrimination.

Pulse-height discrimination with proportional counters affords an easy means for detecting one kind of radiation in the presence of another. For example, to count a combined alpha–beta source with an arrangement like that in Fig. 10.8(a) or (b), one sets the discriminator level so that only pulses above a certain size are reg-

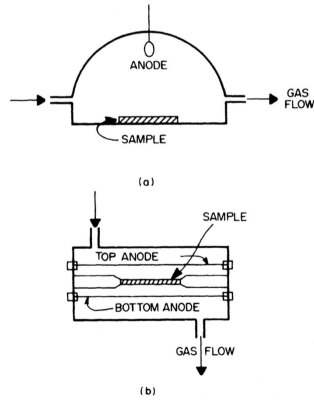

**Fig. 10.8** Diagram of (a) $2\pi$ and (b) $4\pi$ gas-flow proportional counters.

istered. One then measures the count rate at different operating voltages of the tube, leaving the discriminator level set. The resulting count rate from the alpha–beta source will have the general characteristics shown in Fig. 10.10. At low voltages, only the most energetic alpha particles will produce pulses large enough to be counted. Increasing the potential difference causes the count rate to reach a plateau when essentially all of the alpha particles are being counted. With a further increase in voltage, increased gas multiplication enables pulses from the beta particles to surpass the discriminator level and be counted. At still higher voltages, a steeper combined alpha–beta plateau is reached. The use of proportional counters for neutron measurements is described in Section 10.7. Gamma-ray discrimination

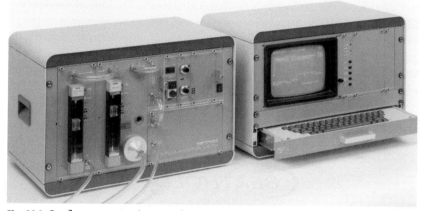

**Fig. 10.9** Gas-flow proportional counter for monitoring tritium activity concentration in air. See text. (Courtesy Berthold Technologies USA, LLC.)

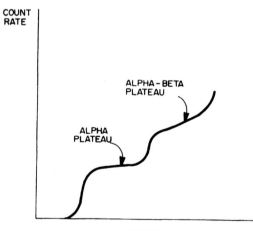

**Fig. 10.10** Count rate vs. operating voltage for a proportional counter used with discriminator for counting mixed alpha–beta sources.

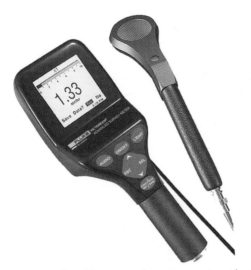

**Fig. 10.11** Portable survey and count-rate meter with optional GM pancake probe. See text. (Courtesy Fluke Biomedical.)

is used to advantage in monitoring for neutrons in mixed gamma-neutron fields. The charged recoil nuclei from which the neutrons scatter generally produce large pulses compared to those from the Compton electrons and photoelectrons produced by the photons.

Geiger–Mueller counters are very convenient and reliable radiation monitors, providing both visual and audible responses. They usually come equipped with a removable shield that covers a thin window to enable the detection of beta and alpha particles in addition to gamma rays. Readout can be, e.g., in counts per minute or in $mR\,h^{-1}$ with $^{137}Cs$ calibration. With the latter, special energy compensation of the probe is needed to flatten the energy response for low-energy photons. Figure 10.11 shows an example of a counter with a pancake GM probe. The instrument is also compatible with other kinds of probes, and can be used to detect alpha, beta, gamma, and neutron radiation. Examples of various GM and scintillation (Section 10.3) probes employed in a variety of applications are shown in Fig. 10.12.

Ideally, after the primary discharge in a GM tube, the positive ions from the counter gas drift to the cathode wall, where they are neutralized. Because of the high potential difference, however, some positive ions can strike the cathode with sufficient energy to release secondary electrons. Since these electrons can initiate another discharge, leading to multiple pulses, some means of quenching the discharge must be used. By one method, called external quenching, a large resistance between the anode and high-voltage supply reduces the potential difference after each pulse. This method has the disadvantage of making the tube slow ($\sim 10^{-3}$ s) in returning to its original voltage. Internal quenching of a GM tube by addition of an appropriate gas is more common. The quenching gas is chosen with a lower ionization potential and a more complex molecular structure than the counter gas. When a positive ion of the counter gas collides with a molecule of the quench-

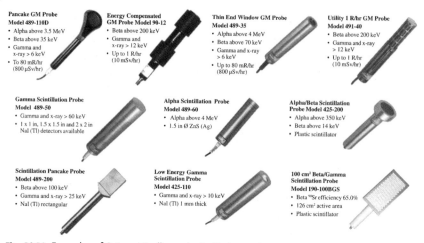

**Pancake GM Probe
Model 489-110D**
- Alpha above 3.5 MeV
- Beta above 35 keV
- Gamma and
  x-ray > 6 keV
- To 80 mR/hr
  (800 μSv/hr)

**Energy Compensated
GM Probe Model 90-12**
- Beta above 200 keV
- Gamma and
  x-ray > 12 keV
- Up to 1 R/hr
  (10 mSv/hr)

**Thin End Window GM Probe
Model 489-35**
- Alpha above 4 MeV
- Beta above 70 keV
- Gamma and x-ray
  > 6 keV
- Up to 80 mR/hr
  (800 μSv/hr)

**Utility 1 R/hr GM Probe
Model 491-40**
- Beta above 200 keV
- Gamma and x-ray
  > 12 keV
- Up to 1 R/hr
  (10 mSv/hr)

**Gamma Scintillation Probe
Model 489-50**
- Gamma and x-ray > 60 keV
- 1 x 1 in, 1.5 x 1.5 in and 2 x 2 in
  NaI (Tl) detectors available

**Alpha Scintillation Probe
Model 489-60**
- Alpha above 4 MeV
- 1.5 in Ø ZnS (Ag)

**Alpha/Beta Scintillation
Probe Model 425-200**
- Alpha above 350 keV
- Beta above 14 keV
- Plastic scintillator

**Scintillation Pancake Probe
Model 489-200**
- Beta above 100 keV
- Gamma and x-ray > 25 keV
- NaI (Tl) rectangular

**Low Energy Gamma
Scintillation Probe
Model 425-110**
- Gamma and x-ray > 10 keV
- NaI (Tl) 1 mm thick

**100 cm² Beta/Gamma
Scintillation Probe
Model 190-100BGS**
- Beta ⁹⁰Sr efficiency 65.0%
- 126 cm² active area
- Plastic scintillator

**Fig. 10.12** Examples of Geiger–Mueller and scintillation probes with specifications. (Courtesy Fluke Biomedical.)

ing gas, the latter, because of its lower ionization potential, can transfer an electron to the counter gas, thereby neutralizing it. Positive ions of the quenching gas, reaching the cathode wall, spend their energy in dissociating rather than producing secondary electrons. A number of organic molecules (e.g., ethyl alcohol) are suitable for internal quenching. Since the molecules are consumed by the dissociation process, organically quenched GM tubes have limited lifetimes ($\sim 10^9$ counts). Alternatively, the halogens chlorine and bromine are used for quenching. Although they dissociate, they later recombine. Halogen-quenched GM tubes are often preferred for extended use, although other factors limit their lifetimes.

## 10.2
## Ionization in Semiconductors

### Band Theory of Solids

Section 2.9 briefly described crystalline solids and the origin of the band structure of their electronic energy levels. In addition to the forces that act on an electron to produce the discrete bound states in an isolated atom, neighboring atoms in the condensed phase can also affect its behavior. The influence is greatest on the motion of the most loosely bound, valence electrons in the atoms and least on the more tightly-bound, inner-shell electrons. As depicted schematically in Fig. 2.7, with many atoms present, coalescing of the discrete states into the two bands of allowed energies with a forbidden gap between them depends on the size of $R_0$, the orderly spacing of atoms in the crystal. The figure indicates that bands are not formed if $R_0$ is very large. By the same token, the bands would overlap into a single continuum with no forbidden gap if $R_0$ is small.

While the study of crystals is a complex subject, some insight into their electrical properties can be gained by pursuing these simple physical concepts. With isolated bands present, the upper hatched area on the far right in Fig. 2.7 is called the *conduction band*, and the lower hatched area, the *valence band*. Electrically, solids can be classified as *insulators, semiconductors,* or *conductors*. The distinction is manifest in the mobility of electrons in response to an electric field applied to the solid. This response, in turn, is intimately connected with the particular atomic/molecular crystal structure and resultant band gap. The three classifications are represented in Fig. 10.13. Like most ionic solids (Section 2.9), NaCl, for example, is an insulator. The pairing of the univalent sodium and chlorine atoms completely fills the valence band, leaving no vacancies into which electrons are free to move. In principle, thermal agitation is possible and could provide sufficient energy to promote some valence electrons into the unoccupied conduction band, where they would be mobile. However, the probability for this random occurrence is exceedingly small,

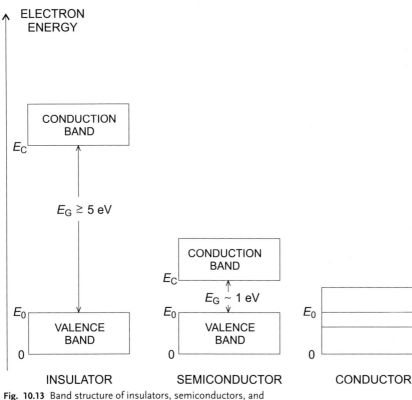

**Fig. 10.13** Band structure of insulators, semiconductors, and conductors. Starting from zero at the bottom of the valence band, the vertical scale shows schematically the energies spanned by the valence and conduction bands and the forbidden gap. The energy at the top of the valence band is denoted by $E_0$. The energy at the bottom of the conduction band is $E_C = E_0 + E_G$, where $E_G$ is the size of the band gap.

because the excitation energy required to span the forbidden gap is of the order of 8.5 eV. Generally, when the gap is greater than about 5 eV, the material is an insulator. While the details are different, a similar situation describes the covalent solids. The width of the band gap in carbon is 5.4 eV, making it an insulator. In silicon the gap is 1.14 eV and in germanium, 0.67 eV. At absolute zero temperature the valence bands in these two metals are completely filled and the conduction band is empty. They are insulators. At room temperatures ($kT \sim 0.025$ eV), a small fraction of their electrons are thermally excited into the conduction band, giving them some conductivity. Covalent solids having an energy gap $\sim 1$ eV are called intrinsic semiconductors. In conductors, the valence and conduction bands merge, as indicated in Fig. 10.13, providing mobility to the valence electrons. Sodium, with its single atomic ground-state 3s electron is a conductor in the solid phase.

We focus now on semiconductors and the properties that underlie their importance for radiation detection and measurement. One can treat conduction electrons in the material as a system of free, identical spin-$\frac{1}{2}$ particles (Sections 2.5, 2.6). They can exchange energy with one another, but otherwise act independently, like molecules in an ideal gas, except that they also obey the Pauli exclusion principle. Under these conditions, it is shown in statistical mechanics that the average number $N(E)$ of electrons per quantum state of energy $E$ is given by the Fermi distribution,

$$N(E) = \frac{1}{e^{(E-E_F)/kT} + 1}.$$  (10.6)

Here $k$ is the Boltzmann constant, $T$ is the absolute temperature, and $E_F$ is called the *Fermi energy*. At any given time, each quantum state in the system is either empty or occupied by a single electron. The value of $N(E)$ is the probability that a state with energy $E$ is occupied. To help understand the significance of the Fermi energy, we consider the distribution at the temperature of absolute zero, $T = 0$. For states with energies $E > E_F$ above the Fermi energy, the exponential term in the denominator of (10.6) in infinite; and so $N(E) = 0$. For states with energies below $E_F$, the exponential term is zero; and so $N(E) = 1$. Thus, all states in the system below $E_F$ are singly occupied, while all above $E_F$ are empty. This configuration has the lowest energy possible, as expected at absolute zero. At temperatures $T > 0$, the Fermi energy is defined as that energy for which the average, or probable, number of electrons is $\frac{1}{2}$.

It is instructive to diagram the relative number of free electrons as a function of energy in various types of solids at different temperatures. Figure 10.14(a) shows the energy distribution of electrons in the conduction band of a conductor at a temperature above absolute zero ($T > 0$). Electrons occupy states with a thermal distribution of energies above $E_C$. The lower energy levels are filled, but unoccupied states are available for conduction near the top of the band. A diagram for the same conductor at $T = 0$ is shown in Fig. 10.14(b). All levels with $E < E_F$ are occupied and all with $E > E_F$ are unoccupied.

Figure 10.15(a) shows the electron energy distribution in an insulator with $T > 0$. The valence band is full and the forbidden-gap energy $E_G$ ($\sim 5$ eV) is so wide that the electrons cannot reach the conduction band at ordinary temperatures. Fig-

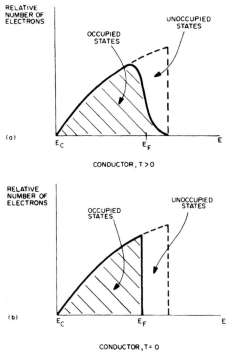

**Fig. 10.14** Relative number of electrons in the conduction band
of a conductor at absolute temperatures (a) $T > 0$ and (b) $T = 0$.

ure 10.15(b) shows a semiconductor in which $E_G \sim 1$ eV is considerably narrower
than in the insulator. The tail of the thermal distribution [Eq. (10.6)] permits a relatively small number of electrons to have energies in the conduction band. In this
case, the number of occupied states in the conduction band is equal to the number
of vacant states in the valence band and $E_F$ lies midway in the forbidden gap at an
energy $E_0 + E_G/2$. At room temperature, the density of electrons in the conduction
band is $1.5 \times 10^{10}$ cm$^{-3}$ in Si and $2.4 \times 10^{13}$ cm$^{-3}$ in Ge.

### Semiconductors

Figure 10.16 schematically represents the occupation of energy states for the semiconductor from Fig. 10.15(b) with $T > 0$. The relatively small number of electrons in
the conduction band are denoted with minus signs and the equal number of positive ions they leave in the valence band with plus signs. The combination of two
charges is called an electron–hole pair, roughly analogous to an ion pair in a gas.
Under the influence of an applied electric field, electrons in the conduction band
will move. In addition, electrons in the valence band move to fill the holes, leaving
other holes in their place, which in turn are filled by other electrons, and so on.
This, in effect, causes the holes to migrate in the direction opposite to that of the
electrons. The motions of both the conduction-band electrons and the valence-band

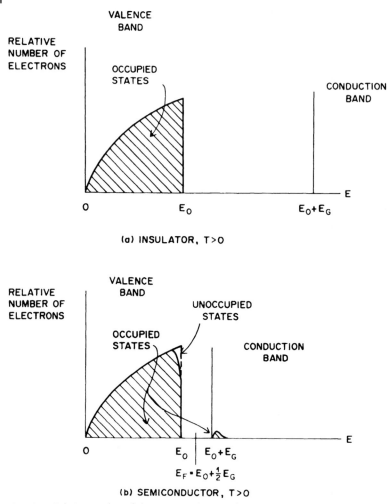

**Fig. 10.15** Relative number of electrons in valence and conduction bands of (a) an insulator and (b) a semiconductor with $T > 0$.

holes contribute to the observed conductivity. The diagram in Fig. 10.16 represents an intrinsic (pure) semiconductor. Its inherent conductivity at room temperature is restricted by the small number of electron–hole pairs, which, in turn, is limited by the size of the gap compared with $kT$.

The conductivity of a semiconductor can be greatly enhanced by doping the crystal with atoms from a neighboring group in the periodic system. As an example, we consider the addition of a small amount of arsenic to germanium.[1] When a crystal is formed from the molten mixture, the arsenic impurity occupies a substitutional

---

1   The reader is referred to the Periodic Table at the back of the book.

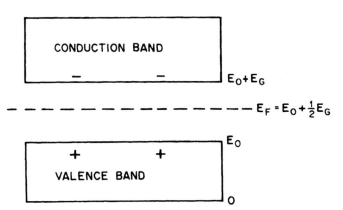

Fig. 10.16 Occupation of energy states in an intrinsic semiconductor at room temperature. A relatively small number of electrons (−) are thermally excited into the conduction band, leaving an equal number of holes (+) in the valence band. The Fermi energy $E_F$ lies at the middle of the forbidden gap.

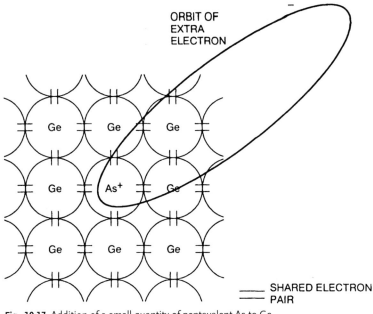

Fig. 10.17 Addition of a small quantity of pentavalent As to Ge crystal lattice provides very loosely bound "extra" electrons that have a high probability of being thermally excited into the conduction band at room temperatures. Arsenic is called a donor impurity and the resulting semiconductor, n-type.

position in the germanium lattice, as indicated schematically in Fig. 10.17. (The As atom has a radius of 1.39 Å compared with 1.37 Å for Ge.) Since As has five valence electrons, there is one electron left over after all of the eight covalent bonds have been formed with the neighboring Ge atoms. In Fig. 10.17, two short straight lines are used to represent a pair of electrons shared covalently by neighboring atoms and the loop represents, very schematically, the orbit of the extra electron contributed by $As^+$, which is in the crystal lattice. There is no state for the extra electron to occupy in the filled valence band. Since it is only very loosely bound to the $As^+$ ion (its orbit can extend over several tens of atomic diameters), this electron has a high probability of being thermally excited into the conduction band at room temperature. The conductivity of the doped semiconductor is thus greatly increased over its value as an intrinsic semiconductor. The amount of increase can be controlled by regulating the amount of arsenic added, which can be as little as a few parts per million. An impurity such as As that contributes extra electrons is called a donor and the resulting semiconductor is called n-type (negative).

Since little energy is needed to excite the extra electrons of an n-type semiconductor into the conduction band, the energy levels of the donor impurity atoms must lie in the forbidden gap just below the bottom of the conduction band. The energy-level diagram for Ge doped with As is shown in Fig. 10.18. The donor states are found to lie 0.013 eV below the bottom of the conduction band, as compared with the total gap energy, $E_G = 0.67$ eV, for Ge. At absolute zero all of the donor states are occupied and no electrons are in the conduction band. The Fermi energy lies between the donor levels and the bottom of the conduction band. As $T$ is increased, thermally excited electrons enter the conduction band from the donor states, greatly increasing the conductivity. Antimony can also be used as a donor impurity in Ge or Si to make an n-type semiconductor.

Another type of semiconductor is formed when Ge or Si is doped with gallium or indium, which occur in the adjacent column to their left in the periodic system.

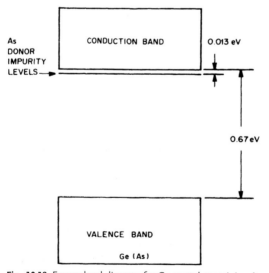

**Fig. 10.18** Energy-level diagram for Ge crystal containing As donor atoms (n-type semiconductor).

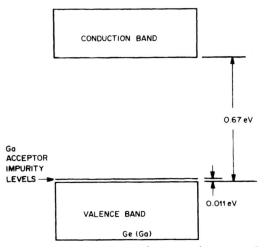

**Fig. 10.19** Energy-level diagram for Ge crystal containing Ga acceptor atoms (p-type semiconductor).

In this case, the valence shell of the interposed impurity atom has one less electron than the number needed to form the regular covalent crystal. Thus, the doped crystal contains positively charged "holes," which can accept electrons. The dopant is then called an acceptor impurity and the resulting semiconductor, p-type (positive). Holes in the valence band move like positive charges as electrons from neighboring atoms fill them. Because of the ease with which valence-band electrons can move and leave holes with the impurity present, the effect of the acceptor impurity is to introduce electron energy levels in the forbidden gap slightly above the top of the valence band. Figure 10.19 shows the energy-level diagram for a p-type Ge semiconductor with Ga acceptor atoms added. The action of the p-type semiconductor is analogous to that of the n-type. At absolute zero all of the electrons are in the valence band, the Fermi energy lying just above. When $T > 0$, thermally excited electrons occupy acceptor-level states, giving enhanced conductivity to the doped crystal.

**Semiconductor Junctions**

The usefulness of semiconductors as electronic circuit elements and for radiation measurements stems from the special properties created at a diode junction where n- and p-type semiconductors are brought into good thermodynamic contact. Figure 10.20 shows an electron energy-level diagram for an n–p junction. The two semiconductor types in contact form a single system with its own characteristic Fermi energy $E_F$. Because $E_F$ lies just below the conduction band in the isolated n region and just above the valence band in the isolated p region, the bands must become deformed over the junction region, as shown in the figure. When the n- and p-type semiconductors are initially brought into contact, electrons flow from the donor impurity levels on the n side over to the lower-energy acceptor sites on

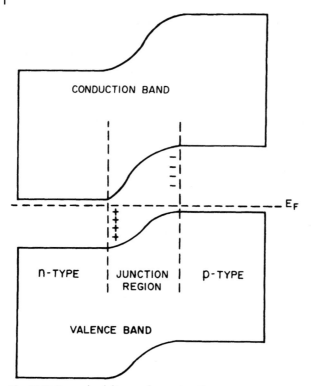

**Fig. 10.20** Energy-level diagram for n–p junction.

the p side. This process accumulates negative charge on the p side of the junction region and leaves behind immobile positive charges on the n side in the form of ionized donor impurity atoms. The net effect at equilibrium is the separation of charge across the junction region (as indicated by the + and – symbols in Fig. 10.20) and the maintenance of the deformed bands.

Even with thermal equilibrium, electrons move both ways through the junction region. With reference to Fig. 10.20, thermal agitation will cause some electrons to get randomly promoted into the conduction band in the p region, at the same time leaving holes in the valence band. A promoted electron can then travel freely to the junction region, where it will be drawn into the n region. This process gives rise to a spontaneous *thermal current* of electrons in the direction from the p to the n side of the junction. Also, some conduction-band electrons in the n region randomly receive enough energy to be able to move into the p region. There an electron can combine with a vacated hole in the valence band. This process provides a *recombination current* of electrons from the n to the p side. It balances the thermal current, so that no net charge flows through the device. Holes in the valence band appear to migrate by being successively filled by neighboring electrons, thus acting like positive charge carriers. Silicon devices (1.14 eV band gap) operate at room temperatures. The smaller gap of germanium (0.67 eV) necessitates operation at low temperatures (e.g., 77 K, liquid nitrogen) to suppress thermal noise.

The junction region over which the charge imbalance occurs is also called the depletion region, because any mobile charges initially there moved out when the two sides were joined. The depletion region acts, therefore, like a high-resistivity parallel-plate ionization chamber, making it feasible to use it for radiation detection. Ion pairs produced there will migrate out, their motion giving rise to an electrical signal. The performance of such a device is greatly improved by using a bias voltage to alleviate recombination and noise problems. The biased junction becomes a good rectifier, as described next.

Consider the n–p junction device in Fig. 10.21(a) with the negative side of an external bias voltage $V$ applied to the n side. When compared with Fig. 10.20, it is seen that the applied voltage in this direction lowers the potential difference across the junction region and causes a relatively large current $I$ to flow in the circuit. Bias in this direction is called forward, and a typical current–voltage curve is shown at the right in Fig. 10.21(a). One obtains a relatively large current with a small bias voltage. When a reverse bias is applied in Fig. 10.21(b), comparison with Fig. 10.20 shows that the potential difference across the junction region increases. Therefore, a much smaller current flows—and in the opposite direction—under reverse bias, as illustrated on the right in Fig. 10.21(b). Note the vastly different voltage and current scales on the two curves in the figure. Such n–p junction devices are rectifiers, passing current readily in one direction but not the other.

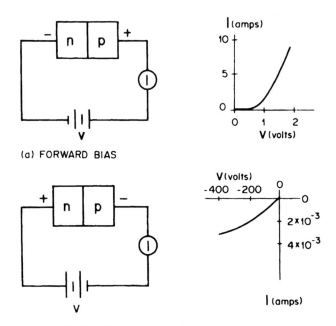

(a) FORWARD BIAS

(b) REVERSE BIAS

**Fig. 10.21** (a) Forward- and (b) reverse-biased n–p junctions and typical curves of current vs. voltage. Note the very different scales used for the two curves. Such an n–p junction is a good rectifier.

**Radiation Measuring Devices**

The reverse-biased n–p junction constitutes an attractive radiation detector. The depletion region, which is the active volume, has high resistivity, and ions produced there by radiation can be collected swiftly and efficiently. It can serve as a rate meter or to analyze pulses. The number of electron–hole pairs produced in a pulse is proportional to the energy absorbed in the active volume, and so the junction can be used as a spectrometer. The "W values" for Si and Ge are, respectively, 3.6 eV and 3.0 eV per electron–hole pair, as compared with the corresponding figure of ~30 eV per ion pair in gases. Statistically, the relatively large number of charge carriers produced per unit energy absorbed in semiconductors endows them with much better energy resolution than other detectors. Unique among detectors is the fact that the physical size of the depletion region can be varied by changing the bias voltage. For measuring alpha and beta radiation, junctions are fabricated with a very thin surface barrier between the outside of the device and the depletion region. Several examples of semiconductor radiation instruments will be briefly described.

Electronic dosimeters have advanced rapidly in recent years. They are in widespread use, particularly in nuclear power-plant, home-security, and military applications. Figure 10.22 shows a silicon diode personal electronic dosimeter for photons in the energy range 50 keV to 6 MeV. Other models in the series measure photons down to 20 keV and beta radiation. The unit operates as both a passive and an active dosimeter. It has adjustable dose and dose-rate warning and alarm levels. It can be used either in an autonomous or in a satellite mode, with remote computer interfacing with exposure-records management software. The device performs regular internal operating checks and reports dose increments, date, and time at specified intervals. A microprocessor counts pulses, converts them to dose, and calculates dose rate.

High-purity germanium (HPGe) detectors are available for a wide variety of tasks (Fig. 10.23). The crystals are housed in a vacuum-tight cryostat unit, which typically contains the preamplifier in a cylindrical package. Depending on the intended application, germanium detectors come in a number of different planar and coaxial configurations.

A light-weight, rugged, portable multichannel analyzer for gamma spectra is shown in Fig. 10.24. It is controlled from a key pad connected through an interface module to the HPGe detector. The unit has 16k channels and gives a live display of data being acquired. It holds 23 spectra in its internal memory. Nuclide ID and activity calculations are performed by using stored calibration information. The instrument can also interface with a computer to utilize other software applications.

CZT is the name given to *cadmium zinc telluride* semiconductors, which operate at normal temperatures. The relatively high-density crystal is advantageous for stopping secondary electrons. The resolution is intermediate between Ge and Si. With its high sensitivity for gamma detection, CTZ applications include homeland security, waste effluent monitoring, and first-responder technology. Detectors, such as the one shown in Fig. 10.25, are small, rugged, programmable, and operate with very low energy consumption.

## General Features

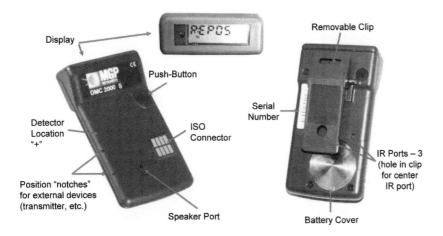

### DMC 2000S DETECTOR BOARD

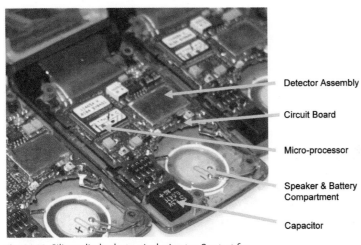

**Fig. 10.22** Silicon diode electronic dosimeter. See text for description. (Courtesy MGP Instruments, Inc.)

The sensitive lung counter in Fig. 10.26 employs three coaxial HPGe crystal detectors. It is specially designed to measure the relatively low-energy (15 keV–400 keV) gamma and X rays emitted by uranium and other actinides in the lung (e.g., $^{241}$Am, $^{239}$Pu, $^{238}$Pu, $^{237}$Np). If there is insoluable material in the organ, it is not amenable to bioassay monitoring. External measurements to identify the weak photons in the presence of background entail considerable technical challenges. The counter is calibrated with torso phantoms, employing a series of chest-wall plates, like the one shown in the figure. Calibration phantoms, having differing thicknesses of overlay and fat-to-muscle ratios are used to better represent a wide

**Fig. 10.23** Examples of HPGe detectors. (Courtesy Canberra, Inc.)

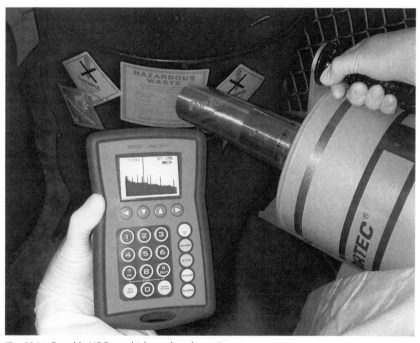

**Fig. 10.24** Portable HPGe multichannel analyzer. (Image provided courtesy of ORTEC, a brand of Advanced Measurement Technology, AMETEK.)

**Fig. 10.25** Cadmium zinc telluride (CZT) crystal detector. (Courtesy RFTrax, Inc.)

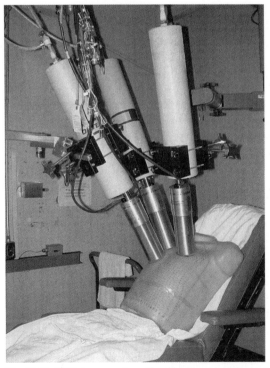

**Fig. 10.26** ORNL lung counter with calibration torso. See text. (Courtesy Robert L. Coleman, Oak Ridge National Laboratory, managed by UT-Battelle, LLC, for the U.S. Department of Energy.)

DETECTORS

**Fig. 10.27** Particle identifier.

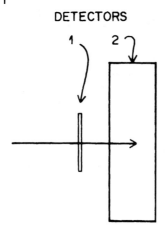

range of male and female subjects. Ultrasound measurements are used to correlate calibration parameters with the height and weight of persons being measured. For reduced background, the facility is shielded by 14-inch thick irreplaceable steel from pre-World-War II battleships, free of radioisotopes of cobalt, cesium, and other nuclides of the atomic age.

The *particle identifier* utilizes two detectors in the configuration shown in Fig. 10.27. A heavy particle passes through a thin detector (1) and is stopped in a thick detector (2). The pulse height from 1 is proportional to the stopping power, $-dE/dx$, and that from 2 is proportional to the kinetic energy, $E = Mv^2/2$, with which the particle enters it, where $M$ is the particle's mass and $v$ is its velocity. The signals from 1 and 2 can be combined electronically in coincidence to form the product $E(-dE/dx)$. Since the ln term in the stopping-power formula [Eq. (5.23)] varies slowly with the particle energy $E$, it follows that

$$E\left(-\frac{dE}{dx}\right) = kz^2 M,$$ (10.7)

approximately, where $z$ is the particle's charge and $k$ is a constant of proportionality. The quantity $z^2 M$, which is thus determined by the measurement, is characteristic of a particular heavy particle, which can then be identified.

## 10.3
## Scintillation

### General

Scintillation was the first method used to detect ionizing radiation (Roentgen having observed the fluorescence of a screen when he discovered X rays). When radiation loses energy in a luminescent material, called a scintillator or phosphor, it causes electronic transitions to excited states in the material. The excited states decay by emitting photons, which can be observed and related quantitatively to the action of the radiation. If the decay of the excited state is rapid ($10^{-8}$ or $10^{-9}$ s), the process is called fluorescence; if it is slower, the process is called phosphorescence.

Scintillators employed for radiation detection are usually surrounded by reflecting surfaces to trap as much light as possible. The light is fed into a photomultiplier tube for generation of an electrical signal. There a photosensitive cathode converts a fraction of the photons into photoelectrons, which are accelerated through an electric field toward another electrode, called a dynode. In striking the dynode, each electron ejects a number of secondary electrons, giving rise to electron multiplication. These secondary electrons are then accelerated through a number of additional dynode stages (e.g., 10), achieving electron multiplication in the range $10^7$–$10^{10}$. The magnitude of the final signal is proportional to the scintillator light output, which, under the right conditions, is proportional to the energy loss that produced the scintillation.

Since materials emit and absorb photons of the same wavelength, impurities are usually added to scintillators to trap energy at levels such that the wavelength of the emitted light will not fall into a self-absorption region. Furthermore, because many substances, especially organic compounds, emit fluorescent radiation in the ultraviolet range, impurities are also added as wavelength shifters. These lead to the emission of photons of visible light, for which glass is transparent and for which the most sensitive photomultiplier tubes are available.

Good scintillator materials should have a number of characteristics. They should efficiently convert the energy deposited by a charged particle or photon into detectable light. The efficiency of a scintillator is defined as the fraction of the energy deposited that is converted into visible light. The highest efficiency, about 13%, is obtained with sodium iodide. A good scintillator should also have a linear energy response; that is, the constant of proportionality between the light yield and the energy deposited should be independent of the particle or photon energy. The luminescence should be rapid, so that pulses are generated quickly and high count rates can be resolved. The scintillator should also be transparent to its own emitted light. Finally, it should have good optical quality for coupling to a light pipe or photomultiplier tube. The choice of a particular scintillation detector represents a balancing of these factors for a given application.

Two types of scintillators, organic and inorganic, are used in radiation detection. The luminescence mechanism is different in the two.

## Organic Scintillators

Fluorescence in organic materials results from transitions in individual molecules. Incident radiation causes electronic excitations of molecules into discrete states, from which they decay by photon emission. Since the process is molecular, the same fluorescence can occur with the organic scintillator in the solid, liquid, or vapor state. Fluorescence in an inorganic scintillator, on the other hand, depends on the existence of a regular crystalline lattice, as described in the next section.

Organic scintillators are available in a variety of forms. Anthracene and stilbene are the most common organic crystalline scintillators, anthracene having the highest efficiency of any organic material. Organic scintillators can be polymerized into plastics. Liquid scintillators (e.g., xylene, toluene) are often used and are practical when large volumes are required. Radioactive samples can be dissolved or suspended in them for high-efficiency counting. Liquid scintillators are especially suited for measuring soft beta rays, such as those from $^{14}$C or $^3$H. High-$Z$ elements (e.g., lead or tin) are sometimes added to organic scintillator materials to achieve greater photoelectric conversion, but usually at the cost of decreased efficiency.

Compared with inorganic scintillators, organic materials have much faster response, but generally yield less light. Because of their low-$Z$ constituents, there are little or no photoelectric peaks in gamma-ray pulse-height spectra without the addition of high-$Z$ elements. Organic scintillators are generally most useful for measuring alpha and beta rays and for detecting fast neutrons through the recoil protons produced.

### Inorganic Scintillators

Inorganic scintillator crystals are made with small amounts of activator impurities to increase the fluorescence efficiency and to produce photons in the visible region. As shown in Fig. 10.28, the crystal is characterized by valence and conduction bands, as described in Section 10.2. The activator provides electron energy levels in the forbidden gap of the pure crystal. When a charged particle interacts with the crystal, it promotes electrons from the valence band into the conduction band, leaving behind positively charged holes. A hole can drift to an activator site and ionize it. An electron can then drop into the ionized site and form an excited neutral impurity complex, which then decays with the emission of a visible photon. Because the photon energies are less than the width of the forbidden gap, the crystal does not absorb them.

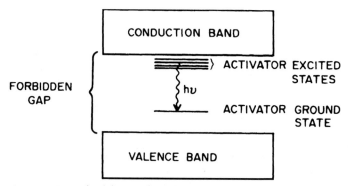

**Fig. 10.28** Energy-level diagram for activated crystal scintillator. Because the energy levels of the activator complex are in the forbidden gap, the crystal is transparent at the fluorescent photon energies $h\nu$.

The alkali halides are good scintillators. In addition to its efficient light yield, sodium iodide doped with thallium [NaI(Tl)] is almost linear in its energy response. It can be machined into a variety of sizes and shapes. Disadvantages are that it is hygroscopic and somewhat fragile. NaI has become a standard scintillator material for gamma-ray spectroscopy. CsI(Na), CsI(Tl), and LiI(Eu) are examples of other inorganic scintillators. Silver-activated zinc sulfide is also commonly used. It is available only as a polycrystalline powder, from which thin films and screens can be made. The use of ZnS, therefore, is limited primarily to the detection of heavy charged particles. (Rutherford used ZnS detectors in his alpha-particle scattering experiments.) Glass scintillators are also widely used.

Two examples of scintillator probes are displayed in Fig. 10.29. In addition to the detector material, each contains a photomultiplier tube, which is reflected in its size. The unit on the left is used for gamma surveys. It has a cylindrical NaI(Tl) crystal with a height and diameter of 2.5 cm. It operates between 500 V and 1200 V. The unit on the right uses ZnS and is suitable for alpha/beta surveys. It operates in the same voltage range. The window area is approximately 100 cm$^2$.

Specialized scintillation devices have been designed for other specific purposes. One example is the phoswich (= phosphor sandwich) detector, which can be used to count beta particles or low-energy photons in the presence of high-energy photons. It consists of a thin NaI(Tl) crystal in front coupled to a larger scintillator of another material, often CsI(Tl), having a different fluorescence time. Signals that come from the photomultiplier tube can be distinguished electronically on the basis of the different decay times of the two phosphors to tell whether the light came only from the thin front crystal or from both crystals. In this way, the low-energy radiation can be counted in the presence of a high-energy gamma-ray background.

Figure 10.30 shows a pulse-height spectrum measured with a 4 × 4 in. NaI(Tl) scintillator exposed to 662-keV gamma rays from $^{137}$Cs. Several features should be noted. Only those photons that lose all of their energy in the crystal contribute to the total-energy peak, also called the photopeak. These include incident photons that produce a photoelectron directly and those that undergo one or more Compton

**Fig. 10.29** Examples of scintillation probes: (left) NaI(Tl) for gamma surveys; (right) ZnS for alpha/beta monitoring. (Courtesy Ludlum Instruments, Inc.)

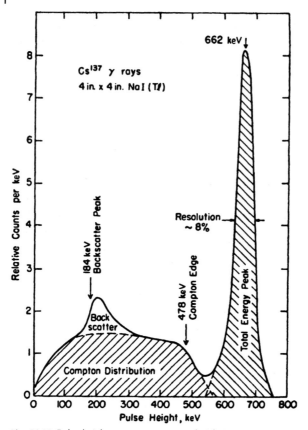

**Fig. 10.30** Pulse-height spectrum measured with 4 × 4 in. NaI(Tl) scintillator exposed to 662-keV gamma rays from $^{137}$Cs. The resolution is about 8% of the peak energy. The maximum Compton-electron energy is 478 keV. [Reprinted with permission from R. D. Evans, "Gamma Rays," in *American Institute of Physics Handbook*, 3d ed., p. 8-210, McGraw-Hill, New York (1972). Copyright 1972 by McGraw-Hill Book Company.]

scatterings and then produce a photoelectron. In the latter case, the light produced by the Compton recoil electrons and that produced by the final photoelectron combine to yield a single scintillation pulse around 662 keV. (The light produced by Auger electrons and characteristic X rays absorbed in the crystal is also included in the same pulse, these processes taking place rapidly.) Other photons escape from the crystal after one or more Compton scatterings, and therefore do not deposit all of their energy in producing the scintillation. These events give rise to the continuous Compton distribution at lower pulse heights, as shown in the figure. The Compton edge at 478 keV is the maximum Compton-electron recoil energy, $T_{max}$, given by Eq. (8.20). The pulses that exceed $T_{max}$ in magnitude come from two or more Compton recoil electrons produced by the photon before it escapes from the crystal. The backscatter peak is caused by photons that are scattered into the scintillator from surrounding materials. The energy of a $^{137}$Cs gamma ray that is scattered at 180° is $662 - T_{max} = 662 - 478 = 184$ keV. The backscattered radiation peaks at an energy slightly above this value, as shown.

The relative area under the total-energy peak and the Compton distribution in Fig. 10.30 depends on the size of the scintillator crystal. If the crystal is very large, then relatively few photons escape. Most of the pulses occur around 662 keV. If it is small, then only single interactions are likely and the Compton continuum is large. In fact, the ratio of the areas under the total-energy peak and the Compton distribution in a small detector is equal to the ratio of the photoelectric and Compton cross sections in the crystal material.

The occurrence of so-called escape peaks is accentuated when a detector is small, whether it be a scintillator or semiconductor. In NaI, for example, when a K-shell vacancy in iodine following photoelectric absorption is filled by an L-shell electron, a 28-keV characteristic X ray is emitted. The X ray will likely escape if the crystal is small, and a pulse size of $h\nu_0 - 28$ keV will be registered, where $h\nu_0$ is the energy of the incident photon. When the incident photons are monoenergetic, the pulse-height spectrum shows the escape peak, as illustrated in Fig. 10.31. The relative size of the peaks at the two energies $h\nu_0$ and $h\nu_0 - 28$ keV depends on the physical dimensions of the NaI crystal. Germanium has an X-ray escape peak 11 keV below the photopeak. Another kind of escape peak can occur in the pulse-height spectra of monoenergetic high-energy photons of energy $h\nu_0$, which produce electron–positron pairs in the detector. The positron quickly stops and annihilates with an

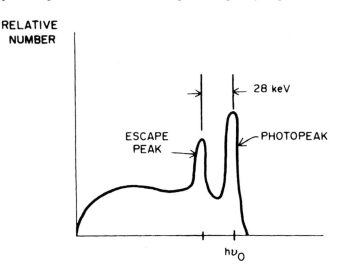

Fig. 10.31 When characteristic X rays from iodine escape from a NaI scintillator, an escape peak appears 28 keV below the photopeak.

atomic electron, producing two 0.511-MeV photons. In small detectors one or both annihilation photons can escape without interacting, leading to escape peaks at the energies $h\nu_0 - 0.511$ and $h\nu_0 - 1.022$ MeV.

An example of a portable NaI gamma analyzer for laboratory and field use is shown in Fig. 10.32. It has 512 channels and can store 30 spectra in memory. The instrument displays spectra and dose rates, as seen in the figure. Software includes peak analysis, nuclide identification search, and capability for data transfer. Adjustable alarm thresholds with an audible signal can be set.

### Example

Monoenergetic 450-keV gamma rays are absorbed in a NaI(Tl) crystal having an efficiency of 12%. Seventy-five percent of the scintillation photons, which have an average energy of 2.8 eV, reach the cathode of a photomultiplier tube, which converts 20% of the incident photons into photoelectrons. Assume that variations in the pulse heights from different gamma photons are due entirely to statistical fluctuations in the number of visible photons per pulse that reach the cathode. (a) Calculate the average number of scintillation photons produced per absorbed gamma photon. (b) How many photoelectrons are produced, on the average, per gamma photon? (c) What is the average energy expended by the incident photon to produce a photoelectron from the cathode of the photomultiplier tube (the "$W$ value")? (d) Compare this value with the average energy needed to produce an ion pair in a gas or a semiconductor.

### Solution

(a) The total energy of the visible light produced with 12% efficiency is 450 keV $\times$ 0.12 = 54.0 keV. The average number of scintillation photons is therefore 54,000/2.8 = 19,300. (b) The average number of photons that reach the photomultiplier cathode is 0.75 $\times$ 19,300 = 14,500, and so the average number of photoelectrons that produce a pulse is 0.20 $\times$ 14,500 = 2900. (c) Since one 450-keV incident gamma photon produces an average of 2900 photoelectrons that initiate the signal, the "$W$ value" for the scintillator is 450,000/2900 = 155 eV/photoelectron. (d) For gases, $W \sim 30$ eV ip$^{-1}$; and so the average number of electrons produced by absorption of a photon would be about 450,000/30 = 15,000. For a semiconductor, $W \sim 3$ eV ip$^{-1}$ and the corresponding number of electrons would be 150,000. A "$W$ value" of several hundred eV per electron produced at the photocathode is typical for scintillation detectors. Energy resolution is discussed in Section 11.11.

The energy resolution of a spectrometer depends on several factors, such as the efficiency of light or charge collection and electronic noise. Resolution is also inherently limited by random fluctuations in the number of charge carriers collected when a given amount of energy is absorbed in the detector. As the last example shows, the number of charge carriers produced in germanium is substantially larger than that in a scintillator. Therefore, as discussed in the next chapter, the *relative* fluctuation about the mean—the inherent resolution—is much better for germanium. The upper panel in the example presented in Fig. 10.33 compares pulse-height spectra measured with a NaI detector for pure $^{133}$Ba and for a mixture of $^{133}$Ba and $^{239}$Pu. The soft Pu gamma photons, at energies marked by the arrows, are not distinguished.

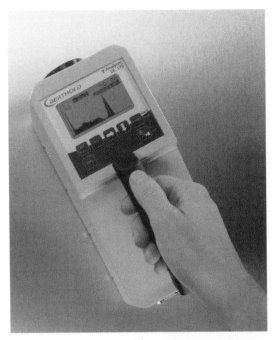

(a)

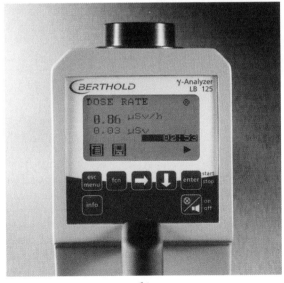

(b)

**Fig. 10.32** (a) Hand-held NaI gamma analyzer and dose-rate meter with (b) close-up of display. (Courtesy Berthold Technologies USA, LLC.)

*NaI Detector* (arrows indicate undetected Pu peaks)

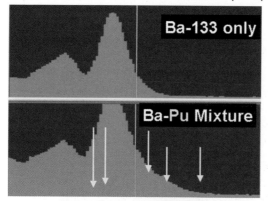

Energy Range 220–480 keV

*HPGe Detector:* Pu peaks are easy to see

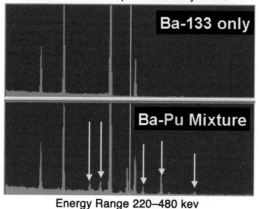

Energy Range 220–480 kev

**Fig. 10.33** Comparison of spectra of $^{133}$Ba and $^{133}$Ba-$^{239}$Pu mixture made with NaI and HPGe detectors. (Image provided courtesy of ORTEC, a brand of Advanced Measurement Technology, AMETEK.)

The lower panel compares the same spectra measured with HPGe. With this much greater resolution, the presence of both nuclides is unmistakable.

Research has accelerated on the development of new scintillation as well as other methods of radiation detection, driven by concerns for homeland security. Large inorganic scintillators provide the most sensitive means for general detection of gamma rays. Promising new approaches in this direction include the investigation of cerium-doped lanthanum halides, $LaCl_3 : Ce^{3+}$ and $LaBr_3 : Ce^{3+}$. Large crystals can be produced with relatively high light yield and good energy resolution. Figure 10.34 shows radiation monitors in place at a typical border crossing into the United States.

**Fig. 10.34** Sensitive monitors installed at portals in passenger lanes of border crossings into the U.S. can detect small amounts of radioactive material. Alarms are sometimes triggered by naturally radioactive agricultural products, ceramic tiles, and occasionally by a passenger, recently administered radioactive thalium, technetium, or iodine by a physician. (Courtesy Joseph C. McDonald. Reprinted with permission from "Detecting Illicit Radioactive Sources," by Joseph C. McDonald, Bert M. Coursey, and Michael Carter in Physics Today, November 2004. Copyright 2004, American Institute of Physics.)

## 10.4
## Photographic Film

Since the early days of experience with ionizing radiation, films have been used extensively for detection and measurement. (Recall from Section 1.3 how Becquerel discovered radioactivity.) Film emulsions contain small crystals of a silver halide (e.g., AgBr), suspended in a gelatine layer spread over a plastic or glass surface, wrapped in light-tight packaging. Under the action of ionizing radiation, some secondary electrons released in the emulsion become trapped in the crystalline lattice, reducing silver ions to atomic silver. Continued trapping leads to the formation of microscopic aggregates of silver atoms, which comprise the latent image. When developed, the latent images are converted into metallic silver, which appears to the eye as darkening of the film. The degree of darkening, called the optical density, increases with the amount of radiation absorbed. An optical densitometer can be used to measure light transmission through the developed film.

Badges containing X-ray film packets have been worn, clipped to the clothing, for beta–gamma personnel radiation-dose monitoring and for worker identification. As discussed below, doses from gamma and beta radiation can be inferred by comparing densitometer readings from exposed film badges with readings from a calibrated set of films given different, known doses under the same conditions. The darkening response of film to neutrons, on the other hand, is too weak to be used in this way for neutron personnel monitoring. Instead, special fine-grain, nuclear track emulsions are employed. The tracks of individual recoil protons that neutrons produce in the emulsion are then observed and counted under a microscope. The work is tedious, and the technique is limited by the fact that the ranges of recoil protons with energies less than about 2.5 MeV are too short to produce recognizable tracks.

Film calibration and the use of densitometer readings to obtain dose would appear, in principle, to be straightforward. In practice, however, the procedure is complicated by a number of factors. First, the density produced in film from a given dose of radiation depends on the emulsion type and the particular lot of the manufacturer. Second, firm is affected by environmental conditions, such as exposure to moisture, and by general aging. Elevated temperatures contribute to base fog in an emulsion before development. Third, significant variations in density are introduced by the steps inherent in the film-development process itself. These include the type, concentration, and age of the developing solution as well as the development time and handling through agitation, rinsing, and fixing. Variations from these sources are significantly reduced by applying the following procedure to both the film dosimeters worn by workers and those used for the calibration of the dosimeters. All units should be from the same manufacturer's production lot, stored and handled in similar fashion, developed at the same time under the same conditions, and read with a single densitometer, and even by a single operator. Experience shows that an acceptable degree of reproducibility can be thus attained.

A serious problem of a different nature for dose determination is presented by the strong response of film to low-energy photons. The upper curve in Fig. 10.35 illustrates the relative response (darkening) of film enclosed in thin plastic to a fixed dose of monoenergetic photons as a function of their energy. From about 5 MeV down to 200 keV, the relative response, set at unity in the figure, is flat. Below about 200 keV, the rising photoelectric absorption cross section of the silver in the film leads increasingly to more blackening at the fixed dose than would occur if film were tissue- or air-equivalent. The relative response peaks at around 40 keV and then drops off at still lower energies because of absorption of the photons in the packaging material around the film.

The lower curve in Fig. 10.35 shows the relative response when the incident radiation passes through a cadmium absorber of suitable thickness placed over the film. The absorption of photons in the cadmium filter tends to compensate for the over-response of the film at low energies, while having little effect at high energies, thus extending the usefulness of the badge to lower-energy photons.

Film badges are also used for personnel monitoring of beta radiation, for which there is usually negligible energy dependence of the response. For mixed beta–gamma radiation exposures, the separate contribution of the beta particles is assessed by comparing (1) the optical density behind a suitable filter that absorbs them and (2) the density through a neighboring "open window." The latter consists only of the structural material enclosing the film. Since beta particles have short ranges, a badge that has been exposed to them alone will be darkened behind the open window, but not behind the absorbing filter. Such a finding would also result from exposure to low-energy photons. To distinguish these from beta particles, one

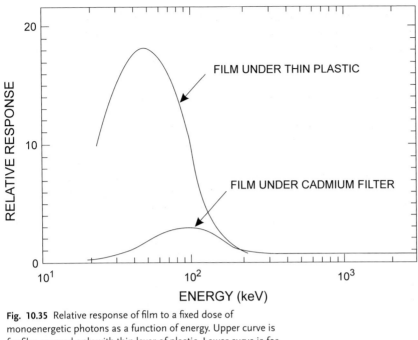

**Fig. 10.35** Relative response of film to a fixed dose of monoenergetic photons as a function of energy. Upper curve is for film covered only with thin layer of plastic. Lower curve is for film covered with a cadmium filter to compensate for the over-response to low-energy photons.

can employ two additional filters, one of high and the other of low atomic number, such as silver and aluminum. They should have the same density thickness, so as to be equivalent beta-particle absorbers. The high-$Z$ filter will strongly absorb low-energy photons, which are attenuated less by the low-$Z$ material. The presence of low-energy photons will contribute to a difference in darkening behind the two.

Figure 10.36 shows an exploded view of a multi-purpose film badge used at several sites from about 1960 to 1980. It was designed for routine beta–gamma and neutron personnel monitoring; criticality applications; assessment of a large, accidental gamma dose; and for personal security identification. The laminated picture front, identifying the wearer, was of low-$Z$ material. An assembly, comprised of filters and other units, was placed behind the picture and in front of the film packs, which included both X-ray and nuclear track emulsions. There were four filter areas in the assembly through which radiation could pass to reach the firm. (1) At the window position, the only material traversed in addition to the laminated picture (52 mg cm$^{-2}$) was the paper wrapper around the film (28 mg cm$^{-2}$), giving a total density thickness of 80 mg cm$^{-2}$. (2) The (low-$Z$) plastic filter had a thickness of about 215 mg cm$^{-2}$. (3) A gold foil was sandwiched between two pieces of cadmium, each 0.042 cm thick. The combination presented a Cd-Au-Cd absorber thickness of about 1000 mg cm$^{-2}$. (4) The aluminum filter (275 mg cm$^{-2}$) was provisionally included at the time for eventual help in determining the effective energy of photon exposures.

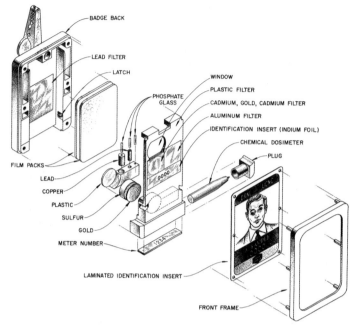

ORNL Badge-Meter Model II

**Fig. 10.36** Multi-element film badge in use during the period 1960–1980. [From W. T. Thornton, D. M. Davis, and E. D. Gupton, *The ORNL Badge Dosimeter and its Personnel Monitoring Applications*, Report ORNL-3126, Oak Ridge, TN (1961).]

Readings from badges worn by personnel were analyzed to provide a number of dose quantities, as mandated by regulations, basically in the following ways. The optical density behind the Cd-Au-Cd filter served as a measure of "deep dose" to tissues inside the body. The thickness of the plastic filter plus the picture and film wrapper (300 $mg\,cm^{-2}$) corresponded to the 3-mm depth specified for the lens of the eye. Assessment of "skin dose" (specified at a depth of 7 $mg\,cm^{-2}$) was based on the Cd-Au-Cd reading and the difference between the densities behind the window and the plastic filter. Depending on the particular beta fields anticipated, an empirical constant was worked out for weighting this difference in the evaluation of the skin dose. The numerical value of the constant was determined from calibration with a particular beta source, often natural uranium. Considerable uncertainty attended determination of skin dose from a badge exposed to an unknown beta–gamma radiation field. The fluence of fast neutrons was proportional to the number of recoil proton tracks observed in the film behind the Cd-Au-Cd filter, which strongly absorbed thermal neutrons. Thermal neutrons produced protons of 0.524 MeV energy in the $^{14}N(n,p)^{14}C$ reaction with nitrogen (Section 9.7) in the emulsions. These were observed in the window portion of the film, and their number was proportional to the fluence of thermal neutrons. These data plus knowledge of tissue composition and neutron cross sections were applied to estimate neutron doses.

As mentioned, the badge in Fig. 10.36 served other purposes in addition to personal identification and routine radiation monitoring. As an adjunct, three silver meta-phosphate glass rods could measure gamma doses in excess of 100 rad (Section 12.2) and also responded to thermal neutrons. The three rods were surrounded by different shields of lead, copper, and plastic. Comparing their relative responses gave an indication of the effective energy of the photons. The response of the glass rods would be potentially important for accidental exposures to high-level radiation. The chemical dosimeter could enable a swift visual indication of persons exposed in an incident. Several elemental foils with different neutron-activation energy thresholds were included to provide data about the neutron energy spectrum in case of a criticality accident (Section 10.9). A 0.5-g sulfur pellet is activated effectively only by neutrons with energies greater than about 3.2 MeV (Section 9.7). There were two gold foils, which are activated by thermal neutrons. One was enclosed in the Cd-Au-Cd filter and the other was inserted bare behind the sulfur pellet. Significant exposure to neutrons in a criticality accident can be readily identified by the induced activity in the indium foil contained in the badge (Section 9.7). Activity induced in the gold could also serve the same purpose. Although less sensitive than indium, the longer half-life of $^{198}$Au (2.696 d) compared with $^{116m}$In (54.15 min) is an advantage.

Multi-element film dosimeters for personnel monitoring became largely replaced by thermoluminescent dosimeters (next section) during the 1980s. However, film badges still serve in many applications, such as hospitals, universities, and small laboratories, where radiation fields are relatively uncomplicated and well known.

## 10.5
### Thermoluminescence

In connection with Fig. 10.28, we described how ionizing radiation can produce electron–hole pairs in an inorganic crystal. These lead to the formation of excited states with energies that lie in the forbidden gap when particular added activator impurities are present. In a scintillation detector, it is desirable for the excited states to decay quickly to the ground states, so that prompt fluorescence results. In another class of inorganic crystals, called thermoluminescent dosimeters (TLDs), the crystal material and impurities are chosen so that the electrons and holes remain trapped at the activator sites at room temperature. Placed in a radiation field, a TLD crystal serves as a passive integrating detector, in which the number of trapped electrons and holes depends on its radiation exposure history.

After exposure, the TLD material is heated. As the temperature rises, trapped electrons and holes migrate and combine, with the accompanying emission of photons with energies of a few eV. Some of the photons enter a photomultiplier tube and produce an electronic signal. The sample is commonly processed in a TLD

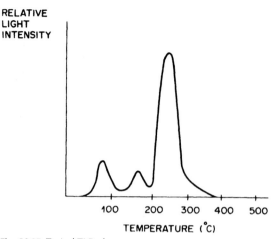

**Fig. 10.37** Typical TLD glow curve.

reader, which automatically heats the material, measures the light yield as a function of temperature, and records the information in the form of a glow curve, such as that shown in Fig. 10.37. Typically, several peaks occur as traps at different energy levels are emptied. The total light output or the area under the glow curve can be compared with that from calibrated TLDs to infer radiation dose. All traps can be emptied by heating to sufficiently high temperature, and the crystal reused.

A number of TLD materials are in use. Manganese-activated calcium sulfate, $CaSO_4$ : Mn, is sensitive enough to measure doses of a few tens of $\mu$rad. Its traps are relatively shallow, however, and it has the disadvantage of "fading" significantly in 24 h. $CaSO_4$ : Dy is better. Another popular TLD crystal is LiF, which has inherent defects and impurities and needs no added activator. It exhibits negligible fading and is close to tissue in atomic composition. It can be used to measure gamma-ray doses in the range of about 0.01–1000 rad. Other TLD materials include $CaF_2$ : Mn, $CaF_2$ : Dy, and $Li_2B_4O_7$ : Mn.

Figure 10.38 shows schematic drawings of two TLD personnel dosimeters. The beta–gamma system on the left has four LiF chips. Elements 1, 2, and 3 are Harshaw TLD-700 material, which is essentially pure $^7$LiF and, therefore, insensitive to neutrons.[2] The first chip has a thickness of 0.015 in = 0.38 mm, and is situated behind 1,000 mg cm$^{-2}$ of Teflon and plastic to measure the regulatory "deep dose" (Section 14.9). Chip 2 is set behind a thin absorber and a 0.004 in = 0.10 mm layer of copper, giving a total thickness of 333 mg cm$^{-2}$. The copper filters out low-energy photons while transmitting some beta particles. Its response, compared with that of chip 3 behind a thin absorber ("open window") for low-energy photon discrimination, is used for the assessment of shallow dose (Section 14.9). The remaining element 4 consists of TLD-600, which is enriched to about 96% in the isotope $^6$Li. This chip is sensitive to thermal neutrons and is at the regulatory depth of the lens of the eye (300 mg cm$^{-2}$). If a person wearing the dosimeter is exposed to fast neu-

2   Natural lithium is 92.5% $^7$Li and 7.5% $^6$Li.

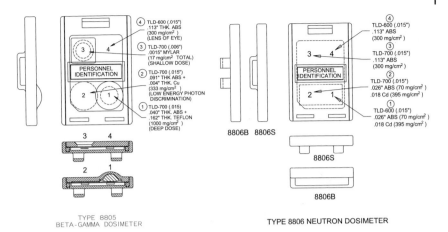

**Fig. 10.38** Schematic views of beta–gamma (left) and neutron (right) TLD dosimeters. (Courtesy Thermo Electron Corp.)

trons, some will be moderated by the body and detected as slow neutrons only in chip 4 of the dosimeter, thus furnishing evidence of neutron exposure.

When a potential for exposure to neutrons exists, a special TLD dosimeter, such as that shown on the right in Fig. 10.38, should be employed. Readings from the pairs of TLD-600 and TLD-700 elements, one sensitive and the other insensitive to neutrons, can be compared. Since their responses to gamma rays are identical, differences can be attributed to neutrons. The cadmium filters for chips 1 and 2 absorb incident thermal neutrons. Differences in their readings, therefore, are associated with fast neutrons. Without the cadmium filters, differences between elements 3 and 4 indicate total (fast-plus-thermal) neutron exposure.

The beta–gamma dosimeter of Fig. 10.38 and its system of filters is displayed in Fig. 10.39. Thermoluminesent dosimeters can be processed by automated readout systems, which can transfer results to a central computer system for dosimetry records. Computer algorithms have been written to unfold the required dose assessments for individuals from the readings obtained from the different chips. Laboratory accreditation is provided through the National Voluntary Laboratory Accreditation Program (NVLAP).

## 10.6
## Other Methods

### Particle Track Registration

A number of techniques have been devised for directly observing the tracks of individual charged particles. Neutron dosimetry with the film badge described in Section 10.4 utilized neutron-sensitive emulsions in which the tracks of recoil pro-

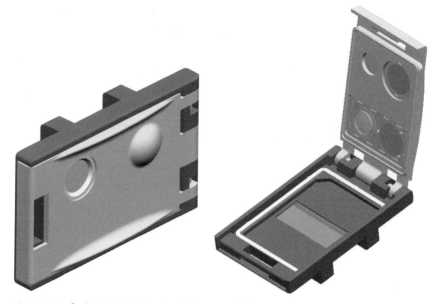

**Fig. 10.39** Left: closed TLD dosimeter. Right: open dosimeter, showing four filter areas, chips removed. (Courtesy Thermo Electron Corp.)

tons from the elastic scattering of fast neutrons could be counted and analyzed. Figure 5.1 shows an example of alpha- and beta-particle tracks in photographic film. In the cloud chamber, moisture from a supersaturated vapor condenses on the ions left in the wake of a passing charged particle, rendering the track visible. In the bubble chamber, tiny bubbles are formed as a superheated liquid starts to boil along a charged particle's track. Another device, the spark chamber, utilizes a potential difference between a stack of plates to cause a discharge along the ionized path of a charged particle that passes through the stack.

Track etching is possible in some organic polymers and in several types of glasses. A charged particle causes radiation damage along its path in the material. When treated chemically or electrochemically, the damaged sites are attacked preferentially and made visible, either with a microscope or the unaided eye. Track etching is feasible only for particles of high LET. The technique is widely used in neutron dosimetry (e.g., CR-39 detectors). Although neutral particles do not produce a trail of ions, the tracks of the charged recoil particles they produce can be registered by techniques discussed here.

**Optically Stimulated Luminescence**

*Optically stimulated luminescence* (OSL) shares some similarities and some contrasts with thermoluminescence. A number of materials exhibit both phenomena. Under irradiation, electrons become trapped in long-lived excited states of doped crystals. With TLDs, dose is inferred from the amount of light emitted under thermal stimulation. With OSL, the light emission is caused by optical stimulation. Reading a TLD empties all of the trapped-electron states, erasing the primary record and returning the dosimeter to its original condition for reuse. Reading with OSL, on the other hand, depletes relatively little of the stored charge, essentially preserving the primary record and enabling the dosimeter to be read again. The variable stimulation power with OSL can be used to advantage to achieve sensitivity over a wide range of doses.

Although a decades-old idea, practical use of OSL for dosimetry became a reality with the development of the Luxel® personnel dosimeters by Landau, Inc. in the late 1990s. The detector material is aluminum oxide, grown in the presence of carbon, $Al_2O_3 : C$. (Crystals with different dopants can be fabricated for specialized applications.) Figure 10.40 displays a Luxel® dosimeter. A thin $Al_2O_3$ strip is sandwiched between a multi-element, sealed filter pack. As with film and TLD, the different filters are used to provide specific information about mixed radiation fields for personnel dose assessment. The individual read outs are fed into a computer algorithm that estimates the regulatory deep and shallow doses. Neutron dose assessment can be added by the inclusion of an optional CR-39 detector in the dosimeter, which is analyzed by track etching and counting.

Landauer employs two dosimeter read-out methods. Since the induced light emitted from the detector must be measured in the presence of the stimulating light, it is essential that the two light sources not be mixed. In one method, the stimulation is caused by a pulsed laser and the emission signal is read between pulses. The other method employs continuous stimulation by light-emitting diodes (LEDs) or CW (continuous-wave) laser, and the measurement of the light emitted from the detector at wavelengths outside the LED or laser spectrum. The pulsed system is more expensive and more complex, but considerably faster than the continuous-stimulation method.

### Direct Ion Storage (DIS)

*Direct Ion Storage* (DIS) has been recently developed into an important basis for a personal dosimeter. A predetermined amount of electric charge is placed on the floating plate of a nonvolatile solid-state DIS memory cell. The charge is tunneled onto the gate through oxide-silicon material that surrounds it. At normal temperatures, the stored charge is trapped permanently on the gate because of the extremely low probability of thermal excitations of electrons through the adjacent material. The amount of charge stored on the gate can be "read" without disturbing it by making conductivity measurements with the cell. In this configuration the device is not sensitive to ionizing radiation, because the low mobility of charge carriers in the oxide prevents neutralization of charge on the gate before recombination.

To make a dosimeter, the memory cell is enclosed within a conductive wall, forming in a small ($\sim$10 cm$^3$) ionization chamber, containing air or other gas. The oxide layer is provided with a small opening to make contact between the floating plate and the chamber gas. Radiation now produces ion pairs with high mobility in the

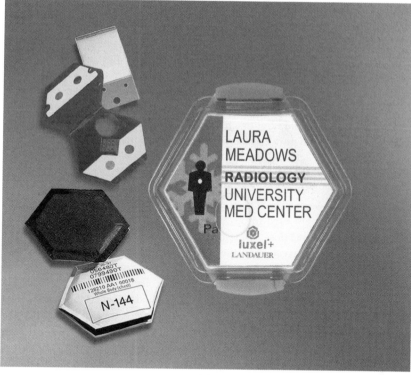

(a)

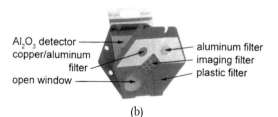

$Al_2O_3$ detector ──────
copper/aluminum
filter ──────
open window ──────

────── aluminum filter
────── imaging filter
────── plastic filter

(b)

**Fig. 10.40** Luxel® personnel dosimeter. (b) Filter pack, showing detector, filters, and open window. (Courtesy Landauer, Inc.)

gas. Electric fields direct the charge carriers through the hole in the oxide layer to reduce the charge initially stored on the floating plate. DIS dosimeters are calibrated to provide dose assessment from the measured loss of charge. Depending on the radiation fields to be monitored, characteristics of the wall material and thickness as well as other factors can be varied to fit the application.

The basic design of the direct ion storage dosimeter gives it the flat energy response of an ionization chamber. It has instant, non-destructive read-out. The dosimeter shown in Fig. 10.41 with its reader is rugged, light (20 g without holder),

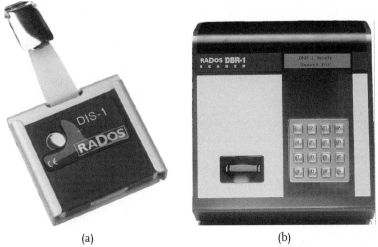

**Fig. 10.41** (a) Direct ion storage personal dosimeter and
(b) reader. (Courtesy RADOS Technology.)

and waterproof. It measures photon dose equivalent in the range 1–40 $\mu$Sv and
beta dose equivalent from 10 $\mu$Sv–40 Sv.

### Radiophotoluminescence

The film badge in Fig. 10.36 contained three silver meta-phosphate glass rods to
measure large photon doses ($\geq$100 rad = 1 Gy), as might occur in an accident. Energy absorbed from the ionizing radiation leads to the migration of electrons to
permanent sites associated with the silver in the glass. As a result, new absorption
frequencies are produced, and the glass will fluoresce under exposure to ultraviolet light. The fluorescence yield can then be compared with calibrated standards to
infer dose. Since the fluorescence does not change the glass, the read-out is nonde-
structive. Although radiophotoluminescence has been used for routine personnel
dosimetry, it has generally been limited to high-dose applications.

### Chemical Dosimeters

Radiation produces chemical changes. One of the most widely studied chemical
detection systems is the Fricke dosimeter, in which ferrous ions in a sulfate solution are oxidized by the action of radiation. As in all aqueous chemical dosimeters,
radiation interacts with water to produce free radicals (e.g., H and OH), which are
highly reactive. The OH radical, for example, can oxidize the ferrous ion directly:
$Fe^{2+} + OH \rightarrow Fe^{3+} + OH^-$. After irradiation, aqueous chemical dosimeters can be
analyzed by titration or light absorption. The useful range of the Fricke dosimeter is from about 40 to 400 Gray (Gy). The dose measurements are accurate and
absolute. The aqueous system approximates soft tissue.

Other chemical-dosimetry systems are based on ceric sulfate, oxalic acid, or a combination of ferrous sulfate and cupric sulfate. Doses of the order of 0.1 Gy can be measured chemically with some chlorinated hydrocarbons, such as chloroform. Higher doses result in visible color changes in some systems.

### Calorimetry

The energy imparted to matter from radiation is usually efficiently converted into heat. (Radiation energy can also be expended in nuclear transformations and chemical changes.) If the absorber is thermally insulated, as in a calorimeter, then the temperature rise can be used to infer absorbed dose absolutely. However, a relatively large amount of radiation is required for calorimetric measurements. An absorbed energy of 4180 J kg$^{-1}$ (= 4180 Gy) in water raises the temperature only 1°C (Problem 39). Because they are relatively insensitive, calorimetric methods in dosimetry have been employed primarily for high-intensity radiation beams, such as those used for radiotherapy. Calorimetric methods are also utilized for the absolute calibration of source strength.

### Cerenkov Detectors

When a charged particle travels in a medium faster than light, it emits visible electromagnetic radiation, analogous to the shock wave produced in air at supersonic velocities. The speed of light in a medium with index of refraction $n$ is given by $c/n$, where $c$ is the speed of light in a vacuum. Letting $v = \beta c$ represent the speed of the particle, we can express the condition for the emission of Cerenkov radiation as $\beta c > c/n$, or

$$\beta n > 1. \tag{10.8}$$

The light is emitted preferentially in the direction the particle is traveling and is confined to a cone with vertex angle given by $\cos\theta = 1/\beta n$. It follows from (10.8) that the threshold kinetic energy for emission of Cerenkov light by a particle of rest mass $M$ is given by (Problem 40)

$$T = Mc^2\left(\frac{n}{\sqrt{n^2 - 1}} - 1\right). \tag{10.9}$$

The familiar "blue glow" seen coming from a reactor core [e.g., Figs. 9.1 and 9.2(b)] is Cerenkov radiation, emitted by energetic beta particles traveling faster than light in the water.

Cerenkov detectors are employed to observe high-energy particles. The emitted radiation can also be used to measure high-energy beta-particle activity in aqueous samples.

## 10.7
## Neutron Detection

### Slow Neutrons

Neutrons are detected through the charged particles they produce in nuclear reactions, both inelastic and elastic. In some applications, pulses from the charged
particles are registered simply to infer the presence of neutrons. In other situations, the neutron energy spectrum is sought, and the pulses must be further analyzed. For slow neutrons (kinetic energies $T \lesssim 0.5$ eV), detection is usually the
only requirement. For intermediate (0.5 eV $\lesssim T \lesssim 0.1$ MeV) and fast ($T \gtrsim 0.1$ MeV)
neutrons, spectral measurements are frequently needed. We discuss slow-neutron
detection methods first.

Table 10.2 lists the three most important nuclear reactions for slow-neutron detection. The reaction-product kinetic energies and cross sections are given for capture of thermal neutrons (energy $= 0.025$ eV). Since the incident kinetic energy
of a thermal neutron is negligible, the sum of the kinetic energies of the reaction
products is equal to the $Q$ value itself. Given $Q$, equations analogous to Eqs. (3.18)
and (3.19) can be applied to calculate the discrete energies of the two products,
leading to the values given in Table 10.2. We shall describe slow-neutron detection
by means of these reactions and then briefly discuss detection by fission reactions
and foil activation.

### $^{10}$B(n,$\alpha$)

One of the most widely used slow-neutron detectors is a proportional counter using boron trifluoride ($BF_3$) gas. For increased sensitivity, the boron is usually highly
enriched in $^{10}$B above its 19.7% natural isotopic abundance. If the dimensions of
the tube are large compared with the ranges of the reaction products, then pulse
heights at the $Q$ values of 2.31 MeV and 2.79 MeV should be observed with areas

**Table 10.2** Reactions Used for Slow-Neutron Detection
(Numerical Data Apply to Thermal-Neutron Capture)

| Reaction | $Q$ Value (MeV) | Product Kinetic Energies (MeV) | Cross Section (Barns) |
|---|---|---|---|
| $^{10}_{5}B + ^{1}_{0}n \rightarrow$ $\begin{cases} ^{7}_{3}Li^{*} + ^{4}_{2}He \ (96\%) \end{cases}$ | 2.31 | $T_{Li} = 0.84$ $T_{He} = 1.47$ | 3840 |
| $^{7}_{3}Li + ^{4}_{2}He \ (4\%)$ | 2.79 | $T_{Li} = 1.01$ $T_{He} = 1.78$ | |
| $^{6}_{3}Li + ^{1}_{0}n \rightarrow ^{3}_{1}H + ^{4}_{2}He$ | 4.78 | $T_{H} = 2.73$ $T_{He} = 2.05$ | 940 |
| $^{3}_{2}He + ^{1}_{0}n \rightarrow ^{3}_{1}H + ^{1}_{1}H$ | 0.765 | $T_{3H} = 0.191$ $T_{1H} = 0.574$ | 5330 |

RELATIVE
NUMBER

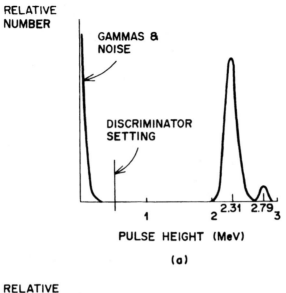

GAMMAS &
NOISE

DISCRIMINATOR
SETTING

2.31  2.79
1        2        3

PULSE HEIGHT (MeV)

(a)

RELATIVE
NUMBER

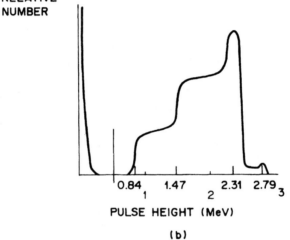

0.84    1.47          2.31  2.79
1                2        3

PULSE HEIGHT (MeV)

(b)

**Fig. 10.42** (a) Idealized thermal-neutron pulse-height spectrum
from large BF$_3$ tube in which reaction products are completely
absorbed in the gas. (b) Spectrum from tube showing wall
effects.

in the ratio 96 : 4, as shown in Fig. 10.42(a). With most practical sizes, however,
a significant number of Li nuclei and alpha particles enter the wall of the tube,
and energy lost there is not registered. Since the two reaction products separate
"back-to-back" to conserve momentum, when one strikes the wall the other is directed away from it. This wall effect introduces continua to the left of the peaks.
As sketched in Fig. 10.42(b), one continuum takes off from the peak at 2.31 MeV
and is approximately flat down to 1.47 MeV. (A similar continuum occurs below

the small peak at 2.79 MeV.) Over this interval, the total energy (1.47 MeV) of the alpha particle is absorbed in the gas while only part of the energy of the Li nucleus is absorbed there, the rest going into the wall. Below 1.47 MeV, the spectrum again drops and is approximately flat down to the energy 0.84 MeV of the Li recoil. Pulses occur here when the Li nucleus stops in the gas and the alpha particle enters the wall.

The $BF_3$ proportional counter can discriminate against gamma rays, which are usually present with neutrons and produce secondary electrons that ionize the gas. Compared with the neutron reaction products, electrons produced by the photons are sparsely ionizing and give much smaller pulses. As indicated in Fig. 10.42(a), amplitude discrimination can be used to eliminate these counts as well as electronic noise if the gamma fluence rate is not too large. In intense gamma fields, however, the pileup of multiple pulses from photons can become a problem.

In other counter designs, a boron compound is used to line the interior walls of the tube, in which another gas, more suitable for proportional counting than $BF_3$, is used. Boron-loaded scintillators (e.g., ZnS) are also employed for slow-neutron detection.

### $^6$Li(n,$\alpha$)

As shown in Table 10.2, this reaction, compared with $^{10}$B(n,$\alpha$), has a higher Q value (potentially better gamma-ray discrimination), but lower cross section (less sensitivity). The isotope $^6$Li is 7.42% abundant in nature, but lithium enriched in $^6$Li is available.

Lithium scintillators are frequently used for slow-neutron detection. Analogous to NaI(Tl), crystals of LiI(Eu) can be employed. They can be made large compared with the ranges of the reaction products, so that the pulse-height spectra are free of wall effects. However, the scintillation efficiency is then comparable for electrons and heavy charged particles, and so gamma-ray discrimination is much poorer than with $BF_3$ gas.

Lithium compounds can be mixed with ZnS to make small detectors. Because secondary electrons produced by gamma rays easily escape, gamma-ray discrimination with such devices is good.

### $^3$He(n,p)

This reaction has the highest cross section of the three in Table 10.2. Like the $BF_3$ tube, the $^3$He proportional counter exhibits wall effects. However, $^3$He is a better counter gas and can be operated at higher pressures with better detection efficiency. Because of the low value of Q, though, gamma discrimination is worse.

### (n,f)

Slow-neutron-induced fission of $^{233}$U, $^{235}$U, or $^{239}$Pu is utilized in fission counters. The Q value of ~200 MeV for each is large. About 165 MeV of this energy is converted directly into kinetic energy of the heavy fission fragments. Fission pulses are extremely large, enabling slow-neutron counting to be done at low levels, even in a high background. Most commonly, the fissile material is coated on the inner surface of an ionization chamber. A disadvantage of fissionable materials is that they are alpha emitters, and one must sometimes contend with the pileup of alpha-particle pulses.

### Activation Foils

Slow neutrons captured by nuclei induce radioactivity in a number of elements, which can be made into foils for neutron detection. The amount of induced activity will depend on a number of factors—the element chosen, the mass of the foil, the neutron energy spectrum, the capture cross section, and the time of irradiation. Examples of thermal-neutron activation-foil materials include Mn, Co, Cu, Ag, In, Dy, and Au.

### Intermediate and Fast Neutrons

Nuclear reactions are also important for measurements with intermediate and fast neutrons. In addition, neutrons at these speeds can, by elastic scattering, transfer detectable kinetic energies to nuclei, especially hydrogen. Elastic recoil energies are negligible for slow neutrons. Detector systems can be conveniently discussed in four groups—those based on neutron moderation, nuclear reactions, elastic scattering alone, and foil activation. Recent developments also include bubble detectors.

### Neutron Moderation

Two principal systems in this category have been developed: the long counter and moderating spheres enclosing a small thermal-neutron detector. A cross section of the cylindrical long counter is shown in Fig. 10.43. This detector, which is one of the oldest still in use, can be constructed to give nearly the same response from a neutron of any energy from about 10 keV to 5 MeV. The long counter contains a $BF_3$ tube surrounded by an inner paraffin moderator, as shown. The instrument is sensitive to neutrons incident from the right. Those from other directions are either reflected or thermalized by the outer paraffin jacket and then absorbed in the $B_2O_3$ layer. Neutrons that enter from the right are slowed down in the inner paraffin moderator, high-energy neutrons reaching greater depths on the average than low-energy ones. With this arrangement, the probability that a moderated neutron will enter the $BF_3$ tube and be registered does not depend strongly on the initial energy with which it entered the counter. Holes on the front face make it easier for neutrons with energies <1 MeV to penetrate past the surface, from which they might otherwise be reflected. The long counter does not measure neutron spectra.

Neutron spectral information can be inferred by the use of polyethylene moderating spheres (Bonner spheres) of different diameters with small lithium iodide scintillators at their centers. A series of five or more spheres, ranging in diameter from 2 to 12 in., is typically used. The different sizes provide varying degrees of moderation for neutrons of different energies. The response of each sphere is calibrated for monoenergetic neutrons from thermal energy to 10 MeV or more.

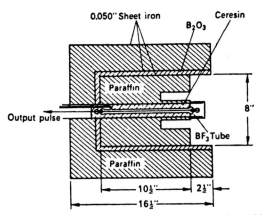

**Fig. 10.43** The original long counter of Hanson and McKibben.
[Reprinted with permission from A. O. Hanson and
M. L. McKibben, "A Neutron Detector Having Uniform
Sensitivity from 10 keV to 5 MeV," *Phys. Rev.* **72**, 673 (1947).
Copyright 1947 by the American Physical Society.]

The spheres are then exposed in an unknown neutron field and the count rates measured. An unfolding procedure is used to infer information about the neutron spectrum from knowledge of the calibration curves and the measured count rates. Because the unfolding procedure does not yield very precise results, this method is not widely used for spectral measurements. With a relatively large sphere, it is found that the response as a function of neutron energy is similar to the dose equivalent per neutron. It therefore serves as a neutron rem meter in many applied health physics operations. Such an instrument is shown in Fig. 10.44.

### Nuclear Reactions

The $^6$Li(n,$\alpha$) and $^3$He(n,p) reactions are the only ones of major importance for neutron spectrometry. Ideally, an incident neutron of energy $T$ that undergoes a reaction causes a detector to register a peak at an energy $Q + T$. In practice, many times another peak also occurs at an energy $Q$ due to neutrons that have been slowed by multiple scattering in building walls and shielding around the detector. Slow-neutron cross sections can be orders of magnitude larger than at higher energies. The additional peak at $Q$ is sometimes called the epithermal peak.

Crystals of LiI(Eu) are used in neutron spectroscopy. However, the nonlinearity of their response with the energy of the reaction products (tritons and alpha particles) is a serious handicap. Lithium-glass scintillators are also in use, principally as fast responding detectors in neutron time-of-flight measurements. In another type of neutron spectrometer, a thin LiF sheet is placed between two semiconductor diodes. At relatively low neutron energies $T$, the recoil products will tend to be ejected back-to-back, giving coincidence counts in both semiconductors with a total pulse height of $Q + T$, from which $T$ can be ascertained.

In the $^3$He(n,p) proportional counter, monoenergetic neutrons of energy $T$ produce a peak at energy $T + 0.765$ MeV, as illustrated in Fig. 10.45. The epithermal

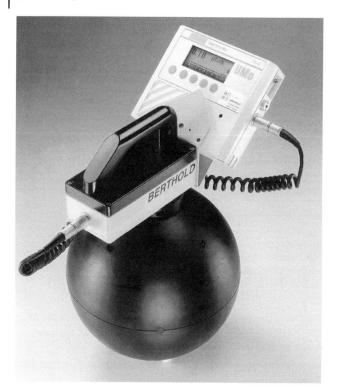

**Fig. 10.44** Instrument for measuring neutron dose-equivalent rate. A $^3$He proportional counter tube is located at the center of a polyethylene moderator sphere. The cylindrical tube has approximately equal height and diameter to minimize directional dependence of the response. Sensitivity is about 3 counts per nSv (3 counts per $\mu$rem) in the energy range 1 to 10 MeV. Good gamma discrimination. (Courtesy Berthold Technologies USA, LLC.)

peak is also shown at the energy $Q = 0.765$ MeV. In addition to these peaks from the reaction products, one finds a continuum of pulse heights from recoil $^3$He nuclei that elastically scatter incident neutrons. It follows from Eq. (9.3) that the maximum kinetic energy that a neutron of mass $m = 1$ and kinetic energy $T$ can transfer to a helium nucleus of mass $M = 3$ is

$$T_{max} = \frac{4mM}{(m+M)^2} T = \frac{4 \times 1 \times 3}{(1+3)^2} T = \frac{3}{4} T. \tag{10.10}$$

The elastic continuum in Fig. 10.45 thus extends up to the energy $0.75T$. Wall effects can be reduced in the $^3$He proportional counter by using gas pressures of several atmospheres and also by adding a heavier gas (e.g., Kr).

### Elastic Scattering
A number of instruments are based on elastic scattering alone, especially from hydrogen. As discussed in Section 9.5, a neutron can lose all of its kinetic energy $T$ in a single head-on collision with a proton. Also, since n–p scattering is isotropic

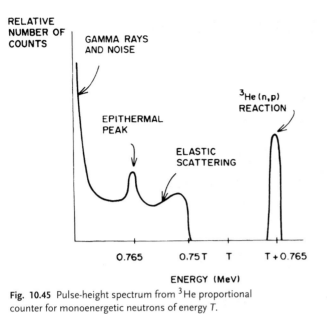

**Fig. 10.45** Pulse-height spectrum from $^3$He proportional counter for monoenergetic neutrons of energy $T$.

in the center-of-mass system for neutron laboratory energies up to ~10 MeV, the average energy imparted to protons by neutrons in this energy range is $T/2$ (Section 9.6).

Organic proton-recoil scintillators are available for neutron spectrometry in a variety of crystal, plastic, and liquid materials. The full proton recoil energies can be caught in these scintillators. Complications in the use of proton-recoil scintillators include nonlinearity of response, multiple neutron scattering, and competing nuclear reactions. For applications in mixed fields, the gamma response can, in principle, be separated electronically from the neutron response on the basis of quicker scintillation.

Proportional counters have been designed with hydrocarbon gases, such as $CH_4$. These have inherently lower detection efficiencies than solid-state devices, but offer the potential for better gamma discrimination. Wall effects can be important. Proportional counters have also been constructed with polyethylene or other hydrogenous material surrounding the tube. One such device, based on the Bragg–Gray principle, will be discussed in Section 12.6.

A proton-recoil telescope, illustrated in Fig. 10.46, can be used to accurately measure the spectrum of neutrons in a collimated beam. At an angle $\theta$, the energy $T_p$ of a recoil proton from a thin target struck by a neutron of incident energy $T$ is, by Eq. (9.5),

$$T_p = T\cos^2\theta. \tag{10.11}$$

The $E(-dE/dx)$ coincidence particle identifier (Fig. 10.27) can be used to reduce background, eliminate competing events, and measure $T_p$.

**INCIDENT
NEUTRON**

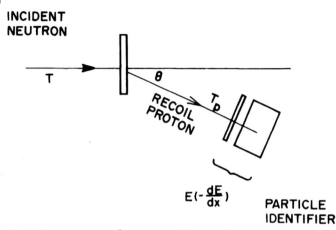

**Fig. 10.46** Arrangement of proton-recoil telescope for measuring spectrum of a neutron beam.

Neutron spectra can also be inferred from the observed range distribution of recoil protons in nuclear track emulsions. Neutrons with at least several hundred keV of energy are needed to produce protons with recognizable tracks.

### Threshold Foil Activation

Like low-energy neutrons, intermediate and fast neutrons can be detected by the radioactivity they induce in various elements. With many nuclides, a threshold energy exists for the required nuclear reaction. When foils of several nuclides are simultaneously exposed to a neutron field, differences in the induced activity between them can be used to obtain information about the neutron energy spectrum as well as the fluence.

As described at the end of Section 9.8, the activity induced in a foil or other target is a combined result of (1) the neutron fluence (and fluence rate) at energies above threshold and (2) the energy-dependent cross section for the reaction. Activation provides an estimate of neutron fluence at an effective threshold energy above the minimum given by Eq. (9.29). The effective threshold energy is thus only an approximate concept; for a given material, different specific values can be found in the literature. Table 10.3 lists some reactions and their effective threshold energies used for fast-neutron detection. As an example, if an exposed aluminum foil shows induced activity from $^{27}$Mg and a simultaneously exposed cobalt foil shows no induced activity from $^{56}$Mn, then one can infer that neutrons with energies 3.8 MeV < $T$ < 5.2 MeV were present. To obtain accurate spectral data from threshold-detector systems, one must take into account such factors as the masses of the particular isotopes in the foils, their neutron cross sections as functions of energy, the exposure history of the foils, and the half-lives of the induced radioisotopes.

### Bubble Detectors

The popular bubble detector is a unique and important personal neutron dosimeter. Figure 10.47 shows a pair of detectors, one exposed to neutrons and the other unexposed. The basic dosimeter consists of 8 cm$^3$ of a clear polymer in which tens

**Table 10.3** Reactions for Threshold Activation Detectors of Neutrons

| Reaction | Effective Threshold (MeV) |
|---|---|
| $^{115}In(n,n')^{115m}In$ | 0.5 |
| $^{58}Ni(n,p)^{58}Co$ | 1.9 |
| $^{27}Al(n,p)^{27}Mg$ | 3.8 |
| $^{56}Fe(n,p)^{56}Mn$ | 4.9 |
| $^{59}Co(n,\alpha)^{56}Mn$ | 5.2 |
| $^{24}Mg(n,p)^{24}Na$ | 6.0 |
| $^{197}Au(n,2n)^{196}Au$ | 8.6 |

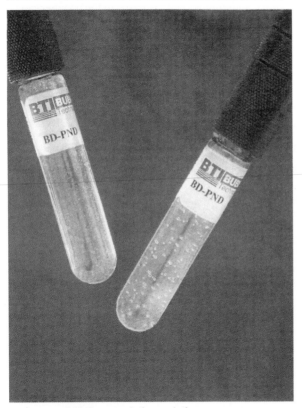

**Fig. 10.47** Bubble detectors before and after exposure to neutrons. (Courtesy H. Ing, Bubble Technology Industries, Inc.)

of thousands of microscopic droplets of a superheated liquid (Freon-12, for example) are dispersed. A liquid that continues to exist as such at temperatures above its normal boiling point is said to be superheated. Under this condition, a sudden disturbance in the droplet, such as the passage of a charged particle generated from a neutron interaction, can produce a boiling, explosive phase transition into a vapor. Droplets are instantly transformed into small, visible bubbles in the dosimeter.

They are fixed indefinitely in the polymer, and can be counted visually or with an automatic reader. Dosimeter properties can be adjusted to meet different objectives by varying the polymer and the detecting liquid. After reading, the bubbles can be made to disappear by recompression through a screw-cap assembly on the unit, thus restoring it to the unexposed state.

The bubble detector is a sensitive, passive neutron dosimeter. Although commonly manufactured to have about one bubble per mrem of fast-neutron dose equivalent (Section 12.2), it has been produced with up to three orders of magnitude higher sensitivity. Its threshold neutron energy of about 100 keV is lower than that of nuclear-track film. The tissue-equivalent dose response is flat from approximately 200 keV to more than 15 MeV. It is isotropic and completely insensitive to gamma radiation. A compound containing $^6$Li and dispersed in the polymer can be used to monitor thermal neutrons. Sets of bubble detectors, fabricated with different neutron-energy thresholds, have been employed to obtain spectral information for dosimetry.

## 10.8
## Suggested Reading

The best sources of information in the diverse and rapidly expanding field of radiation detection and instrumentation are on the World Wide Web. Detailed data can be found on virtually any current or historic topic. The following publications are suggested as supplements to this chapter.

1  Frame, Paul W., "A History of Radiation Detection Instrumentation," *Health Phys.* **88**, 613–637 (2005). [This important publication appears in the issue commemorating the 50th anniversary of the Health Physics Society. It provides a comprehensive, in-depth review of the history of radiation detection from early days through modern technology. Extensive bibliography. The numerous photographs in the article are of instruments in Oak Ridge Associated Universities' Historical Instrumentation Collection, which is managed by Dr. Frame. The collection can be accessed on-line at http://www.orau.org/ptp/museumdirectory.htm.]

2  ICRU Report 31, *Average Energy Required to Produce an Ion Pair*, International Commission on Radiation Units and Measurements, Washington, DC (1979).

3  Knoll, Glenn F., *Radiation Detection and Measurement*, 3rd Ed., Wiley, New York (2000). [This authoritative textbook covers many of the subjects of this chapter in detail.]

4  Poston, John W., Sr., "External Dosimetry and Personnel Monitoring," *Health Phys.* **88**, 557–564 (2005). [This review is another in the issue commemorating the 50th anniversary of the Health Physics Society. Radiation detection is discussed with emphasis on personnel dosimetry. Accompanying historical information is included. Bibliography.]

## 10.9
## Problems

1. How many electrons are collected per second in an ionization chamber when the current is $5 \times 10^{-14}$ A? What is the rate of energy absorption if $W = 29.9$ eV ip$^{-1}$?

2. How many ion pairs does a 5.6-MeV alpha particle produce in $N_2$ (Fig. 10.2)?

3. Why do the $W$ values for heavy charged particles increase at low energies (Fig. 10.2)?

4. A beam of alpha particles produced a current of $10^{-14}$ A in a parallel-plate ionization chamber for 8 s. The chamber contained air at STP.
   (a) How many ion pairs were produced?
   (b) How much energy did the beam deposit in the chamber?
   (c) If the chamber volume was 240 cm$^3$, what was the energy absorbed per unit mass in the chamber gas (1 gray absorbed dose = 1 J kg$^{-1}$)?

5. A 10-cm$^2$ beam of charged particles is totally absorbed in an ionization chamber, producing a saturation current of $10^{-6}$ A. If $W = 30$ eV ip$^{-1}$, what is the average beam intensity in units of eV cm$^{-2}$ s$^{-1}$?

6. A 5-MeV alpha-particle beam of cross-sectional area 2 cm$^2$ is stopped completely in an ionization chamber, producing a current of 10 $\mu$A under voltage saturation conditions.
   (a) If $W = 32$ eV ip$^{-1}$, what is the intensity of the beam?
   (b) What is the fluence rate?

7. A thin radioactive source placed in an ionization chamber emits $10^6$ alpha particles per second with energy 3.81 MeV. The particles are completely stopped in the gas, for which $W = 36$ eV ip$^{-1}$. Calculate
   (a) the average number of ion pairs produced per second
   (b) the current that flows under saturation conditions
   (c) the amount of charge collected in 1 h.

8. Assume that the $W$ values for protons and carbon-recoil nuclei are both 30 eV ip$^{-1}$ in $C_2H_4$ gas. What is the maximum number of ion pairs that can be produced by a 3-MeV neutron interacting elastically with (a) H or (b) C?

9. An alpha-particle source is fabricated into a thin foil. Placed first in a $2\pi$ gas-flow proportional counter, it shows only a single pulse height and registers 7080 counts min$^{-1}$ (background negligible). The source is next placed in $4\pi$ geometry in an air ionization chamber operated under saturation conditions, where it produces a current of $5.56 \times 10^{-12}$ A. Assume that the foil stops the recoil nuclei following alpha decay but absorbs a negligible amount of energy from the alpha particles.
   (a) What is the activity of the source?
   (b) What is the alpha-particle energy?
   (c) Assume that the atomic mass number of the daughter nucleus is 206 and calculate its recoil energy.

10. A $^{210}$Po source is placed in an air ionization chamber, and a saturation current of $8 \times 10^{-12}$ A is observed. Assume that the ionization is due entirely to 5.30-MeV alpha particles stopping in the chamber. How many stop per second?

11. An ionization chamber is simultaneously bombarded by a beam of $8 \times 10^6$ helium ions per second and a beam of $1 \times 10^8$ carbon ions per second. The helium ions have an initial energy of 5 MeV, and the carbon ions have an initial energy of 100 keV. All ions stop in the chamber gas. For the helium ions, $W = 36$ eV ip$^{-1}$; for the carbon ions, $W = 48$ eV ip$^{-1}$. Calculate the saturation current.

12. A source emits 5.16-MeV alpha particles, which are absorbed at a rate of 842 per minute in the gas of a parallel-plate ionization chamber. The saturation current is $3.2 \times 10^{-13}$ A. Calculate the $W$ value for the alpha particles in the gas.

13. Show that $1$ eV ip$^{-1} = 1$ J C$^{-1}$.

14. A saturation current of $2.70 \times 10^{-14}$ A is measured with a parallel-plate ionization chamber in a radiation field. The chamber contains air ($W = 34$ eV ip$^{-1}$) at 20°C and 752 torr.
    (a) What is the rate of energy absorption in the chamber?
    (b) If the chamber has a sensitive volume of 750 cm$^3$, what is the dose rate in the air in J kg$^{-1}$ s$^{-1}$ ($=$ Gy s$^{-1}$)?

15. A parallel-plate ionization chamber is being designed to work with air ($W = 34$ eV ip$^{-1}$) at STP. When the dose rate in the chamber is 10.0 mGy h$^{-1}$, the saturation current is to be $10^{-11}$ A. What volume must the chamber have?

16. Where does most of the gas multiplication occur inside a cylindrical proportional-counter tube?

17. What is the ratio of the pulse heights from a 1-MeV proton ($W = 30$ eV ip$^{-1}$) and a 1-MeV carbon nucleus ($W = 40$ eV ip$^{-1}$) absorbed in a proportional counter?

18. Why is a GM counter not useful for determining the absorbed energy in a gas?

19. Show that, at high energies, where the average number of electrons per quantum state is small, the quantum-mechanical distribution Eq. (10.6) approaches the classical Boltzmann distribution, $N = \exp(-E/kT)$.

20. Figure 10.15(b) shows the relative number of electrons at energies $E$ in an intrinsic semiconductor when $T > 0$. Make such a sketch for the As-doped Ge semiconductor shown in Fig. 10.18.

21. What type of semiconductor results when Ge is doped with (a) Sb or (b) In?

22. Make a sketch like that in Fig. 10.17 for Ge doped with Ga (p-type semiconductor).

23. Use the nonrelativistic stopping-power formula and show that Eq. (10.7) holds.

24. What are the relative magnitudes of the responses of an $E(-dE/dx)$ particle identifier to a proton, an alpha particle, and a stripped carbon nucleus?

25. The $W$ value for silicon is 3.6 eV ip$^{-1}$. Calculate the mean number of ion pairs produced by a 300-keV beta particle absorbed in Si.

26. Calculate the number of ion pairs produced by a 4-MeV alpha particle in
    (a) an ionization chamber filled with air ($W = 36$ eV ip$^{-1}$)
    (b) a silicon surface-barrier detector ($W = 3.6$ eV ip$^{-1}$).

27. At a temperature of absolute zero, some electrons occupy states at the donor impurity levels of an n-type semiconductor (Fig. 10.18). True or false?

28. Why is reverse, rather than forward, bias used for semiconductor junctions in radiation measurements?

29. A 1.27-MeV photon loses 540 keV and 210 keV in successive Compton scattering events in the sensitive volume of a Ge detector before escaping.
    (a) Estimate the total number of secondary electrons produced by the events.
    (b) Would the device register the passage of the photon as a single event or as two events?

30. A silicon semiconductor detector has a dead layer of 1 $\mu$m followed by a depletion region 250 $\mu$m in depth.
    (a) Are these detector dimensions suitable for alpha-particle spectroscopy up to 4 MeV?
    (b) For beta-particle spectroscopy up to 500 keV?

31. If 2.1 MeV of absorbed alpha-particle energy produces 41,100 scintillation photons of average wavelength 4800 Å in a scintillator, calculate its efficiency.

32. A 600-keV photon is absorbed in a NaI(Tl) crystal having an efficiency of 11.2%. The average wavelength of the scintillation photons produced is 5340 Å, and 11% of them produce a signal at the cathode of the photomultiplier tube. Calculate the average energy from the incident radiation that produces one photoelectron at the cathode (the "$W$ value").

33. Figure 10.30 was obtained for a 4 × 4 in. NaI(Tl) crystal scintillator. Sketch the observed spectrum if measurements were made with the same source, but with a NaI crystal that was
    (a) very large
    (b) very small.

(c) In (b), interpret the relative areas under the photopeak and the Compton continuum.

34. In Fig. 10.30, what physical process gives rise to pulses with energies between the Compton edge and the total energy peak?

35. A 1.17-MeV gamma ray is Compton scattered once at an angle of 48° in a scintillator and again at an angle of 112° before escaping.

    (a) What (average) pulse height is registered?

    (b) If the photon were scattered once at 48° and then photoelectrically absorbed, what pulse height would be registered?

36. The mass attenuation coefficient of NaI (density = 3.67 g cm$^{-3}$) for 500-keV photons is 0.090 cm$^2$ g$^{-1}$. What percentage of normally incident photons interact in a crystal 4 cm thick?

37. Calculate the energy of the backscatter peak for 3-MeV gamma photons.

38. What sequence of events produces the escape peak seen at 11 keV below the total-energy peak when pulse heights from monoenergetic photons are measured with a germanium detector? (See Fig. 10.31.)

39. Calculate the temperature rise in a calorimetric water dosimeter that absorbs 10 J kg$^{-1}$ from a radiation beam (= an absorbed dose of 10 Gy, a lethal dose if given acutely over the whole body). What absorbed dose is required to raise the temperature 1°C?

40. Prove Eq. (10.9).

41. (a) Calculate the threshold kinetic energy for an electron to produce Cerenkov radiation in water (index of refraction = 1.33).

    (b) What is the threshold energy for a proton?

42. An electron enters a water shield (index of refraction 1.33) with a speed $v = 0.90c$, where $c$ is the speed of light in a vacuum. Assuming that 0.1% of its energy loss is due to Cerenkov radiation as long as this is possible, calculate the number of photons emitted if their average wavelength is 4200 Å.

43. Given the value $Q = 4.78$ MeV for the $^6$Li(n,$\alpha$)$^3$H reaction in Table 10.2, calculate $T_H$ and $T_{He}$.

44. Sketch the pulse-height spectrum from a $^3$He proportional counter exposed to thermal neutrons. Include wall effects at the appropriate energies.

45. (a) What is the maximum energy that a 3-MeV neutron can transfer to a $^3$He nucleus by elastic scattering?

    (b) What is the maximum energy that can be registered in a $^3$He proportional counter from a 3-MeV neutron?

46. Sketch the pulse-height spectrum from a thin layered boron-lined proportional counter tube for thermal neutrons. Repeat for a layer thicker than the ranges of the reaction products.

47. A proportional-counter tube is lined with a thin coating of a lithium compound (containing $^6$Li). The tube size and gas pressure are sufficiently large to stop any charged particle that enters the gas. The counter gas is insensitive to thermal neutrons.
    (a) Sketch the pulse-height spectrum when the tube is exposed to thermal neutrons. Show the relevant energies and indicate what kind of events produce the signal there.
    (b) Repeat for a thick layer of the lithium compound.

48. At what neutron energy is the cross section for the $^3$He(n,p) reaction equal to the thermal-neutron cross section for the $^6$Li(n,$\alpha$) reaction?

49. In the proton-recoil telescope (Fig. 10.46), what is the energy of the scattered neutron if $\theta = 27°$ and $T_p = 1.18$ MeV?

50. Calculate the threshold energy (a) neglecting and (b) including nuclear recoil for the reaction $^{24}$Mg(n,p)$^{24}$Na shown in Table 10.3. The $^{24}$Na nucleus is produced in an excited state from which it emits a 1.369-MeV gamma photon and goes to its ground state. (See Appendix D.)

## 10.10
## Answers

1. $3.12 \times 10^5$ s$^{-1}$;
   9.33 MeV s$^{-1}$
2. $1.54 \times 10^5$ on average
4. (a) $5.00 \times 10^5$
   (b) 18.0 MeV
   (c) $9.29 \times 10^{-9}$ Gy
5. $1.87 \times 10^{13}$ eV cm$^{-2}$ s$^{-1}$
9. (a) 236 Bq
   (b) 5.29 MeV
   (c) 0.101 MeV
11. $2.11 \times 10^{-7}$ A
15. 95.1 cm$^3$
24. $1 : 16 : 432$

30. (a) Yes
    (b) No
31. 5.1%
32. 190 eV
35. (a) 0.933 MeV
    (b) 1.17 MeV
36. 73.3%
39. 0.00239°C; 4180 Gy
41. (a) 264 keV
    (b) 485 MeV
42. 135
49. 0.31 MeV

# 11
# Statistics

## 11.1
## The Statistical World of Atoms and Radiation

As we have seen throughout this book, the mathematical descriptions used to make quantitative predictions about nature on the atomic scale do so on a statistical basis. How much energy will a 1-MeV proton lose in its next collision with an atomic electron? Will a 400-keV photon penetrate a 2-mm lead shield without interacting? How many disintegrations will occur during the next minute with a given radioactive source? These questions can be answered precisely, but only in statistical terms. Whether we measure a number of counts to infer the activity of a source or the number of electrons produced in a proportional counter to infer the energy of a photon, there is an irreducible uncertainty due to statistical fluctuations in the physical processes that occur. Repetition of the measurement results in a spread of values. How certain, then, is a measurement?

In this chapter we formalize some statistical concepts in order to place confidence limits on the measured values of quantities. The uncertainties we address here are the inherent ones due to quantum physics. Other sources of error or uncertainty, such as the precision with which the position of a pointer on a dial can be read, are not considered here. We begin by analyzing the determination of the activity of a radioactive sample by counting.

## 11.2
## Radioactive Disintegration—Exponential Decay

To determine the activity of a long-lived radionuclide in a sample, the sample can be counted for a specified length of time. Knowledge of the counter efficiency—the ratio of the number of counts and the number of disintegrations—then yields the sample activity. If the counting experiment is repeated many times, the numbers of counts observed in a fixed length of time will be found to be distributed about their mean value, which represents the best estimate of the true activity. The spread of the distribution about its mean is a measure of the uncertainty of the determination.

*Atoms, Radiation, and Radiation Protection.* James E. Turner
Copyright © 2007 WILEY-VCH Verlag GmbH & Co. KGaA, Weinheim
ISBN: 978-3-527-40606-7

In the following sections, we shall calculate the probability for a given number of disintegrations to occur during time $t$ when $N$ atoms of a radionuclide with decay constant $\lambda$ are initially present. The half-life of the nuclide need not be long compared with the observation time. We consider the probability of decay for each atom, assuming that all atoms are identical and independent and that the decay process is spontaneous and random.

Equation (4.7) shows that the relative number of atoms that have not decayed in a sample in time $t$ is $e^{-\lambda t}$. We can interpret exponential decay as implying that the probability that an atom survives a time $t$ without disintegrating is

$$q = \text{probability of survival} = e^{-\lambda t}. \tag{11.1}$$

For decay during the time $t$, it follows that

$$p = \text{probability of decay} = 1 - q = 1 - e^{-\lambda t}. \tag{11.2}$$

Note that there are only two alternatives for a given atom in the time $t$, since $p + q = 1$.

In Section 4.2 we treated the number of atoms $N$ present at time $t$ as a continuous variable. It is, of course, a discrete variable, taking on only non-negative integral values. In the context of the discussion there, we tacitly regarded $N$ as being very large. Fluctuations were not considered, and the decay of a sample over time was represented smoothly by the exponential function. In what follows, the number of radioactive atoms that decay in a sample will be treated properly as a discrete variable.

## 11.3
## Radioactive Disintegration—a Bernoulli Process

Consider a set of $N$ identical atoms, characterized by a decay constant $\lambda$, at time $t = 0$. We ask, "What is the probability that exactly $n$ atoms will decay between $t = 0$ and a specified later time $t$?" The integer $n$ can have any value in the range $0 \le n \le N$. During $t$, each atom can be regarded as "trying" to decay (success) or not decay (failure). Observation of a set of $N$ atoms from time 0 to time $t$ can thus be regarded as an experiment that meets the following conditions:

1. It consists of $N$ trials (i.e., $N$ atoms each having a chance to decay).
2. Each trial has a binary outcome: success or failure (decay or not).
3. The probability of success (decay) is constant from trial to trial (all atoms have an equal chance to decay).
4. The trials are independent.

In statistics, these four conditions characterize a Bernoulli process. The number of successes, $n$, from $N$ trials—called Bernoulli trials—is a binomial random variable, and the probability distribution of this discrete random variable is called the binomial distribution.

The preceding conditions are met for a set of $N$ identical radioactive atoms, observed for a time $t$. Therefore, if many such sets of $N$ atoms are prepared and observed for time $t$, the numbers of atoms that decay from each set are expected to be represented by the binomial distribution. After the next example, illustrating radioactive decay as a Bernoulli process, we derive this distribution and show how it and its spread depend upon $N$, $n$, $p$, $q$, and the observation time $t$.

*Example*

A sample of $N = 10$ atoms of $^{42}K$ (half-life $= 12.4$ h) is prepared and observed for a time $t = 3$ h.

(a) What is the probability that atoms number 1, 3, and 8 will decay during this time?

(b) What is the probability that atoms 1, 3, and 8 decay, while none of the others decay?

(c) What is the probability that exactly three atoms (any three) decay during the 3 hours?

(d) What is the probability that exactly six atoms will decay in the 3 hours?

(e) What is the chance that no atoms will decay in 3 hours?

(f) What is the general formula for the probability that exactly $n$ atoms will decay, where $0 \le n \le 10$?

(g) What is the sum of all possible probabilities from (f)?

(h) If the original sample consisted of $N = 100$ atoms, what would be the chance that no atoms decay in 3 hours?

*Solution*

(a) The decay constant for $^{42}K$ is $\lambda = 0.693/(12.4 \text{ h}) = 0.0559 \text{ h}^{-1}$. The probability that a given atom survives the time $t = 3$ h without decaying is, by Eq. (11.1),

$$q = e^{-\lambda t} = e^{-0.0559 \times 3} = e^{-0.168} = 0.846. \tag{11.3}$$

The probability that a given atom will decay is

$$p = 1 - q = 0.154. \tag{11.4}$$

The probability that atoms 1, 3, and 8 decay in this time is

$$p^3 = (0.154)^3 = 0.00365. \tag{11.5}$$

(b) The answer to (a) is independent of the fate of the other seven atoms. The probability that none of the others decay in the 3 hours is $q^7 = (0.846)^7 = 0.310$. The probability that only atoms 1, 3, and 8 decay while the others survive is therefore

$$p^3 q^7 = (0.00365)(0.310) = 0.00113. \tag{11.6}$$

(c) The last answer, $p^3 q^7$, gives the probability that a particular, designated three atoms decay—and only those three—in the specified time. The probability that exactly three atoms (any three) decay is $p^3 q^7$ times the number of ways that a group of three can be chosen from among the $N = 10$ atoms. To make such a group, there are 10 choices for the first member, 9 choices for the second member, and 8 choices for the third member. Thus, there are $N(N-1)(N-2) = 10 \times 9 \times 8 = 720$ ways of selecting the three atoms for decay. However, these groups are not all distinct. In a counting experiment, the decay of atoms 1, 3, and 8 in that order is not distinguished from decay in the order 1, 8, and 3. The number of ways of ordering the three atoms that decay is $3 \times 2 \times 1 = 3! = n! = 6$. Therefore, the number of ways that any $n = 3$ atoms can be chosen from among $N = 10$ is given by the binomial coefficient[1]

$$\binom{N}{n} = \binom{10}{3} \equiv \frac{10!}{3!7!} = \frac{10 \times 9 \times 8}{3!} = \frac{720}{6} = 120. \tag{11.7}$$

The probability that exactly 3 atoms decay is

$$P_3 = \binom{10}{3} p^3 q^7 = 120 \times 0.00113 = 0.136. \tag{11.8}$$

(d) The probability that exactly 6 atoms decay in the 3 hours is

$$P_6 = \binom{10}{6} p^6 q^4 = \frac{10!}{6!4!} (0.154)^6 (0.846)^4 = 0.00143. \tag{11.9}$$

(e) The probability that no atom decays is

$$P_0 = \binom{10}{0} p^0 q^{10} = \frac{10!}{0!10!} (0.846)^{10} = 0.188. \tag{11.10}$$

Thus, there is a probability of 0.188 that no disintegrations occur in the 3-hour period, which is approximately one-fourth of the half-life.

(f) We can see from (c) and (d) that the general expression for the probability that exactly $n$ of the 10 atoms decay is

$$P_n = \binom{10}{n} p^n q^{10-n}. \tag{11.11}$$

(g) The sum of the probabilities for all possible numbers of disintegrations, $n = 0$ to $n = 10$, should be unity. From Eq. (11.11), and with the help of the last footnote for the binomial expansion, we write

$$\sum_{n=0}^{10} P_n = \sum_{n=0}^{10} \binom{10}{n} p^n q^{10-n} = (p+q)^{10}. \tag{11.12}$$

Since $p + q = 1$, the total probability (11.12) is unity.

(h) With $N = 100$ atoms in the sample, the probability that none would decay in the 3 h is $q^{100} = (0.846)^{100} = 5.46 \times 10^{-8}$. This is a much smaller probability than in (e), where there are only 10 atoms in the initial sample. Seeing no atoms decay in a sample of size 100 is a rare event.

---

1 The general expansion of a binomial to an integral power $N$ is given by

$$(p+q)^N = \sum_{n=0}^{N} \binom{N}{n} p^n q^{N-n},$$

where

$$\binom{N}{n} = \frac{N!}{n!(N-n)!} = \frac{N(N-1)\cdots(N-n+1)}{n!}.$$

## 11.4
## The Binomial Distribution

We summarize the results that describe the Bernoulli process of radioactive decay and generalize Eq. (11.11) for any initial number $N$ of identical radioactive atoms. The probability that exactly $n$ will disintegrate in time $t$ is

$$P_n = \binom{N}{n} p^n q^{N-n}. \tag{11.13}$$

Here $p$ and $q$ are defined by Eqs. (11.1) and (11.2). Since the $P_n$ are just the terms in the binomial expansion and since $p + q = 1$, the probability distribution represented by Eq. (11.13) is normalized; that is,

$$\sum_{n=0}^{N} P_n = (p+q)^N = 1. \tag{11.14}$$

The function defined by Eq. (11.13) with $p + q = 1$ is called the binomial distribution and applies to any Bernoulli process. Besides radioactive decay, other familiar examples of binomial distributions include the number of times "heads" occurs when a coin is tossed $N$ times and the frequency with which exactly $n$ sixes occur when five dice are rolled. The binomial distribution finds widespread industrial applications in product sampling and quality control.

The expected, or mean, number of disintegrations in time $t$ is given by the average value $\mu$ of the binomial distribution (11.13):

$$\mu \equiv \sum_{n=0}^{N} n P_n = \sum_{n=0}^{N} n \binom{N}{n} p^n q^{N-n}. \tag{11.15}$$

This sum is evaluated in Appendix E. The result, given by Eq. (E.4), is

$$\mu = Np. \tag{11.16}$$

Thus, the mean is just the product of the total number of trials and the probability of the success of a single trial.

Repeated observations of many sets of $N$ identical atoms for time $t$ is expected to give the binomial probability distribution $P_n$ for the number of disintegrations $n$. The scatter, or spread, of the distribution of $n$ is characterized quantitatively by its variance $\sigma^2$ or standard deviation $\sigma$, defined as the positive square root of the variance. The variance is defined as the expected value of the squared deviation from the mean of all values of $n$:

$$\sigma^2 \equiv \sum_{n=0}^{N} (n - \mu)^2 P_n. \tag{11.17}$$

As shown in Appendix E, [Eq. (E.14)], the standard deviation is given by

$$\sigma = \sqrt{Npq}. \tag{11.18}$$

Although we have been discussing the number of disintegrations of a radio-nuclide, the results can also be applied to the number of counts registered from disintegrations in an experiment. Unless the efficiency $\epsilon$ of a counter is 100%, the number of counts will be less than the number of disintegrations. With $N$ atoms initially present, the probability that a given atom will disintegrate in time $t$ and be registered as a count is, in place of Eq. (11.2),

$$p^* = \text{prob. of a count} = \epsilon p = \epsilon(1 - e^{-\lambda t}). \tag{11.19}$$

The probability that the given atom will not give a count, either by not decaying or by decaying but not being registered, is

$$q^* = \text{prob. of no count} = 1 - p^* = 1 - \epsilon + \epsilon e^{-\lambda t}. \tag{11.20}$$

The formalism developed thus far for the number of disintegrations can be applied to the number of counts by using $p^*$ and $q^*$ in place of $p$ and $q$. The binomial distribution function $P_n$ then applies to the number of counts, rather than disintegrations, obtained in time $t$. If $\epsilon = 1$, then Eqs. (11.19) and (11.20) are identical with (11.2) and (11.1), respectively. Also, one can divide the number of disintegrations by the time $t$ and then apply the above formalism to obtain the average disintegration and count *rates* over the observation time.

### Example

An experimenter repeatedly prepares a large number of samples identical to that in the example from the last section: $N = 10$ atoms of $^{42}$K at time $t = 0$. He does not know how much activity is initially present, but he wants to estimate it by determining the mean number of disintegrations that occur in a given time. To this end, each new sample is placed in a counter, having an efficiency $\epsilon = 32\%$, and observed for 3 h, the same time period as before.

(a) What is the probability that exactly 3 counts will be observed?
(b) What is the expected number of counts in 3 h?
(c) What is the expected count rate, averaged over the 3 h?
(d) What is the expected disintegration rate, averaged over the 3 h?
(e) What is the standard deviation of the count rate over the 3 h?
(f) What is the standard deviation of the disintegration rate over the 3 h?
(g) If $\epsilon = 100\%$, the count rate would be equal to the disintegration rate. What would then be the expected value and standard deviation of the disintegration rate?

### Solution

(a) In the previous example, Eq. (11.8) gave a probability $P_3 = 0.136$ for the occurrence of exactly 3 disintegrations. This, of course, remains true here. However, with $\epsilon = 0.32$, the probability of observing exactly 3 counts in the time $t = 3$ h is smaller, because obtaining 3 counts will generally require more man 3 atoms to decay, with a correspondingly lower probability. We let $n^*$ represent the number of disintegrations detected by the counter. The probability for exactly 3 counts, that is, for $n^* = 3$, is given by Eq. (11.8) with $p$ and $q$ replaced by $p^*$ and $q^*$. From Eqs. (11.19) and (11.4), we have $p^* = \epsilon p = 0.32 \times 0.154 = 0.049$; and, from Eq. (11.20), $q^* = 0.951$. The probability of observing exactly 3 counts is, from Eq. (11.8),

$$P_3^* = \binom{10}{3} p^{*3} q^{*7} = 120(0.049)^3 (0.951)^7 = 0.00993. \tag{11.21}$$

(b) The expected number of counts $\mu^*$ is given by Eq. (11.16) with $p$ replaced by $p^*$:

$$\mu^* = Np^* = 10 \times 0.049 = 0.490. \tag{11.22}$$

This is the expected number of counts in 3 h. [With $\epsilon = 1.00$, $\mu^* = \mu = Np = 1.54$. The ratio $\mu^*/\mu$ is the average fraction of disintegrating atoms that get counted, that is, the counter efficiency, $\epsilon$.]

(c) The expected count rate for $t = 3$ h is $r_c = \mu^*/t = 0.490/(3\text{ h}) = 0.163\text{ h}^{-1}$.

(d) The average disintegration rate for $t = 3$ h is $r_d = \mu^*/t\epsilon = (0.163\text{ h}^{-1})/0.32 = 0.509\text{ h}^{-1}$. [Because the observation time is not small compared with the half-life, the average disintegration rate over 3 h is less than the initial activity. The latter is $\lambda N = (0.0559\text{ h}^{-1})(10) = 0.559\text{ h}^{-1}$.]

(e) The standard deviation of the count rate is, from Eq. (11.18),

$$\sigma_{cr} = \frac{\sqrt{Np^* q^*}}{t} = \frac{\sqrt{10 \times 0.049 \times 0.951}}{3\text{ h}} = 0.228\text{ h}^{-1}. \tag{11.23}$$

(f) The standard deviation of the disintegration rate is $\sigma_{cr} = \sigma_{cr}/\epsilon = 0.713\text{ h}^{-1}$.

(g) The actual disintegration rate, which would be equal to the count rate if $\epsilon = 1.00$, is

$$r_d = \frac{Np}{t} = \frac{10 \times 0.154}{3\text{ h}} = 0.513\text{ h}^{-1}. \tag{11.24}$$

The standard deviation is then

$$\sigma_{dr} = \frac{\sqrt{Npq}}{t} = \frac{\sqrt{10 \times 0.154 \times 0.846}}{3} = 0.380\text{ h}^{-1}. \tag{11.25}$$

The disintegration rate $r_d$, Eq. (11.24), is the same as that found in (d), except for round-off ($\epsilon$ was given to only two significant figures). The ratio $r_c/r_d$ is just equal to the counter efficiency, since $\mu^*/\mu = \epsilon$. It follows that a large number of repeated measurements would be expected to give a distribution of values for the disintegration rate about its true mean, independently of the counter efficiency. Note, however, that the standard deviation (11.25) obtained for the disintegration rate is considerably smaller here than in (f). Thus $r_d$ is determined with greater precision here. Improving counter efficiency results in a larger number of counts in a given time period, the increased sample size (number of counts) giving greater precision in the statistical results.

This example has dealt with a radioactive source that is unrealistically weak to measure. However, it illustrates the basic way in which the random process of radioactive decay can be registered by an instrument that counts events. Information needed to characterize a source (e.g., activity, half-life) is obtained in this manner on a statistical basis. In practice, it is usually desirable to obtain estimates with high statistical precision, acquired by sampling large numbers of events.

### Example

More realistically, consider a $^{42}$K source with an activity of 37 Bq ($= 1$ nCi). The source is placed in a counter, having an efficiency of 100%, and the numbers of counts in one-second intervals are registered.

(a) What is the mean disintegration rate?
(b) Calculate the standard deviation of the disintegration rate.
(c) What is the probability that exactly 40 counts will be observed in any second?

### Solution

(a) The mean disintegration rate is the given activity, $r_d = 37$ s$^{-1}$.

(b) The standard deviation of the disintegration rate is given by Eq. (11.18). We work with the time interval, $t = 1$ s. Since the decay constant is $\lambda = 0.0559$ h$^{-1} = 1.55 \times 10^{-5}$ s$^{-1}$, we have

$$q = e^{-\lambda t} = e^{-1.55 \times 10^{-5} \times 1} = 0.9999845 \tag{11.26}$$

and $p = 1 - q = 0.0000155$.[2] The number of atoms present is

$$N = \frac{r_d}{\lambda} = \frac{37 \text{ s}^{-1}}{1.55 \times 10^{-5} \text{ s}^{-1}} = 2.39 \times 10^6. \tag{11.27}$$

From Eq. (11.18), we obtain for the standard deviation of the disintegration rate

$$\sigma_{dr} = \frac{\sqrt{Npq}}{t} = \frac{\sqrt{2.39 \times 10^6 \times 0.0000155 \times 0.9999845}}{1 \text{ s}} = 6.09 \text{ s}^{-1}, \tag{11.28}$$

which is about 16% of the mean disintegration rate.

(c) The probability of observing exactly $n = 40$ counts in 1 s is given by Eq. (11.13). However, the factors quickly become unwieldy when $N$ is not small (e.g., $69! = 1.71 \times 10^{98}$). For large $N$ and small $n$, as we have here, we can write for the binomial coefficient

$$\binom{N}{n} \equiv \frac{N(N-1)\cdots(N-n+1)}{n!} \cong \frac{N^n}{n!}, \tag{11.29}$$

since each of the $n$ factors in the numerator is negligibly different from $N$. Equation (11.13) then gives

$$P_{40} = \frac{(2.39 \times 10^6)^{40}}{40!}(0.0000155)^{40}(0.9999845)^{2.39 \times 10^6 - 40} \tag{11.30}$$

$$= \frac{(2.39)^{40}(10^{240})(0.0000155)^{40}(0.9999845)^{2.39 \times 10^6}}{40!}, \tag{11.31}$$

where $n = 40 \ll N$ has been dropped from the last exponent. The right-hand side can be conveniently evaluated with the help of logarithms. To reduce round-off errors, we

---

2  Note that, for small $\lambda t$,
$\quad p = 1 - e^{-\lambda t} \cong 1 - (1 - \lambda t) = \lambda t.$

use four decimal places:

$$\log(2.39)^{40} = 15.1359$$

$$\log(10)^{240} = 240.0000$$

$$\log(0.0000155)^{40} = -192.3867$$

$$\log(0.9999845)^{2.39 \times 10^6} = -16.0886$$

$$-\log 40! = \underline{-47.9116}$$

$$\log P_{40} = -1.251. \tag{11.32}$$

Thus, $P_{40} = 10^{-1.251} = 0.0561$.

This example shows that computations employing the basic binomial distribution that describes the Bernoulli process of radioactive decay can get quite cumbersome. Precise numerical evaluations can present formidable problems, especially for large $N$, large $n$, and small $p$. These are the conditions typically met in practice, as this example illustrates. Fortunately, very good approximations can be made to the binomial distribution for just these cases, greatly simplifying numerical work. We consider in the next two sections the Poisson and then the normal, or Gaussian, distributions. These distributions also arise naturally in describing many phenomena.

## 11.5
## The Poisson Distribution

We consider the conditions $N \gg 1$, $N \gg n$, and $p \ll 1$, as we did in the last example. As seen from Eqs. (11.1) and (11.2), small $p$ implies that $q$ is near unity and therefore $\lambda t \ll 1$. Under these conditions, as shown in Appendix E, the terms of the binomial distribution (11.13) are given to a very good approximation by the Poisson distribution [Eq. (E.23)]:

$$P_n = \frac{\mu^n e^{-\mu}}{n!}. \tag{11.33}$$

It is also shown in Appendix E that $\mu = Np$ is the mean of the Poisson distribution, as it is for the binomial. The standard deviation of the Poisson distribution [Eq. (E.28)] is the square root of the mean:

$$\sigma = \sqrt{\mu}. \tag{11.34}$$

Whereas the binomial distribution is characterized by two independent parameters, $N$ and $p$ (or $q$), the Poisson distribution has the single parameter $\mu$.

As the next example shows, computations for radioactive decay can be simplified considerably when the conditions for representation by Poisson statistics are satisfied.

*Example*

Repeat the last example by using Poisson statistics to approximate the binomial distribution.

*Solution*

(a) As before, the mean disintegration rate is the given activity, which we write as $\mu = 37 \text{ s}^{-1}$.

(b) The standard deviation [Eq. (11.34)] is

$$\sigma = \sqrt{\mu} = \sqrt{37} = 6.08 \text{ s}^{-1}, \tag{11.35}$$

as compared with $6.09 \text{ s}^{-1}$ found before [Eq. (11.28)].

(c) The probability of exactly 40 disintegrations occurring in a given second is, by Eq. (11.33),

$$P_{40} = \frac{37^{40} e^{-37}}{40!} = 0.0559, \tag{11.36}$$

in close agreement with the value 0.0561 found before. (Lack of exact agreement to several significant figures between the results found here with the binomial and Poisson distributions can be attributed to round-off.)

Like the binomial distribution, the distribution (11.33) can be derived in its own right for a Poisson process.[3] The conditions required are the following:

1. The number of successes in any one time interval is independent of the number in any other disjoint time interval. (The Poisson process has no memory.)
2. The probability that a single success occurs in a very short time interval is proportional to the length of the interval.
3. The probability that more than one success will occur in a very short time interval is negligible.

The Poisson process can describe such diverse phenomena as the number of traffic accidents that occur during August in a certain county, the number of eggs laid daily by a brood of hens, and the number of cosmic rays registered hourly in a counter. The events occur at random, but at an expected average rate. Generally, the Poisson distribution describes the number of successes for any random process whose probability is small ($p \ll 1$) and constant.

Figure 11.1 shows a comparison of the binomial and Poisson distributions. In all panels, the mean, $\mu = 10$, of both distributions is kept fixed; the probability of success $p$ and sample size $N$ are varied between panels. Since the mean is the same, the Poisson distribution is the same throughout the figure. Both distributions are asymmetric, favoring values of $n \leq \mu$. As pointed out after Eq. (E.19), although the binomial probability $P_n = 0$ when $n > N$, the Poisson $P_n$ are never exactly zero. (In the upper left-hand panel of Fig. 11.1 when $n > 15$, for example, it can be seen

---

3 However, we shall not carry out the derivation of Eq. (11.33) from the postulates.

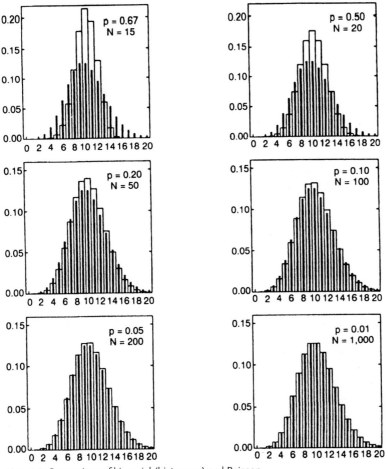

**Fig. 11.1** Comparison of binomial (histogram) and Poisson (solid bars) distributions, having the same mean, $\mu = 10$, but different values of the probability of success p and sample size N. The ordinate in each panel shows the probability $P_n$ of exactly n successes, shown on the abscissa. With fixed $\mu$, the Poisson distribution is the same throughout. (Courtesy James S. Bogard.)

that $P_n = 0$ for the binomial, but not the Poisson, distribution.) As p gets smaller in Fig. 11.1, the Poisson distribution approaches the binomial (having the same mean) more and more closely, as we have discussed. The two distributions are close to one another when $p = 0.10$ and become virtually indistinguishable when $p = 0.01$. Both distributions also become progressively more bell-shaped with decreasing p, suggesting a relationship with the normal distribution (next section).

In the typical counting of radioactive samples with significant activity, the approximation of using Poisson statistics in place of the exact, but often cumbersome, binomial distribution is usually warranted. However, the justification for doing so

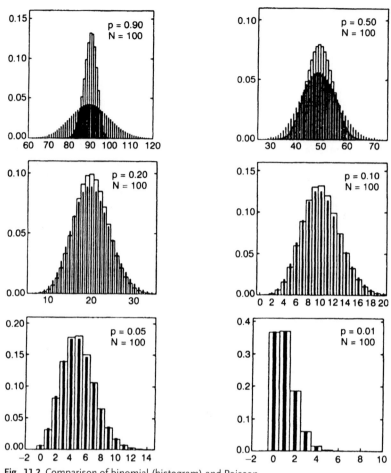

**Fig. 11.2** Comparison of binomial (histogram) and Poisson (solid bars) distributions for fixed N and different p. The ordinate shows $P_n$ and the abscissa, n. The mean of the two distributions in a given panel is the same. (Courtesy James S. Bogard.)

depends on the extent to which $p \ll 1$ and $N \gg 1$. In the last two examples, involving 37 Bq of $^{42}$K and 1-s counting intervals, the Poisson description was seen to be extremely accurate. We had for the probability of success (decay) $p = 1.55 \times 10^{-5}$ and $N = 2.39 \times 10^6$ trials (atoms).

Figure 11.2 shows a comparison of another sort, in which $N = 100$ is held constant and $p$ is varied. The middle panel on the right ($p = 0.10$) is the same as in Fig. 11.1. The mean value is the same for both distributions in each panel of Fig. 11.2; but it shifts from panel to panel, becoming progressively smaller as $p$ is decreased. Again, we see that the binomial and Poisson distributions become virtually indistinguishable for small $p$. However, they are not bell-shaped when they become similar, as they were in the last figure.

## 11.6
## The Normal Distribution

Figure 11.1 indicates that the binomial and Poisson distributions become similar and tend to approach the shape of a normal, or Gaussian, distribution as $p$ gets small and $N$ gets large. The latter distribution although defined for a continuous (rather than discrete) variable, will be found to be an extremely useful and accurate approximation to the binomial and Poisson distributions for large $N$.

In Appendix E, it is shown how the normal distribution can be obtained from the Poisson under specific conditions. The distribution, Eq. (E.35), can be written as the probability density for the continuous random variable $x$:

$$f(x) = \frac{1}{\sqrt{2\pi}\sigma} e^{-(x-\mu)^2/2\sigma^2}. \tag{11.37}$$

The probability that $x$ lies between $x$ and $x+dx$ is $f(x)\,dx$. The function is normalized to unit area when integrated over all values of $x$, from $-\infty < x < \infty$. In place of the single Poisson parameter $\mu$, the normal distribution (11.37) is characterized by two independent parameters, its mean $\mu$ and standard deviation $\sigma$. It has inflection points at $x = \mu \pm \sigma$, where it changes from concave downward to concave upward in going away from the mean (Problem 18).

Figure 11.3 shows a comparison of binomial and normal distributions, having the same means and standard deviations. The binomial distributions in the two upper panels appeared in the two previous figures. In the upper left, the binomial and normal distributions are quite different in shape. The normal random variable can be negative, whereas the binomial integer $n$ cannot be negative. Also, the binomial distribution is skewed; the normal is symmetric. As we saw in the lower right-hand panel of Fig. 11.2, the Poisson distribution matches the binomial closely for the conditions in the upper left panel of Fig. 11.3. In fact, the binomial and Poisson distributions match throughout Fig. 11.3. As $N$ increases, $\mu$ becomes larger; and both the binomial and normal distributions shift toward the right. The binomial and Poisson practically match the normal distribution when $\mu = 30$ (lower left panel).

The probability that $x$ has a value between $x_1$ and $x_2$ is equal to the area under the curve $f(x)$ between these two values:

$$P(x_1 \leq x \leq x_2) = \frac{1}{\sqrt{2\pi}\sigma} \int_{x_1}^{x_2} e^{-(x-\mu)^2/2\sigma^2}\,dx. \tag{11.38}$$

For computational purposes, it is convenient to transform the normal distribution, which depends on the two parameters $\mu$ and $\sigma$, into a single, universal form. The

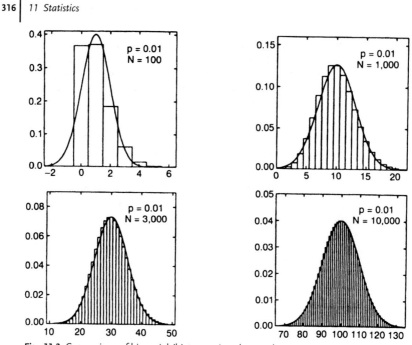

**Fig. 11.3** Comparison of binomial (histogram) and normal (solid line) distributions, having the same means and standard deviations. The ordinate in each panel gives the probability $P_n$ for the former and the density $f(x)$ [Eq. (11.37)] for the latter, the abscissa giving $n$ or $x$. (Courtesy James S. Bogard.)

standard normal distribution, having zero mean and unit standard deviation, is obtained by making the substitution

$$z = \frac{x - \mu}{\sigma}. \tag{11.39}$$

Equation (11.38) then becomes $(dx = \sigma \, dz)$

$$P(z_1 \leq z \leq z_2) = \frac{1}{\sqrt{2\pi}} \int_{z_1}^{z_2} e^{-z^2/2} \, dz. \tag{11.40}$$

Table 11.1 lists values of the integral,

$$P(z \leq z_0) = \frac{1}{\sqrt{2\pi}} \int_{-\infty}^{z_0} e^{-z^2/2} \, dz, \tag{11.41}$$

giving the probability that the normal random variable $z$ has a value less than or equal to $z_0$. This probability is illustrated by the shaded area under the standard normal curve, as indicated at the top of Table 11.1. The following example illustrates the use of the table.

*Example*

Repeated counts are made in 1-min intervals with a long-lived radioactive source. The observed mean value of the number of counts is 813, with a standard deviation of 28.5 counts. (a) What is the probability of observing 800 or fewer counts in a given minute? (b) What is the probability of observing 850 or more counts in 1 min? (c) What is the probability of observing 800 to 850 counts in a minute? (d) What is the symmetric range of values about the mean number of counts within which 90% of the 1-min observations are expected to fall?

*Solution*

The normal distribution is indistinguishable from the binomial and Poisson distributions within the precision given in the problem. Whereas the desired quantities would be tedious to calculate by using either of the latter two, they are readily obtained for the normal distribution.

(a) We use as estimates of the true mean and standard deviation $\mu = 813$ and $\sigma = 28.5$. Let $x$ represent the number of counts observed in 1 min. We have for the standard normal random variable, from Eq. (11.39), $z = (x-813)/28.5$. The probability that $x$ has a value less than or equal to 800 is the same as the probability that $z$ is less than or equal to $(800 - 813)/28.5 = -0.456$. Interpolating in Table 11.1, we find that $P(x \le 800) = P(z \le -0.456) = 0.324$.

(b) For $x = 850$, $z = (850-813)/28.5 = 1.30$. Whereas Table 11.1 gives values $P(z \le z_0)$, we are asked here for a complementary value, $P(z \ge z_0) = 1 - P(z \le z_0)$. Thus, $P(x \ge 850) = P(z \ge 1.30) = 1 - P(z \le 1.30) = 1 - 0.9032 = 0.097$.

(c) It follows from (a) and (b) that the probability that $x$ lies between 800 and 850 is equal to the area under the standard normal distribution in the interval $0.456 < z < 1.30$. Using Table 11.1, we find that

$$P(800 < x < 850) = P(-0.456 < z < 1.30) \tag{11.42}$$

$$= P(z < 1.30) - P(z < -0.456) \tag{11.43}$$

$$= 0.903 - 0.324 = 0.579. \tag{11.44}$$

As a check, the sum of the answers to (a), (b), and (c) must add to give unity (except for possible roundoff): $0.324 + 0.097 + 0.579 = 1.000$.

(d) With respect to the standard normal curve, 90% of the area in the symmetric interval $\pm z$ about $z = 0$ corresponds to the value of $z$ for which $P(z \le z_0) = 0.9500$. From Table 11.1 we see that this occurs when $z_0 = 1.645$. Equation (11.39) implies that the corresponding interval in $x$ is $\pm 1.645$ standard deviations from the mean $\mu$. Thus, 90% of the values observed for $x$ are expected to fall within the range $813 \pm 1.645(28.5) = 813 \pm 46.9$.

Table 11.2 lists some useful values of one-tail areas complementary to the areas given in Table 11.1. To illustrate its use, the answer to part (b) of the last example can be gotten by interpolating in Table 11.2 with $k_\alpha = 1.30$, giving $P(x \ge 850) = 0.0975$. Also, the second entry in the table shows that one-half of the area under a normal distribution lies within the interval $\pm 0.675\sigma$ about the mean.

**Table 11.1** Areas Under the Standard Normal Distribution from −∞ to z

| z | .00 | .01 | .02 | .03 | .04 | .05 | .06 | .07 | .08 | .09 |
|---|---|---|---|---|---|---|---|---|---|---|
| -3.4 | .0003 | .0003 | .0003 | .0003 | .0003 | .0003 | .0003 | .0003 | .0003 | .0002 |
| -3.3 | .0005 | .0005 | .0005 | .0004 | .0004 | .0004 | .0004 | .0004 | .0004 | .0003 |
| -3.2 | .0007 | .0007 | .0006 | .0006 | .0006 | .0006 | .0006 | .0005 | .0005 | .0005 |
| -3.1 | .0010 | .0009 | .0009 | .0009 | .0008 | .0008 | .0008 | .0008 | .0007 | .0007 |
| -3.0 | .0013 | .0013 | .0013 | .0012 | .0012 | .0011 | .0011 | .0011 | .0010 | .0010 |
| -2.9 | .0019 | .0018 | .0017 | .0017 | .0016 | .0016 | .0015 | .0015 | .0014 | .0014 |
| -2.8 | .0026 | .0025 | .0024 | .0023 | .0023 | .0022 | .0021 | .0021 | .0020 | .0019 |
| -2.7 | .0035 | .0034 | .0033 | .0032 | .0031 | .0030 | .0029 | .0028 | .0027 | .0026 |
| -2.6 | .0047 | .0045 | .0044 | .0043 | .0041 | .0040 | .0039 | .0038 | .0037 | .0036 |
| -2.5 | .0062 | .0060 | .0059 | .0057 | .0055 | .0054 | .0052 | .0051 | .0049 | .0048 |
| -2.4 | .0082 | .0080 | .0078 | .0075 | .0073 | .0071 | .0069 | .0068 | .0066 | .0064 |
| -2.3 | .0107 | .0104 | .0102 | .0099 | .0096 | .0094 | .0091 | .0089 | .0087 | .0084 |
| -2.2 | .0139 | .0136 | .0132 | .0129 | .0125 | .0122 | .0119 | .0116 | .0113 | .0110 |
| -2.1 | .0179 | .0174 | .0170 | .0166 | .0162 | .0158 | .0154 | .0150 | .0146 | .0143 |
| -2.0 | .0228 | .0222 | .0217 | .0212 | .0207 | .0202 | .0197 | .0192 | .0188 | .0183 |
| -1.9 | .0287 | .0281 | .0274 | .0268 | .0262 | .0256 | .0250 | .0244 | .0239 | .0233 |
| -1.8 | .0359 | .0352 | .0344 | .0336 | .0329 | .0322 | .0314 | .0307 | .0301 | .0294 |
| -1.7 | .0446 | .0436 | .0427 | .0418 | .0409 | .0401 | .0392 | .0384 | .0375 | .0367 |
| -1.6 | .0548 | .0537 | .0526 | .0516 | .0505 | .0495 | .0485 | .0475 | .0465 | .0455 |
| -1.5 | .0668 | .0655 | .0643 | .0630 | .0618 | .0606 | .0594 | .0582 | .0571 | .0559 |
| -1.4 | .0808 | .0793 | .0778 | .0764 | .0749 | .0735 | .0722 | .0708 | .0694 | .0681 |
| -1.3 | .0968 | .0951 | .0934 | .0918 | .0901 | .0885 | .0869 | .0853 | .0838 | .0823 |
| -1.2 | .1151 | .1131 | .1112 | .1093 | .1075 | .1056 | .1038 | .1020 | .1003 | .0985 |

Area

0   z

**Table 11.1** (Continued)

| z | .00 | .01 | .02 | .03 | .04 | .05 | .06 | .07 | .08 | .09 |
|------|-------|-------|-------|-------|-------|-------|-------|-------|-------|-------|
| −1.1 | .1357 | .1335 | .1314 | .1292 | .1271 | .1251 | .1230 | .1210 | .1190 | .1170 |
| −1.0 | .1587 | .1562 | .1539 | .1515 | .1492 | .1469 | .1446 | .1423 | .1401 | .1379 |
| −0.9 | .1841 | .1814 | .1788 | .1762 | .1736 | .1711 | .1685 | .1660 | .1635 | .1611 |
| −0.8 | .2119 | .2090 | .2061 | .2033 | .2005 | .1977 | .1949 | .1922 | .1894 | .1867 |
| −0.7 | .2420 | .2389 | .2358 | .2327 | .2296 | .2266 | .2236 | .2206 | .2177 | .2148 |
| −0.6 | .2743 | .2709 | .2676 | .2643 | .2611 | .2578 | .2546 | .2514 | .2483 | .2451 |
| −0.5 | .3085 | .3050 | .3015 | .2981 | .2946 | .2912 | .2877 | .2843 | .2810 | .2776 |
| −0.4 | .3446 | .3409 | .3372 | .3336 | .3300 | .3264 | .3228 | .3192 | .3156 | .3121 |
| −0.3 | .3821 | .3783 | .3745 | .3707 | .3669 | .3632 | .3594 | .3557 | .3520 | .3483 |
| −0.2 | .4207 | .4168 | .4129 | .4090 | .4052 | .4013 | .3974 | .3936 | .3897 | .3859 |
| −0.1 | .4602 | .4562 | .4522 | .4483 | .4443 | .4404 | .4364 | .4325 | .4286 | .4247 |
| −0.0 | .5000 | .4960 | .4920 | .4880 | .4840 | .4801 | .4761 | .4721 | .4681 | .4641 |
| 0.0 | .5000 | .5040 | .5080 | .5120 | .5160 | .5199 | .5239 | .5279 | .5319 | .5359 |
| 0.1 | .5398 | .5438 | .5478 | .5517 | .5557 | .5596 | .5636 | .5675 | .5714 | .5753 |
| 0.2 | .5793 | .5832 | .5871 | .5910 | .5948 | .5987 | .6026 | .6064 | .6103 | .6141 |
| 0.3 | .6179 | .6217 | .6255 | .6293 | .6331 | .6368 | .6406 | .6443 | .6480 | .6517 |
| 0.4 | .6554 | .6591 | .6628 | .6664 | .6700 | .6736 | .6772 | .6808 | .6844 | .6879 |
| 0.5 | .6915 | .6950 | .6985 | .7019 | .7054 | .7088 | .7123 | .7157 | .7190 | .7224 |
| 0.6 | .7257 | .7291 | .7324 | .7357 | .7389 | .7422 | .7454 | .7486 | .7517 | .7549 |
| 0.7 | .7580 | .7611 | .7642 | .7673 | .7704 | .7734 | .7764 | .7794 | .7823 | .7852 |
| 0.8 | .7881 | .7910 | .7939 | .7967 | .7995 | .8023 | .8051 | .8078 | .8106 | .8133 |
| 0.9 | .8159 | .8186 | .8212 | .8238 | .8264 | .8289 | .8315 | .8340 | .8365 | .8389 |
| 1.0 | .8413 | .8438 | .8461 | .8485 | .8508 | .8531 | .8554 | .8577 | .8599 | .8621 |
| 1.1 | .8643 | .8665 | .8686 | .8708 | .8729 | .8749 | .8770 | .8790 | .8810 | .8830 |
| 1.2 | .8849 | .8869 | .8888 | .8907 | .8925 | .8944 | .8962 | .8980 | .8997 | .9015 |

**Table 11.1** (Continued)

| z | .00 | .01 | .02 | .03 | .04 | .05 | .06 | .07 | .08 | .09 |
|---|-----|-----|-----|-----|-----|-----|-----|-----|-----|-----|
| 1.3 | .9032 | .9049 | .9066 | .9082 | .9099 | .9115 | .9131 | .9147 | .9162 | .9177 |
| 1.4 | .9192 | .9207 | .9222 | .9236 | .9251 | .9265 | .9278 | .9292 | .9306 | .9319 |
| 1.5 | .9332 | .9345 | .9357 | .9370 | .9382 | .9394 | .9406 | .9418 | .9429 | .9441 |
| 1.6 | .9452 | .9463 | .9474 | .9484 | .9495 | .9505 | .9515 | .9525 | .9535 | .9545 |
| 1.7 | .9554 | .9564 | .9573 | .9582 | .9591 | .9599 | .9608 | .9616 | .9625 | .9633 |
| 1.8 | .9641 | .9649 | .9656 | .9664 | .9671 | .9678 | .9686 | .9693 | .9699 | .9706 |
| 1.9 | .9713 | .9719 | .9726 | .9732 | .9738 | .9744 | .9750 | .9756 | .9761 | .9767 |
| 2.0 | .9772 | .9778 | .9783 | .9788 | .9793 | .9798 | .9803 | .9808 | .9812 | .9817 |
| 2.1 | .9821 | .9826 | .9830 | .9834 | .9838 | .9842 | .9846 | .9850 | .9854 | .9857 |
| 2.2 | .9861 | .9864 | .9868 | .9871 | .9875 | .9878 | .9881 | .9884 | .9887 | .9890 |
| 2.3 | .9893 | .9896 | .9898 | .9901 | .9904 | .9906 | .9909 | .9911 | .9913 | .9916 |
| 2.4 | .9918 | .9920 | .9922 | .9925 | .9927 | .9929 | .9931 | .9932 | .9934 | .9936 |
| 2.5 | .9938 | .9940 | .9941 | .9943 | .9945 | .9946 | .9948 | .9949 | .9951 | .9952 |
| 2.6 | .9953 | .9955 | .9956 | .9957 | .9959 | .9960 | .9961 | .9962 | .9963 | .9964 |
| 2.7 | .9965 | .9966 | .9967 | .9968 | .9969 | .9970 | .9971 | .9972 | .9973 | .9974 |
| 2.8 | .9974 | .9975 | .9976 | .9977 | .9977 | .9978 | .9979 | .9979 | .9980 | .9981 |
| 2.9 | .9981 | .9982 | .9982 | .9983 | .9984 | .9984 | .9985 | .9985 | .9986 | .9986 |
| 3.0 | .9987 | .9987 | .9987 | .9988 | .9988 | .9989 | .9989 | .9989 | .9990 | .9990 |
| 3.1 | .9990 | .9991 | .9991 | .9991 | .9992 | .9992 | .9992 | .9992 | .9993 | .9993 |
| 3.2 | .9993 | .9993 | .9994 | .9994 | .9994 | .9994 | .9994 | .9995 | .9995 | .9995 |
| 3.3 | .9995 | .9995 | .9995 | .9996 | .9996 | .9996 | .9996 | .9996 | .9996 | .9997 |
| 3.4 | .9997 | .9997 | .9997 | .9997 | .9997 | .9997 | .9997 | .9997 | .9997 | .9998 |

**Table 11.2** One-Tail Areas $\alpha$ Under the Standard Normal Distribution from $z = k_\alpha$ to $\infty$

| Area, $\alpha$ | $k_\alpha$ |
|---|---|
| 0.5000 | 0.000 |
| 0.2500 | 0.675 |
| 0.1587 | 1.000 |
| 0.1000 | 1.282 |
| 0.0500 | 1.645 |
| 0.0250 | 1.960 |
| 0.0228 | 2.000 |
| 0.0100 | 2.326 |
| 0.0050 | 2.576 |
| 0.0013 | 3.000 |
| 0.0002 | 3.500 |

## 11.7
## Error and Error Propagation

As we have seen, the standard deviation of the values observed for a random variable provides a measure of the uncertainty in the knowledge of the mean of that variable. The uncertainty is often expressed as the probable error, which is the symmetric range about the mean within which there is a 50% chance that a measurement will fall. For a normal distribution, the probable error is thus $\pm 0.675\sigma$ (Table 11.2).

Another measure of uncertainty is the fractional standard deviation, defined as the ratio of the standard deviation and the mean of a distribution, $\sigma/\mu$. This dimensionless quantity, which is also called the coefficient of variation, expresses the uncertainty in relative terms. For the Poisson distribution, the fractional standard deviation is simply

$$\frac{\sigma}{\mu} = \frac{\sqrt{\mu}}{\mu} = \frac{1}{\sqrt{\mu}}. \tag{11.45}$$

In Fig. 11.1, for example, the standard deviation of the Poisson distribution is $\sqrt{\mu} = \sqrt{10} = 3.16$. The fractional standard deviation is $1/\sqrt{10} = 0.316$.

Often in practice one has only a single measurement of a random variable, such as a number $n$ of counts, and wishes to express an uncertainty associated with it. The best estimate of the mean of the distribution from this single measurement is that result: namely, $n$. If one assumes that the distribution sampled is Poisson or normal, then the best estimate of the standard deviation is $\sqrt{n}$. The significance of the measurement, then, is that the true mean is estimated to lie within the interval $n \pm \sqrt{n}$, with a probability (confidence) of 0.683.

Many measurements involve more than one random variable. For example, the activity of a source can be obtained by counting a sample and then subtracting the number of background counts measured with a blank. Both the number of gross counts with the sample present and the number of background counts are subject to the random statistical fluctuations that we have been discussing. The number of net counts, gross minus background, is ascribed to the activity of the sample. The subject of error propagation deals with estimating the standard deviation of a quantity that depends on fluctuations in more than one independent random variable.

Consider a quantity $Q(x, y)$ that depends on two independent, random variables $x$ and $y$. We can make $N$ repeated measurements of $x$ and $y$, obtaining a set of $N$ pairs of data, $x_i$ and $y_i$, with $i = 1, 2, \ldots, N$. The sample means and standard deviations, $\sigma_x$ and $\sigma_y$, can be computed for the two variables. Also, for each data pair the values $Q_i = Q(x_i, y_i)$ can be calculated. We assume that the scatter of the $x_i$ and $y_i$ about their means is small. As shown in Appendix E, the standard deviation of $Q$ is given by [Eq. (E.41)]

$$\sigma_Q = \sqrt{\left(\frac{\partial Q}{\partial x}\right)^2 \sigma_x^2 + \left(\frac{\partial Q}{\partial y}\right)^2 \sigma_y^2}. \tag{11.46}$$

We shall apply this relation in subsequent sections.

## 11.8
## Counting Radioactive Samples

We turn now to the statistics of count-rate measurements and their associated confidence limits. We discuss gross count rates, net count rates, and optimum counting times for long-lived sources. We conclude the section with a discussion of the counting of short-lived samples.

### Gross Count Rates

To obtain the gross count rate of a long-lived sample-plus-background, for example, one measures a number of counts $n_g$ in a time $t_g$. The gross count rate is then simply $r_g = n_g/t_g$. The standard deviation of this rate is determined by the standard deviation of $n_g$. (We shall assume throughout that time measurements are precise.) Assuming that the number of gross counts is Poisson distributed with mean $\mu_g$, we have for its standard deviation $\sigma_g = \sqrt{\mu_g}$. Therefore, the standard deviation of the gross *count rate* is given by

$$\sigma_{gr} = \frac{\sigma_g}{t_g} = \frac{\sqrt{\mu_g}}{t_g} = \sqrt{\frac{r_g}{t_g}}, \tag{11.47}$$

where $\mu_g = r_g t_g$ has been used in writing the last equality. Since the count rate of a long-lived sample is constant, Eq. (11.47) shows that the standard deviation associated with its measurement decreases as the square root of the counting time.

### Example
The activity of a long-lived sample is measured with 35% efficiency in a counter with negligible background. The sample has a reported activity of 42.0 dpm (disintegra-

tions per minute = min$^{-1}$). To check this value, technician A takes a 1-min reading, that registers 19 counts. His observed rate, 19.0 cpm (counts per minute = min$^{-1}$), differs from that based on the reported activity, namely, $0.35 \times 42.0 = 14.7$ cpm, by 4.3 cpm. Technician B takes a 60-min reading, which registers 1148 counts. His observed count rate, 19.13 cpm, differs from that based on the reported activity by 4.4 cpm, about the same as A.

(a) Does A's check substantiate the reported activity?

(b) What is the estimated activity, based on B's check?

(c) How can A's and B's findings for the measured activity be reconciled?

*Solution*

(a) Assuming that the reported activity is the true activity, one can ask, "What is the probability that technician A's measurement would differ from the expected value by no more than he found?" Based on the reported activity, we assume that the mean number of counts in the 1-min interval is $\mu_A = 14.7$, 4.3 counts less than the observed 19. Also, the assumed standard deviation of the number of counts in this time interval is $\sqrt{\mu_A} = 3.83$. The actual number of counts observed by technician A exceeds the assumed true mean by $4.3/3.83 = 1.12$ standard deviations. Referring to Table 11.1, we find that the interval $\mu \pm 1.12\sigma$ includes 0.737 of the total unit area. A's observation, therefore, is consistent with the stated activity, in that there is a probability of 0.737 that a single measurement would be as close, or closer, to the mean as he found.

(b) Technician B's rate of 19.13 cpm is almost the same as that (19.0 cpm) of A. However, B has a much larger number of counts, and so his result inherently has greater statistical significance, as we now show. Again, we assume that the reported activity is the true activity. Thus, we assume for the mean number of counts in 1 h, $\mu_B = 14.7 \times 60 = 882$, with a standard deviation $\sqrt{\mu_B} = \sqrt{882} = 29.7$. The difference between B's observation $n_B = 1148$ and the expected number of counts is, in multiples of the standard deviation,

$$\frac{n_B - \mu_B}{\sqrt{\mu_B}} = \frac{1148 - 882}{29.7} = 8.96. \tag{11.48}$$

The probability that B's observation is a random occurrence, 9 standard deviations away from the assumed true mean is about $1 \times 10^{-19}$. B's measurement strongly suggests that the true activity is likely in the neighborhood of $(19.13 \text{ min}^{-1})/0.35 = 54.7$ dpm.

(c) Whereas technician A's observation is consistent with the reported activity, B's observation makes it very unlikely that the reported activity is close to the true value. There is nothing inherently inconsistent with the two findings, however. Indeed, an estimate of the true activity, based on A's observation, is $(19.0 \text{ cpm})/0.35 = 54.3$ dpm, virtually the same as B's estimate. It is the *total number of counts* that determines the *statistical significance* of the observation. It is not unlikely that A's single, 1-min observation would differ by 4.3 counts from an expected mean of 14.7. It is highly unlikely, though, that B's observation would differ by $1148 - 882 = 266$ counts from an expected mean of 882.

**Net Count Rates**

As an application of the error-propagation formula, Eq. (11.46), we find the standard deviation of the net count rate of a sample, obtained experimentally as the difference between gross and background count rates, $r_g$ and $r_b$. As with gross counting, one also measures the number $n_b$ of background counts in a time $t_b$. The net count rate ascribed to the sample is then the difference

$$r_n = r_g - r_b = \frac{n_g}{t_g} - \frac{n_b}{t_b}. \tag{11.49}$$

To find the standard deviation of $r_n$, we apply Eq. (11.46) with $Q = r_n$, $x = n_g$, and $y = n_b$. From Eq. (11.49) we have $\partial r_n/\partial n_g = 1/t_g$ and $\partial r_n/\partial n_b = -1/t_b$. Thus, the standard deviation of the net count rate is given by

$$\sigma_{nr} = \sqrt{\frac{\sigma_g^2}{t_g^2} + \frac{\sigma_b^2}{t_b^2}} = \sqrt{\sigma_{gr}^2 + \sigma_{br}^2}. \tag{11.50}$$

Here $\sigma_g$ and $\sigma_b$ are the standard deviations of the numbers of gross and background counts, and $\sigma_{gr}$ and $\sigma_{br}$ are the standard deviations of the gross and background count rates. Equation (11.50) expresses the well-known result for the standard deviation of the sum or difference of two Poisson or normally distributed random variables. Using $n_g$ and $n_b$ as the best estimates of the means of the gross and background distributions and assuming that the numbers of counts obey Poisson statistics, we have $\sigma_g^2 = n_g$ and $\sigma_b^2 = n_b$. Therefore, the last equation can be written

$$\sigma_{nr} = \sqrt{\frac{n_g}{t_g^2} + \frac{n_b}{t_b^2}} = \sqrt{\frac{r_g}{t_g} + \frac{r_b}{t_b}}, \tag{11.51}$$

where the substitutions $r_g = n_g/t_g$ and $r_b = n_b/t_b$ have been made to obtain the last equality. Both expressions in (11.51) are useful in solving problems, depending on the particular information given.

*Example*

A long-lived radioactive sample is placed in a counter for 10 min, and 1426 counts are registered. The sample is then removed, and 2561 background counts are observed in 90 min. (a) What is the net count rate of the sample and its standard deviation? (b) If the counter efficiency with the sample present is 28%, what is the activity of the sample and its standard deviation in Bq? (c) Without repeating the background measurement, how long would the sample have to be counted in order to obtain the net count rate to within ±5% of its true value with 95% confidence? (d) Would the time in (c) also be sufficient to ensure that the *activity* is known to within ±5% with 95% confidence?

*Solution*

(a) We have $n_g = 1426$, $t_g = 10$ min, $n_b = 2561$, and $t_b = 90$ min. The gross and background count rates are $r_g = 1426/10 = 142.6$ cpm and $r_b = 2561/90 = 28.5$ cpm.

Therefore, the net count rate is $r_n = 142.6 - 28.5 = 114$ cpm. The standard deviation can be found from either of the expressions in (11.51). Using the first (which does not depend on the calculated values, $r_g$ and $r_b$), we find

$$\sigma_{nr} = \sqrt{\frac{1426}{(10\ \text{min})^2} + \frac{2561}{(90\ \text{min})^2}} = 3.82\ \text{min}^{-1} = 3.82\ \text{cpm}. \tag{11.52}$$

(b) Since the counter efficiency is $\epsilon = 0.28$, the inferred activity of the sample is $A = r_n/\epsilon = (114\ \text{min}^{-1})/0.28 = 407\ \text{dpm} = 6.78\ \text{Bq}$. The standard deviation of the activity is $\sigma_{nr}/\epsilon = (3.82\ \text{min}^{-1})/0.28 = 13.6\ \text{dpm} = 0.227\ \text{Bq}$.

(c) A 5% uncertainty in the net count rate is $0.05 r_n = 0.05 \times 114 = 5.70$ cpm. For the true net count rate to be within this range of the mean at the 95% confidence level means that $5.70\ \text{cpm} = 1.96\sigma_{nr}$ (Table 11.2), or that $\sigma_{nr} = 2.91$ cpm. Using the second expression in (11.51) with the background rate as before (since we do not yet know the new value of $n_g$), we write

$$\sigma_{nr} = 2.91\ \text{min}^{-1} = \sqrt{\frac{142.6\ \text{min}^{-1}}{t_g} + \frac{28.5\ \text{min}^{-1}}{90\ \text{min}}}. \tag{11.53}$$

Solving, we find that $t_g = 17.5$ min.

(d) Yes. The relative uncertainties remain the same and scale according to the efficiency. If the efficiency were larger and the counting times remained the same, then a larger number of counts and less statistical uncertainty would result.

### Optimum Counting Times

If the total time $T = t_g + t_b$ for making the gross and background counts is fixed, one can partition the individual times $t_g$ and $t_b$ in a certain way to minimize the standard deviation of the net count rate. To find this partitioning, we can write the second equality in (11.51) as a function of either time variable alone and minimize it by differentiation.[4] Substituting $T - t_g$ for $t_b$ and minimizing the mathematical expression for the *variance*, rather than the standard deviation (simpler than dealing with the square root), we write

$$\frac{d}{dt_g} \sigma_{nr}^2 = \frac{d}{dt_g} \left( \frac{r_g}{t_g} + \frac{r_b}{T - t_g} \right) = 0. \tag{11.54}$$

It follows that

$$-\frac{r_g}{t_g^2} + \frac{r_b}{(T - t_g)^2} = 0 \tag{11.55}$$

or

$$-\frac{r_g}{t_g^2} + \frac{r_b}{t_b^2} = 0, \tag{11.56}$$

---

4  Alternatively, the function $f(t_g, t_b) \equiv \sigma_{nr} - \lambda(t_g + t_b)$, involving the Lagrange multiplier $\lambda$, can be minimized by solving the two equations, $\partial f/\partial t_g = 0$ and $\partial f/\partial t_b = 0$. Equation (11.57) results.

giving

$$\frac{t_g}{t_b} = \sqrt{\frac{r_g}{r_b}}. \tag{11.57}$$

The ratio of the optimum counting times is thus equal to the square root of the ratio of the respective count rates.

**Counting Short-Lived Samples**

As we have seen, radioactive decay is a Bernoulli process. The distribution in the number of disintegrations in a given time for identical samples of a pure radionuclide is thus described by the binomial distribution. When $p \ll 1$ and $N \gg 1$, the Poisson and normal distributions give excellent approximations to the binomial. However, for a rapidly decaying radionuclide, or whenever the time of observation is not short compared with the half-life, $p$ will not be small. The formalism presented thus far in this section cannot be applied for counting such samples. As an illustration of dealing statistically with a short-lived radionuclide, the section concludes with an analysis of a counting experiment in which the activity dies away completely and background is zero.

We show how the binomial distribution leads directly to the formulas we have been using when $\lambda t \ll 1$ and then what the distribution implies when $\lambda t$ is large. The expected number of atoms that disintegrate during time $t$ in a sample of size $N$ can be written, with the help of Eqs. (11.16) and (11.2),

$$\mu = N(1 - e^{-\lambda t}). \tag{11.58}$$

For a long-lived sample (or short counting time), $\lambda t \ll 1$, $e^{-\lambda t} \cong 1 - \lambda t$, and so $\mu \cong N\lambda t$. The expected disintegration rate is $\mu / t = \lambda N$, as we had in an earlier chapter [Eq. (4.2)]. With the help of Eqs. (11.18), (11.1), and (11.2), we see that the standard deviation of the number of disintegrations is

$$\sigma = \sqrt{N(1 - e^{-\lambda t})e^{-\lambda t}} = \sqrt{\mu e^{-\lambda t}}. \tag{11.59}$$

Again, for $\lambda t \ll 1$, we obtain $\sigma = \sqrt{\mu}$. This very important property of the binomial distribution is exactly true for the Poisson distribution. Thus, a single observation from a distribution that is expected to be binomial gives estimates of both the mean and the standard deviation of the distribution when $\lambda t \ll 1$. If the number of counts obtained is reasonably large, then it can be used for estimating $\sigma$.

Equations (11.58) and (11.59) also hold for long times ($\lambda t \gg 1$), for which the Poisson description of radioactive decay is not accurate. If we make an observation over a time so much longer than the half-life that the nuclide has decayed away ($\lambda t \to \infty$), then Eqs. (11.58) and (11.59) imply that $\mu = N$ and $\sigma = 0$. The interpretation of this result is straightforward. The expected number of disintegrations is equal to the original number of atoms $N$ in the sample and the standard deviation of this number is zero. We have observed every disintegration and know exactly how many atoms were originally present. Repeating the experiment over and over, one would always obtain the same result.

The situation is different if one registers disintegrations with a counter having an efficiency $\epsilon < 1$. Not observing every decay introduces uncertainty in the number of atoms initially present. In place of Eqs. (11.58) and (11.59) one has for the mean and standard deviation of the number of counts, with the help of Eqs. (11.19) and (11.20),

$$\mu_c = \epsilon N(1 - e^{-\lambda t}) \tag{11.60}$$

and

$$\sigma_c = \sqrt{\epsilon N(1 - e^{-\lambda t})(1 - \epsilon + \epsilon e^{-\lambda t})}. \tag{11.61}$$

When the sample has decayed away completely ($\lambda t \to \infty$), the result is $\mu_c = \epsilon N$ and

$$\sigma_c = \sqrt{\epsilon N(1 - \epsilon)} = \sqrt{\mu_c(1 - \epsilon)}. \tag{11.62}$$

The standard deviation of the number of atoms initially present is $\sigma = \sigma_c/\epsilon$.

*Example*
A sample containing a certain radionuclide registers 91,993 counts before dying completely away. Background is zero. What is the expected value of the initial number of atoms of the radionuclide and the standard deviation of the initial number, if the counter efficiency $\epsilon$ is (a) 100% and (b) 42%?

*Solution*
(a) With $\epsilon = 1$, the numbers of counts and disintegrations are the same. There were exactly $\mu = 91,993$ atoms present initially ($\sigma = 0$).

(b) With $\epsilon = 0.42$, we use the observed number of counts as the estimate of $\mu_c$. The expected number of atoms originally present is then $\mu_c/\epsilon = 91,993/0.42 = 2.19 \times 10^5$. The standard deviation of the number of counts is, by Eq. (11.62),

$$\sigma_c = \sqrt{91993(1 - 0.42)} = 231. \tag{11.63}$$

The standard deviation of the number of atoms initially present is $\sigma = \sigma_c/\epsilon = 231/0.42 = 550$.

## 11.9
## Minimum Significant Measured Activity—Type-I Errors

In many operations that involve counting a sample (e.g., bioassay monitoring, smear counting), a decision has to be made as to whether the sample contains "significant" activity. A high gross count number for a particular sample could be the result of a large random fluctuation in background, or it could be due to the

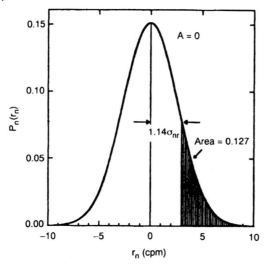

**Fig. 11.4** Probability density $P_n(r_n)$ for measurement of net count rate $r_n$ when no activity is present. See example in text. (Courtesy James S. Bogard.)

radioactivity of the sample. In practice, a critical count level is often established for screening when large numbers of samples must be routinely processed under identical conditions. If a given sample reads more than the critical number of gross counts, it is assumed to have significant activity, and graded action can then be taken. A type-I error is said to occur if it is concluded that activity is present when, in fact, there is none (false positive). A type-II error occurs when it is wrongly concluded that no activity is present (false negative). The two types of error carry different implications. This section and the next develop some statistical procedures that have been formulated to ascertain "minimum significant measured activity" and "minimum detectable true activity." We assume that the distributions of gross and background counts are normal and consider only long-lived radionuclides.

### Example

A sample, counted for 10 min, registers 530 gross counts. A 30-min background reading gives 1500 counts. (a) Does the sample have activity? (b) Without changing the counting times, what minimum number of gross counts can be used as a decision level such that the risk of making a type-I error is no greater than 0.050?

### Solution

(a) The numbers of gross and background counts are $n_g = 530$ and $n_b = 1500$; the respective counting times are $t_g = 10$ min and $t_b = 30$ min. The gross and background count rates are $r_g = n_g/t_g = 53$ cpm and $r_b = n_b/t_b = 50$ cpm, giving a net count rate $r_n = r_g - r_b = 3$ cpm. The question of whether activity is present cannot be answered in an absolute sense from these measurements. The observed net rate could occur randomly with or without activity in the sample. We can, however, compute the *prob-*

*ability* that the result would occur randomly when we assume that the sample has no activity. To do this, we compare the net count rate with its estimated standard deviation $\sigma_{nr}$, given by Eq. (11.51):

$$\sigma_{nr} = \sqrt{\frac{r_g}{t_g} + \frac{r_b}{t_b}} = \sqrt{\frac{53}{10} + \frac{50}{30}} = 2.64 \text{ cpm.} \tag{11.64}$$

The observed net rate differs from 0 by $3/2.64 = 1.14$ standard deviations. As found in Table 11.1, the area under the standard normal curve to the right of this value is 0.127. Assuming that the activity $A$ is zero, as shown in Fig. 11.4, we conclude that an observation giving a net count rate greater than the observed $r_n = 1.14\sigma_{nr} = 3$ cpm would occur randomly with a probability of 0.127. This single set of measurements, gross and background, is thus consistent with the conclusion that the sample likely contains little or no activity. However, one does not know where the bell-shaped curve in Fig. 11.4 should be centered. Based on this single measurement, the most likely place is $r_n = 3$ cpm, with the sample activity corresponding to that value of the net count rate.

(b) Assigning a maximum probability for type-I errors enables one to give a more definitive answer, with that proviso, for reporting the presence or absence of activity in a sample. When $A = 0$, as in Fig. 11.4, the net rate that leaves 5% of the area under the normal curve to its right is $r_1 = 1.65\sigma_{nr}$ (Table 11.2). Using Eq. (11.51), we write

$$r_1 = 1.65\sqrt{\frac{r_g}{t_g} + \frac{r_b}{t_b}} = 1.65\sqrt{\frac{r_1 + 50}{10} + \frac{50}{30}}, \tag{11.65}$$

where the substitution $r_g = r_1 + r_b$ has been made. This equation is quadratic in $r_1$. After some manipulation, one finds that

$$r_1^2 - 0.272r_1 - 18.2 = 0. \tag{11.66}$$

The solution is $r_1 = 4.40$ cpm. The corresponding gross count rate is $r_g = r_1 + r_b = 4.40 + 50 = 54.4$ cpm, and so the critical number of gross counts is $n_g = r_g t_g = (54.4 \text{ min}^{-1}) \times (10 \text{ min}) = 544$. Thus, a sample giving $n_g > 544$ (i.e., a minimum of 545 gross counts) can be reported as having significant activity, with a probability no greater than 0.05 of making a type-I error.

We can generalize the last example to compute decision levels for arbitrary choices of $t_g$, $t_b$, and the maximum acceptable probability $\alpha$ for type-I errors. As in Table 11.2, we let $k_\alpha$ represent the number of standard deviations of the net count rate that gives a one-tail area equal to $\alpha$. Then, like Eq. (11.65), we can write for $r_1$, the minimum significant measured net count rate,

$$r_1 = k_\alpha\sqrt{\sigma_{gr}^2 + \sigma_{br}^2} = k_\alpha\sqrt{\frac{r_1 + r_b}{t_g} + \frac{r_b}{t_b}}. \tag{11.67}$$

Solving for $r_1$, we obtain (Problem 38)

$$r_1 = \frac{k_\alpha^2}{2t_g} + \frac{k_\alpha}{2}\sqrt{\frac{k_\alpha^2}{t_g^2} + 4r_b\left(\frac{t_g + t_b}{t_g t_b}\right)}. \tag{11.68}$$

This general expression leads to other useful formulas when the counting times are equal $(t_g = t_b = t)$. The last term under the radical can then be written $4r_b(2t/t^2) = 8n_b/t^2$, where $n_b = r_b t$ is the number of background counts obtained in the time $t$. The minimum significant count difference (gross minus background) becomes (Problem 39)

$$\Delta_1 = r_1 t = \tfrac{1}{2}k_\alpha^2 + \tfrac{1}{2}k_\alpha\sqrt{k_\alpha^2 + 8n_b},\tag{11.69}$$

$$= k_\alpha\sqrt{2n_b}\left(\frac{k_\alpha}{\sqrt{8n_b}} + \sqrt{1 + \frac{k_\alpha^2}{8n_b}}\right).\tag{11.70}$$

If the counter efficiency is $\epsilon$, then the corresponding net number of disintegrations in the sample during the time $t$ is $\Delta_1/\epsilon$. It follows that the minimum significant measured activity is

$$A_I = \frac{\Delta_1}{\epsilon t}.\tag{11.71}$$

In many instances, $k_\alpha/\sqrt{n_b} \ll 1$. One then has the approximate formula,

$$\Delta_1 \cong k_\alpha\sqrt{2n_b},\tag{11.72}$$

in place of (11.70).

Often, background is stable and the expected number of background counts $B$ in the time $t$ is known with much greater accuracy than that associated with the single measurement $n_b$. In that case, with no activity present, the standard deviation of the number of net counts is simply $\sqrt{B}$. It follows that the minimum significant net count difference is then

$$\Delta_1 = k_\alpha\sqrt{B} \qquad \text{(Background accurately known)}.\tag{11.73}$$

Comparison with Eqs. (11.72) and (11.70) shows that the minimum significant measured net count difference and hence the minimum significant measured activity are lower by a factor of approximately $\sqrt{2}$ when the background is well known. For fixed counting times, $t_g$ and $t_b$, the minimum significant measured activity $A_I$ is determined completely by the choice of $\alpha$, the level of the background, and the accuracy with which the background is known.

*Example*

A 10-min background measurement with a certain counter yields 410 counts. A sample is to be measured for activity by taking a gross count for 10 min. The maximum acceptable risk for making a type-I error is 0.05. The counter efficiency is such that 3.5 disintegrations in a sample result, on average, in one net count.

(a) Calculate the minimum significant net count difference and the minimum significant measured activity in Bq.

(b) How much error is made in (a) by using the approximate formula (11.72) in place of (11.69)?

(c) What is the decision level for type-I errors in terms of the number of gross counts in 10 min?

*Solution*

(a) With equal counting times, $t_g = t_b = t = 10$ min, one can use Eq. (11.69) in place of the general expression (11.68). For $\alpha = 0.05$, $k_\alpha = 1.65$. With $n_b = 410$, we obtain

$$\Delta_1 = \tfrac{1}{2}(1.65)^2 + \tfrac{1}{2}(1.65)\sqrt{(1.65)^2 + 8(410)} = 48.6 = 49 \qquad (11.74)$$

for the minimum significant count difference in 10 min (rounded upward to the nearest integer). The counter efficiency is $\epsilon = 1/3.5 = 0.286$ dpm/cpm. It follows from Eq. (11.71) that the minimum significant measured activity is $A_I = 48.6/(0.286 \times 10 \text{ min}) = 17.0$ dpm $= 0.283$ Bq.

(b) The approximate formula (11.72) gives $\Delta_1 \cong 1.65(2 \times 410)^{1/2} = 47$ net counts. The percent error made by using the approximation in this example is $[(49-47)/49] \times 100 = 4.1\%$. [The criterion for the validity of (11.72) is that $k_\alpha/\sqrt{n_b} \ll 1$. In this example, $k_\alpha/\sqrt{n_b} = 1.65/\sqrt{410} = 0.081$.]

(c) The decision level for gross counts in 10 min is $n_1 = n_b + \Delta_1 = 459$.

The value $n_1 = 459$ in the last example can serve as a decision level for screening samples for the presence of activity by gross counting for 10 min. A sample showing $n_g < 459$ counts can be reported as having less than the "minimum significant measured activity," $A_I = 0.283$ Bq. A sample showing $n_g \geq 459$ counts can be reported as having an activity $(n_g - n_b)/\epsilon t = (n_g - 410)/2.86$ dpm. A blank will read high, on average, one time in twenty. Use of a decision level thus implies acceptance of a certain risk, set by choosing $\alpha$, for making a type-I error.

For samples having zero activity, the probability of making a type-I error is just equal to the value chosen for $\alpha$. For samples having activity, a type-I error cannot occur, by definition. Therefore, when one screens a large collection of samples, some with $A = 0$ and some with $A > 0$, the probability of making a type-I error with any given sample never exceeds $\alpha$.

## 11.10
## Minimum Detectable True Activity—Type-II Errors

We consider next the implications for making a type-II error by using a critical decision level. We denote the maximum acceptable risk for a type-II error by $\beta$. With the critical net count rate set at $r_1$, based on the choice of $\alpha$ for type-I errors, the situation is depicted in Fig. 11.5. For a certain net count rate $r_2$, corresponding to a sample activity $A_{II}$, the area under the curve to the left of $r_1$ in the figure will be $\beta$. When $A = A_{II}$, use of $r_1$ as a screening level thus leads to a type-II error with probability $\beta$. When $A > A_{II}$, the probability of a type-II error is less than $\beta$; and when $A < A_{II}$, the probability is more than $\beta$. The activity $A_{II}$, which is the smallest that will not be missed with frequency greater than $\beta$, is called the minimum detectable true activity. We now show how it can be calculated.

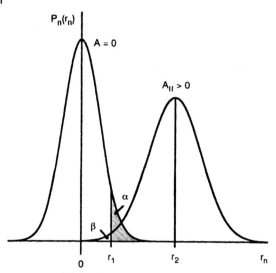

**Fig. 11.5** Probability density $P_n(r_n)$ for net count rate $r_n$. When activity $A = 0$, $r_1$ is fixed by choice of probability $\alpha$ for type-I errors. Use of $r_1$ and choice of probability $\beta$ for type-II errors fixes $r_2$, corresponding to the minimum detectable true activity $A_{II}$. (If $\alpha = \beta$, then the intersection of the two curves occurs at the value $r_n = r_1$.) (Courtesy James S. Bogard.)

Repeated measurements of the gross and background count rates with the activity $A_{II}$ in a sample would give the net count rate, $r_g - r_b$, distributed about $r_2$, as shown in Fig. 11.5. Since the quantity $r_g - r_b - r_2$ is distributed normally about the mean value of zero, we can describe it with the help of the standard normal distribution. Letting $k_\beta$ represent the number of standard deviations that leave an area $\beta$ to its left, we write [analogous to Eq. (11.67)]

$$r_g - r_b - r_2 = -k_\beta \sqrt{\frac{r_g}{t_g} + \frac{r_b}{t_b}} \tag{11.75}$$

$$= -k_\beta \sqrt{\frac{r_g - r_b}{t_g} + r_b \left( \frac{t_g + t_b}{t_g t_b} \right)}, \tag{11.76}$$

where $r_b/t_g$ has been subtracted and added under the radical. On both sides of this equation we set the net count rate $r_g - r_b = r_1$, the decision-level rate established for type-I errors. We can now write

$$r_2 = r_1 + k_\beta \sqrt{\frac{r_1}{t_g} + r_b \left( \frac{t_g + t_b}{t_g t_b} \right)}. \tag{11.77}$$

Substituting for $r_1$ from Eq. (11.68), we obtain

$$r_2 = k_\alpha \left[ \frac{k_\alpha}{2t_g} + \frac{1}{2} \sqrt{\frac{k_\alpha^2}{t_g^2} + 4r_b \left( \frac{t_g + t_b}{t_g t_b} \right)} \right]$$

$$+ k_\beta \left[ \frac{k_\alpha}{t_g} \left( \frac{k_\alpha}{2t_g} + \frac{1}{2} \sqrt{\frac{k_\alpha^2}{t_g^2} + 4 r_b \left( \frac{t_g + t_b}{t_g t_b} \right)} \right) + r_b \left( \frac{t_g + t_b}{t_g t_b} \right) \right]^{1/2}. \tag{11.78}$$

This general result gives the net rate that corresponds to the minimum detectable true activity for a given background rate $r_b$ and arbitrary choices of $\alpha$, $\beta$, and the counting times. When the latter are equal ($t_g = t_b = t$), Eq. (11.78) gives for the number of net counts with the minimum detectable true activity

$$\Delta_2 = r_2 t = \sqrt{2 n_b} \left\{ k_\alpha \left[ \frac{k_\alpha}{\sqrt{8 n_b}} + \sqrt{1 + \frac{k_\alpha^2}{8 n_b}} \right] \right.$$

$$\left. + k_\beta \left[ 1 + \frac{k_\alpha^2}{4 n_b} + \frac{k_\alpha}{\sqrt{2 n_b}} \sqrt{1 + \frac{k_\alpha^2}{8 n_b}} \right]^{1/2} \right\}. \tag{11.79}$$

With the help of Eq. (11.70), we can also write

$$\Delta_2 = \Delta_1 + k_\beta \sqrt{2 n_b} \left[ 1 + \frac{k_\alpha^2}{4 n_b} + \frac{k_\alpha}{\sqrt{2 n_b}} \sqrt{1 + \frac{k_\alpha^2}{8 n_b}} \right]^{1/2}. \tag{11.80}$$

The minimum detectable true activity is given by

$$A_{II} = \frac{\Delta_2}{\epsilon t}, \tag{11.81}$$

where $\epsilon$ is the counter efficiency. As with Eqs. (11.70) and (11.72), we obtain a simple formula when $k_\alpha / \sqrt{n_b} \ll 1$:

$$\Delta_2 \cong (k_\alpha + k_\beta) \sqrt{2 n_b}. \tag{11.82}$$

When the background count $B$ is accurately known, we have seen by Eq. (11.73) that the minimum significant count difference is $\Delta_1 = k_\alpha \sqrt{B}$. If a sample has exactly the minimum detectable true activity, then the expected number of net counts $\Delta_2$ is just $k_\beta$ standard deviations greater than $\Delta_1$. The standard deviation of the net count rate is $\sqrt{(B + \Delta_2)}$. Thus,

$$\Delta_2 = k_\alpha \sqrt{B} + k_\beta \sqrt{B + \Delta_2}. \tag{11.83}$$

Solving for $\Delta_2$, we find

$$\Delta_2 = \sqrt{B} \left( k_\alpha + \frac{k_\beta^2}{2 \sqrt{B}} + k_\beta \sqrt{1 + \frac{k_\alpha}{\sqrt{B}} + \frac{k_\beta^2}{4 B}} \right) \tag{11.84}$$

(Background accurately known).

This expression for $\Delta_2$ is then to be used in Eq. (11.81) to obtain the minimum detectable true activity. When $k_\alpha / \sqrt{B} \ll 1$ and $k_\beta / \sqrt{B} \ll 1$, one has, approximately, in place of Eq. (11.82),

$$\Delta_2 \cong (k_\alpha + k_\beta) \sqrt{B} \qquad \text{(Background accurately known).} \tag{11.85}$$

As with $\Delta_1$ and $A_I$, accurate knowledge of the background lowers $\Delta_2$ and $A_{II}$ by about a factor of $\sqrt{2}$.

*Example*

The counting arrangement ($\alpha = 0.05$, $\epsilon = 0.286$, $t_g = t_b = 10$ min, and $n_b = 410$) and critical gross count number $n_1 = 459$ from the last example are to be used to screen samples for activity. The maximum acceptable probability for making a type-II error is $\beta = 0.025$. (a) Calculate the minimum detectable true activity in Bq. (b) How much error is made by using the approximate formula (11.82) in place of the exact (11.79) or (11.80)?

*Solution*

(a) With equal gross and background counting times, we can use Eqs. (11.80) and (11.81) to find $A_{II}$. For $\beta = 0.025$, $k_\beta = 1.96$ (Table 11.2). With $\Delta_1 = 48.6$ counts from the last example [Eq. (11.74)], Eq. (11.80) gives

$$\Delta_2 = 48.6 + 1.96\sqrt{2(410)}\left[1 + \frac{(1.65)^2}{4(410)} + \frac{1.65}{\sqrt{2(410)}}\sqrt{1 + \frac{(1.65)^2}{8(410)}}\right]^{1/2} = 106$$

(11.86)

net counts. The minimum true detectable activity is, from Eq. (11.81),

$$A_{II} = \frac{106}{0.286 \times 10 \text{ min}} = 37.1 \text{ dpm} = 0.618 \text{ Bq.}$$

(11.87)

(b) The approximate formula (11.82) gives $\Delta_2 \cong (1.65 + 1.96)(2 \times 410)^{1/2} = 103$ net counts. The error made in using the approximation to compute $A_{II}$ in this case is $[(106 - 103)/106] \times 100 = 2.8\%$.

This example illustrates how a protocol can be set up for reporting activity in a series of samples that are otherwise identical. As shown in Fig. 11.6, the decision level for a 10-min gross count is $n_1 = 459$, corresponding to the minimum significant count difference $\Delta_1 = 49$ and the minimum significant measured activity, $A_I = 0.283$ Bq. A sample for which $n_g < 459$ is considered as having no reportable activity. When $n_g \geq 459$, a sample is reported as having an activity

$$A = \frac{n_g - n_b}{\epsilon t} = \frac{n_g - 410}{(0.286)(600 \text{ s})} = \frac{n_g - 410}{172} \text{ Bq.}$$

(11.88)

Note that $A$ will be greater than the minimum significant measured activity, $A_I = 0.283$ Bq. From part (a) in the last example, when $n_g = \Delta_2 + 410 = 516$, the reported value of the activity will be $A_{II} = 0.618$ Bq, the minimum detectable true activity. For a sample of unknown activity, the probability of making a type-I error does not exceed $\alpha = 0.05$. (If $A = 0$, the probability equals $\alpha$.) The probability of making a type-II error with a given sample does not exceed $\beta = 0.025$, as long as the activity is greater than $A_{II} = 0.618$ Bq. (If $A = A_{II}$, the probability equals $\beta$.) When $0 < A < A_{II}$, the probability for a type-II error is greater than $\beta$.

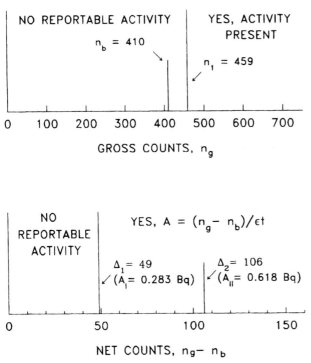

**Fig. 11.6** Gross and net counts for reporting activity in samples. See text after example in Section 11.10. (Courtesy James S. Bogard.)

In practice, one does not usually do a background count with each sample. If the only source of background variation is the random fluctuation in the count rate $r_b$, then the latter can be assessed with a long count. As seen from Eqs. (11.68) and (11.78), increasing $t_g$ alone reduces the decision level and hence $A_I$ and $A_{II}$.[5]

## 11.11
### Criteria for Radiobioassay, HPS NI3.30-1996

The analysis discussed in the last two sections largely follows work presented by Altshuler and Pasternack (1963).[6] The important subject of radiobioassay has received a great deal of attention over the years. Different decision rules have been suggested for activity determinations in low-level radioactive counting [Strom and MacLellan (2001)]. Much of the literature relates back to the seminal papers of Currie (1968, 1984), who carefully defined basic concepts and objectives. He considered

5   D. J. Strom and P. S. Stansbury, "Minimum Detectable Activity when Background is Counted Longer than the Sample," *Health Phys.* **63**, 360–361 (1992).

6   Citations made in this section can be found in Section 11.14, Suggested Reading.

operational quantities such as a level that may be recognized as "detected", one that is expected to lead to detection, and one that has adequate measurement precision for quantitative assessment.

A large effort, involving a number of agencies and many specialists, led to the development and publication of *An American National Standard—Performance Criteria for Radiobioassay*, approved in 1996 by the American National Standards Institute and published by the Health Physics Society as HPS N13.30-1996. The express purpose of the Standard "…is to provide criteria for quality assurance, evaluation of performance, and the accreditation of radiobioassay service laboratories. These criteria include bias, precision, and determination of the MDA [*minimum detectable amount*]."

Building on the original work of Currie and others, the N13.30 Standard presents a protocol that defines a *decision level* and a *minimum detectable amount* (MDA) for measurements of a radioactive analyte in a sample. These quantities play the same roles as their counterparts in the last two sections. However, they differ from the former in the way in which background is assessed. In our earlier treatment, the decision level is applied to net counts above background, measured with a subject under analysis. The background count is typically made with the subject replaced by an *appropriate blank*—e.g., synthetic urine, radiometrically the same as the subject, but having no added radioactivity. In N13.30, an additional measurement is made by the routine procedure, where the subject contains no analyte above that of the appropriate blank. When systematic errors are negligible, the N13.30 decision level replaces the term $\sqrt{2n_b}$ in Eq. (11.72) above by the standard deviation $s_0$ of the *net count* of a subject with no added analyte. It thus includes contributions of both a subject with no added analyte under the routine procedure and the background as measured with the appropriate blank. Instead of comparing the subject count with the single background determination as before, the comparison is made with the specified net count, having standard deviation $s_0$.

The MDA defined in N13.30 is similarly comparable to Eq. (11.82) with $\sqrt{2n_b}$ replaced by $s_0$. However, an additional complication arises when dealing with very low background rates. A well maintained alpha-particle counter might register a single count on average over a long counting time with a blank having no activity. Use of the formalism is based on having good estimates of the mean and standard deviation of the background. A semi-empirical value of three counts is added to $k_\beta s_0$ in the formula for the MDA in order to render $\beta \leq 0.05$ when the background is very low.

The American National Standards Institute has played an important part in the quality of radiation protection in the United States. Their work led to the Nuclear Regulatory Commission's test program administered by the National Voluntary Laboratory Accreditation Program (NVLAP) and to the Department of Energy's Laboratory Accreditation Program (DOELAP). The Institute's activities extend into other areas besides radiobioassay.

## 11.12
### Instrument Response

We treat two common statistical aspects basic to understanding and interpreting certain radiation measurements. The measurements deal with energy resolution in pulse-height analysis and the effects of instrument dead time on count rates.

### Energy Resolution

In Fig. 10.30, the resolution of the total-energy peak, 8%, refers to the relative width of the peak at one-half its maximum value. Called the full width at half maximum (FWHM), one has FWHM = 0.08(662) = 53 keV. For a normal curve, which the peak in Fig. 10.30 approximates well, with standard deviation $\sigma$, it can be shown that FWHM = $2.35\sigma$.

The resolution of a spectrometer depends on several factors. These include noise in the detector and associated electronic systems as well as fluctuations in the physical processes that convert radiation energy into a measured signal. The latter source of variation is dominant in many applications, and we discuss it here. The random fluctuations associated with the statistical nature of energy loss present an irreducible physical limit to the resolution attainable with any energy-proportional device. Assuming that other sources of random noise are small, we can relate the width of the total energy peak in Fig. 10.30 to the distribution of the number of entities that are collected to register an event.

When a 662-keV photon is absorbed in the NaI crystal, one or more Compton electrons can be ejected along with a photoelectron to produce an event registered under the energy peak. A number of low-energy secondary electrons are also produced. Scintillation photons are generated, many of which enter the photomultiplier tube. The tube converts a fraction of these into photoelectrons, whose number is then proportional to the size of the pulse registered. (Electron multiplication in the photomultiplier tube occurs with relatively small standard deviation.) Repeated absorption of 662-keV photons in the crystal produces a distribution in the number of photoelectrons, which is just that shown by the total-energy peak in Fig. 10.30. Applying Poisson statistics, we can express the resolution in terms of the average number $\mu$ of photoelectrons (with standard deviation $\sigma = \sqrt{\mu}$):

$$R = \frac{\text{FWHM}}{\mu} = \frac{2.35\sigma}{\mu} = \frac{2.35}{\sqrt{\mu}}, \tag{11.89}$$

with FWHM now referring to the number, rather than energy, distribution. With $R = 0.08$ for the scintillator, it follows that the average number of photoelectrons collected per pulse is $\mu = 863$.

For different types of detectors, the physical limitation on resolution imposed by the inherent statistical spread in the number of entities collected can be compared in terms of the average energy needed to produce a single entity. For the NaI detector just given, since an event is registered with the expenditure of 662 keV, this average energy is "$W$" = (662000 eV)/(863 photoelectrons) = 767 eV per photoelectron.

**Table 11.3** Comparison of Resolution of Typical NaI, Gas, and Semiconductor Detectors for $^{137}$Cs Photons

| Detector | Average Energy Per Entity Collected (eV) | Resolution (%) |
|---|---|---|
| NaI | 767 | 8.0 |
| Gas | 30 | 1.6 |
| Semiconductor | 3 | 0.50 |

By comparison, for a gas proportional counter $W \cong 30$ eV per ion pair (Table 10.1). The average number of electrons produced by the absorption of a $^{137}$Cs photon in a gas is $662000/30 = 22100$. The resolution of the total-energy peak (other sources of fluctuations being negligible) with a gas counter is $R = 2.35/(22100)^{1/2} = 0.016$. For germanium, $W \cong 3$ eV per ion pair; and the resolution for $^{137}$Cs photons is $R = 2.35/(221,000)^{1/2} = 0.0050$. A comparison of spectra measured with NaI and with Ge was shown in Fig. 10.33.

Resolution improves as the square root of the average number of entities collected. The preceding comparisons are summarized in Table 11.3. Note that the resolution defined by Eq. (11.89) depends on the energy of the photons being detected through the average value $\mu$.

*Example*

For the scintillator analyzed in the example given after Fig. 10.30, it was found that the average energy needed to produce a photoelectron was 155 eV. (a) What is the resolution for the total-energy peak for 450-keV photons? (b) What is the width of the total-energy peak (FWHM) in keV? (c) What is the resolution for 1.2-MeV photons?

*Solution*

(a) The average number of photoelectrons produced by absorption of a 450-keV photon is $450,000/155 = 2900$. The resolution is therefore by Eq. (11.89), $R = 2.35/(2900)^{1/2} = 0.0436$.

(b) For 450-keV photons, it follows that FWHM $= 0.0436 \times 450 = 19.6$ keV.

(c) Equation (11.89) implies that the resolution decreases as the square root of the photon energy. Thus, the resolution for 1.2-MeV photons is $0.0436 (0.450/1.2)^{1/2} = 0.0267$.

The resolution achievable in gas and semiconductor detectors is considerably better (by a factor of about 2 to 4) than the Poisson limit implied by Eq. (11.89). The departure of ionization events from complete randomness is not surprising in view of the energy-loss spectrum for charged particles discussed earlier (Section 5.3). Some energy is spent in excitations, rather than ionizations, and in overcoming electron binding energies. Also, a typical energy loss of several tens of eV gives a secondary electron enough energy to produce several more ion pairs in clusters along a track. The Fano factor has been introduced as a measure of the departure of fluctuations from pure Poisson statistics. It is defined as the ratio of the observed variance and the variance predicted by the latter:

$$F = \frac{\text{Observed variance}}{\text{Poisson variance}}. \tag{11.90}$$

Reported values of Fano factors for gas proportional counters are in the range from about 0.1 to 0.2 and, for semiconductors, from 0.06 to 0.15. For scintillation detectors, $F$ is near unity, indicating a Poisson-limited resolution.

**Dead Time**

An instrument that responds sequentially to individual events requires a certain minimum time to recover from one event before it is ready to respond to the next. This recovery interval, called the *dead time*, can be due to physical processes in the detector and to instrument electronics. When a radioactive sample is counted, there is a possibility that two interactions in the detector will occur too close together in time to be registered as separate events.

Two models have been proposed to approximate the dead-time behavior of counters. Following a count, a *paralyzable detector* is unable to provide a second response until a certain dead time $\tau$ has passed without another event occurring. Another event during $\tau$ causes the insensitive period to be restarted. A *nonparalyzable detector*, on the other hand, simply ignores other events if they occur during $\tau$. Differences in the two models are illustrated in Fig. 11.7.

The top line represents seven events as they occur along the horizontal time axis, and the two axes below show the responses of the two types of detector. Events 1

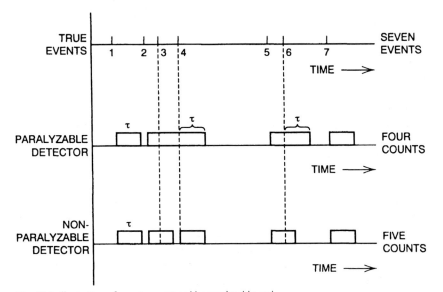

**Fig. 11.7** Illustration of counts registered by paralyzable and nonparalyzable models with dead time $\tau$. See text.

and 2 are registered by both counters. After event 2 is registered, events 3 and then 4 restart the dead period for the paralyzable detector, which misses both. Event 3 is ignored by the nonparalyzable counter, which recovers in time from event 2 to register 4. Events 5 and 7 are recorded by both, but 6 is missed. Of the seven events in this example, the paralyzable counter registers four and the nonparalyzable, five. Such instruments thus actually count the number of *intervals* between events to which they respond, rather than the number of events themselves. In practice, counting systems often exhibit behavior intermediate to the two extremes illustrated in Fig. 11.7.

Dead-time corrections can be made to convert a measured count rate $r_c$ into a true event rate $r_t$. With a nonparalyzable system, the fraction of the time that the instrument is dead is $r_c \tau$. Therefore, the fraction of the time that it is sensitive is $1 - r_c \tau$, which is also the fraction of the number of true events that can be recorded:

$$\frac{r_c}{r_t} = 1 - r_c \tau. \tag{11.91}$$

Thus, the true event rate for a nonparalyzable counter is given in terms of the recorded count rate and the dead time by the relation

$$r_t = \frac{r_c}{1 - r_c \tau} \quad \text{(Nonparalyzable).} \tag{11.92}$$

When the count rate is low or the dead time short ($r_c \tau \ll 1$),

$$r_t \cong r_c (1 + r_c \tau). \tag{11.93}$$

With a paralyzable counter, on the other hand, only intervals longer than $\tau$ are registered. To analyze for the dead time, we need the distribution of time intervals between successive random events that occur at the average rate $r_t$. The average number of events that take place in a time $t$ is $r_t t$. If an event occurs at time $t = 0$, then the probability that no events occur in time $t$ immediately following that event is given by the Poisson term, $P_0 = \exp(-r_t t)$. The probability that an event will occur in the next time interval $dt$ is $r_t \, dt$. Therefore, given an event at time $t = 0$, the probability that the next event will occur between $t$ and $t + dt$ is

$$P(t) \, dt = r_t e^{-r_t t} \, dt. \tag{11.94}$$

The probability that a time interval larger than $t$ will elapse is

$$\int_\tau^\infty r_t e^{-r_t t} \, dt = e^{-r_t \tau}. \tag{11.95}$$

The observed count rate $r_c$ is the product of the true event rate $r_t$ and this probability:

$$r_c = r_t e^{-r_t \tau} \quad \text{(Paralyzable).} \tag{11.96}$$

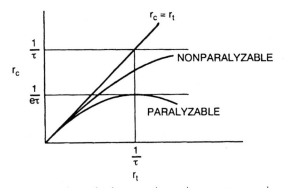

**Fig. 11.8** Relationship between observed count rates $r_c$ and true event rates $r_t$ for nonparalyzable and paralyzable counters with dead time $\tau$.

Unlike Eq. (11.92), one can only solve Eq. (11.96) numerically for $r_t$ in terms of $r_c$ and $\tau$. For low event rates or short dead time ($r_t\tau \ll 1$), Eq. (11.96) gives

$$r_c = r_t(1 - r_t\tau). \tag{11.97}$$

This relationship also leads to the same Eq. (11.93) for nonparalyzable systems when $r_t\tau \ll 1$ (Problem 61).

Figure 11.8 shows plots of the measured count rates $r_c$ as functions of the true event rate $r_t$ for the nonparalyzable and paralyzable models. For small $r_t$, both give nearly the same result (Problem 61). For the nonparalyzable model, Eq. (11.92) shows that $r_c$ cannot exceed $1/\tau$. Therefore, as $r_t$ increases, $r_c$ approaches the asymptotic value $1/\tau$, which is the highest possible count rate. For the paralyzable counter, on the other hand, the behavior at high event rates is quite different. Differentiation of Eq. (11.96) shows that the observed count rate goes through a maximum $1/e\tau$ when $r_t = 1/\tau$ (Problem 63). With increasing event rates, the measured count rate with a paralyzable system will decrease beyond this maximum and approach zero, because of the decreasing opportunity to recover between events. With a paralyzable system, there are generally two possible event rates that correspond to a given count rate.

*Example*

A counting system has a dead time of 1.7 $\mu$s. If a count rate of $9 \times 10^4$ s$^{-1}$ is observed, what is the true event rate if the counter is (a) nonparalyzable or (b) paralyzable?

*Solution*

(a) For the nonparalyzable counter with $r_c = 9 \times 10^4$ s$^{-1}$ and $\tau = 1.7 \times 10^{-6}$ s, Eq. (11.92) gives for the true event rate

$$r_t = \frac{9 \times 10^4 \text{ s}^{-1}}{1 - (9 \times 10^4 \text{ s}^{-1})(1.7 \times 10^{-6} \text{ s})} = 1.06 \times 10^5 \text{ s}^{-1}. \tag{11.98}$$

(b) For the paralyzable case, Eq. (11.96) gives

$$9 \times 10^4 = r_t e^{-1.7 \times 10^{-6} r_t}. \tag{11.99}$$

We expect there to be two solutions for $r_t$, which we find by iteration. We can work directly with Eq. (11.99), but it is more convenient to take the natural logarithm of both sides and rearrange slightly:

$$1.7 \times 10^{-6} r_t = \ln \frac{r_t}{9 \times 10^4}. \tag{11.100}$$

In view of the answer to part (a), we try the solution $r_t = 1.06 \times 10^5$ s$^{-1}$ here. The left-hand side of (11.100) then has the value 0.180, compared with the smaller value on the right-hand side, 0.164. Since the paralyzable counter misses more events than the nonparalyzable, $r_t$ should be larger now that in part (a). Trying $r_t = 1.10 \times 10^5$ s$^{-1}$ gives 0.187 on the left of (11.100) compared with the larger 0.201 on the right. Thus, Eq. (11.100) is satisfied by a value of $r_t$ between these two trial values. Further refinement yields the solution $r_t = 1.08 \times 10^5$ s$^{-1}$. Since this solution is close to the result (11.98) for the nonparalyzable counter, we expect the second solution to be at a higher event rate. We proceed by increasing the event rate in steps by an order of magnitude. For $r_t = 10^{-6}$ s$^{-1}$, the left- and right-hand sides of Eq. (11.100) give, respectively, 1.70 and 2.41. For $r_t = 10^7$ s$^{-1}$, the results are 17.0 and 4.71. Therefore, the solution is between these two values of $r_t$. One finds $r_t = 1.75 \times 10^6$ s$^{-1}$.

## 11.13
## Monte Carlo Simulation of Radiation Transport

We saw in Section 5.4 for charged particles, Section 8.7 for photons, and Section 9.4 for neutrons how radiation transport through matter is governed by attenuation coefficients or cross sections, giving the interaction probabilities. Use of the linear attenuation coefficient to describe the statistical nature of radiation penetration in matter can be illustrated by an example for photons.

Equation (8.43), coupled with the "good-geometry" experiment (Fig. 8.7) that measures $\mu$, implies that the probability that a normally incident photon will reach a depth $x$ in a material without interacting is $P(x) = e^{-\mu x}$. For example, the attenuation coefficient for 500-keV photons in soft tissue is $\mu = 0.097$ cm$^{-1}$. The probabilities that a normally incident, 500-keV photon will reach a depth of 1 cm or 2 cm in tissue without interacting are, respectively, $P(1) = e^{-0.0907 \times 1} = 0.908$ and $P(2) = e^{-0.0907 \times 2} = 0.824$. The probability that an incident photon will have its first interaction somewhere between $x = 1$ cm and $x = 2$ cm is, therefore, $P(1) - P(2) = 0.0839$. The probability of the first interaction's being between 9 cm and 10 cm is $P(9) - P(10) = 0.418 - 0.379 = 0.039$. The last result can be obtained in another way. It is the product of the probability $P(9)$ that a given incident photon

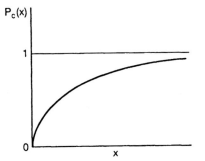

$P_c(x)$

1

0

x

**Fig. 11.9** Cumulative probability $P_c(x) = 1 - e^{-\mu x}$ that a given incident photon has its first interaction before reaching a depth $x$. The linear attenuation coefficient is $\mu$.

will reach a depth of 9 cm and the probability $1 - P(1)$ that it will interact in the first cm thereafter:

$$P(9)[1 - P(1)] = 0.418(1 - 0.908) = 0.039. \tag{11.101}$$

Also, the probability that an incident photon reaches a depth of 9 cm without interacting is just $[P(1)]^9 = (0.908)^9 = 0.420 = P(9)$.

In general, the probability that the first interaction of a normally incident photon will take place at a depth between $x$ and $x + dx$ is given by $P_1(x)\, dx = P(x)\mu\, dx$, $P(x)$ being the probability that it will reach the depth $x$ and $\mu\, dx$ the probability that it will interact in $dx$. The cumulative probability that a normally incident photon will interact before reaching a depth $x$ is

$$P_c(x) = \int_0^x P_1(x)\, dx = \mu \int_0^x e^{-\mu x}\, dx = 1 - e^{-\mu x}. \tag{11.102}$$

(The relative number that have interacted is equal to 1 minus the relative number that have not.) This function is shown in Fig. 11.9.

Probabilities describing radiation transport, such as we have been discussing here, are in agreement with numerous experimental measurements made under specified conditions. From the knowledge of the numerical value of $\mu$, we can also simulate radiation transport on a computer by using Monte Carlo procedures. The Monte Carlo method is a technique of numerical analysis that uses random sampling to construct the solution of a mathematical or physical problem. For example, the numerical value of $\pi$ can be estimated as follows. Figure 11.10 shows a quadrant of a circle enclosed by a square having sides of unit length. Computer programs generate a sequence of random numbers, $0 \leq R < 1$, each number containing a "seed" used to generate the next. Pairs of random numbers can be selected as values $(x_i, y_i)$ that determine points which lie in the square in Fig. 11.10. For each point, one tests whether $x_i^2 + y_i^2 \leq 1$, the radius of the circular arc. If so, then the point lies inside the circle, and is tallied; if not, the point is ignored. After a large number of random trials, one computes the ratio of the number of tallied points and the total number of points tested. With increasingly many trials, this

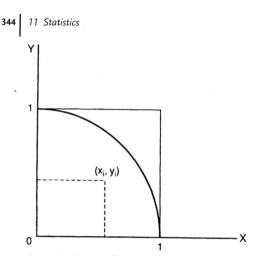

**Fig. 11.10** Diagram for evaluating $\pi$ by a Monte Carlo procedure.

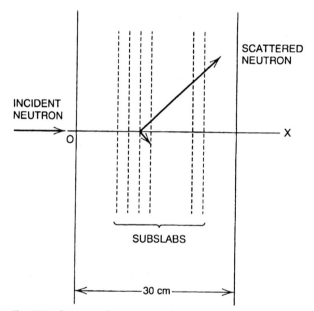

**Fig. 11.11** Geometrical arrangement for computation of
neutron depth–dose curve in tissue slab.

ratio should approach the area $\pi/4$ of the circular arc, thus enabling $\pi$ to be evaluated.

To illustrate the use of Monte Carlo techniques for calculating radiation transport and for dosimetry, we outline the computation of absorbed dose as a function of depth for a uniform, broad beam of monoenergetic neutrons normally incident on a 30-cm soft-tissue slab, infinite in lateral extent. A numerical example then follows. In Fig. 11.11 we consider a neutron incident along an axis $X$ at the origin $O$ on the slab face. We simulate the fate of this (and subsequent) neutrons by Monte

Carlo procedures. We first select a flight distance for the neutron to its first collision, based on the attenuation coefficient $\mu$. for the neutrons of specified energy
in tissue. This is accomplished by setting the cumulative probability for flight distances equal to the first number $R_1$ obtained from the computer random-number
generator:

$$P_c(x) = 1 - e^{-\mu x} = R_1, \tag{11.103}$$

with $0 \leq R_1 < 1$. Solving for $x$, we obtain for the location of the first collision site[7]

$$x = -\frac{1}{\mu} \ln(1 - R_1). \tag{11.104}$$

If $x > 30$ cm, then the neutron traverses the slab completely without interacting,
and we start the simulation again with another neutron. Given a collision site in
the slab, one can next select the type of nucleus struck: H, O, C, or N, the principal
elements in soft tissue. This can be done, as shown in the example that follows, by
partitioning the unit interval into sections with lengths equal to the relative attenuation coefficients for these elements at the given neutron energy. A second random
number $R_2$ then determines the type of nucleus struck. Another random number
can be used to pick the kind of interaction (i.e., elastic or inelastic scattering or
absorption). Additional random numbers then determine all other specifics, such
as the energies and directions of travel of any secondary products. The choices are
made from cumulative probability distributions based on experimental data, theoretical models, or both. The input data to the computer code must contain all the
necessary information to simulate the events as they occur statistically in nature.
As much detailed information and possible alternatives as desired can be included,
though at a cost of increased running time. All secondary products can also be
transported and allowed to interact or escape from the slab. Such a complete history is calculated for each incident neutron and all of its products. For analysis of
results, the slab can be subdivided along the X-axis into parallel subslabs as indicated in Fig. 11.11, for example, at 1-cm intervals. The energies deposited in each
subslab as well as any other details [e.g., values of linear energy transfer (LET)] are
accumulated for a large number of incident neutrons to obtain statistically significant results. (Standard deviations in the compiled statistical quantities are readily
calculated.) The end results can be compiled into a histogram showing the average
energy deposited in each subslab per incident neutron. This energy is proportional
to the average absorbed dose at the depth of the subslab per unit fluence from a
uniform, broad beam of neutrons normally incident on the slab.[8] One thus obtains
a histogram that represents the dose as a function of depth in the slab. Individual
contributions at different values of LET can also be compiled to furnish the dose
equivalent as a function of depth. In fact, the Monte Carlo simulation, by its nature,

---

7   In Fig. 11.9, if we chose a large number of
points randomly along the ordinate between
0 and 1 and use Eq. (11.104) to determine the

resulting values of $x$, the latter will be
distributed as the function $P(x) = e^{-\mu x}$.

8   See last reference listed in Section 11.14.

gives complete details of all physical events that occur, to the extent that they are represented by the input data to the code.

### Example

As in Fig. 11.11, consider a 100-keV neutron normally incident on a soft-tissue slab, having a thickness of 3 cm to simulate a small rodent. Elastic scattering from nuclei is the only important neutron interaction at this energy. The linear attenuation coefficients $\mu_i$ by element for 100-keV neutrons in tissue are given in Table 11.4. Develop a Monte Carlo procedure to calculate the energy deposited at different depths in the slab as a result of 100-keV incident neutrons. Use the "random"-number sequence in Table 11.5 to compute a neutron history.

### Solution

An algorithm for radiation transport is not unique; things can be arranged in a number of ways. Once the algorithm is specified, however, it should be applied in exactly the same manner for each incident neutron. We first select a flight distance for the incident neutron, using Eq. (11.104), the value of $\mu$ given in Table 11.4, and the first random number, $R_1$, in Table 11.5:

$$x = -\frac{1}{0.92315} \ln(1 - 0.87810) = 2.28 \text{ cm.} \tag{11.105}$$

We use the next random number to determine the type of nucleus struck, comparing it with the cumulative probability in the last column of Table 11.4. The second random number, $R_2 = 0.68671$, is in the first interval, assigned to hydrogen. [For $0.842 \leq R_2 < 0.950$ the collision would have been with oxygen, etc.] Thus, the first neutron has a collision with a hydrogen nucleus (proton) at a depth of 2.28 cm. We next select an

**Table 11.4** Data Used to Pick Flight Distances and Struck Nuclei for 100-keV Neutrons in Soft Tissue

| Element | $\mu_i$ (cm$^{-1}$) | $\mu_i/\mu$ | Cumulative $\mu_i/\mu$ |
|---|---|---|---|
| H | 0.777 | 0.842 | 0.842 |
| O | 0.100 | 0.108 | 0.950 |
| C | 0.0406 | 0.044 | 0.994 |
| N | 0.00555 | 0.006 | 1.000 |
| Totals | $\mu = 0.92315$ | 1.000 | |

**Table 11.5** Random Numbers

| $i$ | $R_i$ |
|---|---|
| 1 | 0.87810 |
| 2 | 0.68671 |
| 3 | 0.03621 |
| 4 | 0.10389 |
| 5 | 0.97268 |

energy loss, $Q$, employing the spectrum given in Fig. 9.6. Since the spectrum is flat, a simple choice is $Q = R_3 E_n = 0.03621 \times 100$ keV $= 3.62$ keV. A proton of this energy has a range of less than 0.3 $\mu$m (Table 5.3), and so this amount of energy can be assumed to be absorbed locally at the collision site. [Energy-loss spectra for neutron collisions with the other nuclei, O, C, and N, need to be supplied, either from experimental data or a model. Assuming isotropic scattering in the center-of-mass system would be a reasonable approximation in this example.] The new direction of travel for the neutron after scattering can be determined next. Equation (9.5) gives the polar angle of recoil of the struck proton: $\cos^2 \theta = Q/E_n = 3.62/100 = 0.0362$. Thus, $\theta = 79.0°$; and the neutron is scattered at a polar angle $\xi = 90° - \theta = 11.0°$ with respect to the line along which the neutron was incident. The azimuthal angle of scatter, $\eta$, is completely random between $0°$ and $360°$. We select $\eta = 0°$ to be the vertical direction and use the next random number to pick $\eta$ clockwise about this direction: $\eta = 360 R_4 = 360 \times 0.10389 = 37.4°$.

The scattered neutron is transported next. Typically, tables like Table 11.4 are entered into a Monte Carlo computer code at a number of energies, close enough to allow interpolation between. Because the energy of the scattered neutron in this example is close to 100 keV, we use Table 11.4 to select the next flight distance. Using $R_5$ in place of $R_1$ in Eq. (11.104), we find that the distance of travel to the next collision is 3.90 cm. The remaining distance in the slab from the first collision point to the back surface along the line of travel of the scattered neutron is $(3.00 - 2.28)/\cos 11° = 0.733$ cm. Therefore, the scattered neutron exits the back face of the slab with no further interaction.

The Monte Carlo method finds wide application in calculations for radiation transport, shielding, and dosimetry. Because of its generality and literal simulation of events, special assumptions are not a requirement. Radiation fields need not be uniform, parallel, or monoenergetic. Targets can have any shape or composition. Data describing energy and angular cross sections can be input with any desired degree of fineness in mesh. Subslabs or other subvolumes for analysis can be made smaller for finer detail. Generally, the more versatile a code is made and the greater the detail of the input and output, the longer the running time will be. Smaller analysis subvolumes require a larger number of particle histories for a given degree of statistical confidence in the results. In practice, a thoughtful balance between the demands put on a Monte Carlo code and the reasons for wanting the results can best guide the design of such calculations.

A number of techniques can help streamline Monte Carlo computations and make them more efficient. Stratified sampling can often reduce variance. For example, for the calculations of the first collision of 1000 neutrons in the previous example, we could use Table 11.4 as described. However, instead of selecting the struck nucleus randomly each time from Table 11.4, we could eliminate the variance due to this step entirely by letting the first 842 neutrons collide with H, the next 108 collide with O, the next 44 with C, and the last 6 with N.

Splitting and Russian roulette are two other schemes often employed. In computing neutron penetration of thick shields, for instance, only a small fraction of the incident neutrons get through, limiting the statistical significance of the computed spectra of energies and angles of the emergent particles. The bulk of computer time can be spent in transporting neutrons with many collisions in the shield only to find that they are eventually absorbed before they come out. A series of depths can be assigned such that, when a neutron crosses one, it is replaced by $n$ identical, independent neutrons, each carrying a statistical weight of $1/n$. This splitting gives a larger number of neutrons at the greater depths, each carrying a lesser weight, with total weight preserved. Splitting can be made more efficient by playing Russian roulette with a neutron that tends to become less desirable for the calculations, such as one that tends to return to the entrance surface. At some point in the computations, such a particle is given a chance $p$ to survive Russian roulette or be killed with probability $1-p$. If the neutron is killed, its transport is stopped; otherwise, its statistical weight is increased by the factor $1/p$.

Importance sampling can be used to increase the efficiency of Monte Carlo calculations, depending on what is being sought. In dosimetry, one is often interested in LET distributions of the recoil nuclei produced by neutron collisions. In the last example, the least probable collision is one with nitrogen. One can increase the effective sample size for such a collision by replacing it each time with several that are then assigned reduced statistical weights. Still other techniques used to increase the efficiency of Monte Carlo computations include the use of exponential transformations to artificially increase mean free paths for deep penetration problems, rejection methods, and special sampling functions to represent complicated distributions.

## 11.14
### Suggested Reading

1 Altshuler, B., and Pasternack, B., "Statistical Measures of the Lower Limit of Detection of a Radioactive Counter," *Health Phys.* **9**, 293–298 (1963). [A succinct, very well written paper.]

2 Currie, L. A., "Limits of Quantitative Detection and Quantitative Determination, Applications to Radiochemistry," *Anal. Chem.* **40**, 586–593. [A classic paper in the field.]

3 Currie, L. A., *Lower Limit of Detection: Definition and Elaboration of a Proposed Position for Radiological Effluent and Environmental Measurement.* U.S. Nuclear Regulatory Commission, NUREG/CR-4007, Washington, DC (1984).

4 Evans, R. D., *The Atomic Nucleus,* McGraw-Hill, New York (1955). [Reprinted by Krieger Publishing Co., Malabar, FL (1985). Chapters 26, 27, and 28 are devoted to statistical aspects of nuclear processes and radiation measurements. Excellent, comprehensive treatments and problem sets.]

5 Kase, K. R., Bjarngard, B. E., and Attix, F. H., eds., *The Dosimetry of Ionizing Radiation,* Vol. III, Academic Press, San Diego, CA (1990). [Chapters 5 and 6 provide in-depth treatments of Monte Carlo methods of electron and photon transport for radiation dosimetry and radiotherapy.]

6 Knoll, G. F., *Radiation Detection and Measurement,* 3rd ed., Wiley, New York (2000). [Chapters describe basic counting statistics and error propagation and their applications to various radiation-detection systems and instruments. Fano factor is discussed.]

7 Roberson, P. L., and Carlson, R. D., "Determining the Lower Limit of Detection for Personnel Dosimetry Systems," *Health Phys.* **62**, 2–9 (1992).

8 Strom, D. J., and MacLellan, J. A., "Evaluation of Eight Decision Rules for Low-level Radioactivity Counting," *Health Phys.* **81**, 27–34 (2001). [An important analysis of different decision rules used to decide whether a low-level measurement differs from background.]

9 Tsoulfanidis, N., *Measurement and Detection of Radiation*, McGraw-Hill, New York (1983). [Chapter 2, entitled "Statistics and Errors," is a good survey of basic theory and statistics for radiation measurements, with a number of worked examples.]

10 Turner, J. E., Wright, H. A., and Hamm, R. N., "A Monte Carlo Primer for Health Physicists," *Health Phys.* **48**, 717–733 (1985).

11 Turner, J. E., Zerby, C. D., Woodyard, R. L., Wright, H. A., Kinney, W. E., Snyder, W. S., and Neufeld, J., "Calculation of Radiation Dose from Protons to 400 MeV," *Health Phys.* **10**, 783–808 (1964). [An early example of Monte Carlo calculations of dose from energetic protons. Appendix A gives a derivation of the relation between energy deposited in subslabs per incident particle and the dose per unit fluence from a uniform, broad beam.]

## 11.15
## Problems

1. What is the probability that a normally incident, 400-keV photon will penetrate a 2-mm lead sheet without interacting (Section 8.7)?

2. (a) What is the probability that a given atom of $^{226}$Ra will live 1000 y before decaying?
   (b) What is the probability that it will live 2000 y?
   (c) If the atom is already 10,000 years old, what is the probability that it will live another 1000 y?

3. An unbiased die is rolled 10 times.
   (a) What is the probability that exactly 4 threes will occur?
   (b) What is the probability that exactly 4 of any one number alone will occur?
   (c) What is the probability that two numbers occur exactly 4 times?

4. (a) What is the mean number of threes expected in 10 rolls of a die?
   (b) What would be the probability of observing exactly 4 threes, according to Poisson statistics?
   (c) Why is the answer to part (b) different from the answer to Problem 3(a)?
   (d) Which answer is correct? Why?

5. What is the standard deviation of the number of threes that occur in 10 rolls of a die?

6. In the example in Section 11.3 with $N = 10$, the probabilities $P_3$, $P_6$, and $P_0$ were evaluated.
   (a) Calculate the other values of $P_n$ and show explicitly that the $P_n$ add to give unity.
   (b) Plot the distribution $P_n$ vs. $n$.

7. A sample consists of 16 atoms of $^{222}$Rn.
   (a) What is the probability that exactly one-half of the atoms will decay in 2 d?
   (b) In 3 d?
   (c) Calculate the probability that one week could pass without the decay of a single atom.
   (d) What is the probability that all of the atoms will decay in the first day?

8. (a) In the last problem, what is the mean number of atoms that decay in 2 d?
   (b) What is the mean value of the square of the number of atoms that decay in 2 d?
   (c) What is the standard deviation of the number of atoms that decay in 2 d?

9. Identical samples containing $^{32}$P are measured for a period of 1 h in a counter, having an efficiency of 44% and negligible background. From a large number of observations, the mean number of counts in 1 h is found to be $2.92 \times 10^4$.
   (a) What is the activity of $^{32}$P in a sample?
   (b) Estimate the standard deviation of the number of counts obtained in 1 h.

10. (a) Estimate the number of $^{32}$P atoms in a sample in the last problem.
    (b) Estimate the standard deviation of the sample activity in Bq.
    (c) If the counter efficiency were 100% and the same mean number of counts were observed in 1 h, what would be the standard deviation of the activity in Bq?

11. For the Poisson distribution, show that $P_{n+1} = \mu P_n / (n + 1)$.

12. For the Poisson distribution with $\mu = 4.0$, show that $P_3 = P_4 = 0.195$; and with $\mu = 8.0$, show that $P_7 = P_8 = 0.140$.

13. When the mean value $\mu$ of the Poisson distribution is an integer, show that the probability of observing one less than the mean is equal to the probability of observing the mean (as illustrated by the last problem).

14. Using a counter having an efficiency of 38% and negligible background, a technician records a total of 1812 counts from a long-lived radioactive source in 2 min.
    (a) Estimate the activity of the source in Bq.
    (b) Estimate the standard deviation of the activity.

(c) How long would the source have to be counted in order for the standard deviation of the activity to be 1% of its mean?

15. (a) Estimate the probability that exactly 17 counts would be registered in 1 s in the last problem.

   (b) Repeat for 27 counts.

16. (a) In the last problem, what is the probability that no counts would be registered in 1 s?

   (b) In 2 s?

   (c) How are the probabilities in parts (a) and (b) related?

17. (a) What are the means and standard deviations for the two distributions in the lower left panel of Fig. 11.3?

   (b) For the normal distribution calculate $f(35)$ and $f(40)$.

   (c) Determine the probability $P(35 \leq x \leq 40)$.

18. Show directly from Eq. (11.37) that the normal distribution has inflection points at $x = \mu \pm \sigma$. (The second derivative vanishes at an inflection point.)

19. Use Table 11.1 to verify the entries in Table 11.2.

20. The activities of two sources can be compared by counting them for equal times and then taking the ratio of the two count numbers, $n_1$ and $n_2$. Show that the standard deviation of the ratio $Q = n_1/n_2$ is given by

$$\sigma_Q = Q\left(\frac{\sigma_1^2}{n_1^2} + \frac{\sigma_1^2}{n_1^2}\right)^{1/2},$$

where $\sigma_1$ and $\sigma_2$ are the standard deviations of the two individual count numbers.

21. See the last problem. Two sources are counted for 15 min each and yield $n_1 = 1058$ and $n_2 = 1416$ counts.

   (a) What is the standard deviation of the ratio $n_1/n_2$?

   (b) What is the standard deviation of $n_2/n_1$?

22. The estimated count rate for a long-lived radioactive sample is 93 cpm. Background is negligible. What is the estimated fractional standard deviation of the rate when it is counted for

   (a) 5 min?

   (b) 24 h?

   (c) 1 wk?

23. A sample of long-lived radionuclide gives 939 counts in 3 min.

   (a) What is the probable error in the count rate?

   (b) How long must the sample be counted to determine the count rate to within ±3% with 95% confidence?

24. The count rate in the last problem is 5.22 cps.

   (a) What is the probability that exactly 26 counts would be observed in 5 s?

   (b) Is the use of Poisson statistics warranted here?

   (c) Justify your answer to part (b).

25. According to Poisson statistics, how many counts are needed to obtain a coefficient of variation of 1%?

26. If a 5-min gross count of a sample gives a fractional standard deviation of 2.5%, how much longer should it be counted to reduce the uncertainty to 0.5%?

27. The true count rate of a long-lived radioactive source is 316 cpm. Background is negligible.
    (a) What is the expected number of counts in 5 min?
    (b) What is the standard deviation of the number of counts in 5 min?
    (c) What is the standard deviation of the count rate obtained from a 5-min count?

28. Technician A is asked to make a 5-min measurement with the source in the last problem to determine the count rate and its standard deviation. The measurement yields 1558 counts. Background can be neglected.
    (a) What is the observed count rate?
    (b) What is the standard deviation of the count rate?
    (c) What is the probability that technician B, in an independent 5-min measurement, would obtain a value as close, or closer, to the true count rate as technician A?
    (d) If technician B counts for 1 h, what is the answer to part (c)?

29. A 1-h measurement of background with a certain counter gives 1020 counts. A long-lived sample is placed in the counter, and 120 counts are registered in 5 min.
    (a) What is the standard deviation of the net count rate?
    (b) Without doing additional background counting, how long would the sample have to be counted in order to obtain a standard deviation that is 10% of the net count rate?

30. With a certain counting system, a 6-h background measurement registers 6588 counts. A long-lived sample is then placed in the counter, and 840 gross counts are registered in 30 min.
    (a) What is the net count rate?
    (b) What is the standard deviation of the net count rate?
    (c) Without additional background counting, how many gross counts would be needed in order to obtain the net count rate of the sample to within $\pm 5\%$ of its true value with 90% confidence?
    (d) Without a remeasurement of background, what is the smallest value obtainable for the standard deviation of the net count rate by increasing the gross counting time?

31. In the last problem, a total time of 6 h 30 min was used for background and gross counts. What division of this total time

would give the smallest standard deviation of the net count rate?

32. A certain counter has an efficiency of 35%. A background measurement gives 367 counts in 100 min. When a long-lived sample is placed in the counter, 48 gross counts are registered in 10 min.
    (a) Estimate the standard deviation of the sample activity in Bq.
    (b) For optimum statistical accuracy, how should the total time of 110 min be apportioned between taking background and gross counts?

33. A sample containing a short-lived radionuclide is placed in a counter, which registers 57,912 counts before the activity disappears completely. Background is zero.
    (a) If the counter efficiency is 28%, how many atoms of the radionuclide were present at the beginning of the counting period?
    (b) What is the standard deviation of the number?

34. Repeat the last problem for an efficiency of (a) 45% and (b) 80%.

35. What value $\epsilon < 1$ of the counter efficiency maximizes the standard deviation of the count number, Eq. (11.62)? Interpret your answer.

36. Describe an experiment for obtaining the half-life of a radionuclide when its activity decreases noticeably during the time an observation is made. Include a provision for background counts. How can one minimize the statistical uncertainty in the half-life determination? (See R. D. Evans, *The Atomic Nucleus*, pp. 812–816, cited in Section 11.14.)

37. How are Eqs. (11.58)–(11.62) affected when background is taken into account?

38. Show that Eq. (11.68) follows from (11.67).

39. Show that Eqs. (11.69) and (11.70) follow from (11.67) when $t_g = t_b$.

40. A sample is to be placed in a counter and a gross count taken for 10 min. The maximum acceptable risk for making a type-I error is 0.050. The efficiency of the counter is such that 3.5 disintegrations in a sample result in one net count. A 20-min background measurement yields 820 counts. Under these conditions, what is the minimum significant measured activity in Bq, as inferred from a 10-min gross count?

41. At a certain facility, gross and background counts are made for the same length of time, and $\alpha = 0.050$ and $\beta = 0.025$. The background is not accurately known beforehand. Calibration at the facility shows that one net count corresponds to a sample activity of 260 Bq. The minimum significant measured activity

is 159 Bq. Using approximate formulas, calculate the minimum detectable true activity.

42. You are to establish a procedure for measuring the radioactivity of samples placed in a counter. The calibration of the counter is such that one net count in the allotted time corresponds to 2.27 Bq of sample activity. The expected number of background counts, $B = 1080$, during the allotted time is accurately known. The acceptable risk for making a type-I error is 5% and that for a type-II error, 1%.

(a) What is the minimum significant net sample count?

(b) What is the minimum significant measured activity?

(c) What is the minimum detectable true activity?

(d) A certain sample-plus-background reading registers 1126 gross counts. What is the implied sample activity?

(e) What would be the answer to (d) if one were willing to risk a type-I error in 10% of the measurements?

(f) With the type-I error probability set at its original value of 5%, what would be the minimum detectable true activity if one were to accept a 10% (instead of 1%) risk of a type-II error?

43. What percentage error is made in parts (b), (c), (e), and (f) of the last problem by using approximate formulas?

44. You are shown a counting facility for analyzing samples for radioactivity. The calibration is such that 84 net counts correspond to $3.85 \times 10^4$ Bq of sample activity. The background count of 1270 is stable and accurately known. You are told that the minimum significant net sample count is 70.

(a) What is the minimum significant measured activity for the facility?

(b) What is the maximum risk of making a type-I error?

(c) What is the minimum detectable true activity, if the maximum risk for making a type-II error is 5%?

(d) Does the minimum detectable true activity increase or decrease, if the risk of a type-II error is decreased from 5%?

(e) If the true activity is exactly equal to the minimum detectable true activity $A_{II}$, what is the probability that a single measurement will give a result that implies an activity less than $A_{II}$?

45. Two counting systems are being considered for routine use. The calibration constant for activity for counter 1 is 0.459 Bq per net count; and the expected number of background counts, $B_1 = 7928$, is accurately known. The calibration constant for counter 2 is 0.294 Bq per net count; and the expected number of background counts, $B_2 = 15,160$, is also accurately known.

(a) At a given level of risk for a type-I error, what is the ratio of the minimum significant measured activities for the two counters?

(b) It is decided that the acceptable risks for type-I and type-II errors are both to be 10%. Additional shielding can be placed around counter 1 to reduce its background. If nothing else is changed, what number of background counts, $B_1$, would then be required to achieve a minimum detectable true activity of 30 Bq?

(c) What factors determine the value of the calibration constant?

(d) What difference does it make in the minimum significant measured activity and in the minimum detectable true activity, if the expected number of background counts is not accurately known?

46. At a certain facility, one net count corresponds to an activity of 12.9 Bq in a sample; and the expected number, 816, of background counts is well known.

(a) If the maximum acceptable risk for making a type-I error is 0.050, what is the minimum significant net count?

(b) If the minimum detectable true activity is $A_{II} = 1300$ Bq, what is the risk of making a type-II error when a sample has an activity exactly equal to $A_{II}$? (Use the approximate formula.)

(c) What is the probability of making a type-II error in a single measurement with a sample that has an activity exactly equal to the minimum significant measured activity?

(d) If the minimum detectable true activity $A_{II}$ were taken to be the same as the minimum significant measured activity $A_1$, what would be the probability of making a type-II error with a sample having an activity equal to $A_{II}$?

47. Repeat Problem 42 with the measured number of background counts 1080 not accurately known.

48. Repeat Problem 46(a) and (b) with the measured number of background counts 816 not accurately known.

49. Show that Eqs. (11.79) and (11.80) follow from Eq. (11.78).

50. In a court of law, one judge might tend to be lenient, while another is a "hanging judge." With respect to the possible guilt of a defendant, what relative importance do the two judges place on not making errors of type-I and type-II?

51. The resolution of a scintillation counter is 8.2% for 0.662-MeV gamma photons from $^{137}$Cs. What is its resolution for 1.17-MeV gamma rays from $^{60}$Co?

52. The $W$ value for silicon is 3.6 eV ip$^{-1}$ at 77 K. Calculate the Poisson energy resolution (FWHM in keV) for the absorption of 5.3-MeV alpha particles.

53. The $W$ value for Ge is 3.0 eV ip$^{-1}$ at 77 K. Based on Poisson statistics, calculate the expected energy resolution (FWHM) of an HPGe detector for 0.662-MeV photons from $^{137}$Cs. Assume that charge collection is complete and that electronic noise is negligible.

54. The resolution of a certain HPGe spectrometer ($W = 3.0$ eV ip$^{-1}$) for 662-keV $^{137}$Cs photons is found to be 0.50%. Assume that variations in pulse height are due entirely to statistical fluctuations in the number of ions produced by an absorbed photon. What is the resolution for the 1.332-MeV gamma ray from $^{60}$Co?

55. Using Poisson statistics, calculate the energy resolution (percent) of both detectors for the 4-MeV alpha particles in Problem 26 of the last chapter. How would the resolution of the two compare if the alpha-particle energy were 6 MeV?

56. A 30-keV beta particle is absorbed in a scintillator having an efficiency of 8%. The average scintillation photon energy is 3.4 eV. If 21% of the photons are counted, what is the resolution for the measured beta-particle energy? What would be the resolution in a semiconductor detector ($W = 3$ eV ip$^{-1}$)?

57. What are the pros and cons of using NaI scintillators vs. HPGe detectors for gamma-ray spectroscopy?

58. Give an argument that, because of energy conservation alone, the Fano factor cannot be unity.

59. From the definition (11.90) of the Fano factor, show that the resolution of a counter can be expressed as
$R = \text{FWHM}/\mu = 2.35(F/\mu)^{1/2}$.

60. The resolution for 360-keV photons with a certain gas proportional counter ($W = 30$ eV ip$^{-1}$) is observed to be 0.80%. Determine the Fano factor.

61. Show that Eq. (11.97) leads to (11.93) when $r_t \tau \ll 1$.

62. A nonparalyzable counter with a source in place registers 128,639 counts in 2.5 s. An identical source is added, and 210,649 counts are recorded in 2.5 s. Background is zero, and self-absorption in the sources is negligible. What is the dead time of the counter?

63. Show that the maximum count rate obtainable with a paralyzable system having a dead time $\tau$ is $1/e\tau$.

64. If the maximum count rate attainable with a paralyzable counter is 64,100 s$^{-1}$, what is the dead time?

65. The mass attenuation coefficient for 600-keV photons in aluminum is 0.080 cm$^2$ g$^{-1}$.

(a) What is the mean free path of the photons?
(b) What is the probability that a normally incident photon will penetrate a 10-cm Al slab without interacting?
(c) If the photon penetrates the first 9 cm, what is the probability that it will penetrate the additional 1 cm?
(d) How does the probability in (c) compare with the probability that the photon will penetrate the first cm?
(e) If the photon penetrates the first 9 cm, what is the probability that it would penetrate another 10 cm, for a total of 19 cm?
(f) Do the answers to (b) and (e) imply that a photon can penetrate 10 cm or 19 cm with equal likelihood? Explain.

66. A 2-MeV neutron has a collision with hydrogen.
    (a) What is the probability that it loses an energy between 0.63 MeV and 0.75 MeV?
    (b) If the neutron loses 0.75 MeV, at what angle is it scattered?

67. (a) What is the probability that a 10-MeV neutron will lose 6 MeV or more in a collision with hydrogen?
    (b) When a 10-MeV neutron loses 6 MeV or more in a collision with hydrogen, what is the range of scattering angles that the neutron can make with respect to the original direction?

68. Monoenergetic neutrons are normally incident on a homogeneous slab of thickness 1.70 cm followed by a second homogeneous slab of thickness 2.10 cm. The attenuation coefficients for the two slabs are, respectively, $0.085$ cm$^{-1}$ and $1.10$ cm$^{-1}$.
    (a) Outline a Monte Carlo procedure to determine the first collision depths of successive incident neutrons, based on the use of a random-number sequence.
    (b) Use the sequence in Table 11.5 to find the first collision site of two neutrons.
    (c) What is the probability that a neutron will penetrate both slabs without having an interaction?
    (d) How is the answer to part (c) affected, if the order of the slabs is reversed?

69. Monoenergetic neutrons normally incident on the plane surface S in Fig. 11.12 have a 30% chance of penetrating the surface and a 70% chance of being reflected isotropically back. Write formulas that can be used with a random-number sequence in a Monte Carlo procedure to determine
    (a) whether an incident neutron will be reflected
    (b) the polar and azimuthal angles for a reflected neutron.

70. A Monte Carlo calculation is made with 50,000 monoenergetic neutrons normally incident at a point on a homogeneous tissue slab of unit density, as indicated in Fig. 11.11. The energy

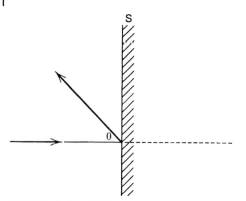

**Fig. 11.12** Problem 69.

deposited in a subslab of thickness 0.25 cm at a certain depth in the slab is found to be 347 MeV. From these data, determine what the energy absorbed per unit mass (absorbed dose) would be at this depth for a uniform, broad beam of normally incident neutrons with unit fluence.

71. The total linear attenuation coefficient for 10-keV electrons in water is 77.6 $\mu m^{-1}$, partitioned as follows:

| | |
|---|---|
| Elastic scattering | 38.2 $\mu m^{-1}$ |
| Ionization | 37.4 |
| Excitation | 2.0 |
| Total | 77.6 $\mu m^{-1}$ |

(a) What is the probability that a 10-keV electron will travel 100 Å without having an interaction?

(b) What is the probability that it will travel 200 Å without interacting and then experience its first collision as an ionization within the next 10 Å?

(c) What is the probability that the first collision will be an excitation at a distance between 60 Å and 80 Å?

72. Attenuation coefficients for 10-keV electrons in water are given in the last problem. Using the sequence of random numbers $R_i$, in Table 11.6, find the flight distance and type of collision for three electrons. Take the event types in the order given as the last problem.

73. In a certain radiation-transport problem, the probability density $P(Q)$ for energy loss $Q$ decreases linearly with $Q$ from $Q = 0$ to the maximum possible value of $Q$, $Q = 800$ eV, where $P(800) = 0$.

(a) Write an analytic function for the normalized distribution $P(Q)$ with $Q$ expressed in eV.

(b) What are the units of $P(Q)$?

**Table 11.6** Problem 72

| $i$ | $R_i$ |
|---|---|
| 1 | 0.1018 |
| 2 | 0.7365 |
| 3 | 0.3248 |
| 4 | 0.4985 |
| 5 | 0.8685 |
| 6 | 0.2789 |

(c) What is the probability that a collision will result in an energy loss $Q$ between 300 eV and 350 eV?

(d) Write an analytic function for the cumulative probability $P_c(Q)$ for energy loss $\leq Q$.

(e) Use $P_c(Q)$ from part (d) to solve part (c).

(f) Calculate the median energy loss.

## 11.16
## Answers

2. **(a)** 0.648
   **(b)** 0.421
   **(c)** 0.648
4. **(a)** 1.67
   **(b)** 0.0610
5. 1.18
8. **(a)** 4.86
   **(b)** 27.0
   **(c)** 1.84
10. **(a)** $3.28 \times 10^7$
    **(b)** 0.108 Bq
    **(c)** 0.0475 Bq
14. **(a)** 39.7 Bq
    **(b)** 0.933 Bq
    **(c)** 11.0 min
15. **(a)** 0.0858
    **(b)** 0.00173
21. **(a)** 0.0303
    **(b)** 0.0543
23. **(a)** $\pm 6.89$ cpm
    **(b)** 13.6 min
24. **(a)** 0.0780
    **(b)** Yes

25. $10^4$
27. **(a)** 1580
    **(b)** 39.7
    **(c)** 7.95 cpm
28. **(a)** 312 cpm
    **(b)** 7.90 cpm
    **(c)** 0.420
    **(d)** 0.918
29. **(a)** 2.25 cpm
    **(b)** 116 min
32. **(a)** 0.0342 Bq
    **(b)** $t_b = 51.4$ min
    $t_g = 58.6$ min
40. 0.247 Bq
41. 348 Bq
42. **(a)** 54
    **(b)** 123 Bq
    **(c)** 307 Bq
    **(d)** No sign. act.
    **(e)** 104 Bq
    **(f)** 222 Bq
46. **(a)** 47
    **(b)** 0.030

(c)  0.50

(d)  0.50

52.  10.3 keV

54.  0.532%

55.  0.71% and 0.22%;

0.58% and 0.18%

60.  0.14

62.  4.30 $\mu s$

65.  (a)  4.63 cm

(b)  0.115

(c)  0.806

(d)  Same

(e)  0.115

(f)  No

67.  (a)  0.40

(b)  50.8° to 90.0°

71.  (a)  0.460

(b)  0.00763

(c)  0.00233

72.  13.8 Å, ionization

50.6 Å, ionization

261 Å, elastic

73.  (a)  $P(Q) = 0.0025 - 3.13 \times 10^{-6} Q$

(b)  $eV^{-1}$

(c)  0.0741

(d)  $P_c(Q) = 0.0025 Q - 1.57 \times 10^{-6} Q^2$

(f)  235 eV

# 12
# Radiation Dosimetry

## 12.1
## Introduction

Radiation dosimetry is the branch of science that attempts to quantitatively relate specific measurements made in a radiation field to physical, chemical, and/or biological changes that the radiation would produce in a target. Dosimetry is essential for quantifying the incidence of various biological changes as a function of the amount of radiation received (dose–effect relationships), for comparing different experiments, for monitoring the radiation exposure of individuals, and for surveillance of the environment. In this chapter we describe the principal concepts upon which radiation dosimetry is based and present methods for their practical utilization.

When radiation interacts with a target it produces excited and ionized atoms and molecules as well as large numbers of secondary electrons. The secondary electrons can produce additional ionizations and excitations until, finally, the energies of all electrons fall below the threshold necessary for exciting the medium. As we shall see in detail in the next chapter, the initial electronic transitions, which produce chemically active species, are completed in very short times ($\lesssim 10^{-15}$ s) in local regions within the path traversed by a charged particle. These changes, which require the direct absorption of energy from the incident radiation by the target, represent the initial physical perturbations from which subsequent radiation effects evolve. It is natural therefore to consider measurements of ionization and energy absorption as the basis for radiation dosimetry.

As experience and knowledge have been gained through the years, basic ideas, philosophy, and concepts behind radiation protection and dosimetry have continually evolved. This process continues today. On a world scale, the recommendations of the International Commission on Radiological Protection (ICRP) have played a major role in establishing protection criteria at many facilities that deal with radiation. In the United States, recommendations of the National Council on Radiation protection and Measurements (NCRP) have provided similar guidance. There is close cooperation among these two bodies and also the International Commission on Radiation Units and Measurements (ICRU). As a practical matter, there is a certain time delay between the publication of the recommendations and the official

*Atoms, Radiation, and Radiation Protection.* James E. Turner
Copyright © 2007 WILEY-VCH Verlag GmbH & Co. KGaA, Weinheim
ISBN: 978-3-527-40606-7

promulgation of statutory regulations by organizations responsible for radiation protection. As a result, at any point in time, some differences might exist among particular procedures in effect at different locations, even though they are based on publications of the ICRP or NCRP. We shall discuss the implementation of radiation protection criteria and exposure limits in Chapter 14. The present chapter will deal with radiation quantities and units of historical and current importance.

## 12.2
## Quantities and Units

### Exposure

Exposure is defined for gamma and X rays in terms of the amount of ionization they produce in air. The unit of exposure is called the roentgen (R) and was introduced at the Radiological Congress in Stockholm in 1928 (Chap. 1). It was originally defined as that amount of gamma or X radiation that produces in air 1 esu of charge of either sign per 0.001293 g of air. (This mass of air occupies 1 cm$^3$ at standard temperature and pressure.) The charge involved in the definition of the roentgen includes both the ions produced directly by the incident photons as well as ions produced by all secondary electrons. Since 1962, exposure has been defined by the International Commission on Radiation Units and Measurements (ICRU) as the quotient $\Delta Q/\Delta m$, where $\Delta Q$ is the sum of all charges of one sign produced in air when all the electrons liberated by photons in a mass $\Delta m$ of air are completely stopped in air. The unit roentgen is now defined as

$$1 \text{ R} = 2.58 \times 10^{-4} \text{ C kg}^{-1}. \tag{12.1}$$

The concept of exposure applies only to electromagnetic radiation; the charge and mass used in its definition, as well as in the definition of the roentgen, refer only to air.

*Example*
Show that 1 esu cm$^{-3}$ in air at STP is equivalent to the definition (12.1) of 1 R of exposure.

*Solution*
Since the density of air at STP is 0.001293 g cm$^{-3}$ and 1 esu $= 3.34 \times 10^{-10}$ C (Appendix B), we have

$$\frac{1 \text{ esu}}{\text{cm}^3} = \frac{3.34 \times 10^{-10} \text{ C}}{0.001293 \text{ g} \times 10^{-3} \text{ kg g}^{-1}} = 2.58 \times 10^{-4} \text{ C kg}^{-1}. \tag{12.2}$$

### Absorbed Dose

The concept of exposure and the definition of the roentgen provide a practical, measurable standard for electromagnetic radiation in air. However, additional concepts are needed to apply to other kinds of radiation and to other materials, particularly tissue. The primary physical quantity used in dosimetry is the absorbed dose. It is defined as the energy absorbed per unit mass from any kind of ionizing radiation in any target. The unit of absorbed dose, $J\,kg^{-1}$, is called the gray (Gy). The older unit, the rad, is defined as $100\ erg\,g^{-1}$. It follows that

$$1\ Gy \equiv \frac{1\ J}{kg} = \frac{10^7\ erg}{10^3\ g} = 10^4\,\frac{erg}{g} = 100\ rad. \tag{12.3}$$

The absorbed dose is often referred to simply as the dose. It is treated as a point function, having a value at every position in an irradiated object.

One can compute the absorbed dose in air when the exposure is 1 R. Photons produce secondary electrons in air, for which the average energy needed to make an ion pair is $W = 34\ eV\,ip^{-1} = 34\ J\,C^{-1}$ (Sect. 10.1). Using a more precise $W$ value,[1] one finds

$$1\ R = \frac{2.58 \times 10^{-4}\ C}{kg} \times \frac{33.97\ J}{C} = 8.76 \times 10^{-3}\ J\,kg^{-1}. \tag{12.4}$$

Thus, an exposure of 1 R gives a dose in air of $8.76 \times 10^{-3}$ Gy (= 0.876 rad). Calculations also show that a radiation exposure of 1 R would produce a dose of $9.5 \times 10^{-3}$ Gy (= 0.95 rad) in soft tissue. This unit is called the rep ("roentgen-equivalent-physical") and was used in early radiation-protection work as a measure of the change produced in living tissue by radiation. The rep is no longer employed.

### Dose Equivalent

It has long been recognized that the absorbed dose needed to achieve a given level of biological damage (e.g., 50% cell killing) is often different for different kinds of radiation. As discussed in the next chapter, radiation with a high linear energy transfer (LET) (Sect. 7.3) is generally more damaging to a biological system per unit dose than radiation with a low LET (for example, cf. Fig. 13.16).

To allow for the different biological effectiveness of different kinds of radiation, the International Commission on Radiological Protection (ICRP), National Council on Radiation Protection and Measurements (NCRP), and ICRU (Chap. 1) introduced the concept of dose equivalent for radiation-protection purposes. The dose equivalent $H$ is defined as the product of the absorbed dose $D$ and a dimensionless quality factor $Q$, which depends on LET:

$$H = QD. \tag{12.5}$$

In principle, other multiplicative modifying factors can be included along with $Q$ to allow for additional considerations (e.g., dose fractionation), but these are not ordinarily used. Until the 1990 recommendations made in ICRP Publication 60, the dependence of $Q$ on LET was defined as given in Table 12.1. Since then, the

---

1 See p. 29 in Attix reference, Section 12.11.

**Table 12.1** Dependence of Quality Factor $Q$ on LET of Radiation as Formerly Recommended by ICRP, NCRP, and ICRU

| LET (keV $\mu$m$^{-1}$ in Water) | $Q$ |
|---|---|
| 3.5 or less | 1 |
| 3.5–7.0 | 1–2 |
| 7.0–23 | 2–5 |
| 23–53 | 5–10 |
| 53–175 | 10–20 |
| Gamma rays, X rays, electrons, positrons of any LET | 1 |

**Table 12.2** Dependence of Quality Factor $Q$ on LET as Currently Recommended by ICRP, NCRP, and ICRU

| LET, $L$ (keV $\mu$m$^{-1}$ in Water) | $Q$ |
|---|---|
| <10 | 1 |
| 10–100 | $0.32L-2.2$ |
| >100 | $300/\sqrt{L}$ |

ICRP, NCRP, and ICRU have defined $Q$ in accordance with Table 12.2. In the context of quality factor, LET is the unrestricted stopping power, $L_\infty$, as discussed in Section 7.3. For incident charged particles, it is the LET of the radiation in water, expressed in keV per $\mu$m of travel. For neutrons, photons, and other uncharged radiation, LET refers to that which the secondary charged particles they generate would have in water. Like absorbed dose, dose equivalent is a point function. When dose is expressed in Gy, the (SI) unit of dose equivalent is the sievert (Sv). With the dose in rad, the older unit of dose equivalent is the rem ("roentgen-equivalent-man"). Since 1 Gy = 100 rad, 1 Sv = 100 rem.

Dose equivalent has been used extensively in protection programs as the quantity in terms of which radiation limits are specified for the exposure of individuals. Dose equivalents from different types of radiation are simply additive.

*Example*

A worker receives a whole-body dose of 0.10 mGy from 2-MeV neutrons. Estimate the dose equivalent, based on Table 12.1.

*Solution*

Most of the absorbed dose is due to the elastic scattering of the neutrons by the hydrogen in tissue (cf. Table 12.6). To make a rough estimate of the quality factor, we first find $Q$ for a 1-MeV proton—the average recoil energy for 2-MeV neutrons. From Table 5.3 we see that the stopping power for a 1-MeV proton in water is 270 MeV cm$^{-1}$ = 27 keV $\mu$m$^{-1}$. Under the current recommendations of the ICRP, NCRP, and ICRU, $Q$ is defined according to Table 12.2. However, the older recommendations, which include Table 12.1, are still in effect at a number of installations. We see from Table 12.1 that an estimate of $Q \sim 6$ should be reasonable for the recoil protons. The recoil O, C, and N nuclei have considerably higher LET values, but do not contribute as much to the dose as H. (LET is proportional to the square of a particle's charge.) Without going into more detail, we take the overall quality factor, $Q \sim 12$, to be twice that for the recoil protons alone. Therefore, the estimated dose equivalent is $H \sim 12 \times 0.10 = 1.2$ mSv. [The value $Q = 10$ is obtained from detailed calculations (cf. Table 12.5).] We note that Table 12.2 implies a comparable value, $Q = 6.4$, for the protons. Changes in the recommendations are discussed more fully in Chapter 14.

By the early 1990s, the ICRP and NCRP replaced the use of LET-dependent quality factors by *radiation weighting factors*, $w$, specified for radiation of a given type and energy. The quantity on the left-hand side of the replacement for Eq. (12.5), $H = wD$, is then called the *equivalent dose*. In some regulations the older terminology, dose equivalent and quality factor, is still employed. However, the latter has come to be specified by radiation type and energy, rather than LET.

## 12.3
## Measurement of Exposure

### Free-Air Ionization Chamber

Based on its definition, exposure can be measured operationally with the "free-air," or "standard," ionization chamber, sketched in Fig. 12.1. X rays emerge from the target T of an X-ray tube and enter the free-air chamber through a circular aperture of area $A$, defining a right circular cone $TBC$ of rays. Parallel plates $Q$ and $Q'$ in the chamber collect the ions produced in the volume of air between them with center $P'$.

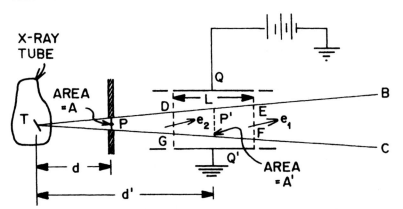

**Fig. 12.1** Schematic diagram of the "free-air" or "standard" ionization chamber.

The exposure in the volume *DEFG* in roentgens would be determined directly if the total ionization produced only by those ions that originate from X-ray interactions in the truncated conical volume *DEFG* could be collected and the resulting charge divided by the mass of air in *DEFG*. This mass is given by $M = \rho A'L$, where $\rho$ is the density of air, $A'$ is the cross-sectional area of the truncated cone at its midpoint $P'$, and $L$, the thickness of the cone, is equal to the length of the collecting plates $Q$ and $Q'$. Unfortunately, the plates collect all of the ions between them, not the particular set that is specified in the definition of the roentgen. Some electrons produced by X-ray interactions in *DEFG* escape this volume and produce ions that are not collected by the plates $Q, Q'$. Also, some ions from electrons originally produced outside *DEFG* are collected. Thus, only part of the ionization of an electron such as $e_1$ in Fig. 12.1 is collected, while ionization from an "outside" electron, such as $e_2$, is collected. When the distance from $P$ to $DG$ is sufficiently large (e.g., $\sim$ 10 cm for 300-keV X rays), electronic equilibrium will be realized; that is, there will be almost exact compensation between ionization lost from the volume *DEFG* by electrons, such as $e_1$, that escape and ionization gained from electrons, such as $e_2$, that enter. The distance from $P$ to $DG$, however, should not be so large as to attenuate the beam significantly between $P$ and $P'$. Under these conditions, when a charge $q$ is collected, the exposure at $P'$ is given by

$$E_{P'} = \frac{q}{\rho A'L}. \tag{12.6}$$

In practice, one prefers to know the exposure $E_P$ at $P$, the location where the entrance port is placed, rather than $E_{P'}$. By the inverse-square law, $E_P = (d'/d)^2 E_{P'}$. Since $A = (d/d')^2 A'$, Eq. (12.6) gives

$$E_P = \left(\frac{d'}{d}\right)^2 \frac{q}{\rho A'L} = \frac{q}{\rho AL}. \tag{12.7}$$

*Example*

The entrance port of a free-air ionization chamber has a diameter of 0.25 cm and the length of the collecting plates is 6 cm. Exposure to an X-ray beam produces a steady current of $2.6 \times 10^{-10}$ A for 30 s. The temperature is 26°C and the pressure is 750 torr. Calculate the exposure rate and the exposure.

*Solution*

We can apply Eq. (12.7) to exposure rates as well as to exposure. The rate of charge collection is $\dot{q} = 2.6 \times 10^{-10}$ A $= 2.6 \times 10^{-10}$ C s$^{-1}$. The density of the air under the stated conditions is $\rho = (0.00129)(273/299)(750/760) = 1.16 \times 10^{-3}$ g cm$^{-3}$. The entrance-port area is $A = \pi(0.125)^2 = 4.91 \times 10^{-2}$ cm$^2$ and $L = 6$ cm. Equation (12.7) implies, for the exposure rate,

$$\dot{E}_P = \frac{\dot{q}}{\rho AL} = \frac{2.6 \times 10^{-10}\ \mathrm{C\,s^{-1}}}{1.16 \times 10^{-3} \times 4.91 \times 10^{-2} \times 6\ \mathrm{g}}$$

$$\times \frac{1\ \mathrm{R}}{2.58 \times 10^{-7}\ \mathrm{C\,g^{-1}}} = 2.95\ \mathrm{R\,s^{-1}}. \tag{12.8}$$

The total exposure is 88.5 R.

Measurement of exposure with the free-air chamber requires some care and attention to details. For example, the collecting plates $Q$ and $Q'$ in Fig. 12.1 must be recessed away from the active volume $DEFG$ by a distance not less than the lateral range of electrons produced there. We have already mentioned minimum and maximum restrictions on the distance from $P$ to $DG$. When the photon energy is increased, the minimum distance required for electronic equilibrium increases rapidly and the dimensions for a free-air chamber become excessively large for photons of high energy. For this and other reasons, the free-air ionization chamber and the roentgen are not used for photon energies above 3 MeV.

**The Air-Wall Chamber**

The free-air ionization chamber is not a practical instrument for measuring routine exposure. It is used chiefly as a primary laboratory standard. For routine use, chambers can be built with walls of a solid material, having photon response properties similar to those of air. Chambers of this type were discussed in Section 10.1.

Such an "air-wall" pocket chamber, built as a capacitor, is shown schematically in Fig. 12.2. A central anode, insulated from the rest of the chamber, is given an initial charge from a charger-reader device to which it is attached before wearing. When exposed to photons, the secondary electrons liberated in the walls and enclosed air tend to neutralize the charge on the anode and lower the potential difference between it and the wall. The change in potential difference is directly proportional to the total ionization produced and hence to the exposure. Thus, after exposure to photons, measurement of the change in potential difference from its original value when the chamber was fully charged can be used to find the exposure. Direct-reading pocket ion chambers are available (Fig. 10.6).

*Example*

A pocket air-wall chamber has a volume of 2.5 cm$^3$ and a capacitance of 7 pF. Initially charged at 200 V, the reader showed a potential difference of 170 V after the chamber was worn. What exposure in roentgens can be inferred?

*Solution*

The charge lost is $\Delta Q = C\Delta V = 7 \times 10^{-12} \times (200 - 170) = 2.10 \times 10^{-10}$ C. The mass of air [we assume standard temperature and pressure (STP)] is $M = 0.00129 \times 2.5 = 3.23 \times 10^{-3}$ g. It follows that the exposure is

$$\frac{2.10 \times 10^{-10} \text{ C}}{3.23 \times 10^{-3} \text{ g}} \times \frac{1 \text{ R}}{2.58 \times 10^{-7} \text{ C g}^{-1}} = 0.252 \text{ R}. \tag{12.9}$$

**PLASTIC**    **ANODE**

**Fig. 12.2** Air-wall pocket ionization chamber, having a plastic wall with approximately the same response to photons as air.

Literally taken, the data given in this problem indicate only that the chamber was partially discharged. Charge loss could occur for reasons other than radiation (e.g., leakage from the central wire). Two pocket ion chambers can be worn simultaneously to improve reliability.

In practice, air-wall ionization chambers involve a number of compromises from an ideal instrument that measures exposure accurately. For example, if the wall is too thin, incident photons will produce insufficient ionization inside the chamber. If the wall is too thick, it will significantly attenuate the incident radiation. The optimal thickness is reached when, for a given photon field, the ionization in the chamber gas is a maximum. This value, called the equilibrium wall thickness, is equal to the range of the most energetic secondary electrons produced in the wall. In addition, a solid wall can be only approximately air equivalent. Air-wall chambers can be made with an almost energy-independent response from a few hundred keV to about 2 MeV—the energy range in which Compton scattering is the dominant photon interaction in air and low-$Z$ wall materials.

## 12.4
### Measurement of Absorbed Dose

One of the primary goals of dosimetry is the determination of the absorbed dose in tissue exposed to radiation. The Bragg–Gray principle provides a means of relating ionization measurements in a gas to the absorbed dose in some convenient material from which a dosimeter can be fabricated. To obtain the tissue dose, either the material can be tissue equivalent or else the ratio of the absorbed dose in the material to that in tissue can be inferred from other information, such as calculations or calibration measurements.

Consider a gas in a walled enclosure irradiated by photons, as illustrated in Fig. 12.3. The photons lose energy in the gas by producing secondary electrons there, and the ratio of the energy deposited and the mass of the gas is the absorbed dose in the gas. This energy is proportional to the amount of ionization in the gas

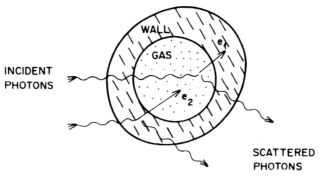

**Fig. 12.3** Gas in cavity enclosed by wall to illustrate Bragg–Gray principle.

when electronic equilibrium exists between the wall and the gas. Then an electron, such as $e_1$ in Fig. 12.3, which is produced by a photon in the gas and enters the wall before losing all of its energy, is compensated by another electron, like $e_2$, which is produced by a photon in the wall and stops in the gas. When the walls and gas have the same atomic composition, then the energy spectra of such electrons will be the same irrespective of their origin, and a high degree of compensation can be realized. The situation is then analogous to the air-wall chamber just discussed. Electronic equilibrium requires that the wall thickness be at least as great as the maximum range of secondary charged particles. However, as with the air-wall chamber, the wall thickness should not be so great that the incident radiation is appreciably attenuated.

The Bragg–Gray principle states that, if a gas is enclosed by a wall of the same atomic composition and if the wall meets the thickness conditions just given, then the energy absorbed per unit mass in the gas is equal to the number of ion pairs produced there times the $W$ value divided by the mass $m$ of the gas. Furthermore, the absorbed dose $D_g$ in the gas is equal to the absorbed dose $D_w$ in the wall. Denoting the number of ions in the gas by $N_g$, we write

$$D_w = D_g = \frac{N_g W}{m}. \tag{12.10}$$

When the wall and gas are of different atomic composition, the absorbed dose in the wall can still be obtained from the ionization in the gas. In this case, the cavity size and gas pressure must be small, so that secondary charged particles lose only a small fraction of their energy in the gas. The absorbed dose then scales as the ratio $S_w/S_g$ of the mass stopping powers of the wall and gas:

$$D_w = \frac{D_g S_w}{S_g} = \frac{N_g W S_w}{m S_g}. \tag{12.11}$$

If neutrons rather than photons are incident, then in order to satisfy the Bragg–Gray principle the wall must be at least as thick as the maximum range of any secondary charged recoil particle that the neutrons produce in it.

As with the air-wall chamber for measuring exposure, condenser-type chambers that satisfy the Bragg–Gray conditions can be used to measure absorbed dose. Prior to exposure, the chamber is charged. The dose can then be inferred from the reduced potential difference across the instrument after it is exposed to radiation.

The determination of dose rate is usually made by measuring the current due to ionization in a chamber that satisfies the Bragg–Gray conditions. As the following example shows, this method is both sensitive and practical.

*Example*
A chamber satisfying the Bragg–Gray conditions contains 0.15 g of gas with a $W$ value of 33 eV ip$^{-1}$. The ratio of the mass stopping power of the wall and the gas is 1.03. What is the current when the absorbed dose rate in the wall is 10 mGy h$^{-1}$?

*Solution*

We apply Eq. (12.11) to the dose rate, with $S_w/S_g = 1.03$. From the given conditions, $\dot{D}_w = 10 \text{ mGy h}^{-1} = (0.010 \text{ J kg}^{-1})/(3600 \text{ s}) = 2.78 \times 10^{-6} \text{ J kg}^{-1}\text{ s}^{-1}$. The rate of ion-pair production in the gas is, from Eq. (12.11),

$$\dot{N}_g = \frac{\dot{D}_w m S_g}{W S_w}$$

$$= \frac{2.78 \times 10^{-6} \text{ J kg}^{-1}\text{ s}^{-1} \times 0.15 \times 10^{-3} \text{ kg}}{33 \text{ eV ip}^{-1} \times 1.60 \times 10^{-19} \text{ J eV}^{-1} \times 1.03} = 7.67 \times 10^{7} \text{ ip s}^{-1}. \quad (12.12)$$

Since the electronic charge is $1.60 \times 10^{-19}$ C, the current is $7.67 \times 10^7 \times 1.60 \times 10^{-19} = 1.23 \times 10^{-11} \text{ C s}^{-1} = 1.23 \times 10^{-11}$ A. Simple electrometer circuits can be used to measure currents smaller than $10^{-14}$ A, corresponding to dose rates much less than $10 \text{ mGy h}^{-1}$ in this example. Note that Eq. (12.12) implies that the current is given by

$$I = \dot{N}_g e = \frac{\dot{D}_w m e S_g}{W S_w}. \quad (12.13)$$

Expressing $W = 33 \text{ J C}^{-1}$ and remembering that the conversion factor from eV to J is numerically equal to the magnitude of the electronic charge $e$, we write, in SI units,

$$I = \frac{\dot{D}_w m S_g}{W S_w} \quad (12.14)$$

$$= \frac{2.78 \times 10^{-6} \text{ J kg}^{-1}\text{ s}^{-1} \times 1.5 \times 10^{-4} \text{ kg}}{33 \text{ J C}^{-1} \times 1.03} = 1.23 \times 10^{-11} \text{ A}. \quad (12.15)$$

The arithmetic is thus shortened somewhat by using Eq. (12.14). However, one must be careful to keep units straight.

## 12.5
## Measurement of X- and Gamma-Ray Dose

Figure 12.4 shows the cross section of a spherical chamber of graphite that encloses $CO_2$ gas. The chamber satisfies the Bragg–Gray conditions for photons over a wide energy range, and so the dose $D_C$ in the carbon wall can be obtained from the measured ionization of the $CO_2$ by means of Eq. (12.11). Since carbon is a major constituent of soft tissue, the wall dose approximates that in soft tissue $D_t$. Calculations show that, for photon energies between 0.2 MeV and 5 MeV, $D_t = 1.1 D_C$ to within 5%. Thus soft-tissue dose can be measured with an accuracy of 5% with the carbon chamber.

Generally, the dose in low-Z wall materials will approximate that in soft tissue over a wide range of photon energies. This fact leads to the widespread use of plastics and a number of other low-Z materials for gamma-dosimeter walls. The ratio of the absorbed dose in many materials relative to that in soft tissue has been calculated. Figure 12.5 shows several important examples. For photon energies from

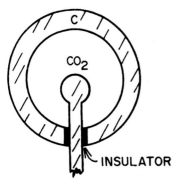

Fig. 12.4 Cross section of graphite-walled $CO_2$ chamber for measuring photon dose.

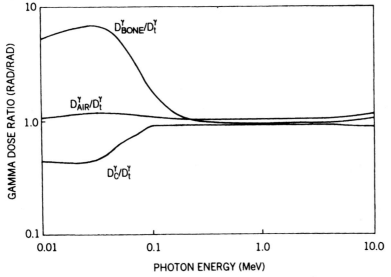

Fig. 12.5 Ratio of absorbed doses in bone, air, and carbon to that in soft tissue, $D_t$.

~0.1 MeV to ~10 MeV, the ratios for all materials of low atomic number are near unity, because Compton scattering dominates. The curve for bone, in contrast to the other two, rises at low energies due to the larger cross section of photoelectric absorption in the heavier elements of bone (e.g., Ca and P).

## 12.6
## Neutron Dosimetry

An ionization device, such as that shown in Fig. 12.4, used for measuring gamma-ray dose will show a reading when exposed to neutrons. The response is due to ionization produced in the gas by the charged recoil nuclei struck by neutrons in the walls and gas. However, the amount of ionization will not be proportional

**Table 12.3** Principal Elements in Soft Tissue of Unit Density

| Element | Atoms cm$^3$ |
|---|---|
| H | $5.98 \times 10^{22}$ |
| O | $2.45 \times 10^{22}$ |
| C | $9.03 \times 10^{21}$ |
| N | $1.29 \times 10^{21}$ |

**Table 12.4** Relative Response of $C-CO_2$ Chamber to Neutrons of Energy $E$ and Photons [Eq. (12.16)]

| Neutron Energy, $E$ (MeV) | $P(E)$ |
|---|---|
| 0.1 | 0.109 |
| 0.5 | 0.149 |
| 1.0 | 0.149 |
| 2.0 | 0.145 |
| 3.0 | 0.151 |
| 4.0 | 0.247 |
| 5.0 | 0.168 |
| 10.0 | 0.341 |
| 20.0 | 0.487 |

to the absorbed dose in tissue unless (1) the walls and gas are tissue equivalent and (2) the Bragg–Gray principle is satisfied for neutrons. As shown in Table 12.3, soft tissue consists chiefly of hydrogen, oxygen, carbon, and nitrogen, all having different cross sections as functions of neutron energy (cf. Fig. 9.2). The carbon wall of the chamber in Fig. 12.4 would respond quite differently from tissue to a field of neutrons of mixed energies, because the three other principal elements of tissue are lacking.

The $C-CO_2$ chamber in Fig. 12.4 and similar devices *can* be used for neutrons of a given energy if the chamber response has been calibrated experimentally as a function of neutron energy. Table 12.4 shows the relative response $P(E)$ of the $C-CO_2$ chamber to photons or to neutrons of a given energy for a fluence that delivers 1 rad to soft tissue. If $D_C^n(E)$ is the absorbed dose in the carbon wall due to 1 tissue rad of neutrons of energy $E$ and $D_C^\gamma$ is the absorbed dose in the wall due to 1 tissue rad of photons, then, approximately,

$$P(E) = \frac{D_C^n(E)}{D_C^\gamma}. \tag{12.16}$$

*Example*

A $C-CO_2$ chamber exposed to 1-MeV neutrons gives the same reading as that obtained when gamma rays deliver an absorbed dose of 2 mGy to the carbon wall. What absorbed dose would the neutrons deliver to soft tissue?

*Solution*

From Table 12.4, the neutron tissue dose $D_n$ would be given approximately by the relation

$$P(E)D_n = 0.149\, D_n = 2\ \text{mGy}, \tag{12.17}$$

or $D_n = 13.4$ mGy.

Tissue-equivalent gases and plastics have been developed for constructing chambers to measure neutron dose directly. These materials are fabricated with the approximate relative atomic abundances shown in Table 12.3. In accordance with the proviso mentioned after Eq. (12.11), the wall of a tissue-equivalent neutron chamber must be at least as thick as the range of a proton having the maximum energy of the neutrons to be monitored.

More often than not, gamma rays are present when neutrons are. In monitoring mixed gamma–neutron radiation fields one generally needs to know the separate contributions that each type of radiation makes to the absorbed dose. One needs this information in order to assign the proper quality factor to the neutron part to obtain the dose equivalent. To this end, two chambers can be exposed—one C—$CO_2$ and one tissue equivalent—and doses determined by a difference method. The response $R_T$ of the tissue-equivalent instrument provides the combined dose, $R_T = D_\gamma + D_n$. The reading $R_C$ of the C—$CO_2$ chamber can be expressed as $R_C = D_\gamma + P(E)D_n$, where $P(E)$ is an appropriate average from Table 12.4 for the neutron field in question. The individual doses $D_\gamma$ and $D_n$ can be inferred from $R_T$ and $R_C$.

*Example*

In an unknown gamma–neutron field, a tissue-equivalent ionization chamber registers 0.082 mGy h$^{-1}$ and a C—$CO_2$ chamber, 0.029 mGy h$^{-1}$. What are the gamma and neutron dose rates?

*Solution*

The instruments' responses can be written in terms of the dose rates as

$$\dot{R}_T = \dot{D}_\gamma + \dot{D}_n = 0.082 \tag{12.18}$$

and

$$\dot{R}_C = \dot{D}_\gamma + P(E)\dot{D}_n = 0.029. \tag{12.19}$$

Since we are not given any information about the neutron energy spectrum, we must assume some value of $P(E)$ in order to go further. We choose $P(E) \sim 0.15$, representative of neutrons in the lower MeV to keV range in Table 12.4. Subtracting both sides of Eq. (12.19) from (12.18) gives $\dot{D}_n = (0.082 - 0.029)/(1 - 0.15) = 0.062$ mGy h$^{-1}$. It follows from (12.18) that $\dot{D}_\gamma = 0.020$ mGy h$^{-1}$.

Very often, as the example illustrates, the neutron energy spectrum is not known and the difference method may not be accurate.

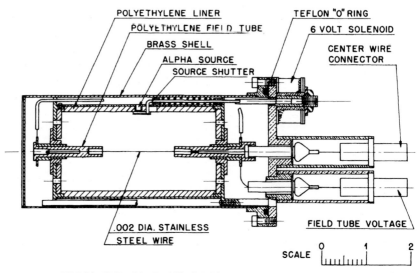

POLYETHYLENE LINER
POLYETHYLENE FIELD TUBE
BRASS SHELL
ALPHA SOURCE
SOURCE SHUTTER
TEFLON "O" RING
6 VOLT SOLENOID
CENTER WIRE
CONNECTOR

.002 DIA. STAINLESS
STEEL WIRE

FIELD TUBE VOLTAGE

SCALE 0 1 2

**ABSOLUTE FAST NEUTRON DOSIMETER**

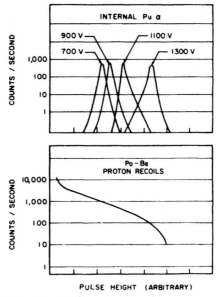

INTERNAL Pu α

900 V 1100 V
700 V 1300 V

COUNTS / SECOND

1,000
100
10
1

Po – Be
PROTON RECOILS

COUNTS / SECOND

10,000
1,000
100
10
1

PULSE HEIGHT (ARBITRARY)

**Fig. 12.6** Hurst fast-neutron proportional counter. Internal
alpha source in wall is used to provide pulses of known size for
energy calibration. (Courtesy Oak Ridge National Laboratory,
operated by Martin Marietta Energy Systems, Inc., for the
Department of Energy.)

As mentioned in Section 10.7, the proportional counter provides a direct method of measuring neutron dose, and it has the advantage of excellent gamma discrimination. The pulse height produced by a charged recoil particle is proportional to the energy that the particle deposits in the gas. The Hurst fast-neutron proportional counter is shown in Fig. 12.6. To satisfy the Bragg–Gray principle, the polyethylene walls are made thicker than the range of a 20-MeV proton. The counter gas can be either ethylene ($C_2H_4$) or cyclopropane ($C_3H_6$), both having the same $H/C = 2$ ratio as the walls. A recoil proton or carbon nucleus from the wall or gas has high LET. Unless only a small portion of its path is in the gas it will deposit much more energy in the gas than a low-LET secondary electron produced by a gamma ray. Rejection of the small gamma pulses can be accomplished by electronic discrimination. Fast-neutron dose rates as low as $10^{-5}$ Gy h$^{-1}$ can be measured in the presence of gamma fields with dose rates up to 1 Gy h$^{-1}$. In very intense fields signals from multiple gamma rays can "pile up" and give pulses comparable in size to those from neutrons.

The LET spectra of the recoil particles produced by neutrons (and hence neutron quality factors) depend on neutron energy. Table 12.5 gives the mean quality factors (based on Table 12.1) and fluence rates for monoenergetic neutrons that give a dose equivalent of 1 mSv in a 40-h work week. The quality factors have been computed by averaging over the LET spectra of all charged recoil nuclei produced by the neutrons. For practical applications, using $\overline{Q} = 3$ for neutrons of energies less than 10 keV and $\overline{Q} = 10$ for higher energies will result in little error. Using $\overline{Q} = 10$ for all neutrons is acceptable, but may be overly conservative. Thus, in monitoring neutrons for radiation-protection purposes, one should generally know or estimate the neutron energy spectrum or LET spectrum (i.e., the LET spectrum of the recoil particles). Measurement of LET spectra is discussed in Section 12.8. Several methods of obtaining neutron energy spectra were described in Section 10.7. The neutron rem meter, shown in Fig. 10.44, was discussed previously.

Figure 12.7 shows an experimental setup for exposing anthropomorphic phantoms, wearing various types of dosimeters, to fission neutrons. A bare reactor was positioned above the circle, drawn on the floor, with an intervening shield placed between it and the phantoms, located 3 m away. In this Health Physics Research Reactor facility, the responses of dosimeters to neutrons with a known energy spectrum and fluence were studied.

Intermediate and fast neutrons incident on the body are subsequently moderated and can be backscattered at slow or epithermal energies through the surface they entered. Exposure to these neutrons can therefore be monitored by wearing a device, such as a thermoluminescent dosimeter (TLD) enriched in $^6$Li, that is sensitive to slow neutrons. Such a device is called an albedo-type neutron dosimeter. (For a medium A that contains a neutron source and an adjoining medium B that does not, the albedo is defined in reactor physics as the fraction of neutrons entering B that are reflected or scattered back into A.)

**Table 12.5** Mean Quality Factors $\overline{Q}$ and Fluence Rates for Monoenergetic Neutrons that Give a Maximum Dose-Equivalent Rate of 1 mSv in 40 h

| Neutron Energy (eV) | $\overline{Q}$ | Fluence Rate ($cm^{-2}\,s^{-1}$) |
|---|---|---|
| 0.025 (thermal) | 2 | 680 |
| 0.1 | 2 | 680 |
| 1.0 | 2 | 560 |
| 10.0 | 2 | 560 |
| $10^2$ | 2 | 580 |
| $10^3$ | 2 | 680 |
| $10^4$ | 2.5 | 700 |
| $10^5$ | 7.5 | 115 |
| $5 \times 10^5$ | 11 | 27 |
| $10^6$ | 11 | 19 |
| $5 \times 10^6$ | 8 | 16 |
| $10^7$ | 6.5 | 17 |
| $1.4 \times 10^7$ | 7.5 | 12 |
| $6 \times 10^7$ | 5.5 | 11 |
| $10^8$ | 4 | 14 |
| $4 \times 10^8$ | 3.5 | 10 |

Source: From *Protection Against Neutron Radiation*, NCRP Report No. 38, National Council on Radiation Protection and Measurements, Washington, D.C. (1971). In its 1987 Report No. 91, the NCRP recommends multiplying the above values of $\overline{Q}$ by two (and reducing the above fluence rates by this factor).

## 12.7
## Dose Measurements for Charged-Particle Beams

For radiotherapy and for radiobiological experiments one needs to measure the dose or dose rate in a beam of charged particles. This is often accomplished by measuring the current from a thin-walled ionization chamber placed at different depths in a water target exposed to the beam, as illustrated in Fig. 12.8. The dose rate is proportional to the current. For monoenergetic particles of a given kind (e.g., protons) the resulting "depth–dose" curve has the reversed shape of the mass stopping-power curves in Fig. 5.6. The dose rate is a maximum in the region of the Bragg peak near the end of the particles' range. In therapeutic applications, absorbers or adjustments in beam energy are employed so that the beam stops at the location of a tumor or other tissue to be irradiated. In this way, the dose there (as well as LET) is largest, while the intervening tissue is relatively spared. To further spare healthy tissue, a tumor can be irradiated from several directions.

If the charged particles are relatively low-energy protons ($\lesssim 400$ MeV), then essentially all of their energy loss is due to electronic collisions. The curve in Fig. 12.8 will then be similar in shape to that for the mass stopping power. Higher-energy

**Fig. 12.7** Anthropomorphic phantoms, wearing a variety of dosimeters in different positions, were exposed to neutrons with known fluence and energy spectra at the Health Physics Research Reactor. (Courtesy Oak Ridge National Laboratory, operated by Martin Marietta Energy Systems, Inc., for the Department of Energy.)

protons undergo significant nuclear reactions, which attenuate the protons and deposit energy by nuclear processes. The depth–dose curve is then different from the mass stopping power. Other particles, such as charged pions, have strong nuclear interactions at all energies, and depth–dose patterns can be quite different.

## 12.8
## Determination of LET

To specify dose equivalent, one needs, in addition to the absorbed dose, the LET of incident charged particles or the LET of the charged recoil particles produced by incident neutral radiation (neutrons or gamma rays). As given in Tables 12.1 and 12.2, the required quality factors are defined in terms of the LET in water, which, for radiation-protection purposes, is the same as the stopping power. Stopping-power values of water for a number of charged particles are available and used in many applications.

Radiation fields more often than not occur with a spectrum of LET values. H. H. Rossi and coworkers developed methods for inferring LET spectra directly

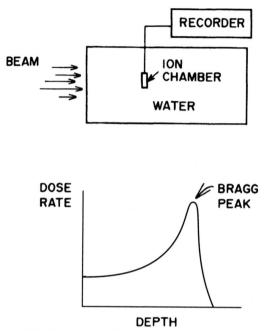

**Fig. 12.8** Measurement of dose or dose rate as a function of depth in water exposed to a beam of charged particles.

from measurements made with a proportional counter.[2] A spherically shaped counter (usually tissue equivalent) is used and a pulse-height spectrum measured in the radiation field. If energy-loss straggling is ignored and the counter gas pressure is low, so that a charged particle from the wall does not lose a large fraction of its energy in traversing the gas, then the pulse size is equal to the product of the LET and the chord length. The distribution of isotropic chord lengths $x$ in a sphere of radius $R$ is given by the simple linear expression

$$P(x)\,dx = \frac{x}{2R^2}\,dx. \tag{12.20}$$

Thus, the probability that a given chord has a length between $x$ and $x+dx$ is $P(x)\,dx$, this function giving unity when integrated from $x = 0$ to $2R$. Using analytic techniques, one can, in principle, unfold the LET spectrum from the measured pulse-height spectrum and the distribution $P(x)$ of track lengths through the gas. However, energy-loss straggling and other factors complicate the practical application of this method.

Precise LET determination presents a difficult technical problem. Usually, practical needs are satisfied by using estimates of the quality factor or radiation weighting factor based on conservative assumptions.

2   Cf. *Microdosimetry*, ICRU Report 36, International Commission on Radiation   Units and Measurements, Bethesda, MD (1983).

## 12.9
## Dose Calculations

Absorbed dose, LET, and dose equivalent can frequently be obtained reliably by calculations. In this section we discuss several examples.

### Alpha and Low-Energy Beta Emitters Distributed in Tissue

When a radionuclide is ingested or inhaled, it can become distributed in various parts of the body. It is then called an internal emitter. Usually a radionuclide entering the body follows certain metabolic pathways and, as a chemical element, preferentially seeks specific body organs. For example, iodine concentrates in the thyroid; radium and strontium are bone seekers. In contrast, tritium (hydrogen) and cesium tend to distribute themselves throughout the whole body. If an internally deposited radionuclide emits particles that have a short range, then their energies will be absorbed in the tissue that contains them. One can then calculate the dose rate in the tissue from the activity concentration there. Such is the case when an alpha or low-energy beta emitter is embedded in tissue. If $A$ denotes the average concentration, in $\mathrm{Bq\,g^{-1}}$, of the radionuclide in the tissue and $\bar{E}$ denotes the average alpha- or beta-particle energy, in MeV per disintegration, then the rate of energy absorption per gram of tissue is $A\bar{E}$ MeV $\mathrm{g^{-1}\,s^{-1}}$. The absorbed dose rate is

$$\dot{D} = A\bar{E}\frac{\mathrm{MeV}}{\mathrm{g\,s}} \times 1.60 \times 10^{-13}\frac{\mathrm{J}}{\mathrm{MeV}} \times 10^{3}\frac{\mathrm{g}}{\mathrm{kg}}$$

$$= 1.60 \times 10^{-10} A\bar{E}\ \mathrm{Gy\,s^{-1}}. \tag{12.21}$$

Note that this procedure gives the average dose rate in the tissue that contains the radionuclide. If the source is not uniformly distributed in the tissue, then the peak dose rate will be higher than that given by Eq. (12.21). The existence of "hot spots" for nonuniformly deposited internal emitters can complicate a meaningful organ-dose evaluation. Nonuniform deposition can occur, for example, when inhaled particulate matter becomes embedded in different regions of the lungs.

*Example*
What is the average dose rate in a 50-g sample of soft tissue that contains $1.20 \times 10^{5}$ Bq of $^{14}$C?

*Solution*
The average energy of $^{14}$C beta particles is $\bar{E} = 0.0495$ MeV (Appendix D). (As a rule of thumb, when not given explicitly, the average beta-particle energy can be assumed to be one-third the maximum energy.) The activity density is $A = (1.20 \times 10^{5}\ \mathrm{s^{-1}})/(50\ \mathrm{g})$. It follows directly from Eq. (12.21) that $\dot{D} = 1.90 \times 10^{-8}\ \mathrm{Gy\,s^{-1}}$.

*Example*

If the tissue sample in the last example has unit density and is spherical in shape and the $^{14}C$ is distributed uniformly, make a rough estimate of the fraction of the beta-particle energy that escapes from the tissue.

*Solution*

We compare the range of a beta particle having the average energy $\bar{E} = 0.0495$ MeV with the radius of the tissue sphere. The sphere radius $r$ is found by writing $50 = 4\pi r^3/3$, which gives $r = 2.29$ cm. From Table 6.1, the range of the beta particle is $R = 0.0042$ cm. Thus a beta particle of average energy emitted no closer than $0.0042$ cm from the surface of the tissue sphere will be absorbed in the sphere. The fraction $F$ of the tissue volume that lies at least this close to the surface can be calculated from the difference in the volumes of spheres with radii $r$ and $r - R$. Alternatively, we can differentiate the expression for the volume, $V = 4\pi r^3/3$:

$$F = \frac{dV}{V} = \frac{3\,dr}{r} = \frac{3 \times 0.0042}{2.29} = 5.50 \times 10^{-3}. \tag{12.22}$$

If we assume that one-half of the average beta-particle energy emitted in this outer layer is absorbed in the sphere and the other half escapes, then the fraction of the emitted beta-particle energy that escapes from the sphere is $F/2 = 2.8 \times 10^{-3}$, a very small amount.

**Charged-Particle Beams**

Figure 12.9 represents a uniform, parallel beam of monoenergetic charged particles of a given kind (e.g., protons) normally incident on a thick tissue slab with fluence rate $\dot{\varphi}$ cm$^{-2}$ s$^{-1}$. To calculate the dose rate at a given depth $x$ in the slab, we consider a thin, disc-shaped volume element with thickness $\Delta x$ in the $x$ direction and area $A$ normal to the beam. The rate of energy deposition in the volume element is $\dot{\varphi}A(-dE/dx)\,\Delta x$, where $-dE/dx$ is the (collisional) stopping power of the beam particles as they traverse the slab at depth $x$. [We ignore energy straggling (Chap. 7).] The dose rate $\dot{D}$ is obtained by dividing by the mass $\rho A\,\Delta x$ of the volume element, where $\rho$ is the density of the tissue:

$$\dot{D} = \frac{\dot{\varphi}A(-dE/dx)\,\Delta x}{\rho A\,\Delta x} = \dot{\varphi}\left(-\frac{dE}{\rho\,dx}\right). \tag{12.23}$$

It follows that the dose per unit fluence at any depth is equal to the mass stopping power for the particles at that depth. If, for example, the mass stopping power is 3 MeV cm$^2$ g$^{-1}$, then the dose per unit fluence can be expressed as 3 MeV g$^{-1}$. This analysis assumes that energy is deposited only by means of electronic collisions (stopping power). As discussed in connection with Fig. 12.8, if significant nuclear interactions occur, for example, as with high-energy protons, then accurate depth–dose curves cannot be calculated from Eq. (12.23). One can then resort to Monte Carlo calculations, in which the fates of individual incident and secondary particles are handled statistically on the basis of the cross sections for the various nuclear interactions that can occur.

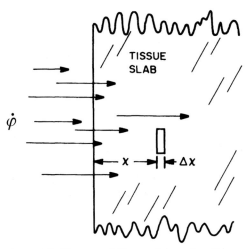

**Fig. 12.9** Uniform, parallel beam of charged particles normally incident on thick tissue slab. Fluence rate $= \dot\varphi$ cm$^{-2}$ s$^{-1}$.

### Point Source of Gamma Rays

We next derive a simple formula for computing the exposure rate in air from a point gamma source of activity $C$ that emits an average photon energy $E$ per disintegration. The rate of energy release in the form of gamma photons escaping from the source is $CE$. Neglecting attenuation in air, we can write for the energy fluence rate, or intensity, through the surface of a sphere of radius $r$ centered about the source $\dot\psi = CE/(4\pi r^2)$. For monoenergetic photons, it follows from Eq. (8.61) that the absorbed dose rate in air at the distance $r$ from the source is

$$\dot D = \dot\psi \frac{\mu_{en}}{\rho} = \frac{CE}{4\pi r^2}\frac{\mu_{en}}{\rho}. \tag{12.24}$$

Here, $\mu_{en}/\rho$ is the mass energy-absorption coefficient of air for the photons. Inspection of Fig. 8.12 shows that this coefficient has roughly the same value for photons with energies between about 60 keV and 2 MeV: $\mu_{en}/\rho \cong 0.027$ cm$^2$ g$^{-1} = 0.0027$ m$^2$ kg$^{-1}$. Therefore, we can apply Eq. (12.24) to any mixture of photons in this energy range, writing

$$\dot D = \frac{CE}{r^2}\frac{0.0027}{4\pi} = \frac{2.15 \times 10^{-4}\,CE}{r^2}. \tag{12.25}$$

With $C$ in Bq (s$^{-1}$), $E$ in J, and $r$ in m, $\dot D$ is in Gy s$^{-1}$. This relationship can be brought into a more convenient form. Expressing the activity $C$ in Ci and the energy $E$ in MeV, we have

$$\dot D = \frac{2.15 \times 10^{-4} \times C \times 3.7 \times 10^{10} \times E \times 1.60 \times 10^{-13}}{r^2}$$

$$= \frac{1.27 \times 10^{-6}\,CE}{r^2}\ \text{Gy s}^{-1}. \tag{12.26}$$

Using hours as the unit of time and changing from dose rate $\dot{D}$ to exposure rate $\dot{X}$ [Eq. (12.4)] gives

$$\dot{X} = \frac{1.27 \times 10^{-6} \, CE \, \text{Gy}}{r^2} \times 3600 \frac{\text{s}}{\text{s}} \times \frac{1 \, \text{R}}{0.0088 \, \text{Gy}}$$

$$= \frac{0.5 CE}{r^2} \, \text{R}\,\text{h}^{-1}. \tag{12.27}$$

This simple formula can be used to estimate the exposure rate from a point source that emits gamma rays.

The specific gamma-ray constant, $\dot{\Gamma}$, for a nuclide is defined by writing

$$\dot{X} = \dot{\Gamma} \frac{C}{r^2}. \tag{12.28}$$

This constant, which numerically gives the exposure rate per unit activity at unit distance, is usually expressed in $\text{R}\,\text{m}^2\,\text{Ci}^{-1}\,\text{h}^{-1}$. Comparison with Eq. (12.27) shows that the specific gamma-ray constant in these units is given approximately by $\dot{\Gamma} = 0.5E$, with $E$ in MeV.

### Example

(a) Estimate the specific gamma-ray constant for $^{137}$Cs. (b) Estimate the exposure rate at a distance of 1.7 m from a 100-mCi point source of $^{137}$Cs.

### Solution

(a) The isotope emits only a 0.662-MeV gamma ray in 85% of its transformations (Appendix D). The average energy per disintegration released as gamma radiation is therefore $0.85 \times 0.662 = 0.563$ MeV. The estimated specific gamma-ray constant for $^{137}$Cs is therefore $\dot{\Gamma} = 0.5E = 0.28 \, \text{R}\,\text{m}^2\,\text{Ci}^{-1}\,\text{h}^{-1}$.

(b) From Eq. (12.28), the exposure rate at a distance $r = 1.7$ m from a point source of activity $C = 100$ mCi $= 0.1$ Ci is

$$\dot{X} = 0.28 \frac{\text{R}\,\text{m}^2}{\text{Ci}\,\text{h}} \times \frac{0.1 \, \text{Ci}}{(1.7 \, \text{m})^2} = 9.7 \times 10^{-3} \, \text{R}\,\text{h}^{-1} = 9.7 \, \text{mR}\,\text{h}^{-1}. \tag{12.29}$$

The accuracy of the approximations leading to Eq. (12.27) varies from nuclide to nuclide. The measured specific gamma-ray constant for $^{137}$Cs, $0.32 \, \text{R}\,\text{m}^2\,\text{Ci}^{-1}\,\text{h}^{-1}$, is somewhat larger than the estimate just obtained. For $^{60}$Co, which emits two gamma photons per disintegration, with energies 1.173 MeV and 1.332 MeV, the estimated specific gamma-ray constant is $0.5(1.173 + 1.332) = 1.3 \, \text{R}\,\text{m}^2\,\text{Ci}^{-1}\,\text{h}^{-1}$, in agreement with the measured value.

In addition to gamma rays, other photons can be emitted from a radionuclide. $^{125}$I, for example, decays by electron capture, giving rise to the emission of characteristic X rays (the major radiation) plus a relatively infrequent, soft (35-keV) gamma photon. Internal bremsstrahlung from a beta particle ($\beta^-$ or $\beta^+$) or captured electron accelerated near the nucleus can also occur, though this contribution is often negligible. The exposure-rate constant, $\dot{\Gamma}_\delta$, of a radionuclide is defined like the specific gamma-ray constant, but includes the exposure rate from *all* photons emitted with energies greater than a specified value $\delta$. In the case of $^{125}$I, the specific gamma-ray constant is 0.0042 R m$^2$ Ci$^{-1}$ h$^{-1}$ and the exposure-rate constant is 0.13 R m$^2$ Ci$^{-1}$ h$^{-1}$ for photons with energies greater than about 10 keV.

**Neutrons**

As discussed in Chapter 9, fast neutrons lose energy primarily by elastic scattering while slow and thermal neutrons have a high probability of being captured. The two principal capture reactions in tissue are $^1$H$(n,\gamma)^2$H and $^{14}$N$(n,p)^{14}$C. Slow neutrons are quickly thermalized by the body. The first capture reaction releases a 2.22-MeV gamma ray, which could deposit a fraction of its energy in escaping the body. In contrast, the nitrogen-capture reaction releases an energy of 0.626 MeV, which is deposited by the proton and recoil carbon nucleus in the immediate vicinity of the capture site. The resulting dose from exposure to thermal neutrons can be calculated, as the next example illustrates.

*Example*
Calculate the dose in a 150-g sample of soft tissue exposed to a fluence of $10^7$ thermal neutrons cm$^{-2}$.

*Solution*
From Table 12.3, the density of nitrogen atoms in soft tissue is $N = 1.29 \times 10^{21}$ cm$^{-3}$, $^{14}$N being over 99.6% abundant. The thermal-neutron capture cross section is $\sigma = 1.70 \times 10^{-24}$ cm$^2$ (Section 9.7). Each capture event by nitrogen results in the deposition of energy $E = 0.626$ MeV, which will be absorbed in the unit-density sample ($\rho = 1$ g cm$^{-3}$). The number of interactions per unit fluence per unit volume of the tissue is $N\sigma$. The dose from the fluence $\varphi = 10^7$ cm$^{-2}$ is therefore

$$D = \frac{\varphi N \sigma E}{\rho}$$

$$= \frac{10^7 \text{ cm}^{-2} \times 1.29 \times 10^{21} \text{ cm}^{-3} \times 1.70 \times 10^{-24} \text{ cm}^2 \times 0.626 \text{ MeV}}{1 \text{ g cm}^{-3}}$$

$$\times \frac{1.6 \times 10^{-13} \text{ J}}{\text{MeV}} \times \frac{1}{10^{-3} \text{ kg g}^{-1}} = 2.20 \times 10^{-6} \text{ Gy}. \tag{12.30}$$

Some additional dose would be deposited by the gamma rays produced by the $^1$H$(n,\gamma)^2$H reaction, for which the cross section is $3.3 \times 10^{-25}$ cm$^2$. However, in a tissue sample as small as 150 g, the contribution of this gamma-ray dose is negligible. It is not negligible in a large target, such as the whole body.

The absorbed dose from fast neutrons is due almost entirely to the energy transferred to the atomic nuclei in tissue by elastic scattering. As discussed in Section 9.6, a fast neutron loses an average of one-half its energy in a single collision with hydrogen. For the other nuclei in soft tissue, the average energy loss is approximately one-half the maximum given by Eq. (9.3). These relationships facilitate the calculation of a "first-collision" dose from fast neutrons in soft tissue. The first-collision dose is that delivered by neutrons that make only a single collision in the target. The first-collision dose closely approximates the actual dose when the mean free path of the neutrons is large compared with the dimensions of the target. A 5-MeV neutron, for example, has a macroscopic cross section in soft tissue of 0.051 cm$^{-1}$, and so its mean free path is $1/0.051 = 20$ cm. Thus, in a target the size of the body, a large fraction of 5-MeV neutrons will not make multiple collisions, and the first-collision dose can be used as a basis for approximating the actual dose. The first-collision dose is, of course, always a lower bound to the actual dose. Moreover, fast neutrons deposit most of their energy in tissue by means of collisions with hydrogen. Therefore, calculating the first-collision dose with tissue hydrogen often provides a simple, lower-bound estimate of fast-neutron dose.

*Example*

Calculate the first-collision dose to tissue hydrogen per unit fluence of 5-MeV neutrons.

*Solution*

The density of H atoms is $N = 5.98 \times 10^{22}$ cm$^{-3}$ (Table 12.3) and the cross section for scattering 5-MeV neutrons is $\sigma = 1.61 \times 10^{-24}$ cm$^2$ (Fig. 9.2). The mean energy loss per collision, $Q_{avg} = 2.5$ MeV, is one-half the incident neutron energy. The dose per unit neutron fluence from collisions with hydrogen is therefore (tissue density $\rho = 1$ g cm$^{-3}$)

$$D = \frac{N\sigma\, Q_{avg}}{\rho} = \frac{5.98 \times 10^{22} \text{ cm}^{-3} \times 1.61 \times 10^{-24} \text{ cm}^2 \times 2.5 \text{ MeV}}{1 \text{ g cm}^{-3}}$$

$$\times \frac{1.6 \times 10^{-13} \text{ J MeV}^{-1}}{10^{-3} \text{ kg g}^{-1}}$$

$$= 3.85 \times 10^{-11} \text{ Gy cm}^2. \tag{12.31}$$

Note that the units of "Gy per (neutron cm$^{-2}$)" are Gy cm$^2$.

Similar calculations of the first-collision doses due to collisions of 5-MeV neutrons with the O, C, and N nuclei in soft tissue give, respectively, contributions of $0.244 \times 10^{-11}$, $0.079 \times 10^{-11}$, and $0.024 \times 10^{-11}$ Gy cm$^2$, representing in total about an additional 10%. Detailed analysis shows that hydrogen recoils contribute approximately 85–95% of the first-collision soft-tissue dose for neutrons with energies between 10 keV and 10 MeV. Table 12.6 shows the analysis of first-collision neutron doses.

Detailed calculations of multiple neutron scattering and energy deposition in slabs and in anthropomorphic phantoms, containing soft tissue, bone, and lungs, have been carried out by Monte Carlo techniques (Section 11.13). Computer programs are available, based on experimental cross-section data and theoretical algorithms, to transport individual neutrons through a target with the same statistical

Table 12.6 Analysis of First-Collision Dose for Neutrons in Soft Tissue

| Neutron Energy (MeV) | First-Collision Dose per Unit Neutron Fluence for Collisions with Various Elements ($10^{-11}$ Gy cm$^2$) | | | | |
|---|---|---|---|---|---|
| | H | O | C | N | Total |
| 0.01 | 0.091 | 0.002 | 0.001 | 0.000 | 0.094 |
| 0.02 | 0.172 | 0.004 | 0.001 | 0.001 | 0.178 |
| 0.03 | 0.244 | 0.005 | 0.002 | 0.001 | 0.252 |
| 0.05 | 0.369 | 0.008 | 0.003 | 0.001 | 0.381 |
| 0.07 | 0.472 | 0.012 | 0.004 | 0.001 | 0.489 |
| 0.10 | 0.603 | 0.017 | 0.006 | 0.002 | 0.628 |
| 0.20 | 0.914 | 0.034 | 0.012 | 0.003 | 0.963 |
| 0.30 | 1.14 | 0.052 | 0.016 | 0.003 | 1.21 |
| 0.50 | 1.47 | 0.122 | 0.023 | 0.004 | 1.62 |
| 0.70 | 1.73 | 0.089 | 0.029 | 0.005 | 1.85 |
| 1.0 | 2.06 | 0.390 | 0.036 | 0.007 | 2.49 |
| 2.0 | 2.78 | 0.156 | 0.047 | 0.012 | 3.00 |
| 3.0 | 3.26 | 0.205 | 0.045 | 0.018 | 3.53 |
| 5.0 | 3.88 | 0.244 | 0.079 | 0.024 | 4.23 |
| 7.0 | 4.22 | 0.485 | 0.094 | 0.032 | 4.83 |
| 10.0 | 4.48 | 0.595 | 0.157 | 0.046 | 5.28 |
| 14.0 | 4.62 | 1.10 | 0.259 | 0.077 | 6.06 |

Source: From "Measurement of Absorbed Dose of Neutrons and Mixtures of Neutrons and Gamma Rays," National Bureau of Standards Handbook 75, Washington, D.C. (1961).

distribution of events that neutrons have in nature. Such Monte Carlo calculations can be made under general conditions of target composition and geometry as well as incident neutron spectra and directions of incidence. Compilations of the results for a large number of neutrons then provide dose and LET distributions as functions of position, as well as any other desired information, to within the statistical fluctuations of the compilations. Using a larger number of neutron histories reduces the variance in the quantities calculated, but increases computer time.

Figure 12.10 shows the results of Monte Carlo calculations carried out for 5-MeV neutrons incident normally on a 30-cm soft-tissue slab, approximating the thickness of the body. (The geometry is identical to that shown for the charged particles in Fig. 12.9.) The curve labeled $E_T$ is the total dose, $E_p$ is the dose due to H recoil nuclei (protons), $E_\gamma$ is the dose from gamma rays from the $^1$H(n,$\gamma$)$^2$H slow-neutron capture reaction, and $E_H$ is the dose from the heavy (O, C, N) recoil nuclei. The total dose builds up somewhat in the first few cm of depth and then decreases as the beam becomes degraded in energy and neutrons are absorbed. The proton and heavy-recoil curves, $E_p$ and $E_H$, show a similar pattern. As the neutrons penetrate, they are moderated and approach thermal energies. This is reflected in the rise of the gamma-dose curve, $E_\gamma$, which has a broad maximum over the region from

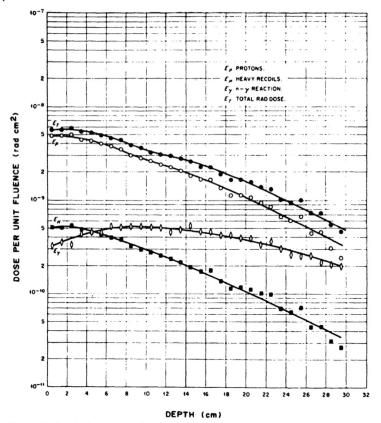

**Fig. 12.10** Depth–dose curves for a broad beam of 5-MeV neutrons incident normally on a soft-tissue slab. Ordinate gives dose per unit fluence at different depths shown by the abscissa. [From "Protection Against Neutron Radiation Up to 30 Million Electron Volts," in *National Bureau of Standards Handbook 63*, p. 44, Washington, D.C. (1957).]

about 6 cm to 14 cm. Note that the total dose decreases by an order of magnitude between the front and back of the slab.

The result of our calculation of the first-collision dose, $D = 3.85 \times 10^{-11}$ Gy cm$^2$ = $3.85 \times 10^{-9}$ rad cm$^2$, due to proton recoils in the last example can be compared with the curve for $E_p$ in Fig. 12.10. At the slab entrance, $E_p = 4.8 \times 10^{-9}$ rad cm$^2$ is greater than $D$, which, as we pointed out, is a lower bound for the actual dose from proton recoils. A number of neutrons are back-scattered from within the slab to add to the first-collision dose deposited directly by the incident 5-MeV neutrons.

## 12.10
## Other Dosimetric Concepts and Quantities

### Kerma

A quantity related to dose for indirectly ionizing radiation (photons and neutrons) is the initial kinetic energy of all charged particles liberated by the radiation per unit mass. This quantity, which has the dimensions of absorbed dose, is called the *kerma* (Kinetic Energy Released per unit MAss). Kerma was discussed briefly for photons in Section 8.9 in connection with the mass energy-transfer coefficient [Eq. (8.62)]. By definition, kerma includes energy that may subsequently appear as bremsstrahlung and it also includes Auger-electron energies. The absorbed dose generally builds up behind a surface irradiated by a beam of neutral particles to a depth comparable with the range of the secondary charged particles generated (cf. Fig. 12.10). The kerma, on the other hand, decreases steadily because of the attenuation of the primary radiation with increasing depth.

The first-collision "dose" calculated for neutrons in the last section is, more precisely stated, the first-collision "kerma." The two are identical as long as all of the initial kinetic energy of the recoil charged particles can be considered as being absorbed locally at the interaction site. Specifically, kerma and absorbed dose at a point in an irradiated target are equal when charged-particle equilibrium exists there and bremsstrahlung losses are negligible.

It is often of interest to consider kerma or kerma rate for a specific material at a point in free space or in another medium. The specific substance itself need not actually be present. Given the photon or neutron fluence and energy spectra at that point, one can calculate the kerma for an imagined small amount of the material placed there. It is thus convenient to describe a given radiation field in terms of the kerma in some relevant, or reference, material. For example, one can specify the air kerma at a point in a water phantom or the tissue kerma in air.

Additional information on kerma can be found in the references listed in Section 12.11.

### Microdosimetry

Absorbed dose is an averaged quantity and, as such, does not specifically reflect the stochastic, or statistical, nature of energy deposition by ionizing radiation in matter. Statistical aspects are especially important when one considers dose in small regions of an irradiated target, such as cell nuclei or other subcellular components. The subject of microdosimetry deals with these phenomena. Consider, for example, cell nuclei having a diameter $\sim$5 $\mu$m. If the whole body receives a uniform dose of 1 mGy of low-LET radiation, then $\frac{2}{3}$ of the nuclei will have no ionizations at all and $\frac{1}{3}$ will receive an average dose of $\sim$3 mGy. If, on the other hand, the whole body receives 1 mGy from fission neutrons, then 99.8% of the nuclei will receive no dose and 0.2% will have a dose of ~500 mGy.[3] The difference arises from the fact that the neutron dose is deposited by recoil nuclei, which have a short range. A proton having an energy of 500 keV has a range of $8 \times 10^{-4}$ cm $= 8$ $\mu$m in soft tissue (Table 5.3), compared with a range of 0.174 cm for a 500-keV electron (Table 6.1). Both particles deposit the same energy. The proton range is comparable to the cell-nucleus diameter; the electron travels the equivalent of ~$1740/5 = 350$ nuclear diameters.

### Specific Energy

When a particle or photon of radiation interacts in a small volume of tissue, one refers to the interaction as an energy-deposition event. The energy deposited by the incident particle and all of the secondary electrons in the volume is called the energy imparted, $\epsilon$. Because of the statistical nature of radiation interaction, the energy imparted is a stochastic quantity. The specific energy (imparted) in a volume of mass $m$ is defined as

$$z = \frac{\epsilon}{m}. \tag{12.32}$$

It has the dimensions of absorbed dose. When the volume is irradiated, it experiences a number of energy-deposition events, which are characterized by the single-event distribution in the values of $z$ that occur. The average absorbed dose in the volume from a number of events is the mean value of $z$. Studies of the distributions in $z$ from different radiations in different-size small volumes of tissue are made in microdosimetry.

Similarly, one can regard an ensemble of identical small volumes throughout an irradiated body and the distribution of specific energy in the volumes due to any number of events. Thus, the example cited from the BEIR-III Report in the next-to-last paragraph can be conveniently described in terms of the distribution of specific energy $z$ in the cell nuclei. For the low-LET radiation, $\frac{2}{3}$ of the nuclei have $z = 0$; in the other $\frac{1}{3}$, $z$ varies widely with a mean value of ~3 mGy. For the fission neutrons, 99.8% of the nuclei have $z = 0$; in the other 0.2%, $z$ varies by many orders of magnitude with a mean value of ~500 mGy.

The stochastic specific energy plays an important role as the microdosimetric analogue of the conventional absorbed dose, which is a nonstochastic quantity.

### Lineal Energy

The lineal energy $y$ is defined as the ratio of the energy imparted $\epsilon$ from a single event in a small volume and the mean length $\bar{x}$ of isotropic chords through the volume:

$$y = \frac{\epsilon}{\bar{x}}. \tag{12.33}$$

---

3   *The Effects on Populations of Exposure to Low Levels of Ionizing Radiation: 1980*, BEIR-III   Report, p. 14, National Academy of Sciences, Washington, D.C. (1980).

Lineal energy, which is a stochastic quantity, has the same dimensions as LET, which is nonstochastic (being the mean value of the linear rate of energy loss). Lineal energy is the microdosimetric analogue of LET. Unlike specific energy, however, it is defined only for single events.

A relationship between lineal energy and LET can be seen as follows. Consider a small volume containing a chord of length $x$ traversed by a charged particle with LET $= L$. We ignore energy-loss straggling and assume that the energy lost by the particle in the volume is absorbed there. We assume further that the chord is so short that the LET is constant over its length. The energy imparted by the single traversal event is $\epsilon = Lx$. For isotropic irradiation of the volume by particles traversing it with LET $= L$, the mean value of the imparted energy is $\bar{\epsilon} = L\bar{x}$. Under these conditions, it follows from the definition (12.33) of the lineal energy that its mean value is the LET: $\bar{y} = \bar{\epsilon}/\bar{x} = L$.

For any convex body, having surface area $S$ and volume $V$, traversed by isotropic chords, the mean chord length is given quite generally by the Cauchy relation, $\bar{x} = 4V/S$. For a sphere of radius $R$, it follows that $\bar{x} = 4R/3$ (Problem 60).

Proposals have been made to use lineal energy instead of LET as a basis for defining quality factors in radiation-protection work. Whereas the measurement of LET spectra is a difficult technical problem, distributions of lineal energy and its frequency- and dose-mean values can be readily measured for many radiation fields. Disadvantages of using lineal energy include the necessity of specifying a universal size for the reference volume, usually assumed to be spherical in shape. There does not appear to be a compelling reason for any particular choice, and the $y$ distributions depend upon this specification. In addition, concepts associated with chords are probably inappropriate for application to the tortuous paths of electrons, especially at low energies.

## 12.11
### Suggested Reading

1 Attix, F. H., *Introduction to Radiological Physics and Radiation Dosimetry*, Wiley, New York (1986). [Clear and rigorous treatments (and in much greater depth) of subjects in this chapter. Dosimetry fundamentals and, especially, instrumentation and measurements are described.]

2 Cember, H., *Introduction to Health Physics*, 3rd Ed., McGraw-Hill, New York (1996).

3 ICRU Report 36, *Microdosimetry*, International Commission on Radiation Units and Measurements, Bethesda, MD (1983).

4 ICRU Report 60, *Fundamental Quantities and Units for Ionizing Radiation*, International Commission on Radiation Units and Measurements, Bethesda, MD (1998). [Provides definitions of fundamental quantities related to radiometry, interaction coefficients, dosimetry, and radioactivity. Gives standardized symbols and units.]

5 Martin, J. E., *Physics for Radiation Protection: A Handbook*, 2nd Ed., Wiley, New York (2006).

6 Rossi, H. H., and M. Zaider, *Microdosimetry and its Applications*, Springer Verlag, Berlin (1996).

**7** Turner, J. E., "An Introduction to Microdosimetry," *Rad. Prot. Management* 9(3), 25–58 (1992).

Section 12.9 showed a sampling of approximate dose calculations that can be performed by hand. A number of sophisticated computational systems can be found on the World Wide Web by simply searching "radiation dose calculations". Valuable links to diverse areas in the profession of health physics are available. Computer codes can be obtained and run on personal computers (e.g., VARSKIN for assessing beta and gamma skin dose from skin and clothing contamination). A useful site is www.doseinfo-radar.com.

## 12.12
## Problems

1.  (a)  What is the average absorbed dose in a 40-cm³ region of a body organ (density $= 0.93$ g cm$^{-3}$) that absorbs $3 \times 10^5$ MeV of energy from a radiation field?
    (b)  If the energy is deposited by ionizing particles with an LET of 10 keV $\mu$m$^{-1}$ in water, what is the dose equivalent according to Table 12.1?
    (c)  Express the answers to (a) and (b) in both rads and rems as well as Gy and Sv.

2.  A portion of the body receives 0.15 mGy from radiation with a quality factor $Q = 6$ and 0.22 mGy from radiation with $Q = 10$.
    (a)  What is the total dose?
    (b)  What is the total dose equivalent?

3.  A beam of X rays produces 4 esu of charge per second in 0.08 g of air. What is the exposure rate in (a) mR s$^{-1}$ and (b) SI units?

4.  If all of the ion pairs are collected in the last example, what is the current?

5.  A free-air ionization chamber operating under saturation conditions has a sensitive volume of 12 cm³. Exposed to a beam of X rays, it gives a reading of $5 \times 10^{-6}$ mA. The temperature is 18°C and the pressure is 756 torr. What is the exposure rate?

6.  A free-air ionization chamber with a sensitive volume of 14 cm³ is exposed to gamma rays. A reading of $10^{-11}$ A is obtained under saturation conditions when the temperature is 20°C and the pressure is 762 torr.
    (a)  Calculate the exposure rate.
    (b)  How would the exposure rate be affected if the temperature dropped 2°C?
    (c)  Would the current change?

7.  A current of $10^{-14}$ A is produced in a free-air chamber operating under saturation conditions. The sensitive volume, in which electronic equilibrium exists, contains 0.022 g of air. What is the exposure rate?

8. A free-air ionization chamber is to be constructed so that a saturation current of $10^{-14}$ A is produced when the exposure rate is 10 mR h$^{-1}$. The temperature is 20°C, and the pressure is 750 torr.
   (a) Calculate the size of the sensitive volume needed.
   (b) What would be the current in this chamber if it were exposed to 20 mR h$^{-1}$ at a temperature of 30°C and a pressure of 750 torr?

9. A free-air ionization chamber has a sensitive volume of 103 cm$^3$ and operates at a temperature of 20°C and a pressure of 750 torr. Placed near a gamma-ray source, it registers a current of $6.60 \times 10^{-11}$ A.
   (a) Calculate the exposure rate in R h$^{-1}$.
   (b) If the temperature changes to 15°C and the pressure to 760 torr, what will be the current? (Source and chamber stay in fixed positions.)
   (c) What will be the exposure rate in (b)?

10. An air-wall pocket chamber contains 7.7 mg of air. Its capacitance is 9.4 pF. With the chamber charged, how much exposure will cause a decrease of 10 V in the potential difference?

11. A pocket air-wall ionization chamber with a capacitance of $10^{-11}$ F contains 2.7 cm$^3$ of air. If it can be charged to 240 V, what is the maximum exposure that it can measure? Assume STP.

12. A pocket dosimeter with a volume of 2.2 cm$^3$ and capacitance of 8 pF was fully charged at a potential difference of 200 V. After being worn during a gamma-ray exposure, the potential difference was found to be 192 V. Assume STP conditions.
   (a) Calculate the exposure.
   (b) What charging voltage would have to be applied to the dosimeter if it is to read exposures up to 3 R?

13. An unsealed air-wall pocket chamber has a volume of 5.7 cm$^3$ and a capacitance of 8.6 pF. The temperature is 25°C and the pressure is 765 torr.
   (a) How much charging voltage is needed for the chamber, if it is to measure exposures up to a maximum of 1.0 R?
   (b) If the same charging voltage is used on another day, when the temperature is 18°C and the pressure is 765 torr, what will be the maximum measurable exposure?

14. A pocket air-wall ionization chamber has a volume of 6.2 cm$^3$ and a capacitance of 8.0 pF. Assume STP conditions.
   (a) If the instrument is to register a range of exposures up to a maximum of 1.0 R, what must the charging voltage be?

(b) If everything else remained the same, what would the range of the instrument be if the capacitance were doubled?

(c) With everything else the same as in (a), what would the range be if the volume were doubled?

15. State the Bragg–Gray principle and the conditions for its validity.

16. What minimum wall thickness of carbon is needed to satisfy the Bragg–Gray principle if the chamber pictured in Fig. 12.4 is to be used to measure absorbed dose from photons with energies up to 5 MeV? (Assume same mass stopping powers for carbon and water and use numerical data given in Chapter 6.)

17. An ionization chamber that satisfies the Bragg–Gray principle contains 0.12 g of $CO_2$ gas ($W = 33$ eV ip$^{-1}$). When exposed to a beam of gamma rays, a saturation current of $4.4 \times 10^{-10}$ A is observed. What is the absorbed-dose rate in the gas?

18. An ionization chamber like that shown in Fig. 12.4 was exposed to gamma rays, and $2.8 \times 10^{12}$ ion pairs were produced per gram of $CO_2$ ($W = 33$ eV ip$^{-1}$). What was the absorbed dose in the carbon walls? The average mass stopping powers of graphite and $CO_2$ for the photons are, respectively, 1.648 and 1.680 MeV cm$^2$ g$^{-1}$.

19. An unsealed air ionization chamber ($W = 34$ eV ip$^{-1}$) is to be designed so that the saturation current is $10^{-11}$ A when the dose rate is 10 mGy h$^{-1}$. The chamber is to be operated at a temperature of 20°C and pressure of 750 torr. What chamber volume is required?

20. An ionization chamber is bombarded simultaneously by a beam of $8 \times 10^6$ helium ions per second and a beam of $1 \times 10^8$ carbon ions per second. The helium ions have an initial energy of 5 MeV and the carbon ions, 100 keV. All ions stop in the chamber gas. The $W$ values for the helium and carbon ions are, respectively, 36 eV ip$^{-1}$ and 48 eV ip$^{-1}$.

(a) Calculate the saturation current.

(b) If the chamber contains 0.876 g of gas, what is the dose rate?

21. What disadvantages are there in using the C–$CO_2$ chamber to monitor neutrons?

22. Why should a neutron dosimeter contain a substantial amount of hydrogen?

23. In a laboratory test, a tissue-equivalent (TE) chamber and a C–$CO_2$ chamber both show correct readings of 10 mGy in response to a gamma-ray exposure that produces 10 mGy in tissue. When exposed to 0.5-MeV neutrons, the TE chamber reads 10 mGy and the C–$CO_2$ chamber shows 1.49 mGy.

When the two chambers are used together to monitor an unknown mixed gamma–neutron field, the TE chamber reads $0.27$ mGy h$^{-1}$ and the C–CO$_2$ chamber reads $0.21$ mGy h$^{-1}$. What are the individual gamma and neutron dose rates? (Assume that the neutrons have an energy of 0.5 MeV.)

24. A C–CO$_2$ chamber like that in Fig. 12.4 is calibrated to read directly in mGy h$^{-1}$ for 1-MeV photons. The chamber satisfies the Bragg–Gray principle for neutrons as well as gamma rays.

    (a) The chamber is exposed to 1-MeV neutrons and gives a reading of $22$ mGy h$^{-1}$. What is the neutron dose rate for soft tissue?

    (b) A 1-MeV gamma source is added, and the reading is increased to $84$ mGy h$^{-1}$. What is the gamma dose rate to tissue?

    (c) What would a tissue-equivalent chamber read in part (b) when exposed to both the neutrons and gamma rays together?

25. In a proportional counter such as the one shown in Fig. 12.6, why, on the average, would pulses produced by gamma photons be much smaller than pulses produced by fast neutrons?

26. If $W = 30$ eV ip$^{-1}$ for a 2-MeV proton and $W = 40$ eV ip$^{-1}$ for a 1-MeV carbon recoil nucleus in a proportional-counter gas, what is the ratio of the pulse heights produced by these two particles if they stop completely in the gas?

27. A beam of fast neutrons is directed toward a proportional counter that operates with methane (CH$_4$) gas. If the maximum pulse height registered is 4.2 MeV, what is the maximum energy of the neutrons?

28. An air ion chamber having a volume of 2.5 cm$^3$ (STP) is placed at a certain depth in a water tank, as shown in Fig. 12.8. An electron beam incident on the tank produces a current of $0.004$ $\mu$A in the chamber. What is the dose rate at that depth?

29. What is the average whole-body dose rate in a 22-g mouse that contains $1.85 \times 10^5$ Bq of $^{14}$C distributed in its body?

30. A patient receives an injection of $1.11 \times 10^8$ Bq of $^{131}$I, 30% of which goes to the thyroid, having a mass of 20 g. What is the average dose rate in the organ?

31. Tritium often gets into body water following an exposure and quickly becomes distributed uniformly throughout the body. What uniform concentration of $^3$H, in Bq g$^{-1}$, would give a dose-equivalent rate of 1 mSv wk$^{-1}$?

32. A 36-g mouse is to be injected with $^{32}$P (half-life $= 14.3$ d; average beta energy $= 0.70$ MeV; no gamma [Appendix D]). Assume that the $^{32}$P distributes itself almost instantaneously throughout the body following injection and that none is lost

from the body for the first few hours thereafter. What activity of $^{32}P$ needs to be administered in order to give the mouse a dose of 10 mGy in the first hour?

33. A 75-$\mu$A parallel beam of 4-MeV electrons passes normally through the flat surface of a sample of soft tissue in the shape of a disc. The diameter of the disc is 2 cm and its thickness is 0.5 cm. Calculate the average absorbed dose rate in the disc.

34. A soft-tissue disc with a radius of 0.5 cm and thickness of 1 mm is irradiated normally on its flat surface by a 6-$\mu$A beam of 100-MeV protons. Calculate the average dose rate in the sample.

35. An experiment is planned in which bean roots are to be placed in a tank of water at a depth of 2.2 cm and irradiated by a parallel beam of 10-MeV electrons incident on the surface of the water. What fluence rate would be needed to expose the roots at a dose rate of 10 Gy min$^{-1}$?

36. A worker inadvertently puts his hand at right angles into a uniform, parallel beam of 50-MeV protons with a fluence rate of $4.6 \times 10^{10}$ protons cm$^{-2}$ s$^{-1}$. His hand was momentarily exposed for an estimated 0.5 s.

    (a) Estimate the dose that the worker received to the skin of his hand.

    (b) If the beam covered an area of 2.7 cm$^2$, what was the beam current?

37. (a) With $C$ expressed in Ci and $E$ in MeV, show that Eq. (12.26) implies that $\dot{X} = 6CE$ R h$^{-1}$ at 1 ft, approximately.

    (b) For the specific gamma-ray constant, show that
    1 R cm$^2$ mCi$^{-1}$ h$^{-1}$ = 0.1 R m$^2$ Ci$^{-1}$ h$^{-1}$.

38. When $^{38}S$ decays, a single 1.88-MeV gamma photon is emitted in 95% of the transformations. Estimate the exposure rate at a distance of 3 m from a point source of $^{38}S$ having an activity of $2.7 \times 10^{12}$ Bq.

39. What is the exposure rate at a distance of 1 ft from a 20-mCi, unshielded point source of $^{60}Co$?

40. What is the activity of an unshielded point source of $^{60}Co$ if the exposure rate at 20 m is 6 R min$^{-1}$?

41. A worker accidently strayed into a room in which a small, bare vial containing 23 Ci of $^{131}I$ was being used to expose a sample. He remained in the room approximately 10 min, standing at a lab bench 5 m away from the source. Estimate the dose that the worker received.

42. A point source consists of a mixture of 4.2 Ci of $^{42}K$ and 1.8 Ci of $^{24}Na$. Estimate the exposure rate at a distance of 40 cm.

43. A parallel beam of monoenergetic photons emerged from a source when the shielding was removed for a short time. The photon energy $h\nu$ and the total fluence $\varphi$ of photons are known.
    (a) Write a formula from which one can calculate the absorbed dose in air in rad from $h\nu$, expressed in MeV, and $\varphi$, expressed in $cm^{-2}$.
    (b) Write a formula for calculating the exposure in R.

44. The thermal-neutron capture cross section for the $^{14}N(n,p)^{14}C$ reaction is 1.70 barns. Calculate
    (a) the $Q$ value for the reaction
    (b) the resulting dose in soft tissue per unit fluence of thermal neutrons.

45. A 100-$cm^3$ sample of water is exposed to 1500 thermal neutrons $cm^{-2}$ $s^{-1}$. How many photons are emitted per second as a result of neutron capture by hydrogen? The cross section for the $^1H(n,\gamma)^2H$ reaction is $3.3 \times 10^{-25}$ $cm^2$.

46. A uniform target with a volume of 5 L is exposed to 100 thermal neutrons $cm^{-2}$ $s^{-1}$. It contains an unknown number of hydrogen atoms. While exposed to the thermal neutrons, it emits $1.11 \times 10^4$ photons $s^{-1}$ as the result of thermal-neutron capture by hydrogen (cross section = 0.33 barn). No other radiation is emitted. What is the density of H atoms in the target? Neglect attenuation of the neutrons and photons as they penetrate the target.

47. (a) Calculate the average recoil energies of a hydrogen nucleus and a carbon nucleus elastically scattered by 4-MeV neutrons.
    (b) What can one say about the relative contributions that these two processes make to absorbed dose and dose equivalent in soft tissue?

48. Using Table 12.6, plot the percentage of the first-collision tissue dose that is due to elastic scattering from hydrogen for neutrons with energies between 0.01 MeV and 14.0 MeV.

49. Calculate the first-collision dose per unit fluence for 14-MeV neutrons based on their interactions with tissue hydrogen alone. Compare the result with Table 12.6.

50. From Table 12.6, the total first-collision dose per unit fluence for 14-MeV neutrons in soft tissue is $6.06 \times 10^{-11}$ Gy $cm^2$. From Table 12.5, the average quality factor for 14-MeV neutrons is 7.5. Use these two values to estimate the constant fluence rate of 14-MeV neutrons that gives a first-collision dose equivalent of 1 mSv in 40 h. How do you account for the lower value, 12 neutrons $cm^{-2}$ $s^{-1}$, given in the last column of Table 12.5?

51. Calculate the first-collision dose per unit fluence that results from the scattering of 10-MeV neutrons by the carbon in soft tissue. Compare your answer with the value given in Table 12.6.

52. In Fig. 12.10, why does the ratio $E_p/E_\gamma$ decrease with increasing depth?

53. A uniform, broad beam of 3.7-MeV neutrons is incident on a 55-g sample of water. The fluence rate is $6.2 \times 10^6$ cm$^{-2}$ s$^{-1}$. Calculate the rate of energy transfer to the sample by collisions with hydrogen only. The cross section is $2.0 \times 10^{-24}$ cm$^2$.

54. A 22-g mouse is irradiated simultaneously by a beam of thermal neutrons, having a fluence rate of $4.2 \times 10^7$ cm$^{-2}$ s$^{-1}$, and a beam of 5-MeV neutrons, having a fluence rate of $9.6 \times 10^6$ cm$^{-2}$ s$^{-1}$.

    (a) Calculate the dose rate to the mouse from the thermal neutrons.

    (b) Calculate the dose rate from the 5-MeV neutrons, interacting with hydrogen only.

    (c) Estimate the total dose rate to the mouse from all interactions, approximating the cross sections of the heavy elements by that of carbon (Fig. 9.2).

55. Figure 12.11 shows two examples of a single collision of a 5-MeV neutron ($E_n = 5$ MeV) with a proton in a 1-g target. In both instances the neutron loses 2 MeV ($E'_n = 3$ MeV) to the

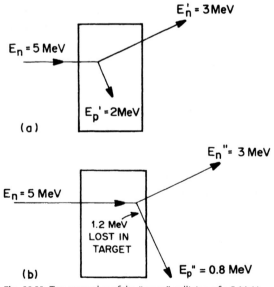

**Fig. 12.11** Two examples of the "same" collision of a 5-MeV neutron with a proton at different locations in a 1-g target. (Problem 55.)

proton ($E'_p = 2$ MeV) and escapes from the target. In Fig.
12.11(a), the recoil proton stops completely in the target. In Fig.
12.11(b), the collision occurs near the back surface and the
proton loses only 1.2 MeV in the target before escaping with an
energy $E''_p = 0.8$ MeV.

(a) What is the average kerma in the target in both instances?
(b) What is the absorbed dose in both cases?

56. What is the tissue kerma rate from the thermal neutrons in
Problem 54?

57. The calculations presented in Section 12.9 for fast neutrons are,
precisely stated, calculations of first-collision *kerma*, rather than
first-collision *dose*, as described there. Explain this distinction.
Give an example in which the neutron first-collision tissue
kerma and first-collision tissue dose would be different.

58. Calculate the specific energies for the examples in Fig. 12.11(a)
and (b).

59. Prove Eq. (12.20). (For a sphere, the distribution for isotropic
chord lengths is the same as that for parallel chords,
distributed uniformly in any direction.)

60. By using (a) the Cauchy relation and (b) Eq. (12.20), show that
the mean isotropic-chord length in a sphere of radius $R$ is $4R/3$.

61. The sensitive volume of the Hurst proportional counter
(Fig. 12.6) is a right circular cylinder, having equal height and
diameter. Show that the mean isotropic-chord length in this
cylinder is the same as that in a sphere, having the same
diameter as the cylinder.

62. Calculate the lineal energies for the examples in Fig. 12.11(a)
and (b), assuming the targets to be cubes of unit density.

63. In microdosimetry, the average number of energy-deposition
events that occur per unit dose in a specified volume in
irradiated tissue is called the *event frequency*. If the event
frequency for cell nuclei in an experiment is 0.37 $Gy^{-1}$, what is
the probability that, for a dose of 0.30 Gy, the number of
energy-loss events in a cell nucleus is (a) 0, (b) 1, (c) 2 or more?

64. Mice are irradiated with 600-keV neutrons in an experiment. It
is hypothesized that the sensitive targets for cancer induction
are the cell nuclei, which are spherical in shape with a diameter
of 5 $\mu$m. For analysis, consider only the first-collision dose
from elastic scattering with tissue hydrogen ($6.0 \times 10^{22}$
atoms $cm^{-3}$; cross section, 5.9 barns).

(a) Show that the average dose delivered per neutron collision
per $cm^3$ is $4.8 \times 10^{-11}$ Gy.
(b) What fluence of 600-keV neutrons is needed to deliver
0.10 Gy?

(c) What is the average number of neutron interactions per cell nucleus when the dose is 0.10 Gy?

(d) What is the probability that a given cell nucleus has exactly one neutron interaction when the dose is 0.10 Gy?

(e) What is the probability that a nucleus has at least one interaction at 0.10 Gy?

(f) At what dose is the average number of interactions per cell nucleus equal to unity?

## 12.13
## Answers

1. (a) $1.29\ \mu$Gy
   (b) $3.30\ \mu$Sv
   (c) 0.129 mrad,
       0.330 mrem
3. (a) $64.7\ \mathrm{mR\,s^{-1}}$
   (b) $1.67 \times$
       $10^{-5}\ \mathrm{C\,kg^{-1}\,s^{-1}}$
5. $1.34\ \mathrm{R\,s^{-1}}$
6. (a) $2.29\ \mathrm{mR\,s^{-1}}$
   (b) No change
   (c) Yes
9. (a) $7.52\ \mathrm{R\,h^{-1}}$
   (b) $6.80 \times 10^{-11}$ A
   (c) Same as (a)
10. 47.3 mR
12. (a) 87.2 mR
    (b) 275 V
16. $\sim$0.8 cm
17. $0.121\ \mathrm{mGy\,s^{-1}}$
18. 14.5 mGy
20. (a) $2.11 \times 10^{-7}$ A
    (b) $9.13\ \mathrm{mGy\,s^{-1}}$
23. $0.20\ \mathrm{mGy\,h^{-1}}$ ($\gamma$)
    $0.07\ \mathrm{mGy\,h^{-1}}$ (n)
27. 4.2 MeV
28. $42.1\ \mathrm{mGy\,s^{-1}}$
30. $49\ \mu\mathrm{Gy\,s^{-1}}$
32. $8.93 \times 10^5$ Bq

33. $4.56 \times 10^4\ \mathrm{Gy\,s^{-1}}$
35. $5.42 \times 10^8\ \mathrm{cm^{-2}\,s^{-1}}$
36. (a) 45.6 Gy; (b) 20 nA
40. $4.3 \times 10^{15}$ Bq
41. $\sim$0.30 mGy
42. $27\ \mathrm{R\,h^{-1}}$
45. $3.31 \times 10^3\ \mathrm{s^{-1}}$
50. $15.3\ \mathrm{cm^{-2}\,s^{-1}}$
51. $2.46 \times 10^{-12}\ \mathrm{Gy\,cm^2}$
53. $8.43 \times 10^7\ \mathrm{MeV\,s^{-1}}$
55. (a) $3.20 \times 10^{-10}$ Gy
        in both
    (b) $3.20 \times 10^{-10}$ Gy
        in Fig. 12.11(a);
        $1.92 \times 10^{-10}$ Gy
        in Fig. 12.11(b)
56. $9.23 \times 10^{-6}\ \mathrm{Gy\,s^{-1}}$
58. $3.20 \times 10^{-10}$ Gy in (a);
    $1.92 \times 10^{-10}$ Gy in (b)
63. (a) 0.895
    (b) 0.0993
    (c) 0.0057
64. (a) $4.80 \times 10^{-11}$ Gy
    (b) $5.89 \times 10^9\ \mathrm{cm^{-2}}$
    (c) 0.137
    (d) 0.119
    (e) 0.128
    (f) 0.733 Gy

# 13
# Chemical and Biological Effects of Radiation

## 13.1
## Time Frame for Radiation Effects

To be specific, we describe the chemical changes produced by ionizing radiation in liquid water, which are relevant to understanding biological effects. Mammalian cells are typically $\sim$70–85% water, $\sim$10–20% proteins, $\sim$10% carbohydrates, and $\sim$2–3% lipids.

Ionizing radiation produces abundant secondary electrons in matter. As discussed in Section 5.3, most secondary electrons are produced in water with energies in the range $\sim$10–70 eV. The secondaries slow down very quickly ($\lesssim10^{-15}$ s) to subexcitation energies; that is, energies below the threshold required to produce electronic transitions ($\sim$7.4 eV for liquid water). Various temporal stages of radiation action can be identified, as we now discuss. The time scale for some important radiation effects, summarized in Table 13.1, covers over 20 orders of magnitude.

## 13.2
## Physical and Prechemical Chances in Irradiated Water

The initial changes produced by radiation in water are the creation of ionized and excited molecules, $H_2O^+$ and $H_2O^*$, and free, subexcitation electrons. These species are produced in $\lesssim10^{-15}$ s in local regions of a track. Although an energetic charged particle may take longer to stop (Sections 5.11 and 6.6), we shall see that portions of the same track that are separated by more than $\sim$0.1 $\mu$m develop independently. Thus we say that the initial physical processes are over in $\lesssim10^{-15}$ s in local track regions.

The water begins to adjust to the sudden physical appearance of the three species even before the molecules can more appreciably in their normal thermal agitation. At room temperature, a water molecule can move an average distance of $\sim$1–2 Å, roughly equal to its diameter (2.9 Å), in $\sim10^{-12}$ s. Thus, $10^{-12}$ s after passage of a charged particle marks the beginning of the ordinary, diffusion-controlled chemical reactions that take place within and around the particle's path. During this prechemical stage, from $\sim10^{-15}$ s to $\sim10^{-12}$ s, the three initial species produced

*Atoms, Radiation, and Radiation Protection.* James E. Turner
Copyright © 2007 WILEY-VCH Verlag GmbH & Co. KGaA, Weinheim
ISBN: 978-3-527-40606-7

**Table 13.1** Time Frame for Effects of Ionizing Radiation

| Times | Events |
|---|---|
| Physical stage<br>$\lesssim 10^{-15}$ s | Formation of $H_2O^+$, $H_2O^*$, and subexcitation electrons, $e^-$, in local track regions ($\lesssim 0.1 \ \mu m$) |
| Prechemical stage<br>$\sim 10^{-15}$ s to $\sim 10^{-12}$ s | Three initial species replaced by $H_3O^+$, OH, $e_{aq}^-$, H, and $H_2$ |
| Chemical stage<br>$\sim 10^{-12}$ s to $\sim 10^{-6}$ s | The four species $H_3O^+$, OH, $e_{aq}^-$, and H diffuse and either react with one another or become widely separated. Intratrack reactions essentially complete by $\sim 10^{-6}$ s |
| Biological stages<br>$\lesssim 10^{-3}$ s<br>$\lesssim 1$ s<br>Minutes<br>Days<br>Weeks<br>Years | <br>Radical reactions with biological molecules complete<br>Biochemical changes<br>Cell division affected<br>Gastrointestinal and central nervous system changes<br>Lung fibrosis develops<br>Cataracts and cancer may appear; genetic effects in offspring |

by the radiation induce changes as follows. First, in about $10^{-14}$ s, an ionized water molecule reacts with a neighboring molecule, forming a hydronium ion and a hydroxyl radical:

$$H_2O^+ + H_2O \rightarrow H_3O^+ + OH. \tag{13.1}$$

Second, an excited water molecule gets rid of its energy either by losing an electron, thus becoming an ion and proceeding according to the reaction (13.1), or by molecular dissociation:

$$H_2O^* \rightarrow \begin{cases} H_2O^+ + e^- \\ H + OH \end{cases}. \tag{13.2}$$

The vibrational periods of the water molecule are $\sim 10^{-14}$ s, which is the time that characterizes the dissociation process. Third, the subexcitation electrons migrate, losing energy by vibrational and rotational excitation of water molecules, and become thermalized by times $\sim 10^{-12}$ s. Moreover, the thermalized electrons orient the permanent dipole moments of neighboring water molecules, forming a cluster, called a hydrated electron. We denote the thermalization–hydration process symbolically by writing

$$e^- \rightarrow e_{aq}^-, \tag{13.3}$$

where the subscript aq refers to the fact that the electron is hydrated (aqueous solution). These changes are summarized for the prechemical stage in Table 13.1. Of the five species formed, $H_2$ does not react further.

## 13.3
## Chemical Stage

At $\sim 10^{-12}$ s after passage of a charged particle in water, the four chemically active species $H_2O^+$, OH, $e_{aq}^-$, and H are located near the positions of the original $H_2O^+$, $H_2O^*$, and $e^-$ that triggered their formation. Three of the new reactants, OH, $e_{aq}^-$, and H, are free radicals, that is, chemical species with unpaired electrons. The reactants begin to migrate randomly about their initial positions in thermal motion. As their diffusion in the water proceeds, individual pairs can come close enough to react chemically. The principal reactions that occur in the track of a charged particle in water during this stage are the following:

$$OH + OH \rightarrow H_2O_2, \tag{13.4}$$

$$OH + e_{aq}^- \rightarrow OH^-, \tag{13.5}$$

$$OH + H \rightarrow H_2O, \tag{13.6}$$

$$H_3O^+ + e_{aq}^- \rightarrow H + H_2O, \tag{13.7}$$

$$e_{aq}^- + e_{aq}^- + 2H_2O \rightarrow H_2 + 2OH^-, \tag{13.8}$$

$$e_{aq}^- + H + H_2O \rightarrow H_2 + OH^-, \tag{13.9}$$

$$H + H \rightarrow H_2. \tag{13.10}$$

With the exception of (13.7), all of these reactions remove chemically active species, since none of the products on the right-hand sides except H will consume additional reactants. As time passes, the reactions (13.4)–(13.10) proceed until the remaining reactants diffuse so far away from one another that the probability for additional reactions is small. This occurs by $\sim 10^{-6}$ s, and the chemical development of the track in pure water then is essentially over.

The motion of the reactants during this diffusion-controlled chemical stage can be viewed as a random walk, in which a reactant makes a sequence of small steps in random directions beginning at its initial position. If the measured diffusion constant for a species is $D$, then, on the average, it will move a small distance $\lambda$ in a time $\tau$ such that

$$\frac{\lambda^2}{6\tau} = D. \tag{13.11}$$

Each type of reactive species can be regarded as having a reaction radius $R$. Two species that approach each other closer than the sum of the their reactive radii have a chance to interact according to Eqs. (13.4)–(13.10). Diffusion constants and reaction radii for the four reactants in irradiated water are shown in Table 13.2.

*Example*
Estimate how far a hydroxyl radical will diffuse in $10^{-12}$ s.

**Table 13.2** Diffusion Constants $D$ and Reaction Radii $R$ for Reactive Species

| Species | $D\ (10^{-5}\ cm^2\ s^{-1})$ | $R$ (Å) |
|---------|---------|---------|
| OH | 2 | 2.4 |
| $e_{aq}^-$ | 5 | 2.1 |
| $H_3O^+$ | 8 | 0.30 |
| H | 8 | 0.42 |

*Solution*

From Eq. (13.11) with $\tau = 10^{-12}$ s and from Table 13.2, we find

$$\lambda = (6\tau D)^{1/2} = (6 \times 10^{-12}\ s \times 2 \times 10^{-5}\ cm^2\ s^{-1})^{1/2}$$

$$= 1.10 \times 10^{-8}\ cm = 1.10\ Å. \tag{13.12}$$

For comparison, the diameter of the water molecule is 2.9 Å. The answer (13.12) is compatible with our taking the time $\sim 10^{-12}$ s as marking the beginning of the chemical stage of charged-particle track development.

### 13.4
### Examples of Calculated Charged-Particle Tracks in Water

Before discussing the biological effects of radiation we present some examples of detailed calculations of charged-particle tracks in water. The calculations have been made from the beginning of the physical stage through the end of the chemical stage.

Monte Carlo computer codes have been developed for calculating the passage of a charged particle and its secondaries in liquid water. In such computations, an individual particle is allowed to lose energy and generate secondary electrons on a statistical basis, as it does in nature. Where available, experimental values of the energy-loss cross sections are used in the computations. The secondary electrons are similarly transported and are allowed to produce other secondary electrons until the energies of all secondaries reach subexcitation levels (<7.4 eV). Such calculations give in complete detail the position and identity of every reactant $H_2O^+$, $H_2O^*$, and subexcitation electron present along the track. These species are allowed to develop according to (13.1), (13.2), and (13.3) to obtain the positions and identities of every one of the reactive species OH, $H_3O^+$, $e_{aq}^-$, and H at $10^{-12}$ s. The computations then carry out a random-walk simulation of diffusion by letting each reactant take a small jump in a random direction and then checking all pairs to see which are closer than the sum of their reaction radii. Those that can react do so and are removed from further consideration [except when H is produced by (13.7)]. The remainder are jumped again from their new positions and the procedure is repeated to develop the track to later times. The data in Table 13.2 and the reaction

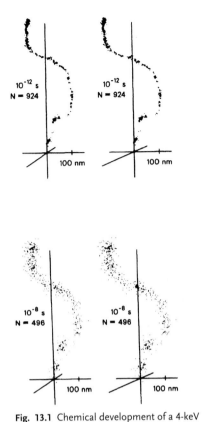
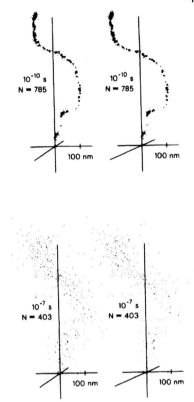
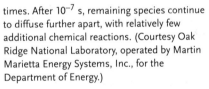

**Fig. 13.1** Chemical development of a 4-keV electron track in liquid water, calculated by Monte Carlo simulation. Each dot in these stereo views gives the location of one of the active radiolytic species, OH, $H_3O^+$, $e_{aq}^-$, or H, at the times shown. Note structure of track with spurs, or clusters of species, at early times. After $10^{-7}$ s, remaining species continue to diffuse further apart, with relatively few additional chemical reactions. (Courtesy Oak Ridge National Laboratory, operated by Martin Marietta Energy Systems, Inc., for the Department of Energy.)

schemes (13.4)–(13.10) can thus be used to carry out the chemical development of a track.

Three examples of calculated electron tracks at $10^{-12}$ s in liquid water were shown in Fig. 6.5. The upper left-hand panel in Fig. 13.1 presents a stereoscopic view of another such track, for a 4-keV electron, starting at the origin in the upward direction. Each dot represents the location of one of the active radiolytic species, OH, $H_3O^+$, $e_{aq}^-$, or H, shown in Table 13.2, at $10^{-12}$ s. There are 924 species present initially. The electron stops in the upper region of the panel, where its higher linear energy transfer (LET) is evidenced by the increased density of dots. The occurrence of species in clusters, or spurs, along the electron's path is seen. As discussed in Section 6.7, this important phenomenon for the subsequent chemical action of ionizing radiation is a result of the particular shape and universality of the energy-loss spectrum for charged particles (Fig. 5.3). The passage of time and the chemical reactions within the track are simulated by the procedures described in the last paragraph. The track is shown at three later stages in Fig. 13.1. At $10^{-7}$ s, the number of reactive species has decreased to 403, and the original structure of the track itself has largely disappeared. Relatively few subsequent reactions take place as the remaining species diffuse ever more widely apart.

A 1-$\mu$m segment of the track of a 2-MeV proton, traveling from left to right in liquid water, is shown in Fig. 13.2, calculated to $2.8 \times 10^{-7}$ s. In contrast to the 4-keV electron in the last figure, the proton track is virtually straight and its high LET leads to a dense formation of reactants along its path. The relative reduction in the number of reactants and the disappearance of the details of the original track structure by $2.8 \times 10^{-7}$ s are, however, comparable. This similarity is due to the fact that intratrack chemical reactions occur only on a local scale of a few hundred angstroms or less, as can be inferred from Figs. 13.1 and 13.2. Separate track segments of this size develop independently of other parts of the track.

These descriptions are borne out by closer examination of the tracks. The middle one-third of the proton track at $10^{-11}$ s in Fig. 13.2 is reproduced on a blown-up scale in the upper line of Fig. 13.3. The second line in this figure shows this segment at $2.8 \times 10^{-9}$ s, as it develops independently of the rest of the track. On an even more expanded scale, the third and fourth lines in Fig. 13.3 show the last third of the track segment from the top line of the figure at $10^{-11}$ s and $2.8 \times 10^{-9}$ s. The scale 0.01 $\mu$m $=$ 100 Å indicates that most of the chemical development of charged-particle tracks takes place within local regions of a few hundred angstroms or less.

Figure 7.1 showed four examples of 0.7-$\mu$m segments of the tracks of protons and alpha particles, having the same velocities, at $10^{-11}$ s. Fast heavy ions of the same velocity have almost the same energy-loss spectrum. Because it has two units of charge, the linear rate of energy loss (stopping power) for an alpha particle is four times that of a proton at the same speed (cf. Section 5.6). Thus the LET of the alpha particles is about four times that of the protons at each energy.

## 13.5
### Chemical Yields in Water

When performing such calculations for a track, the numbers of various chemical species present (e.g., OH, $e_{aq}^-$, $H_2O_2$, etc.) can be tabulated as functions of time. These chemical yields are conveniently expressed in terms of G values—that is, the number of a given species produced per 100 eV of energy loss by the original charged particle and its secondaries, on the average, when it stops in the water. Calculated chemical yields can be compared with experimental measurements. To obtain adequate statistics, computations are repeated for a number of different, independent tracks and the average G values are compiled. As seen from reactions (13.4)–(13.10), G values for the reactant species decrease with time. For example, hydroxyl radicals and hydrated electrons are continually used up, while G values for

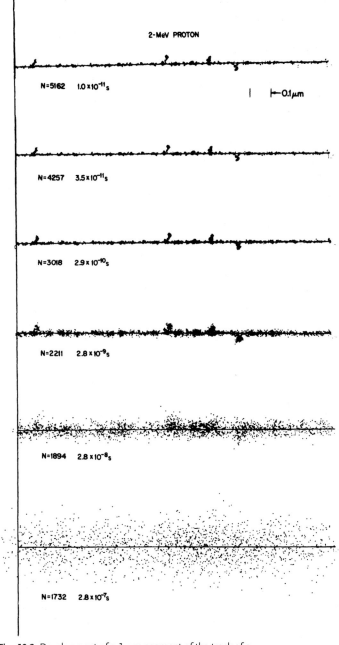

**Fig. 13.2** Development of a 1-$\mu$m segment of the track of a 2-MeV proton, traveling from left to right, in liquid water. (Courtesy Oak Ridge National Laboratory, operated by Martin Marietta Energy Systems, Inc., for the Department of Energy.)

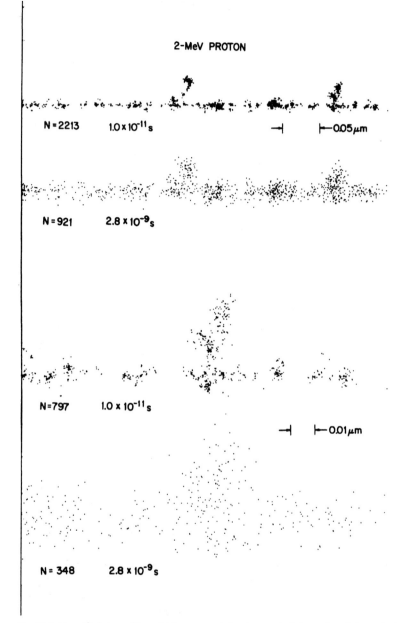

**Fig. 13.3** Magnified view of the middle one-third of the track segment from Fig. 13.2 at $10^{-11}$ s and at $2.8 \times 10^{-9}$ s is shown in the upper two lines. The two lower lines show the right-hand third of this segment at these times under still greater magnification. The figures illustrate how most of the chemical development of charged-particle tracks in pure water takes place in local regions of a few hundred angstroms or less in a track. (Courtesy Oak Ridge National Laboratory, operated by Martin Marietta Energy Systems, Inc., for the Department of Energy.)

Table 13.3 G Values (Number per 100 eV) for Various Species in Water at 0.28 $\mu$s for Electrons at Several Energies

| Species | Electron Energy (eV) | | | | | | | |
|---|---|---|---|---|---|---|---|---|
| | 100 | 200 | 500 | 750 | 1000 | 5000 | 10,000 | 20,000 |
| OH | 1.17 | 0.72 | 0.46 | 0.39 | 0.39 | 0.74 | 1.05 | 1.10 |
| $H_3O^+$ | 4.97 | 5.01 | 4.88 | 4.97 | 4.86 | 5.03 | 5.19 | 5.13 |
| $e_{aq}^-$ | 1.87 | 1.44 | 0.82 | 0.71 | 0.62 | 0.89 | 1.18 | 1.13 |
| H | 2.52 | 2.12 | 1.96 | 1.91 | 1.96 | 1.93 | 1.90 | 1.99 |
| $H_2$ | 0.74 | 0.86 | 0.99 | 0.95 | 0.93 | 0.84 | 0.81 | 0.80 |
| $H_2O_2$ | 1.84 | 2.04 | 2.04 | 2.00 | 1.97 | 1.86 | 1.81 | 1.80 |
| $Fe^{3+}$ | 17.9 | 15.5 | 12.7 | 12.3 | 12.6 | 12.9 | 13.9 | 14.1 |

the other species, such as $H_2O_2$ and $H_2$, increase with time. As mentioned earlier, by about $10^{-6}$ s the reactive species remaining in a track have moved so far apart that additional reactions are unlikely. As functions of time, therefore, the G values change little after $10^{-6}$ s.

Calculated yields for the principal species produced by electrons of various initial energies are given in Table 13.3. The G values are determined by averaging the product yields over the entire tracks of a number of electrons at each energy. [The last line, for $Fe^{3+}$, applies to the Fricke dosimeter (Section 10.6). The measured G value for the Fricke dosimeter for tritium beta rays (average energy 5.6 keV), is 12.9.] The table indicates how subsequent changes induced by radiation can be partially understood on the basis of track structure—an important objective in radiation chemistry and radiation biology. One sees that the G values for the four reactive species (the first four lines) are smallest for electrons in the energy range 750–1000 eV. In other words, the intratrack chemical reactions go most nearly to completion for electrons at these initial energies. At lower energies, the number of initial reactants at $10^{-12}$ s is smaller and diffusion is more favorable compared with reaction. At higher energies, the LET is less and the reactants at $10^{-12}$ s are more spread out than at 750–1000 eV, and thus have a smaller probability of subsequently reacting.

Similar calculations have been carried out for the track segments of protons and alpha particles. The results are shown in Table 13.4. As in Fig. 7.1, pairs of ions have the same speed, and so the alpha particles have four times the LET of the protons in each case. Several findings can be pointed out. First, for either type of particle, the LET is smaller at the higher energies and hence the initial density of reactants at $10^{-12}$ s is smaller. Therefore, the efficiency of the chemical development of the track should get progressively smaller at the higher energies. This decreased efficiency is reflected in the increasing G values for the reactant species in the first four lines (more are left at $10^{-7}$ s) and in the decreasing G values for the reaction products in the fifth and sixth lines (fewer are produced). Second, at a given velocity, the reaction efficiency is considerably greater in the track of an alpha particle than in the track of a proton. Third, comparison of Tables 13.3 and 13.4 shows some

**Table 13.4** G Values (Number per 100 eV) for Various Species at $10^{-7}$ s for Protons of Several Energies and for Alpha Particles of the Same Velocities

| Species Type | Protons (MeV) | | | | Alpha Particles (MeV) | | | |
|---|---|---|---|---|---|---|---|---|
| | 1 | 2 | 5 | 10 | 4 | 8 | 20 | 40 |
| OH | 1.05 | 1.44 | 2.00 | 2.49 | 0.35 | 0.66 | 1.15 | 1.54 |
| $H_3O^+$ | 3.53 | 3.70 | 3.90 | 4.11 | 3.29 | 3.41 | 3.55 | 3.70 |
| $e_{aq}^-$ | 0.19 | 0.40 | 0.83 | 1.19 | 0.02 | 0.08 | 0.25 | 0.46 |
| H | 1.37 | 1.53 | 1.66 | 1.81 | 0.79 | 1.03 | 1.33 | 1.57 |
| $H_2$ | 1.22 | 1.13 | 1.02 | 0.93 | 1.41 | 1.32 | 1.19 | 1.10 |
| $H_2O_2$ | 1.48 | 1.37 | 1.27 | 1.18 | 1.64 | 1.54 | 1.41 | 1.33 |
| $Fe^{3+}$ | 8.69 | 9.97 | 12.01 | 13.86 | 6.07 | 7.06 | 8.72 | 10.31 |

overlap and some differences in yields between electron tracks and heavy-ion track segments. At the highest LET, the reaction efficiency in the heavy-ion track is much greater than that for electrons of any energy.

Electrons, protons, and alpha particles all produce the same species in local track regions at $10^{-15}$ s: $H_2O^+$, $H_2O^*$, and subexcitation electrons. The chemical differences that result at later times are presumably due to the different spatial patterns of initial energy deposition that the particles have.

## 13.6
## Biological Effects

It is generally assumed that biological effects on the cell result from both direct and indirect action of radiation. Direct effects are produced by the initial action of the radiation itself and indirect effects are caused by the later chemical action of free radicals and other radiation products. An example of a direct effect is a strand break in DNA caused by an ionization in the molecule itself. An example of an indirect effect is a strand break that results when an OH radical attacks a DNA sugar at a later time (between ~$10^{-12}$ s and ~$10^{-9}$ s). The difference between direct and indirect effects is illustrated by Fig. 13.4. The dots in the helical configuration schematically represent the location of sugars and bases on a straight segment of DNA 200 Å in length in water. The cluster of dots mostly to the right of the helix gives the positions of the reactants at $10^{-11}$ s and the subsequent times shown after passage of a 5-keV electron along a line perpendicular to the page 50 Å from the center of the axis of the helix.

In addition to any transitions produced by the initial passage of the electron or one of its secondaries (direct effects), the reactants produced in the water can attack the helix at later times (indirect effects). In these computations, the electron was made to travel in a straight line. Also, unreacted radicals were assigned a fixed probability per unit time of simply disappearing, in order to simulate scavenging

1 × 10⁻¹¹ s

3 × 10⁻⁹ s

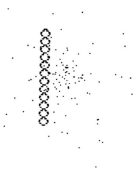

3 × 10⁻¹¹ s

3 × 10⁻⁸ s

3 × 10⁻¹⁰ s

**Fig. 13.4** Direct and indirect action of radiation. Double-helical array of dots represents positions of bases and sugars on a 200-Å straight segment of double-stranded DNA. The other dots show the positions of reactants formed in neighboring water from $10^{-11}$ s to $3 \times 10^{-8}$ s after passage of a 5-keV electron perpendicular to the page in a straight line 50 Å from the center of the helix. In addition to any direct action (i.e., quantum transitions) produced in the DNA by passage of the electron, indirect action also occurs later when the reactants diffuse to the DNA and react with it. Reactants can also disappear by scavenging in this example, crudely simulating a cellular environment. See text. (Courtesy H. A. Wright and R. N. Hamm, Oak Ridge National Laboratory, operated by Martin Marietta Energy Systems, Inc., for the Department of Energy.)

in a cellular environment. Thus, reactants disappear at a much faster rate here than in the previous examples for pure water.

Depending on the dose, kind of radiation, and observed endpoint, the biological effects of radiation can differ widely. Some occur relatively rapidly while others may take years to become evident. Table 13.1 includes a summary of the time scale for some important biological effects caused by ionizing radiation. Probably by about $10^{-3}$ s, radicals produced by a charged-particle track in a biological system have all reacted. Some biochemical processes are altered almost immediately, in less than about 1 s. Cell division can be affected in a matter of minutes. In higher organisms, the time at which cellular killing becomes expressed as a clinical syndrome is related to the rate of cell renewal. Following a large, acute, whole-body dose of radiation, hematopoietic death of an individual might occur in about a month. A higher dose could result in earlier death (1 to 2 wk) from damage to the gastrointestinal tract. At still higher doses, in the range of 100 Gy, damage to membranes and to blood vessels in the brain leads to the cerebrovascular syndrome and death within a day or two. Other kinds of damage, such as lung fibrosis, for example, may take several months to develop. Cataracts and cancer occur years after exposure to radiation. Genetic effects, by definition, are first seen in the next or subsequent generations of an exposed individual.

The biological effects of radiation can be divided into two general categories, stochastic and deterministic, or nonstochastic. As the name implies, stochastic effects are those that occur in a statistical manner. Cancer is one example. If a large population is exposed to a significant amount of a carcinogen, such as radiation, then an elevated incidence of cancer can be expected. Although we might be able to predict the magnitude of the increased incidence, we cannot say which particular individuals in the population will contract the disease and which will not. Also, since there is a certain natural incidence of cancer without specific exposure to radiation, we will not be completely certain whether a given case was induced or would have occurred without the exposure. In addition, although the expected incidence of cancer increases with dose, the severity of the disease in a stricken individual is not a function of dose. In contrast, deterministic effects are those that show a clear causal relationship between dose and effect in a given individual. Usually there is a threshold below which no effect is observed, and the severity increases with dose. Skin reddening is an example of a deterministic effect of radiation.

Stochastic effects of radiation have been demonstrated in man and in other organisms only at relatively high doses, where the observed incidence of an effect is not likely due to a statistical fluctuation in the normal level of occurrence. At low doses, one cannot say with certainty what the risk is to an individual. As a practical hypothesis, one usually assumes that any amount of radiation, no matter how small, entails some risk. However, there is no agreement among experts on just how risk varies as a function of dose at low doses. We shall return to this subject in Section 13.13 in discussing dose–response relationships.

We outline next some of the principal sources of data on the effects of radiation on humans and then describe the effects themselves. This collective body of information, which we only briefly survey here, represents the underlying scientific basis for the radiation-protection standards, criteria, and limits that have been developed. Additional information can be obtained from the references listed in Section 13.15. Virtually all aspects of standards setting are under continuing evaluation and review.

## 13.7
## Sources of Human Data

A considerable body of data exists on radiation effects on man. Risks for certain deleterious effects are reasonably well established at high doses, well above recommended limits. Without attempting to be complete, we mention some of the important sources of data on humans to indicate their scope and the kinds of effects encountered. For many years (into the 1950s), the genetic effects of radiation were considered to pose the greatest danger for human populations exposed to low levels of radiation. Today, the major concern is cancer.

### The Life Span Study

The most important source of information on the effects of ionizing radiation on humans is the continuing Life Span Study of long-term health effects in the atomic-bomb survivors at Hiroshima and Nagasaki. The work is conducted by the joint Japanese/United-States Radiation Effects Research Foundation[1] (RERF). Its objectives include the assessment and characterization of differences in life span and causes of death among the atomic-bomb survivors compared with unexposed persons. Incidence and mortality data are obtained from vital-statistics surveys, death certificates, and other sources. The original sample for the study consisted of about 120,000 persons from among approximately 280,000 identified at the time of the 1950 census as having been exposed to the weapons. Included were a core group of survivors exposed within 2 km of ground zero, other survivors exposed out to distances where little radiation was received, and non-exposed individuals. The sample was eventually constructed by sub-sampling to include all members of the core group and equal-sized samples from the other two, matched by age and sex. Various special cohorts have been formed to study particular questions.

A major task was undertaken to assign doses retrospectively to organs of each individual survivor. Doses were based on analysis of what was known about the weapons' output and the location and shielding of the individual. A number of measurements were conducted at the Nevada Test Site and elsewhere in support of this work. By 1965, a tentative dosimetry system, T65D, was in place for estimating individual doses. This system was substantially updated by the 1986 revision, DS86. The basic quantities determined included the gamma and neutron contributions to the free-in-air kerma and the shielded kerma as functions of the ground distance from the detonations. Doses to different tissues and organs were estimated for individual survivors.

Certain discrepancies persisted between some DS86 predictions when compared with important markers. For example, differences were found between calculated neutron activation products and the activity measured in materials actually exposed at different distances to the bomb radiation. (Activation products include

---

1 Formally called the Atomic Bomb Casualty Commission.

$^{152}$Eu and $^{154}$Eu in rock and concrete, $^{60}$Co in steel and granite, $^{36}$Cl in granite and concrete, $^{63}$Ni in steel, and others). A number of improvements were made in all aspects of the DS86 radiation computations for Hiroshima and Nagasaki. Calculations with newer cross-section values were made of the bomb-released radiation and its air-over-land transport. The greatly advanced capabilities of computers permitted three-dimensional calculations of the detonations and radiation transport. Shielding by terrain and large buildings was upgraded. Differences between predicted and measured activations were resolved under the new dosimetry system, as were other issues. The estimated yield of the Hiroshima weapon (uranium) was revised from 15 to 16 kilotons (TNT), and the epicenter was relocated 20 m higher than before and 15 m to the west. The 21-kiloton yield of the Nagasaki weapon (plutonium) was confirmed with detonation close to its previously assigned site. The new RERF dosimetry system, DS02, has effectively resolved all discrepancies that existed with DS86. Results are now within expected uncertainties for this kind of work. Analysis indicates that the major contribution to the error in doses determined for an individual are the uncertainty in his or her position and orientation at the time of the explosion and the attenuation by surrounding structures. The development of the DS02 system represents a major contribution to the Life Span Study.

Statistically significant excess cancer deaths of the following types have appeared among the atomic-bomb survivors: leukemia; all cancers except leukemia; and cancers of the stomach, colon, lung, female breast, esophagus, ovary, bladder; and multiple myeloma. Mortality data on solid cancer and leukemia were analyzed by using both DS86 and DS02 dose estimates. The new dosimetry system led to only slight revisions in the effects of risk-modifying factors, such as sex, age at exposure, and time since exposure. The risk per unit dose for solid cancers was decreased by about 10%. Leukemia was the first cancer to be linked to radiation exposure among the Japanese survivors. It also has the highest relative risk. The following findings have appeared in some of the approximately 3,000 survivors exposed *in utero*: reduction in IQ with increasing dose, higher incidence of mental retardation among the highly exposed, and some impairment in rate of growth and development.

Statistically significant radiation-related mortality is also seen for non-neoplastic diseases, such as those associated with the heart, respiratory, digestive, and hematopoietic systems. The effects of both cancer and non-cancer mortality are reflected in a general life shortening. The median loss of life in one cohort with estimated doses in the range 0.005 Gy to 1.0 Gy was about 2 months. With doses of 1 Gy or more, the median was about 2.6 years.

Careful searches have been made for genetic effects in the exposed population. Demonstration of such effects is made difficult by the background of naturally occurring spontaneous mutations. Chromosome abnormalities, blood proteins, and other factors have been studied in children born to one or both exposed parents. No significant differences are found in still births, birth weight, sex ratio, infant mortality, or major congenital abnormalities. The Japanese studies indicate, "… that at low doses the genetic risks are small compared to the baseline risks of genetic diseases."[2]

With its enormous scope and scientific value, the studies of the Japanese atomic-bomb survivors have certain drawbacks. The numbers of persons in the lower dose ranges are not sufficiently large to provide direct evidence for radiation effects in man below about 0.2 Gy. The findings nevertheless furnish important estimates of upper limits for the risks for certain effects. In the context of radiation-protection limits, the Japanese exposures were acute and provide no information on how responses might differ for protracted exposures over long times at low dose rates. The exposed populations are also lacking in healthy males of military age. Additional confounding factors in the studies include the possible effects of blast and thermal injuries and poor nutrition and medical care following the attacks on the two cities. In addition, a number of the survivors are still alive, these persons being, of course, the youngest at the time of the exposure. Lifetime risk estimates based on the Japanese data thus still reflect projections of what will happen in this group.

Figure 13.5 presents an example of risk estimation for bomb survivors from the Life Span Study. The excess risk (relative to that at zero dose) for solid cancer is shown as a function of dose. These particular data are averaged over sex and standardized to represent survivors exposed at age 30 who have attained age 60. The doses are grouped into ten intervals and plotted as points at the interval midpoints. The error bars through the points approximate 95% confidence intervals. Two fitted curves are shown as alternative mathematical representations of the risk-*vs.*-dose relationship. The inset shows, for comparison, a linear-quadratic fit for leukemia, which shows greater curvature than solid cancer.

For the purpose of establishing radiation-protection criteria for workers and the public, assessments of risk at low doses and dose rates are of primary concern. Experimentally, it is found that a given large dose of radiation, delivered acutely, is generally more damaging biologically than the same dose delivered over an extended period of time (cf., e.g., Fig. 13.17). In the Life Span Study, therefore, the application of *dose and dose-rate effectiveness factors* (DDREFs) are suggested in order to reduce risks as numerically found in the bomb survivors to values deemed more appropriate for exposure at low doses and dose rates. DDREFs for adjusting linear risk estimates are judged to be in the range 1.1 to 2.3, with a median of 1.5 often being applied.

For a comprehensive review and assessment of health risks from exposure to low levels of ionizing radiation, the reader is referred to the 2006 BEIR VII Phase 2 Report (see references in Section 13.15). Virtually all sources of information on human exposures are addressed.

### Medical Radiation

Studies have been made of populations exposed to therapeutic and diagnostic radiation. While often lacking the sample size and quality of dosimetry that char-

---

2   BEIR VII Report, p. 118 (see references,
    Sect. 13.15).

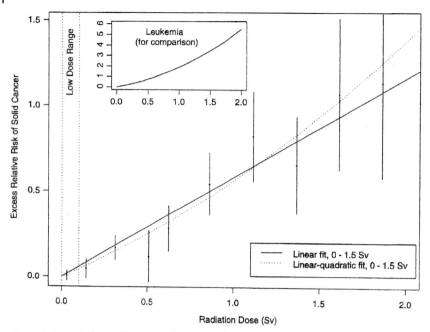

**Fig. 13.5** Example from Life Span Study. Excess relative risks of solid cancer for Japanese atomic-bomb survivors exposed at age 30 who attained age 60. Inset shows the fit of a linear-quadratic model for leukemia, to illustrate the greater degree of curvature observed for that cancer. See text. [Reprinted with permission from Health Risks from Exposure to Low Levels of Ionizing Radiation: BEIR VII Phase 2, © (2006) by the National Academy of Sciences, courtesy of the National Academies Press, Washington, DC.]

acterize the Life Span Study, such investigations can provide some insight into issues outside the scope of the Japanese data—for example, protracted exposures. The following three examples illustrate some findings from medical exposures of humans.

First, X rays were used in the 1930s and 1940s to shrink enlarged thymus glands in children. Treatments could deliver a substantial incidental dose to a child's thyroid, one of the most sensitive tissues for cancer induction by radiation. An abnormally large number of benign and malignant thyroid tumors developed later in life among individuals that underwent this procedure as children for treating the thymus.

Second, it was also common in the 1940s and 1950s to use X rays to treat ringworm of the scalp *(tinea capitis)* in children. A dose of several Gy was administered to the scalp to cause (temporary) epilation, so that the hair follicles could be more effectively treated with medicines. This procedure also resulted in a substantial thyroid dose. Following the establishment of the State of Israel, ringworm of the scalp reached epidemic proportions among immigrants coming there from North Africa. Israeli physicians treated over 10,000 immigrating children, who later showed about a sixfold increase in the incidence of malignant thyroid tumors, compared with unirradiated controls. A survey of 2215 patients similarly treated in New

York yielded excess numbers of thyroid adenomas, leukemia, and brain cancer, but no excess thyroid cancer.

A third example of information obtained on radiation effects from medical exposures is derived from the study of some 14,000 patients treated during the 1930s and early 1940s in Great Britain for ankylosing spondylitis. Large doses of X rays were given to the spine to relieve pain caused by this disease. Retrospective examination of patients' records revealed a small, but statistically significant, increase in leukemia as the cause of death. Doses to the active bone marrow and organs in the treatment field were of the order of several Gy. In addition to uncertainties in the dosimetry, the study lacks a satisfactory cohort of controls—patients having the same disease and receiving similar treatment, but without X-ray therapy.

### Radium-Dial Painters

Radioluminescent paints, made by combining radium with fluorescent materials, were popular in the 1920s. They were used in the production of watch and clock dials, gun sights, and other applications. The industry was widespread. A hundred or more firms purchased the paint, which was applied, almost exclusively by women, to the dials with small brushes. One company reported turning out about 4,300 dials each day. Figure 13.6 shows a typical dial-painting studio of the time in Illinois. Each painter had her materials on the desk top in front of her, and the finished dials can be seen placed to the right of where she sat.

**Fig. 13.6** A studio with radium dial painters, *cir.* 1920s. The proximity of the painted dials at the workers' sides and the supply of radium paint on their desks added an external gamma dose to the internal dose from the ingested radium. [From R. E. Rowland, *Radium in Humans, a Review of U.S. Studies*, Report ANL/ER-3, Argonne National Laboratory, Argonne, IL (1994).]

By the 1920s it was apparent that radiation-related diseases and fatalities were occurring in the industry. The common practice of tipping brushes with the tongue was causing the ingestion of radium, a bone-seeking element, by hundreds of workers. An extensive registry of individual dial painters was subsequently compiled, with information on exposure history. More than 1,000 individuals had their radium body content measured. Bone samples were taken after death and analyzed. An occupational guide of 0.1 $\mu$g for the maximum permissible amount of $^{226}$Ra in the body was later established, based on the findings of the worker studies. It was estimated that this level corresponds to an average dose rate of 0.6 mGy wk$^{-1}$ and a dose equivalent rate of perhaps between $\sim$1 and $\sim$6 mSv wk$^{-1}$. We shall return to this baseline level in the next chapter on radiation-protection limits.

### Uranium Miners

The experience with uranium miners provides another important body of information on radiation effects in human beings. The data are particularly pertinent to the ubiquitous exposure of persons to the naturally occurring daughters of radon. Dating back to the Middle Ages, it was recognized that miners in some parts of Czechoslovakia and southern Germany had abnormally large numbers of lung disorders, referred to as mountain sickness (Bergkrankheit). Well into the twentieth century, miners were exposed to high concentrations of dusts, containing ores of arsenic, uranium, and other metals. The incidence of lung cancer was elevated—in some locations, 50% of the miners died of this disease. Recognition of the role of radon and its daughters was slow in coming. It has been generally accepted as the principal causative agent for lung cancer among uranium miners for only about the last 60 years.

In 1999 the National Research Council published its comprehensive BEIR VI Report, *Health Effects of Exposure to Radon*, updating the 1988 BEIR IV Report on radon and alpha emitters (see references in Section 13. 15). The records of thousands of uranium miners have been examined and analyzed by many investigators in terms of lung-cancer incidence among workers exposed at various levels throughout the world. Supplemented with extensive laboratory work, models have been developed to compute doses to lung tissues per working level month (WLM, Section 4.6) of exposure to radon daughters. Depending on the particular assumptions made and the lung tissue in question, values are in the range 0.2 to 3.0 mGy (WLM)$^{-1}$.[3] A cohort of eleven studies involved 60,606 relatively highly exposed miners worldwide. The mean exposure was 164.4 WLM and the mean duration, 5.7 y. There were 2,674 lung-cancer deaths among the exposed individuals. The data were compared with other cohorts, having successively smaller mean exposures, more applicable to estimating risks at low doses and dose rates. Exposures in the epidemiologic studies of miners are about an order of magni-

3   BEIR VI Report, p. 202.

**Table 13.5** Estimated Number of Lung-cancer Deaths in 1995 in the U.S. Attributable to Indoor Residential Radon*

| Population | Lung-cancer Deaths | Number of Deaths Attributable to Indoor Rn | |
|---|---|---|---|
| | | Model 1 | Model 2 |
| Total Persons | 157,400 | 21,800 | 15,400 |
| Ever Smokers | 146,400 | 18,900 | 13,300 |
| Never Smokers | 11,000 | 2,900 | 2,100 |
| Male | 95,400 | 12,500 | 8,800 |
| Ever Smokers | 90,600 | 11,300 | 7,900 |
| Never Smokers | 4,800 | 1,200 | 900 |
| Female | 62,000 | 9,300 | 6,600 |
| Ever Smokers | 55,800 | 7,600 | 5,400 |
| Never Smokers | 6,200 | 1,700 | 1,200 |

* From the BEIR VI Report.

tude higher than average indoor radon-daughter exposures, although there is some overlap.

In addition to the disparity in dose levels, other factors complicate the application of the uranium-miner experience to an assessment of lung-cancer risk from radon at the relatively low levels in the general population. There are differences between inhaled particle sizes, equilibrium factors, and unattached fractions. There are differences between the breathing rates and physiological characteristics of the male miners and members of all ages and the two sexes in the public. Cigarette smoking is the greatest cause of lung cancer in the world, and most uranium miners were smokers. Synergistic effects occur with the two carcinogens, radon daughters and cigarette smoke.

For estimation of the risk to the general public due to radon, BEIR VI focused on that fraction of the total lung-cancer burden that could presumably be prevented if all radon population exposures were reduced to the background levels of ambient outside air. Compared with outdoors, indoor levels can be considerably higher. Table 13.5 shows an analysis of lung-cancer deaths in the United States for the year 1995, based on data in the BEIR VI Report. The total number of 157,400 deaths, given in the second column, is divided between persons who ever smoked and those who never smoked. A further subdivision is made for males and females. Most of the cases occurred in smokers. Under assumptions used in two preferred risk models, which deal differently with the influence of cigarette smoking, the number of deaths attributable to indoor residential radon daughters was estimated. Model 1 projected 21,800 and Model 2, 15,400 deaths due to residential radon exposures. Thus, the estimates of the two models in Table 13.5 imply that about 1 in 7, or 1 in 10, of all lung-cancer deaths in the U.S. are due indoor residential radon. The BEIR Committee suggested that the number could range from 3,000 to 30,000. As a public-health problem, this assessment clearly identifies indoor radon as the second leading cause of lung cancer after cigarette smoking.

### Accidents

Accidents provide yet another source of information on radiation effects on man, particularly acute effects at high doses. Several fatal accidents have happened with critical assemblies. Serious accidents have occurred with particle accelerators. A larger number of accidental or unknowingly high exposures have resulted from handling radiation devices (X-ray machines and sealed sources) and radioisotopes. Other examples can be cited. In March 1954, high-level fallout from the BRAVO nuclear weapons test reached several Bikini atolls, resulting in substantial doses to some weather-station personnel and Marshallese natives, who were then evacuated. Thyroid abnormalities, including cancer, developed subsequently. In addition, a Japanese fishing vessel *(The Lucky Dragon)* received a large amount of (visible) fallout. The twenty-three men on board suffered massive skin burns and other damage, which could have been lessened considerably by simply rinsing the skin.

On April 26, 2006 the world marked the 20th anniversary of the Chernobyl power-reactor accident in Ukraine, just south of the border with Belarus. It was the most severe accident ever in the nuclear industry. Some 50 persons died within days or weeks, some from the consequences of radiation exposure. (By comparison, the highest individual dose from the 1979 accident at Three Mile Island in the United States was less than 1 mSv.[4]) Enormous quantities of radioactive material were spewed into the atmosphere over a period of days, spreading a cloud of radionuclides over Europe. The resulting contamination of large areas in Belarus, Ukraine, and the Russian Federation led to the relocation of several hundred thousand individuals.

The accident occurred during a low-power test as the result of procedural violations, failure to understand the reactor's behavior, and poor communication between the responsible parties on site. The reactor was being operated with too few control rods, some safety systems shut off, and the emergency cooling system disabled. Even at low power, excess steam pockets (voids) could form in the light-water coolant, thus reducing neutron absorption and increasing the power output, resulting in more voids (positive void coefficient). Reactor control could be quickly lost, as apparently happened.

The consequences were devastating. The acute radiation syndrome (Section 13.8) was confirmed in more than 100 plant employees and first responders, some resulting in death. Severe skin burns from beta radiation occurred. Measurements of blood $^{24}$Na activation indicated that neutrons contributed little to individual doses. Epidemiologic studies have been carried out and are continuing. There is a registry of medical and dosimetric information on hundreds of thousands of individuals. Significant data showing health effects in terms of increased incidence of leukemia and thyroid cancer are well documented. Elevated incidence of thyroid cancer in

---

4  NCRP Report No. 93, *Ionizing Radiation Exposure of the Population of the United States,* p. 28, National Council on Radiation  Protection and Measurements, Bethesda, MD (1987).

children and adolescents is a major effect from the Chernobyl accident. Figure 13.7 shows an example of findings from one study.

Additional information on what has been learned from radiation accidents can be found in several of the references listed in Section 13.15 and on the World Wide Web. Experience with the medical and logistic management of radiation accidents also has important lessons for dealing with potential terrorist attacks that might involve radiation.[5]

## 13.8
## The Acute Radiation Syndrome

If a person receives a single, large, short-term, whole-body dose of radiation, a number of vital tissues and organs are damaged simultaneously. Radiosensitive cells become depleted because their reproduction is impeded. The effects and their severity will depend on the dose and the particular conditions of the exposure. Also, specific responses can be expected to differ from person to person. The complex of clinical symptoms that develop in an individual plus the results of laboratory and bioassay findings are known, collectively, as the acute radiation syndrome.

The acute radiation syndrome can be characterized by four sequential stages. In the initial, or prodromal, period, which lasts until about 48 h after the exposure, an individual is apt to feel tired and nauseous, with loss of appetite (anorexia) and sweating. The remission of these symptoms marks the beginning of the second, or latent, stage. This period, from about 48 h to 2 or 3 wk postexposure, is characterized by a general feeling of well being. Then in the third, or manifest illness, stage, which lasts until 6 or 8 wk postexposure, a number of symptoms develop within a short time. Damage to the radiosensitive hematologic system will be evident through hemorrhaging and infection. At high doses, gastrointestinal symptoms will occur. Other symptoms include fever, loss of hair (epilation), lethargy, and disturbances in perception. If the individual survives, then a fourth, or recovery, stage lasts several additional weeks or months.

Depending on the dose received, the acute radiation syndrome can appear in a mild to very severe form. Table 13.6 summarizes typical expectations for different doses of gamma radiation, which, because of its penetrating power, gives an approximately uniform whole-body dose.

An acute, whole-body, gamma-ray dose of about 4 Gy without treatment would probably be fatal to about 50% of the persons exposed. This dose is known as the LD50—that is, the dose that is lethal to 50% of a population. More specifically, it is also sometimes called the LD50/30, indicating that the fatalities occur within 30 days.

5  See the Proceedings of the 40th Annual Meeting of the NCRP: Advances in Consequence Management for Radiological Terrorism Events, *Health Phys.* **89**, 415–588 (2005).

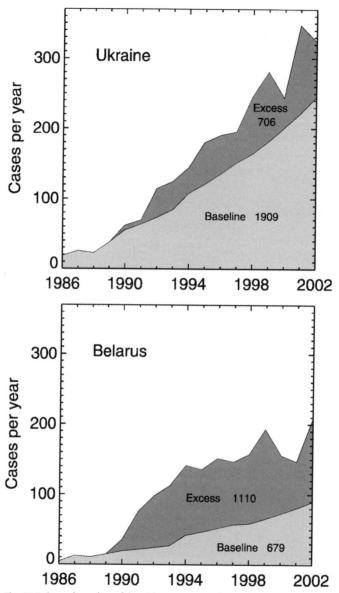

**Fig. 13.7** Annual number of thyroid cancer cases among the birth-year cohorts 1968 to 1985 in Ukraine and Belarus. The total number of observed cases is split into spontaneous (baseline) and excess cases due to $^{131}$I exposures after the Chernobyl accident. The baseline number increases with calendar year, because of aging of the cohort (i.e., the baseline increases with age) and because of intensified surveillance of the thyroid in the aftermath of the accident. After P. Jacob, T. I. Bogdanova, E. Buglova, M. Chepurniy, Y. Demidchik, Y. Gavrilin, J. Kenigsberg, J. Kruk, C. Schotola, S. Shinkarev, M. D. Tronko, and S. Vavilov, "Thyroid Cancer among Ukranians and Belarusians who were Children or Adolescents at the Time of the Chernobyl Accident," *J. Radiol. Ptcn.* **26**, 51–67 (2006). (Courtesy Peter Jacob, GSF National Research Center, Neuherberg, Germany.)

**Table 13.6** Acute Radiation Syndrome for Gamma Radiation

| Dose (Gy) | Symptoms | Remarks |
|---|---|---|
| 0–0.25 | None | No clinically significant effects. |
| 0.25–1 | Mostly none. A few persons may exhibit mild prodromal symptoms, such as nausea and anorexia. | Bone marrow damaged; decrease in red and white blood-cell counts and platelet count. Lymph nodes and spleen injured; lymphocyte count decreases. |
| 1–3 | Mild to severe nausea, malaise, anorexia, infection. | Hematologic damage more severe. Recovery probable, though not assured. |
| 3–6 | Severe effects as above, plus hemorrhaging, infection, diarrhea, epilation, temporary sterility. | Fatalities will occur in the range 3.5 Gy without treatment. |
| More than 6 | Above symptoms plus impairment of central nervous system; incapacitation at doses above ~10 Gy. | Death expected. |

## 13.9
## Delayed Somatic Effects

As indicated in Table 13.1, some biological effects of radiation, administered either acutely or over an extended period, may take a long time to develop and become evident. Such changes are called delayed, or late, somatic effects. In contrast to genetic effects, which are manifested in the offspring of an irradiated parent or parents, late somatic effects occur in the exposed individual. Documentation of late somatic effects due to radiation and estimations of their risks, especially at low doses, are complicated by the fact that the same effects occur spontaneously. The human data on which we focus in this section are supported and expanded by extensive animal experiments.

### Cancer

The risk of getting cancer from radiation depends on many factors, such as the dose and how it is administered over time; the site and particular type of cancer; and a person's age, sex, and genetic background. Additional factors, such as exposure to other carcinogens and promoters, are also important. Cancer causes almost 20% of all deaths in the United States. The relatively small contribution made by low levels of radiation to this large total is not statistically evident in epidemiological studies. In addition, radiogenic cancers are not distinguishable from other cancers. As stated in the BEIR VII Report, "At doses less than 40 times the aver-

**Table 13.7** Lifetime Risk for Incidence and Mortality for All Solid Cancers and for Leukemia from a Dose of 0.1 Gy to 100,000 Persons in a Population Similar to that of the U.S.*

|  | All Solid Cancers | | Leukemia | |
|---|---|---|---|---|
|  | Male | Female | Male | Female |
| Excess cases | 800 | 1,300 | 100 | 70 |
| Number cases without dose | 45,500 | 36,900 | 830 | 590 |
| Excess deaths | 410 | 610 | 70 | 50 |
| Number deaths without dose | 22,100 | 17,500 | 710 | 530 |

\* Adapted from BEIR VII Report (see references, Section 13.15).

age yearly background exposure (100 mSv), statistical limitations make it difficult to evaluate cancer risk in humans." Thus, cancer risk at low doses can at present only be estimated by extrapolation from human data at high doses, where excess incidence is statistically detectable.

Probably the most reliable risk estimates for cancer due to low-LET radiation are those for leukemia and for the thyroid and breast. The minimum latent period of about 2 y for leukemia is shorter than that for solid cancers. Excess incidence of leukemia peaked in the Japanese survivor population around 10 y post-exposure and decreased markedly by about 25 y. These observations are consistent with leukemia experience from other sources, such as patients treated for ankylosing spondylitis and for carcinoma of the uterine cervix. Solid tumors induced by radiation require considerably longer to develop than leukemia. Radiogenic cancers can occur at many sites in the body. We mentioned bone cancers in the radium-dial painters and lung cancers in the uranium miners. The BEIR VII Report provides extensive, detailed information on a wide variety of radiogenic cancers.

The BEIR VII Committee undertook the task of developing models for estimating risks between exposure to low doses of low-LET radiation and adverse health effects. They derived models for both cancer incidence and cancer mortality, allowing for dependence on sex, age at exposure, and time since exposure. Estimates are presented for all solid cancers, leukemia, and a number of site-specific cancers. Special assumptions (e.g., a DDREF) were applied when estimates in lifetime risks for the U.S. population were made from data in the Life Span Study. As an example, Table 13.7 gives a summary from the BEIR VII Report for lifetime risks for all solid cancers and for leukemia. The Committee considered a *linear-no-threshold* model as the most reasonable for describing solid cancers and a *linear-quadratic* model for leukemia (cf., Fig. 13.5). The first line in the table shows the excess number of cancer cases for males and females that would be expected if a population of 100,000 persons, having an age distribution similar to that of the U.S., were to receive a dose of 0.1 Gy of low-LET radiation. The number of cases in the absence of this exposure is shown in the next line. The third and fourth lines display the corresponding information for cancer deaths.

## Life Shortening

Numerous experiments have been carried out in which animals are given sublethal doses of whole-body radiation at various levels. The animals apparently recover, but are subsequently observed to die sooner than controls. This decreased life expectancy was originally described as nonspecific radiation life shortening or as radiation aging. More thorough studies of the effects of low doses of radiation, particularly with careful autopsy examinations, showed that the life shortening due to radiation in animal populations can be attributed to an excess of neoplasia rather than a generally earlier onset of all causes of death. The preponderance of evidence indicates that radiation life shortening at low doses is highly specific, being primarily the result of an increased incidence of leukemia and cancer.

Some investigations have reported a longer average life expectancy in animals exposed to low levels of whole-body radiation than in unexposed controls. Such reports are offered by some as evidence of *radiation hormesis*—that is, the beneficial effect of small doses of radiation. Radiation hormesis has also been extensively investigated in plants, insects, algae, and other systems. As with other low-dose studies of biological effects of radiation, one deals with relatively small effects in a large statistical background of naturally occurring endpoints. Theoretical grounds can be offered in support of low-level radiation hormesis—e.g., stimulation of DNA repair mechanisms that reduce both radiation-induced and spontaneous damage. Evidence for hormesis has been reviewed by the BEIR VII Committee and other bodies. (See references in Section 13.15.) The BEIR VII Report summarizes its judgement in stating, "... the assumption that any stimulatory hormetic effects from low doses of ionizing radiation will have a significant health benefit to humans that exceeds potential detrimental effects from the radiation exposure is unwarranted at this time."

## Cataracts

The biological effects discussed thus far in this section are stochastic. In contrast, a radiogenic cataract is a deterministic effect. There is a practical threshold dose below which cataracts are not produced; and their severity, when they occur, is related to the magnitude of the dose and the time over which it is administered. A cataract is an opacification of the lens of the eye. The threshold for ophthalmologically detectable lens opacification, as observed in patients treated with X rays to the eye, ranges from about 2 Gy for a single exposure to more than 5 Gy for multiple exposures given over several weeks. This level is also consistent with data from Hiroshima and Nagasaki. The threshold for neutrons appears to be lower than for gamma rays. The latent period for radiogenic cataracts is several years, depending on the dose and its fractionation.

Among the biological effects of radiation, a unique feature of a radiogenic cataract is that it can usually be distinguished from other cataracts. The site of the initial detectable opacity on the posterior pole of the lens and its subsequent developmental stages are specific to many radiation cataracts.

## 13.10
### Irradiation of Mammalian Embryo and Fetus

Rapidly dividing cells and tissues in which cells are continually being replaced are among the most radiosensitive: the gonads, gastrointestinal tract, blood-forming organs, lymphatic system, and skin. The developing embryo and fetus, in particular, are highly vulnerable to adverse radiogenic effects, which have been documented in man and in experimental animals.

The principal effects of *in-utero* irradiation are prenatal death, growth retardation, and congenital malformations (teratogenesis). The degree of such effects varies markedly with the stage of development at the time of irradiation. Three such stages can be identified: (1) preimplantation, the time between fertilization of the egg and its implantation in the uterine lining; (2) maximum organogenesis, the time during maximal formation of new organs; and (3) fetal, the final stage, with growth of preformed organs and minimum organogenesis. In humans, these periods are approximately, 0 to 9 d, 10 d to 6 wk, and 6 wk to term.

The unborn is considerably more sensitive to being killed when in the preimplantation stage than later. However, growth retardation and teratogenesis are not generally found as a result of exposure during this stage. Presumably, changes before implantation that predispose the multicellular embryo to such later effects also induce its death. The unborn is most susceptible to teratogenesis when irradiated during the stage of maximum organogenesis. Figure 13.8 shows an example of deformities in a calf whose mother was given 4 Gy of whole-body gamma radiation on the 32nd day after its conception. Calves irradiated similarly, but two additional days after conception, showed little or no damage of this kind. Irradiation during the fetal, or final, stage also produces the greatest degree of permanent growth retardation.

Other types of biological damage have been seen in animals irradiated *in utero* at high doses. However, many of these effects do not appear to occur to the same degree in man, with the exception of damage to the central nervous system. The latter provides, in fact, the most definitive data for an effect of prenatal irradiation in man. The increased prevalence of mental retardation and of microcephaly (small head size), for example, have been documented among the prenatally exposed Japanese survivors.

As described in the next chapter (Section 14.6), special restrictions are recommended by the ICRP and NCRP for the occupational exposure of women of childbearing age and, especially, pregnant women.

## 13.11
### Genetic Effects

Mueller discovered the mutagenic property of ionizing radiation in 1927. Like a number of chemical substances, radiation can alter the genetic information con-

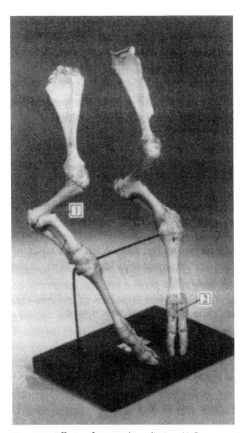

**Fig. 13.8** Effects of prenatal irradiation (4 Gy, whole-body gamma, on 32d day of gestation) on anatomical development of a calf are seen in severe deformities of the forelimbs at birth: (1) bony ankylosis of the humero-radial joints and (2) deformities of the phalanges. In addition, the posterior surfaces of the limbs are turned inward. Such effects are dose- and time-specific. Other fetal calves irradiated two days later suffered only minor damage to the phalanges. (Courtesy G. R. Eisele and W. W. Burr, Jr., Medical and Health Sciences Division, Oak Ridge Associated Universities, Oak Ridge, TN.)

tained in a germ cell or zygote (fertilized ovum). Although mutations can be produced in any cell of the body, only these can transmit the alterations to future generations. Genetic changes may be inconsequential to an individual of a later generation or they may pose a serious handicap.

In the adult human male, the development of mature sperm from the spermatogonial stem cells takes about 10 weeks. Mature sperm cells are produced continually, having passed through several distinct stages. The postspermatogonal cells are relatively resistant to radiation, compared with the stem cells. Thus, an adult male who receives a moderate dose of radiation will not experience an immediate decrease in fertility. However, as his mature sperm cells are depleted, a decrease in fertility, or even sterility, will occur. Depending on the magnitude of the dose and how it is fractionated in time, sterility can be temporary or permanent in males.

The 1990 BEIR V Report (see Section 13.15) states than an acute X-ray dose of 0.15 Gy to the human testes interrupts spermatozoa production to the extent that temporary infertility results. An X-ray dose of 3 to 5 Gy, either acute of fractionated over several weeks, can cause permanent sterility.

In the adult human female, all germ cells are present as ooctyes soon after birth. There are no (oogonial) stem cells, and there is no cell division. The BEIR V Report states than an acute dose of 0.65 to 1.5 Gy to the human ovary impairs fertility temporarily. Fractionation of the dose to the ovaries over several weeks considerably increases the tolerance to radiation. The threshold for permanent sterility in the adult human female for X irradiation of the ovaries is in the range from 2.5 to 6 Gy for acute exposure and is about 6 Gy for protracted exposure.

Every normal cell in the human has 46 paired chromosomes, half derived from the father and half from the mother. Each chromosome contains genes that code for functional characteristics or traits of an individual. The genes, which are segments of deoxyribonucleic acid (DNA), are ordered in linear fashion along a chromosome. The DNA itself is a macromolecule whose structure is a linear array of four varieties of bases, hydrogen bonded in pairs into a double-helical structure. The particular sequence of bases in the DNA encodes the entire genetic information for an individual. The human genome contains about $6 \times 10^9$ base pairs and perhaps 50,000 to 100,000 genes.

Mutations occur naturally and spontaneously among living things. Various estimates indicate that no more than about 5% of all natural mutations in man are ascribable to background radiation. Radicals produced by metabolism, random thermal agitation, chemicals, and drugs, for example, contribute more.

A useful, quantitative benchmark for characterizoing radiation-induced mutation rates is the *doubling dose*. It is defined as the amount of radiation that produces in a generation as many mutations as arise spontaneously. For low-dose-rate, low-LET radiation, the BEIR V Report estimated the doubling dose for mice to be about 1 Gy for various genetic endpoints. It noted that this level is not inconsistent with what might be inferred for man from the atomic-bomb survivors. The BEIR VII Report reviews and discusses doubling-dose estimates, which have been almost exclusively based on both spontaneous and radiation induced rates in mice. The Committee concludes that extrapolation of the doubling dose based on mice for risk estimation in humans should be made with the *human* spontaneous rate. It reports a revised estimate of $0.82 \pm 0.29$ Gy, and suggests retaining the value 1 Gy for the doubling dose as an average rate for mutations.

Radiation-induced genetic changes can result from gene mutations and from chromosome alterations. A gene mutation occurs when the DNA is altered, even by a loss or substitution of a single base. The mutation is called a point mutation when there is a change at a single gene locus. Radiation can also cause breakage and other damage to chromosomes. Some mutations involve a deletion of a portion of a chromosome. Broken chromosomes can rejoin in various ways, introducing errors into the normal arrangement. Figure 13.9 shows two examples of chromosome aberrations induced in human lymphocytes by radiation. Chromosome aberrations

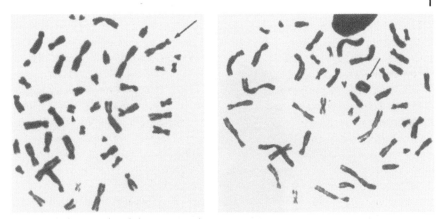

**Fig. 13.9** Radiation-induced chromosome aberrations in
human lymphocytes. Left: chromosome-type dicentric (↙) and
accompanying acentric fragment (▶). Right: chromosome-type
centric ring (↙). The accompanying acentric fragment is not
included in the metaphase spread. (Courtesy H. E. Luippold
and R. J. Preston, Oak Ridge National Laboratory, operated by
Martin Marietta Energy Systems, Inc., for the Department of
Energy.)

occur in somatic cells. Figure 13.10 illustrates genetic effects of radiation in the
fruit fly.

The most extensive studies of the genetic effects of radiation on mammals have
been carried out with mice by W. L. Russell and L. B. Russell. Using literally mil-
lions of mice, they investigated specific locus mutation rates under a variety of conditions of dose, dose rate, and dose fractionation. When compared with the limited
amount of data available for humans, it appears that the data for genetic effects in
the mouse can be applied to man with some degree of confidence. These data play
an important part in assessing the genetic risk and impact on man associated with
the recommended radiation limits to be discussed in the next chapter. We mentioned earlier the doubling dose for mutations in mice, which was established by
the Russells' work. They also measured substantial dose-rate effects on mutations
in the mouse. Protraction of a given dose over time results in fewer mutations
than when the same dose is given acutely, indicating that repair processes come
into play. Males are much more sensitive than females for the induction of genetic
damage by radiation. The latter show little, if any, increased mutation frequency
at low dose rates, even for total accumulated doses of several Gy. Also, mutagenic
effects are lowered when mating is delayed after irradiation.

Radiation does not induce any kinds of mutations that do not occur naturally.
As with other biological endpoints, genetic effects due to radiation are added to
an existing spontaneous pool, thus obscuring their quantitative assessment. An
additional complication arises, because genetic effects are expressed only in the
immediate or later offspring of the irradiated individuals.

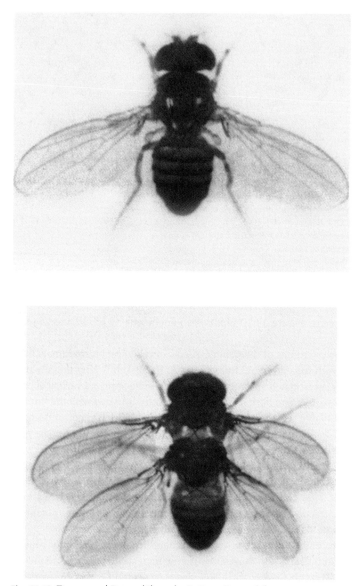

**Fig. 13.10** Top: normal Drosophila male. Bottom: Drosophila male with four wings resulting from one spontaneous and two X-ray induced mutations. [Source: E. B. Lewis, California Institute of Technology. Reprinted with permission from J. Marx, "Genes that Control Development," *Science* **213**, 1485–1488 (1981). Copyright 1981 by the American Association for the Advancement of Science.]

# 13.12
## Radiation Biology

Radiation biology is rapidly advancing our knowledge about the biological effects of radiation. It is beyond the scope of this book to attempt any meaningful review of the varied research being carried out in this exciting field. Studies are directed at discovering and understanding fundamental mechanisms of molecular and cellular responses to radiation.

The complex types of DNA damage produced by radiation can be broadly classified as single-strand breaks, double-strand breaks, and base damages. These structural changes and errors in their repair can lead to gene mutations and chromosomal alterations. A great deal is understood about the molecular details of DNA damage repair and misrepair and its relation to potential tumor induction and other adverse health effects. How a cell operates to deter or prevent the transmission of genetic damage to its progeny is still an unfolding story. Intricate controls exercised by molecular checkpoint genes at specific stages of the cell reproductive cycle appear to recognize and react to the management and repair of damaged DNA.

In order to cause genetic alterations in a cell, it has generally been assumed (or taken for granted) that the cell nucleus must be traversed by a charged-particle track. Research has revealed, however, that nearby cells—called *bystanders*—can also sustain genetic damage, even though no tracks pass through them and hence they presumably receive little or no radiation dose. Studies have been conducted at very low flunece and also with micro-beams directed at individual cells in a target. Mechanisms responsible for producing bystander effects are under investigation. Evidence appears to indicate that mutations in bystander cells with some systems are induced by a different mechanism than those in the directly traversed cells.

*Genomic instability*, which describes the increased rate of accumulation of new genetic changes after irradiation, is observed in some of the progeny of both directly irradiated and bystander cells. The underlying mechanisms for the induction and persistence of genomic instability, which is particularly relevant to tumor development, are poorly understood at present.

In some systems, a small dose of radiation (e.g., several mGy) triggers a cellular response that protects the cells from a large dose of the radiation given subsequently. This phenomenon, which is not universal in all test systems, is called the *adaptive response*. The bacterium *Escherichia coli*, for example, shows a definite adaptive response to oxidative stress. Exposure to a low dose of radiation induces cellular transcription reprogramming. A result is the increased expression of entities that inactivate reactive oxygen species and that repair oxidative DNA damage. For a finite length of time after the initial small priming dose, the bacterial cells are more resistant to a large dose of the radiation than they would be otherwise. Human cells do not show such an adaptive response to oxidative damage. One should note that adaptive response is not the same as hormesis, which ascribes an overall benefit from a small dose of radiation dose.

As more is learned in radiation biology, greater confidence can be placed in the assessment of risk estimates for exposure of persons to ionizing radiation, particularly at low doses. Understanding the basic molecular mechanisms of radiation damage in cells will greatly facilitate the task. We turn next to the subject of dose–response relationships, which underlie radiation-protection regulations in use today.

## 13.13
## Dose–Response Relationships

Biological effects of radiation can be quantitatively described in terms of dose–response relationships, that is, the incidence or severity of a given effect, expressed as a function of dose. These relationships are conveniently represented by plotting a dose–response curve, such as that shown in Fig. 13.11. The ordinate gives the observed degree of some biological effect under consideration (e.g., the incidence of certain cancers in animals per 100,000 population per year) at the dose level given by the abscissa. The circles show data points with error bars that represent a specified confidence level (e.g., 90%). At zero dose, one typically has a natural, or spontaneous, level of incidence, which is known from a large population of unexposed individuals. Often the numbers of individuals exposed at higher dose levels are relatively small, and so the error bars there are large. As a result, although the trend of increasing incidence with dose may be clearly evident, there is no unique dose–response curve that describes the data. In the figure, a solid straight line, consistent with the observations, has been drawn at high doses. The line is constructed in such a way that it intersects the ordinate at the level of natural incidence when a linear extension (dashed curve A) to zero dose is made. In this case, we say that a linear dose–response curve, extrapolated down to zero dose, is used to represent the effect.

Curves with other shapes can usually be drawn through biological dose–effect data. An example of this kind of response is found for leukemia in the atomic-bomb survivors, shown by the inset in Fig. 13.5. Also, extrapolations to low doses can be made in a number of ways. Sometimes there are theoretical reasons for assuming a particular dose dependence, particularly at low doses. The dashed curve B in Fig. 13.11 shows a nonlinear dependence. Both curves A and B imply that there is always some increased incidence of the effect due to radiation, no matter how small the dose. In contrast, the extrapolation shown by the curve C implies that there is a threshold of about 0.75 Gy for inducing the effect.

For many endpoints of carcinogenesis, mutagenesis, and other effects, dose–response functions at low doses and low dose rates can be analyzed in the following way, contrasting high- and low-LET radiations. With low doses of high-LET radiation, the effect is presumably due to individual charged-particle tracks: their spatial density is small, and there is a negligible overlap of different tracks. Since the density of tracks is proportional to the dose, the incidence $E(D)$ (above controls) should also be proportional to the dose $D$ at low doses. This general behavior of the dose

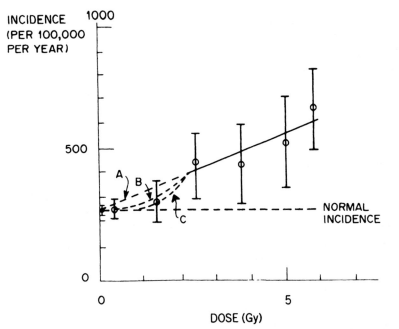

**Fig. 13.11** Example of a dose–response curve, showing the incidence of an effect (e.g., certain cancers per 100,000 population per year) as a function of dose. Circles show measured values with associated error bars. Solid line at high doses is drawn to extrapolate linearly (dashed curve A) to the level of normal incidence at zero dose. Dashed curve B shows a nonlinear extrapolation to zero dose. Dashed curve C corresponds to having a threshold of about 0.75 Gy.

response at low doses for high-LET radiation is shown in Fig. 13.12 by the curve $H$ (which may even begin to decrease in slope at high doses).

For low-LET radiation, dose–response curves in many cases appear to bend upward as the dose increases at low doses and low dose rates, as indicated by the curve $L_1$ in Fig. 13.12. Such behavior is consistent with a quadratic dependence of the magnitude $E(D)$ of the effect as a function of the dose $D$:

$$E(D) = \alpha D + \beta D^2. \tag{13.13}$$

Here $\alpha$ and $\beta$ are constants whose values depend on the biological effect under study, the type of radiation, the dose rate, and other factors. This mathematical form of response, which is commonly referred to as "linear-quadratic" (a misnomer), has a theoretical basis in association with a requirement that *two* interacting lesions are needed to produce the biological damage observed. (It dates back to the 1930s, when it was employed to describe the dose response for some chromosome aberrations, which result from interactions between breaks in two separate chromatids.) As with high-LET radiation, the effect at very low doses must be due to individual tracks. As the dose is increased, the chance for two tracks to overlap soon becomes appreciable at low LET. The response for two-track events should increase as the square of the dose. The initial linear component of the dose–response

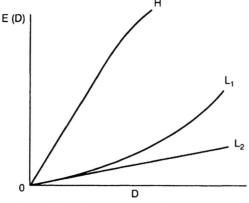

**Fig. 13.12** Schematic representation of dose–response function $E(D)$ at low doses $D$ for high-LET (curve $H$) and low-LET (curve $L_1$) radiations. $L_2$ is the extension of the linear beginning of $L_1$.

function for the low-LET radiation is shown by the curve $L_2$ in Fig. 13.12. To deal with stochastic effects of radiation, the setting of occupational dose limits has been done in a manner consistent with $L_2$. This *linear-nonthreshold (LNT)* dose–response model will be discussed in the next chapter.

Linear-quadratic dose–response relationships are often used to analyze and fit various biological data. However, interpretations other than one- and two-track events can be made to explain their shape. It can be argued, for example, that only single-track damage occurs and that biological repair comes into play, but saturates at high doses. Such a model predicts an upward bend in the dose–response curve with increasing dose.

Another important kind of dose–response relationship is illustrated by the survival of cells exposed to different doses of radiation. The endpoint studied is cell inactivation, or killing, in the sense of cellular reproductive death, or loss of a cell's ability to proliferate indefinitely. Large cell populations can be irradiated and then diluted and tested for colony formation. Cell survival can be measured over three and sometimes four orders of magnitude. It provides a clear, quantitative example of a cause-and-effect relationship for the biological effects of radiation. We next consider cell survival and use it as an example for dose–response modeling.

Cell inactivation is conveniently represented by plotting the natural logarithm of the surviving fraction of irradiated cells as a function of the dose they receive. A linear semilog survival curve, such as that shown in Fig. 13.13, implies exponential survival of the form

$$\frac{S}{S_0} = e^{-D/D_0}. \tag{13.14}$$

Here $S$ is the number of surviving cells at dose $D$, $S_0$ is the original number of cells irradiated, and $D_0$ is the negative reciprocal of the slope of the curve in Fig. 13.13. Analogous to the reciprocals of $\lambda$ in Eq. (4.22) and $\mu$ in Eq. (8.43), it is called the

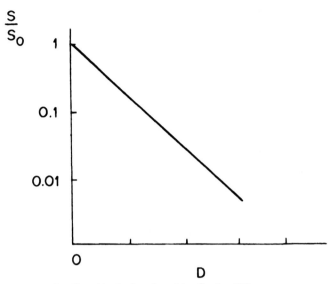

**Fig. 13.13** Semilogarithmic plot of surviving fraction $S/S_0$ as a function of dose $D$, showing exponential survival characterized by straight line.

mean lethal dose; $D_0$ is therefore the average dose absorbed by each cell before it is killed. The surviving fraction when $D = D_0$ is, from Eq. (13.14),

$$\frac{S}{S_0} = e^{-1} = 0.37. \tag{13.15}$$

For this reason, $D_0$ is also called the "D-37" dose.

Exponential behavior can be accounted for by a "single-target," "single-hit" model of cell survival. We consider a sample of $S_0$ identical cells and postulate that each cell has a single target of cross section $\sigma$. We postulate further that whenever radiation produces an event, or "hit," in a cellular target, then that cell is inactivated and does not survive. The biological target itself and the actual physical event that is called a hit need not be specified explicitly. On the other hand, one is free to associate the target and its size with cellular DNA or other components and a hit with an energy-loss event in the target, such as a neutron collision or traversal by a charged particle. When the sample of cells is exposed uniformly to radiation with fluence $\varphi$, then the total number of hits in cellular targets is $\varphi S_0 \sigma$. Dividing by the number of cells $S_0$ gives the average number of hits per target in the cellular population: $\bar{k} = \varphi \sigma$. The distribution of the number of hits per target in the population is Poisson (Problem 27). The probability of there being exactly $k$ hits in the target of a given cell is therefore $P_k = \bar{k}^k e^{-k}/k!$. The probability that a given cell survives the irradiation is given by the probability that its target has no hits: $P_0 = e^{-\bar{k}} = e^{-\varphi \sigma}$.

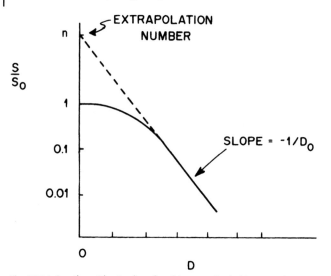

**Fig. 13.14** Semilogarithmic plot of multitarget, single-hit survival.

Thus, the single-target, single-hit model predicts exponential cell survival. Since $P_0 = S/S_0$, we can extend Eq. (13.14) by writing

$$\frac{S}{S_0} = e^{-D/D_0} = e^{-\varphi\sigma}. \tag{13.16}$$

In terms of the model, the inactivation cross section gives the slope of the survival curve on the semilog plot in Fig. 13.13.

A model that yields a survival curve with a different shape is multitarget, single-hit. In this case, $n$ identical targets with cross section $\sigma$ are ascribed to a cell; and all targets in a given cell must be hit at least once in order to inactivate it. As before, we apply Poisson statistics with $\varphi\sigma = D/D_0$ denoting the average number of hits in a given cell target with fluence $\varphi$. The probability that a given target in a cell is hit (one or more times) is equal to one minus the probability that it has not been hit: $1 - e^{-D/D_0}$. The probability that all $n$ targets in a cell are hit is $(1 - e^{-D/D_0})^n$, in which case the cell is inactivated. The survival probability for the cell is therefore

$$\frac{S}{S_0} = 1 - (1 - e^{-D/D_0})^n. \tag{13.17}$$

When $n = 1$, this equation reduces to the single-target, single-hit result. For $n > 1$ the survival curve has the shape shown in Fig. 13.14. There is a shoulder that begins with zero slope at zero dose, reflecting the fact that more than one target must be hit in a cell to inactivate it. As the dose increases, cells accumulate additional struck targets; and so the slope steadily increases. At sufficiently high doses, surviving cells are unlikely to have more than one remaining unhit target. Their response then takes on the characteristics of single-target, single-hit survival, and additional dose produces an exponential decrease with slope $-1/D_0$ on the semilog plot. When

$D$ is large, $e^{-D/D_0}$ is small, and one can use the binomial expansion[6] to write, in place of Eq. (13.17),

$$\frac{S}{S_0} \cong 1 - (1 - ne^{-D/D_0}) = ne^{-D/D_0}. \tag{13.18}$$

The straight line represented by this equation on a semilog plot intercepts the ordinate $(D = 0)$ at the value $S/S_0 = n$, which is called the extrapolation number. As shown in Fig. 13.14, the number of cellular targets $n$ is thus obtained by extrapolating the linear portion of the survival curve back to zero dose.

Many experiments with mammalian cells yield survival curves with shoulders. However, literal interpretation of such data in terms of the elements of a multi-target, single-hit model is not necessarily warranted. Cells in a population are not usually identical. Some might be in different stages of the cell cycle, with different sensitivity to radiation. Repair of initial radiation damage can also lead to the existence of a shoulder on a survival curve.

Still other models of cell survival have been investigated. The multitarget, single-hit model can be modified by postulating that only any $m < n$ of the cellular targets need to be hit in order to produce inactivation. Single-target, multihit models have been proposed, in which more than one hit in a single cellular target is needed for killing. In addition to these target models, other theories of cell survival are based on different concepts.

## 13.14
## Factors Affecting Dose Response

### Relative Biological Effectiveness

Generally, dose–response curves depend on the type of radiation used and on the biological endpoint studied. As a rule, radiation of high LET is more effective biologically than radiation of low LET. Different radiations can be contrasted in terms of their relative biological effectiveness (RBE) compared with X rays. If a dose $D$ of a given type of radiation produces a specific biological endpoint, then RBE is defined as the ratio

$$\text{RBE} = \frac{D_x}{D}, \tag{13.19}$$

where $D_x$ is the X-ray dose needed under the same conditions to produce the same endpoint. As an example, irradiation of *Tradescantia* (spiderwort) produces in stamen hairs pink mutant events that can be counted and scored quantitatively. In experiments with 680-keV neutrons and 250-kVp X rays, it is observed that 0.030 pink events per hair (minus control) are produced by a dose of 16.5 mGy with the

---

6   For small $x$, $(1 - x)^n \cong 1 - nx$.

neutrons and 270 mGy with the X rays. It follows that the RBE for this specific effect is $270/16.5 = 16.4$.[7]

Figure 13.15 shows examples of dose–response curves for irradiation of Sprague-Dawley rats with X rays and with 430-keV neutrons.[8] Groups of rats at age 60 d were given X-ray doses of 0.28, 0.56, and 0.85 Gy and neutron doses of 0.001, 0.004, 0.016, and 0.064 Gy. The straight lines fit the measured data, indicated by the dots in the figure. RBE values at two levels of response for each of the four effects are shown for illustration. One sees that the RBE is different numerically for the four effects and that it also depends on the level of the effect. Its values span the range between 13 and 190 and beyond. As found here and in many experiments, RBE values are largest for small levels of effect. Generally, relative biological effectiveness is observed to depend on the radiation quality (e.g., the LET), dose rate, and dose fractionation, as well as the type and magnitude of the biological endpoint measured. RBE values vary markedly, depending upon these conditions.

The dependence of relative biological effectiveness on radiation quality is often discussed in terms of the LET of the radiation, or the LET of the secondary charged particles produced in the case of photons and neutrons. As a general rule, RBE increases with increasing LET, as illustrated in Fig. 13.15, up to a point. Figure 13.16 represents schematically the RBE for cell killing as a function of the LET of charged particles. Starting at low LET, the efficiency of killing increases with LET, evidently because of the increasing density of ionizations, excitations, and radicals produced in critical targets of the cell along the particle tracks. As the LET is increased further, an optimum range around 100 to 130 keV $\mu m^{-1}$ is reached for the most efficient pattern of energy deposition by a particle for killing a cell. A still further increase in LET results in the deposition of more energy than needed for killing, and the RBE decreases. Energy is wasted in this regime of overkill at very high LET.

The most relevant values of RBE for purposes of radiation protection are those for low doses and low dose rates. For most endpoints, the RBE increases with decreasing dose, as seen in Fig. 13.15, and dose rate. In the context of the linear-quadratic dose–response model illustrated in Fig. 13.12, this increase in the RBE ratio as defined by Eq. (13.19) is associated almost entirely with the decrease in the slope of the curve for the low-LET reference radiation. The maximum values of the RBE determined in this region are denoted by $RBE_M$.[9] For a given radiation and endpoint, $RBE_M$ is thus equal to the ratio of the slope of the dose–response curve $H$ in Fig. 13.12 for the radiation and the slope of $L_2$ from the linear portion of the low-LET reference radiation (e.g., X rays). We shall return to the subject of $RBE_M$ in the next chapter on exposure limits. Table 13.8 summarizes estimates of $RBE_M$ for fission neutrons relative to X rays.

7   NCRP Report No. 104, p. 27 (see references, Sect. 13.15).

8   C. J. Shellabarger, D. Chmelevsky, and A. M. Kellerer, "Induction of Mammary Neoplasms in the Sprague-Dawley Rat by 430 keV Neutrons and X Rays," *J. Nat. Cancer Inst.* **64**, 821 (1980).

9   For stochastic effects. For deterministic effects, the maximum is denoted by $RBE_m$. See ICRP Publication 58, *RBE for Deterministic Effects*, Pergamon Press, Elmsford, N.Y. (1990).

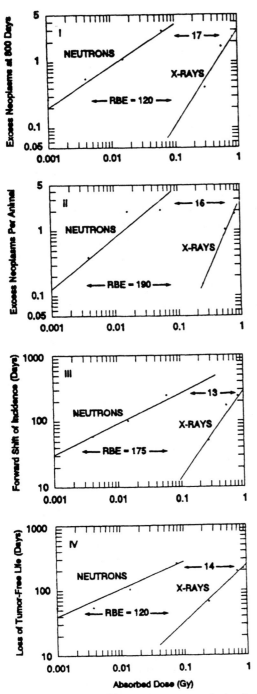

**Fig. 13.15** Examples of dose–response curves for irradiation of Sprague-Dawley rats by X rays and 430-keV neutrons (see text).

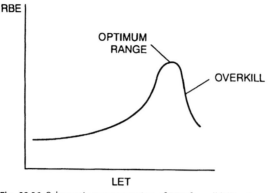

**Fig. 13.16** Schematic representation of RBE for cell killing by charged particles as a function of their LET.

**Table 13.8** Estimated $RBE_M$ Values for Fission Neutrons and X Rays

| Endpoint | Range |
|---|---|
| Cytogenic studies, human lymphocytes in culture | 34–53 |
| Transformation | 3–80 |
| Genetic endpoints in mammalian systems | 5–70 |
| Genetic endpoints in plant systems | 2–100 |
| Life shortening, mouse | 10–46 |
| Tumor induction | 16–59 |

*Source*: From NCRP Report No. 104 (see references, Section 13.15).

## Dose Rate

The dependence of dose–response relationships on dose rate has been demonstrated for a large number of biological effects. In Section 13.11 we mentioned the role of repair mechanisms in reducing the mutation frequency per Gy in mice when the dose rate is lowered. Another example of dose-rate dependence is shown in Fig. 13.17. Mice were irradiated with $^{60}Co$ gamma rays at dose rates ranging up to several tens of $Gy\,h^{-1}$ and the LD50 determined. It was found that LD50 = 8 Gy when the dose rate was several $Gy\,h^{-1}$ or more. At lower dose rates the LD50 increased steadily, reaching approximately 16 Gy at a rate of 0.1 $Gy\,h^{-1}$. Evidently, animal cells and tissues can repair enough of the damage caused by radiation at low dose rates to survive what would be lethal doses if received in a shorter period of time.

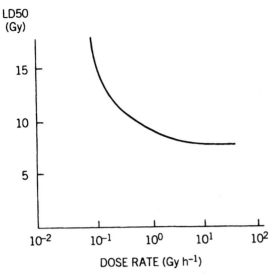

**Fig. 13.17** Dependence of LD50 on dose rate for mice irradiated with $^{60}$Co gamma rays. [Based on J. F. Thomson and W. W. Tourtellotte, *Am. J. Roentg. Rad. Ther. Nucl. Med.* **69**, 826 (1953).]

### Oxygen Enhancement Ratio

Dissolved oxygen in tissue acts as a radio-sensitizing agent. This so-called oxygen effect, which is invariably observed in radiobiology, is illustrated in Fig. 13.18. The curves show the survival of cells irradiated under identical conditions, except that one culture contains dissolved $O_2$ (e.g., from the air) and the other is purged with $N_2$. The effect of oxygen can be expressed quantitatively by means of the oxygen enhancement ratio (OER), defined as the ratio of the dose required under conditions of hypoxia and that under conditions in air to produce the same level of effect. According to this definition, one would obtain the OER from Fig. 13.18 by taking the ratio of doses at a given survival level. OER values are typically 2–3 for X rays, gamma rays, and fast electrons; around 1.7 for fast neutrons; and close to unity for alpha particles.

The existence of the oxygen effect provides strong evidence of the importance of indirect action in producing biological lesions (Section 13.6). Dissolved oxygen is most effective with low- rather than high-LET radiation, because intratrack reactions compete to a lesser extent for the initial reaction product.

### Chemical Modifiers

Chemicals which, like oxygen, have a strong affinity for electrons can make cells more sensitive to radiation. A number of radiosensitizing chemicals and drugs are known. Some sensitize hypoxic cells, but have little or no effect on normally aerated cells. Other agents act as radioprotectors, reducing biological effectiveness.

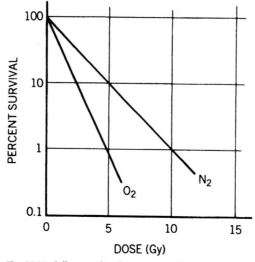

**Fig. 13.18** Cell survival in the presence of dissolved oxygen ($O_2$) and after purging with nitrogen ($N_2$).

The most notable of these are sulfhydryl compounds (e.g., cysteine and cystamine), which scavenge free radicals. Still other chemical modifiers have little effect on cell killing, but substantially enhance some multistep processes, such as oncogenic cell transformation. For carcinogenesis or transformation, for example, such biological promoters can dwarf the effects of physical factors, such as LET and dose rate, on dose–response relationships.

Chemical radiosensitizers for use in radiation therapy are under investigation. Some have the potential to specifically affect resistant hypoxic cells, which are common in tumors. Chemical radioprotectors have been developed for potential military use in a nuclear war.

### Dose Fractionation and Radiotherapy

The goal of treating a malignant tumor with radiation is to destroy it without damaging normal tissues to an intolerable degree. By and large, normal cells and tumor cells have comparable resistance to killing by radiation. Thus, other factors must come into play in radiotherapy. It is found empirically that the most advantageous results are obtained when the radiation is delivered to a patient in fractions, administered perhaps over a period of weeks, rather than all at once.

To understand how the fractionation of dose affects tumor cells more adversely than normal ones in a patient, there are basically four factors to consider at the cellular level: repair, repopulation, redistribution, and reoxygenation. Administering a dose in fractions with adequate time between applications allows the repair of sublethal damage and the repopulation of tissue cells. These processes generally occur on different time scales and to different degrees in the normal and tumor cells. The therapeutic protocol considers optimization of normal-tissue sparing to the detriment of the tumor cells in prescribing the total dose, the number of fractions, the dose per fraction, and the total treatment time.

Because cells exhibit different degrees of radiosensitivity in different phases of the cell cycle, an asynchronous cell population will become partially synchronized by irradiation. The surviving cells will generally be those in the more resistant phases. As the population continues to grow following exposure, the partially synchronized surviving cells become redistributed over the complete cycle, including the more sensitive phases. This process of redistribution, combined with repeated irradiation at intervals, tends to result in increased cell killing relative to that achieved with a single dose.

The oxygen effect is extremely important in radiotherapy. Tumors often have poorly developed blood vessels, intermittent blood flow, and clonogenic cells with greatly reduced oxygen tension. They contain regions with viable cells, which are, however, hypoxic and therefore relatively resistant to radiation (cf., Fig. 13.18). Delivering radiation to a tumor in fractions allows reoxygenation of some hypoxic cells to occur between doses. As an added factor, sensitization increases rapidly with oxygen tension. The result is greater killing of tumor cells than with a comparable single dose. The response of the normal, oxygenated cells is unchanged by this procedure.

## 13.15
## Suggested Reading

A vast amount of information is readily available on the World Wide Web, as well as complete copies (some searchable) of several of the works listed below.

1 Baverstock, K. F., and Stather, J. W., eds., *Low Dose Radiation, Biological Bases of Risk Assessment*, Taylor and Francis, London (1989). [This valuable book of 606 pages contains 54 contributions, covering all essential aspects of its subject. The articles are from scientists in a number of disciplines.]

2 BEIR V, *Health Effects of Exposure to Low Levels of Ionizing Radiation: BEIR V*, Committee on the Biological Effects of Ionizing Radiation, National Research Council, National Academy Press, Washington, DC (1990).

3 BEIR VI, *The Health Effects of Exposure to Indoor Radon: BEIR VI*, Committee on the Biological Effects of Ionizing Radiation, National Research Council, National Academy Press, Washington, DC (1999).

4 BEIR VII, *Health Risks from Exposure to Low Levels of Ionizing Radiation: BEIR VII–Phase 2*, Committee on the Biological Effects of Ionizing Radiation, National Research Council, National Academy Press, Washington, DC (2006). [Reviews health risks from exposure to low levels of low-LET ionizing radiation. Updates BEIR V. Detailed risk estimates for both cancer incidence and cancer mortality are given. Required reading.]

5 Glass, W. A., and Varma, M. N., eds., *Physical and Chemical Mechanisms in Molecular Radiation Biology*, Plenum Press, New York, NY (1991). [A collection of 18 papers at an invited workshop on the subject. Contributions are made from fields of radiological physics, radiation chemistry, and radiation-effects modeling.]

6 Gusev, Igor A., Guskova, Angelina, and Mettler, Fred A., eds., *Medical Management of Radiation Accidents,*

2nd Ed., CRC Press, Boca Raton, FL (2001).

**7** Hall, Eric J., and Giaccia, Amato, J., *Radiobiology for the Radiologist*, 6th Ed., Lippincott Williams and Wilkins, Philadelphia, PA (2005).

**8** Johns, H. E., and Cunningham, J. R., *The Physics of Radiology*, 4th Ed., Charles C. Thomas, Springfield, IL (1983). [Chapter 17, on radiobiology, is highly recommended.]

**9** Mozumda, A., *Fundamentals of Radiation Chemistry*, Academic Press, San Diego, CA (1999).

**10** NCRP Report No. 78, *Evaluation of Occupational and Environmental Exposures to Radon and Radon Daughters in the United States*, National Council on Radiation Protection and Measurements, Washington, DC (1984).

**11** NCRP Report No. 103, *Control of Radon in Houses*, National Council on Radiation Protection and Measurements, Washington, DC (1989).

**12** NCRP Report No. 104, *The Relative Biological Effectiveness of Radiations of Different Quality*, National Council on Radiation Protection and Measurements, Washington, DC (1990).

**13** NCRP Proceedings 42nd Annual Meeting, April 3–4, 2006, *Chernobyl at Twenty*, Health Phys. in press.

**14** Preston, R. Julian, "Radiation Biology: Concepts for Radiation Protection," Health Phys. 88, 545–556 (2005). [An important review article, one of the invited contributions in the volume commemorating the 50th anniversary of the Health Physics Society.]

**15** *Radiation Research* **137** (Supplement), S1–S112 (1994). [A special issue devoted to cancer incidence in atomic-bomb survivors, Radiation Effects Research Foundation, Hiroshima, and Nagasaki, Japan.]

**16** *Radiation Research* **161**, No. 4, April 2004, "A Cohort Study of Thyroid Cancer and Other Thyroid Diseases after the Chornobyl Accident: Objectives, Design and Methods."

**17** Ricks, R. C., Berger, Mary Ellen, and O'Hara, Frederick M., Jr., Eds., *Proceedings of the Fourth International REAC/TS Conference on the Basis for Radiation Accident Preparedness*, CRC Press, Boca Raton, FL (2001).

**18** Sagan, L. A., Guest Editor, *Health Phys.* **52**, 250-680 (1987). [Special issue on radiation hormesis.]

**19** Smith, Jim T., and Beresford, Nicholas A., *Chernobyl—Catastrophe and Consequences*, Springer, Secaucas, NJ (2005).

**20** Sugahara, T., Sagan, L. A., and Aoyama, T., eds., *Low Dose Irradiation and Biological Defense Mechanisms*, Excerpta Medica International Congress Series 1013, 526 pp., Pergamon Press, Elmsford, NY (1992). [Proceedings of the subject conference, held in Kyoto, Japan in July, 1992. The presentations furnish an overview of the state of knowledge regarding effects of low levels of radiation and radiation hormesis.]

**21** United Nations Scientific Committee on the Effects of Atomic Radiation (UNSCEAR), *Sources and Effects of Ionizing Radiation*, UNSCEAR 2000 Report to the General Assembly, United Nations Publications, New York (2000). [In two volumes, 1,300 pp.: Vol. 1, Sources; Vol. 2, Effects.]

## 13.16
## Problems

1. What initial changes are produced directly by ionizing radiation in water (at $\sim 10^{-15}$ s)?

2. What reactive species exist in pure water at times $>10^{-12}$ s after irradiation?

3. Do all of the reactive species (Problem 2) interact with one another?

4. Estimate how far an $H_3O^+$ ion will diffuse, on the average, in water in $5 \times 10^{-12}$ s.

5. Estimate the average time it takes for an OH radical to diffuse 400 Å in water.

6. If an OH radical in water diffuses an average distance of 3.5 Å in $10^{-11}$ s, what is its diffusion constant?

7. Estimate how close an $H_3O^+$ ion and a hydrated electron must be to interact.

8. How far would a water molecule with thermal energy (0.025 eV) travel in $10^{-12}$ s in a vacuum?

9. If a 20-keV electron stops in water and an average of 352 molecules of $H_2O_2$ are produced, what is the G value for $H_2O_2$ for electrons of this energy?

10. If the G value for hydrated electrons produced by 20-keV electrons is 1.13, how many of them are produced, on the average, when a 20-keV electron stops in water?

11. What is the G value for ionization in a gas if $W = 30$ eV ip$^{-1}$ (Section 10.1)?

12. Use Table 13.3 to find the average number of OH radicals produced by a 500-eV electron in water.

13. For what physical reason is the G value for $H_2$ in Table 13.3 smaller for 20-keV electrons than for 1-keV electrons?

14. Why do the G values for the reactant species $H_3O^+$, OH, H, and $e_{aq}^-$ decrease between $10^{-12}$ s and $10^{-6}$ s? Are they constant after $10^{-6}$ s? Explain.

15. For 5-keV electrons, the G value for hydrated electrons is 8.4 at $10^{-12}$ s and 0.89 at $2.8 \times 10^{-7}$ s. What fraction of the hydrated electrons react during this period of time?

16. (a) Why are the yields for the reactive species in Table 13.4 for protons greater than those for alpha particles of the same speed?

    (b) Why are the relative yields of $H_2$ and $H_2O_2$ smaller?

17. A 50-cm$^3$ sample of water is given a dose of 50 mGy from 10-keV electrons. If the yield of $H_2O_2$ is $G = 1.81$ per 100 eV, how many molecules of $H_2O_2$ are produced in the sample?

18. Assume that the annual exposure of a person in the United States to radon daughters is 0.2 WLM. Use the BEIR-IV estimated lifetime risk of 350 excess cancer deaths per $10^6$ person WLM to predict the annual number of such deaths in a population of 250 million people.

19. Distinguish between the "direct" and "indirect" effects of radiation. Give a physical example of each.

20. Give examples of two stochastic and two deterministic biological effects of radiation.

21. What are the major symptoms of the acute radiation syndrome?

22. Given the tenet that the most rapidly dividing cells of the body are the most radiosensitive, show how it is reflected in the information given in Table 13.6 for the acute radiation syndrome.

23. What are the principal late somatic effects of radiation? Are they stochastic or deterministic?

24. According to the BEIR V Report, an acute, whole-body, gamma-ray dose of 0.1 Gy to 100,000 persons would be expected to cause about 800 extra cancer deaths in addition to the 20,000 expected naturally.
    (a) If an "experiment" could be carried out to test this risk estimate for a dose of 0.1 Gy, would a population of 10,000 individuals be sufficiently large to obtain statistically significant results?
    (b) A population of 100,000?
    (c) What kind of statistical distribution describes this problem?

25. Show that $D_0$ in Eq. (13.16) is the mean lethal dose.

26. Survival of a certain cell line exposed to a beam of helium ions is described by the single-target, single-hit model and Eq. (13.16). If 25% of the cells survive a fluence of $4.2 \times 10^7$ cm$^{-2}$, what is the single-target area?

27. Justify the use of Poisson statistics in arriving at Eqs. (13.16) and (13.17).

28. Why do experiments that seek to quantify dose–effect relationships at low doses require large exposed and control populations?

29. Cell survival in a certain set of experiments is described by the function $S/S_0 = e^{-3.1D}$, where $D$ is the dose in Gy.
    (a) What is the mean lethal dose?
    (b) What is the LD50?
    (c) What is the difference between LD50 and mean lethal dose?

30. If 41 Gy reduces the exponential survival of cells to a level of 1%, what is the mean lethal dose?

31. For multitarget, single-hit survival with $D_0 = 7.5$ Gy and an extrapolation number $n = 4$, what fraction of cells survive a dose of 10 Gy?

32. Repeat Problem 31 for $D_0 = 7.5$ Gy and $n = 3$.

33. Repeat Problem 31 for $D_0 = 5.0$ Gy and $n = 4$.

34. What interrelationships do the extrapolation number, the magnitude of $D_0$, and the size of the shoulder have in a multitarget, single-hit cell-survival model?

35. Why does survival in a multitarget, single-hit model become exponential at high doses?

36. **(a)** Sketch a linear plot of the exponential survival curve from Fig. 13.13.
    **(b)** Sketch a linear plot for the multitarget, single-hit curve from Fig. 13.14. What form of curve is it?

37. A multitarget, single-hit survival model requires hitting $n$ targets in a cell at least once each to cause inactivation. A single-target, multihit model requires hitting a single target in a cell $n$ times to produce inactivation. Show that these two models are inherently different in their response. (For example, at high dose consider the probability that hitting a target will contribute to the endpoint.)

38. One can describe the exponential survival fraction, $S/S_0$, by writing $S/S_0 = e^{-pD}$, where $D$ is the number of "hits" per unit volume (proportional to dose) and $p$ is a constant, having the dimensions of volume. Show how $p$ can be interpreted as the target size (or, more rigorously, as an upper limit to the target size in a single-hit model).

39. The cell-survival data in Table 13.9 fit a multitarget, single-hit survival curve. Find the slope at high doses and the extrapolation number. Write the equation that describes the data.

40. Cell survival is described in a certain experiment by the single-target, single-hit response function, $S/S_0 = e^{-1.6D}$, where $D$ is in Gy. At a dose of 1 Gy, what is the probability of there being
    **(a)** no hits
    **(b)** exactly two hits in a given target?

**Table 13.9** Data for Problem 39

| Dose (Gy) | Surviving Fraction |
|-----------|--------------------|
| 0.10 | 0.993 |
| 0.25 | 0.933 |
| 0.50 | 0.729 |
| 1.00 | 0.329 |
| 2.00 | 0.0458 |
| 3.00 | 0.00578 |
| 4.00 | 0.00072 |

41. A colony of identical cells (unit density) is irradiated with neutrons, which deposit an average of 125 keV of energy in a collision. A single neutron collision in a sensitive volume of a cell inactivates the cell.

    (a) If a dose of 0.50 Gy inactivates 11 % of the cells, what is the average number of neutron collisions in the sensitive volume of a cell in the colony?

    (b) What is the size of the sensitive volume of a cell in $\mu m^3$?

    (c) What fraction of the cells are expected to be inactivated by a dose of 2.0 Gy?

42. The survival of a certain cell line when exposed to X rays is found experimentally to be described by the equation

$$\frac{S}{S_0} = 1 - (1 - e^{-0.92D})^2,$$

where $D$ is in Gy. Survival of the same cell line exposed to neutrons is described by

$$\frac{S}{S_0} = e^{-0.92D},$$

with $D$ in Gy.

    (a) What is the RBE for the neutrons (relative to the X rays) for 10% survival of the cells?

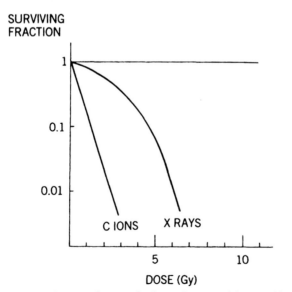

**Fig. 13.19** Surviving fraction of cells as a function of dose (Problem 44).

   (b) At a higher level of survival (lower dose), is the RBE larger
      or smaller?

   (c) Give a reason to explain your answer to (b).

43. What factors can modify dose–effect relationships?
44. Figure 13.19 shows the surviving fraction of cells as a function
    of dose when exposed to either X rays or carbon ions in an
    experiment. From the curves, estimate the RBE of the carbon
    ions for 1% survival and for 50% survival. What appears to
    happen to the RBE as one goes to lower and lower doses?
45. Explain why radiation is used in cancer therapy, even though it
    kills normal cells.
46. Estimate the oxygen enhancement ratio from the cell-survival
    curves in Fig. 13.18.
47. Are the curves in Fig. 13.18 more typical of results expected
    with high-LET or low-LET radiation? Why?

## 13.17
## Answers

| | |
|---|---|
| 4. $4.9 \text{ Å}$ | 26. $3.3 \ \mu\text{m}^2$ |
| 5. $1.3 \times 1^{-7}$ s | 31. $0.706$ |
| 7. $2.4 \text{ Å}$ | 41. (a) $0.117$ |
| 8. $5.2 \text{ Å}$ |     (b) $4.68 \ \mu\text{m}^3$ |
| 11. $3.3$ |     (c) $0.373$ |
| 12. $2.3$ | 42. (a) $1.3$ |
| 15. $0.89$ |     (b) Larger |
| 17. $2.83 \times 10^{14}$ | 44. $2.7, 5.3$ |
| 18. $17{,}500$ | 46. $2.0$ |

# 14
# Radiation-Protection Criteria and Exposure Limits

## 14.1
## Objective of Radiation Protection

Man benefits greatly from the use of X rays, radioisotopes, and fissionable materials in medicine, industry, research, and power generation. However, the realization of these gains entails the routine exposure of persons to radiation in the procurement and normal use of sources as well as exposures from accidents that might occur. Since any radiation exposure presumably involves some risk to the individuals involved, the levels of exposures allowed should be worth the result that is achieved. In principle, therefore, the overall objective of radiation protection is to balance the risks and benefits from activities that involve radiation. If the standards are too lax, the risks may be unacceptably large; if the standards are too stringent, the activities may be prohibitively expensive or impractical, to the overall detriment to society.

The balancing of risks and benefits in radiation protection cannot be carried out in an exact manner. The risks from radiation are not precisely known, particularly at the low levels of allowed exposures, and the benefits are usually not easily measurable and often involve matters that are personal value judgments. Because of the existence of legal radiation-protection standards, in use everywhere, their acceptance rests with society as a whole rather than with particular individuals or groups. Even if the risks from low-level radiation were established quantitatively on a firm scientific basis, the setting of limits would still represent a social judgment in deciding how great a risk to allow. The setting of highway speed limits is an example of such a societal decision—one for which extensive quantitative data are available at the levels of risk actually permitted and accepted.

## 14.2
## Elements of Radiation-Protection Programs

Different uses of ionizing radiation warrant the consideration of different exposure guidelines. Medical X rays, for example, are generally under the control of the physician, who makes a medical judgment as to their being warranted. Specific

*Atoms, Radiation, and Radiation Protection.* James E. Turner
Copyright © 2007 WILEY-VCH Verlag GmbH & Co. KGaA, Weinheim
ISBN: 978-3-527-40606-7

radiation-protection standards, such as those recommended by the International Commission on Radiological Protection (ICRP) and the National Council on Radiation Protection and Measurements (NCRP), have been traditionally applied to the "peaceful uses of atomic energy," the theory being that these activities justify the exposure limits being specified. In contrast, different exposure criteria might be appropriate for military or national-defense purposes or for space exploration, where the risks involved and the objectives are of an entirely different nature than those for other uses of radiation.

We shall concentrate principally on the radiation-protection recommendations of the ICRP and NCRP, which are very similar. The maximum levels of exposure permitted are deemed acceptable in view of the benefits to mankind, as judged by various authorities and agencies who, in the end, have the legal responsibility for radiation safety. Since, in principle, the benefits justify the exposures, the limits apply to an individual worker or member of the public independently of any medical, dental, or background radiation exposure he or she might receive.

Different permissible exposure criteria are usually applied to different groups of persons. Certain levels are permitted for persons who work with radiation. These guidelines are referred to as "occupational" or "on-site" radiation-protection standards. Other levels, often one-tenth of the allowable occupational values, apply to members of the general public. These are referred to as "non-occupational" or "off-site" guides. Several philosophical distinctions can be drawn in setting occupational and nonoccupational standards. In routine operations, radiation workers are exposed in ways that they and their employers have some control over. The workers are also compensated for their jobs and are free to seek other employment. Members of the public, in contrast, are exposed involuntarily to the gaseous and liquid effluents that are permitted to escape from a site where radioactive materials are handled. In addition, off-site exposures usually involve a larger number of persons as well as individuals in special categories of concern, such as children and pregnant women. (Special provisions are also made for occupational radiation exposure of women of child-bearing age.)

On a worldwide scale, the potential genetic effects of radiation have been addressed in setting radiation standards. Exposure of a large fraction of the world's population to even a small amount of radiation represents a genetic risk to mankind that can be passed on indefinitely to succeeding generations. In contrast, the somatic risks are confined to the persons actually exposed.

An essential facet of the application of maximum permissible exposure levels to radiation-protection practices is the ALARA (as low as reasonably achievable) philosophy. The ALARA concept gives primary importance to the principle that exposures should always be kept as low as practicable. The maximum permissible levels are not to be considered as "acceptable," but, instead, they represent the levels that should not be exceeded.

Another consideration in setting radiation-protection standards is the degree of control or specificity that the criteria may require. The ICRP and NCRP have generally made recommendations for the limits for individual workers or members of other groups in a certain length of time, for example, a year or three months. Without requiring the specific means to achieve this end, the recommendations allow maximum flexibility in their application. Many federal and international agencies, however, have very specific regulations that must be met in complying with the ICRP and NCRP limits.

## 14.3
## The NCRP and ICRP

The National Council on Radiation Protection and Measurements is a nonprofit corporation chartered by the U.S. Congress in 1964. One of its most important charges is the dissemination of information and recommendations on radiation in the public interest. It is also charged with the scientific development, evaluation, and application of basic radiation concepts, measurements, and units. The NCRP maintains close working relationships with a large number of organizations, nationally and internationally, that are dedicated to various facets of radiation research, protection, and administration. The Council has approximately 100 members, who serve six-year terms. It has a number of scientific committees, representing virtually all areas of any significance related to radiological protection. The committees are composed of selected experts, who draft recommendations. All recommendations are submitted to the full Council for review, comment, and approval before publication.

The International Commission on Radiological Protection was established in 1928. It has close official relationships with a number of international organizations that include the International Commission on Radiation Units and Measurements (ICRU), the International Atomic Energy Agency, and the World Health Organization. The Commission consists of a Chairman and twelve members. It draws upon a wide spectrum of scientific expertise from outside as well as from its own committees and task groups. Like the NCRP, the ICRP has no legal authority. Recommendations of the two bodies—one for the United States and the other internationally—are made to provide guidance for the setting of radiation-protection criteria, standards, practices, and limits by other (regulatory) agencies. The NCRP and ICRP maintain a close, but independent, relationship. A number of scientists are active in both groups. As we shall see, both the NCRP and ICRP have adopted similar recommendations.

The development and promulgation of recommended radiation-protection criteria is an active and continuing responsibility of both organizations. As more is learned about radiation in its various aspects and about the ever changing needs of society, the basic premises of their work remain under constant study and evaluation. At the time of this writing, the latest recommendations of the ICRP for exposure limits are given in its Publication 60, issued in 1991. Those of the NCRP are contained in its Report No. 116, issued in 1993. These two documents are very similar, though not identical. Both introduce almost the same revised dosimetric concepts, which replace a number of those in the dose-equivalent system then in general use (Section 12.2). Whereas current radiation protection in the U.S. continues to be regulated under the dose-equivalent system, the newer ICRP/NCRP dose quantities are largely employed elsewhere throughout the world today. In practical terms, both systems work in maintaining exposures not only well below acceptable limits, but at low levels in keeping with the ALARA principle. We shall describe both systems in turn (Sections 14.6 and 14.7).

At the time of this writing (2007), the ICRP has before it a major new draft statement, the "2007 ICRP Recommendations." While the numerical limits in Publication 60 continue to be indorsed as providing an appropriate level of protection for normal operations, fundamental changes are proposed in certain concepts and approaches to radiation protection. The 2007 Recommendations will be considered in Section 14.8.

## 14.4
## NCRP/ICRP Dosimetric Quantities

### Equivalent Dose

The equivalent dose, $H_{T,R}$, in a tissue or organ T due to radiation R, is defined as the product of the average absorbed dose, $D_{T,R}$, in T from R and a dimensionless radiation weighting factor, $w_R$, for each radiation:

$$H_{T,R} = w_R D_{T,R}. \tag{14.1}$$

The values of $w_R$ specified by the NCRP are shown in Table 14.1. (The values recommended by the ICRP are the same, except that $w_R = 5$ for protons in the next-to-last entry.) When the radiation consists of components with different $w_R$, then the equivalent dose in T is given by summing all contributions:

$$H_T = \sum_R w_R D_{T,R}. \tag{14.2}$$

With $D_{T,R}$ expressed in Gy (1 Gy = 1 J kg$^{-1}$), $H_{T,R}$ and $H_T$ are in Sv (1 Sv = 1 J kg$^{-1}$).

The equivalent dose replaces the dose equivalent for a tissue or organ, defined in Section 12.2. The two are conceptually different. Whereas dose equivalent in an organ is defined as a point function in terms of the absorbed dose weighted by a quality factor everywhere, equivalent dose in the organ is given simply by the *average* absorbed dose weighted by the factor $w_R$.

For radiation types and energies not included in Table 14.1, the ICRP and NCRP give a prescription for calculating an approximate value of $w_R$ as an average quality factor, $\overline{Q}$. For this purpose, the quality factor Q is defined in terms of the linear energy transfer L by means of Table 12.2, given earlier in the text. One computes the dose–average value of Q at a depth of 10 mm in the standard tissue sphere of

**Table 14.1** Radiation Weighting Factors, $w_R$, from NCRP Report No. 116

| Radiation | $w_R$ |
|---|---|
| X and $\gamma$ rays, electrons, positrons, and muons | 1 |
| Neutrons, energy <10 keV | 5 |
| 10 keV to 100 keV | 10 |
| >100 keV to 2 MeV | 20 |
| >2 MeV to 20 MeV | 10 |
| >20 MeV | 5 |
| Protons, other than recoil protons and energy | |
| >2 MeV | $2^a$ |
| Alpha particles, fission fragments, and nonrelativistic heavy nuclei | 20 |

[a] ICRP Publication 60 recommends $w_R = 5$.

diameter 30 cm specified by the ICRU.[1] Specifically, at the prescribed depth, one calculates

$$w_R \cong \overline{Q} = \frac{1}{D} \int_0^\infty Q(L) D(L) \, dL, \tag{14.3}$$

where $D(L) \, dL$ is the absorbed dose at linear energy transfer (LET) between $L$ and $L + dL$.

The radiation weighting factors and the relationship between $Q$ and $L$ are based on the limiting $\mathrm{RBE_M}$ (relative biological effectiveness) values, such as those in Table 13.6, which are included in NCRP Report No. 116 and ICRP Publication 60.

**Effective Dose**

Since different tissues of the body respond differently to radiation, the probability for stochastic effects that result from a given equivalent dose will generally depend upon the particular tissue or organ irradiated. To take such differences into account, the ICRP and NCRP have assigned dimensionless tissue weighting factors $w_T$, shown in Table 14.2, which add to unity when summed over all tissues T. The equivalent dose $H_T$ in a given tissue, weighted by $w_T$, gives a quantity that is intended to correlate with the overall detriment to an individual, independently of T. The detriment includes the different mortality and morbidity risks for cancers, severe genetic effects, and the associated length of life lost. Table 14.2 implies, for example, that an equivalent dose of 1 mSv to the lung entails the same overall detriment for stochastic effects as an equivalent dose to the thyroid of $(0.12/0.05) \times (1 \text{ mSv}) = 2.4 \text{ mSv}$.

1   See Section 14.9.

**Table 14.2** Tissue Weighting Factors, $w_T$

| Tissue or Organ | $w_T$ |
|---|---|
| Gonads | 0.20 |
| Bone marrow (red) | 0.12 |
| Colon | 0.12 |
| Lung | 0.12 |
| Stomach | 0.12 |
| Bladder | 0.05 |
| Breast | 0.05 |
| Liver | 0.05 |
| Esophagus | 0.05 |
| Thyroid | 0.05 |
| Skin | 0.01 |
| Bone surface | 0.01 |
| Remainder* | 0.05 |

\* Note: The data refer to a reference population of equal numbers of both sexes and a wide range of ages. In the definition of effective dose, they apply to workers, to the whole population, and to either sex. The $w_T$ are based on rounded values of the organ's contribution to the total detriment.

The risk for all stochastic effects for an irradiated individual is represented by the effective dose, $E$, defined as the sum of the weighted equivalent doses over all tissues:

$$E = \sum_T w_T H_T. \tag{14.4}$$

Like $H_T$, $E$ is expressed in Sv. The risk for all stochastic effects is dependent only on the value of the effective dose, whether or not the body is irradiated uniformly. In the case of uniform, whole-body irradiation, $H_T$ is the same throughout the body. Then, since the tissue weighting factors sum to unity,

$$E = \sum_T w_T H_T = H_T \sum_T w_T = H_T, \tag{14.5}$$

the value of the equivalent dose everywhere. The effective dose replaces the earlier effective dose equivalent. The latter quantity was defined the same way as $E$ in Eq. (14.4), with $H_T$ being the organ or tissue dose equivalent.

It should be understood that the procedures embodied in Eq. (14.4) have been set up for use in radiological protection. As the note to Table 14.2 specifies, the values of $w_T$ are simplified and rounded for a reference population of equal numbers of males and females over a wide range of ages. As stated in NCRP Report No. 116 (p. 22), they "should not be used to obtain specific estimates of potential health effects for a given individual."

## Committed Equivalent Dose

When a radionuclide is taken into the body, it can become distributed in various tissues and organs and irradiate them for some time. For the single intake of a radionuclide at time $t_0$, the committed equivalent dose over a subrequent time $\tau$ in an organ or tissue T is defined as

$$H_T(\tau) = \int_{t_0}^{t_0+\tau} \dot{H}_T \, dt, \tag{14.6}$$

where $\dot{H}_T$ is the equivalent-dose rate in T at time $t$. Unless otherwise indicated, an integration time $\tau = 50$ y after intake is implied for occupational use and 70 y for members of the public.[2]

## Committed Effective Dose

By extension, the committed effective dose $E(\tau)$ following the intake of a radionuclide is the weighted sum of the committed equivalent doses in the various tissues T:

$$E(\tau) = \sum_T w_T H_T(\tau). \tag{14.7}$$

As we shall see in more detail in Chapter 16, the effective half-life of a radionuclide in a tissue is determined by its radiological half-life and its metabolic turnover rate. For radionuclides with effective half-lives of no more than a few months, the committed quantities, Eqs. (14.6) and (14.7), are practically realized within one year after intake. If a radionuclide is retained in the body for a long time, then the annual equivalent and effective doses it delivers will be considerably less than the committed quantities.

The committed effective dose replaces the earlier committed effective dose equivalent. The latter is defined like Eq. (14.7), with $H_T$ representing the committed dose equivalent in the organ or tissue T.

## Collective Quantities

The quantities just defined relate to the exposure of an individual person. The ICRP has defined other dosimetric quantities that apply to the exposure of groups or populations to radiation. The *collective equivalent dose* and the *collective effective dose* are obtained by multiplying the average value of these quantities in a population or group by the number of persons therein. The collective quantities are then expressed in the unit, "person-sievert," and can be associated with the total consequences of a given exposure of the population or group. The Commission ad-

2   According to NCRP Report No. 116. ICRP Publication 60 (p. 9) specifies an implied "50 y for adults and from intake to age 70 y for children."

ditionally defines collective dose commitments as the integrals over infinite time of the average individual $\dot{H}_T$ and $\dot{E}$ due to a specified event, either for a critical population group or for the world population.

### Limits on Intake

NCRP Report No. 116 introduces the annual reference level of intake (ARLI). It is defined (p. 59) as "the activity of a radionuclide that, taken into the body during a year, would provide a committed effective dose to a person, represented by Reference Man, equal to 20 mSv. The ARLI is expressed in Becquerels (Bq)." (Reference Man, the ICRP model for dose calculations from the intake of radionuclides, will be described in Chapter 16.) As mentioned after Eq. (14.7), a radionuclide will be retained with an effective half-life in the body that depends on both the radiological half-life and the metabolic turnover rate. By definition, the committed effective dose is delivered over the 50 y following an intake. If the radionuclide has an effective half-life that is short compared with 1 y, then an intake of 1 ARLI during a year will result in an effective dose close to 20 mSv during that year. That is, the effective dose in the year of the intake will be about the same as the committed effective dose. On the other hand, if the effective half-life is comparable to or longer than 1 y, then the effective dose during the year of intake will be less than 20 mSv. For a very long effective half-life, the effective dose each year will be considerably less than 20 mSv, averaging $(20\,\text{mSv})/(50\,\text{y}) = 0.40\,\text{mSv}\,\text{y}^{-1}$. The ARLI is keyed to a cumulated effective dose of 20 mSv over the next 50 y after an intake. ARLI values are computed for both inhalation and ingestion of a radionuclide. The NCRP and ICRP specify that the annual effective *dose limit* (given below in Table 14.4) applies to the sum of the effective dose from external radiation and the committed effective dose from intakes during the year.

Theoretically, continuous intake annually at the level of the ARLI should maintain a worker's exposure within the basic annual and cumulative limits given in Table 14.4. If several radionuclides are involved or if external radiation is a factor, then appropriate reductions among the different ARLIs are to be made to assure compliance with the basic recommendations set forth in the table. The limitation of annual effective dose through application of the ARLI will also protect individual tissues against the likelihood of deterministic effects.

The same quantity, also based on a committed effective dose of 20 mSv, is called the annual limit on intake (ALI) in ICRP Publication 60. Values of the ALI are given in ICRP Publication 68. Prior to ICRP Publication 60 and NCRP Report No. 116, the term ALI was used by both organizations. However, it was then based on a 50-mSv committed effective dose equivalent, rather than the 20-mSv committed effective dose, and other criteria for deterministic effects, as described in Section 14.7.

The NCRP defines the derived reference air concentration (DRAC) as "that concentration of a radionuclide which, if breathed by Reference Man, inspiring $0.02\,\text{m}^3$ per min for a working year, would result in an intake of one ARLI." The working year consists of 50 wk of 40 h each, or 2000 h. Thus,

$$\text{DRAC} = \frac{\text{ARLI}}{0.02 \text{ m}^3 \text{ min}^{-1} \times 60 \text{ min h}^{-1} \times 2000 \text{ h}} = \frac{\text{ARLI}}{2400} \text{ Bq m}^{-3}. \qquad (14.8)$$

The DRAC is introduced to provide a reference level for controlling airborne radionuclides to the level of the ARLI for inhalation.

Prior to NCRP Report No. 116, the derived air concentration (DAC) was used by the NCRP and ICRP. It is defined by the same equation, (14.8), as the DRAC, except that the ARLI is replaced by the ALI.

## 14.5
### Risk Estimates for Radiation Protection

Risk estimates for cancer and genetic effects from radiation have been studied by a number of organizations, which include the ICRP, NCRP, the Radiation Effects Research Foundation, the United Nations Scientific Committee on the Effects of Atomic Radiation (UNSCEAR), the National Radiological Protection Board of the United Kingdom, and the National Academy of Sciences–National Research Council in the United States. Based on these studies, the ICRP in Publication 60 and the NCRP in Report No. 116 concluded that it is appropriate to use for the nominal lifetime fatal cancer risks for low-dose and low-dose-rate exposure the values $4.0 \times 10^{-2}$ Sv$^{-1}$ for an adult worker population and $5.0 \times 10^{-2}$ Sv$^{-1}$ for a population of all ages. The implied unit of the stated risks is "per sievert equivalent dose." These numbers reflect risk estimates that are lower by about a factor of 2 compared with data from high doses and high dose rates. That is, a dose and dose-rate effectiveness factor (DDREF) of about 2 was used.[3]

The detriment from radiation exposure must include other deleterious effects in addition to fatal cancer. The detriments from nonfatal cancers were estimated by the ICRP and NCRP to be $0.8 \times 10^{-2}$ Sv$^{-1}$ for workers and $1.0 \times 10^{-2}$ Sv$^{-1}$ for the whole population. Those for severe genetic effects were, respectively, $0.8 \times 10^{-2}$ Sv$^{-1}$ and $1.3 \times 10^{-2}$ Sv$^{-1}$. The total detriments (equivalent fatal cancer risks) were then $5.6 \times 10^{-2}$ Sv$^{-1}$ for a working population and $7.3 \times 10^{-2}$ Sv$^{-1}$ for a population of all ages. A summary of these figures is given in Table 14.3. The ICRP and NCRP referred to these quantities as "probability coefficients," preferring to employ the term "risk" for the abstract concept rather than a numerical value of the quantity.

The data embodied in Table 14.3 provided the foundation for the protection limits recommended in ICRP Publication 60 and NCRP Report No. 116. They reflect cancer incidence, adjusted for lethality, and heritable effects. Subsequent studies by the ICRP have led to some revision in the totals given in the last line of Table 14.3 to values of $4.9 \times 10^{-2}$ Sv$^{-1}$ and $6.5 \times 10^{-2}$ Sv$^{-1}$, respectively, for workers and for the general population, in place of $5.6 \times 10^{-2}$ Sv$^{-1}$ and $7.3 \times 10^{-2}$ Sv$^{-1}$. The overall estimated probability coefficients for workers and for the public are thus about 10%

3   Section 13.7.

**Table 14.3** Probability Coefficients for Stochastic Effects (per Sv effective dose)

| Detriment | Adult Workers ($10^{-2}$ Sv$^{-1}$) | Whole Population ($10^{-2}$ Sv$^{-1}$) |
|---|---|---|
| Fatal cancer | 4.0 | 5.0 |
| Nonfatal cancer | 0.8 | 1.0 |
| Severe genetic effects | 0.8 | 1.3 |
| Total | 5.6 | 7.3 |

Source: ICRP Publication 60 and NCRP Report No. 116.

lower than before. According to the latest ICRP pronouncement (Section 14.8), the Commission continues to endorse the numerical dose limits recommended in Publication 60 as providing an appropriate level of protection.

## 14.6
## Current Exposure Limits of the NCRP and ICRP

The exposure limits of the NCRP and ICRP embrace the following philosophy, as stated in NCRP Report No. 116 (p. 9):

> The specific objectives of radiation protection are:
> (1) to prevent the occurrence of clinically significant radiation-induced deterministic effects by adhering to dose limits that are below the apparent threshold levels and
> (2) to limit the risk of stochastic effects, cancer and genetic effects, to a reasonable level in relation to societal needs, values, benefits gained and economic factors.

The Council goes on to include the principle of ALARA in its philosophy. It states, further, that for radiation-protection purposes, the risk of stochastic effects is assumed to be proportional to dose without threshold throughout the dose range of relevance in routine radiation protection.

### Occupational Limits

The Council states that the total lifetime detriment incurred each year from radiation by a worker exposed near the limits over his or her lifetime should be no greater than the annual risk of accidental death in a "safe" industry. The annual rate of fatal accidents in 1991 varied from about $0.2 \times 10^{-4}$ to $5 \times 10^{-4}$, being lowest for trade, manufacturing, and service industries and highest for mining and agriculture. The Council cites the 1980 average annual dose equivalent of 2.1 mSv for monitored radiation workers with measurable exposures. Using the total probability coefficient for workers in Table 14.3, one finds for the average total detriment incurred by a worker $(2.1 \times 10^{-3} \text{ Sv y}^{-1}) (5.6 \times 10^{-2} \text{ Sv}^{-1}) = 1.2 \times 10^{-4} \text{ y}^{-1}$. This level is in the range of the average annual risk for accidental death for all industries.

The following recommendation is made by the NCRP for lifetime occupational exposure to radiation:

> The Council ... recommends that the numerical value of the individual worker's lifetime effective dose in tens of mSv be limited to the value of his or her age in years (not including medical and natural background exposure).

To control the distribution of exposure over a working career,

> The Council recommends that the annual occupational effective dose be limited to 50 mSv (not including medical and background exposure).

It is stipulated, further, that the annual effective-dose limit is to be applied to the sum of (1) the relevant effective doses from external radiation in the specified time period and (2) the committed effective doses from intakes during that period.

Under a worst-case scenario, workers near the end of their careers at age 64 with an accumulated occupational effective dose of 640 mSv would not technically have exceeded the lifetime limit just stated. Their lifetime total detriment, from Table 14.3, would be $(0.64 \text{ Sv}) (5.6 \times 10^{-2} \text{ Sv}^{-1}) = 3.6 \times 10^{-2}$. The worst-case scenario for their lifetime risk of a fatal accident in 50 y of working in industry is about $(50 \text{ y})$ $(5 \times 10^{-4} \text{ y}^{-1}) = 2.5 \times 10^{-2}$, comparable with the estimate for radiation.

The ICRP recommends the same occupational annual effective-dose limit of 50 mSv. However, its cumulative limit is different, being simply 100 mSv in any consecutive 5-y period. Over a 50-y working career, the ICRP lifetime limit would be 1000 mSv, compared with 700 mSv at age 70 y for the NCRP. The NCRP recommendations allow somewhat greater flexibility, but require maintaining cumulative lifetime exposure records for an individual. Technically, the ICRP recommendations require exposure records only over 5-y periods.

For preventing deterministic effects, both the NCRP and ICRP recommend the following annual occupational equivalent-dose limits:

> 150 mSv for the crystalline lens of the eye and 500 mSv for localized areas of the skin, the hands, and feet.

The limits for deterministic effects apply irrespective of whether one or several areas or tissues are exposed.

As mentioned at the end of Section 13.10, exposure of the embryo-fetus entails special risks. For occupational exposures,

> the NCRP recommends a monthly equivalent dose limit of 0.5 mSv to the embryo-fetus (excluding medical and natural background radiation) once the pregnancy is known.

The Council no longer recommends specific controls for occupationally exposed women not known to be pregnant. However, the NCRP adopts the ICRP recommendation of reducing the limits on intake of radionuclides, once pregnancy is known. ICRP Publication 60 also recommends an equivalent-dose limit of 2 mSv to a woman's abdomen, once a pregnancy is declared.

### Nonoccupational Limits

In Section 14.2 we mentioned briefly some of the considerations that apply to establishing exposure limits to members of the public. Historically, limits for nonoccupational exposures have been one-tenth those for occupational exposures. That practice continues. The NCRP makes the following recommendations for the exposure of an individual to man-made sources (natural background and medical exposures are not to be included):

> For continuous (or frequent) exposure, it is recommended
> that the annual effective dose not exceed 1 mSv ...
> Furthermore, a maximum annual effective dose limit of
> 5 mSv is recommended to provide for infrequent annual
> exposures. ...

For deterministic effects, the NCRP recommends an annual equivalent dose limit of 50 mSv for the hands, feet, and skin and 15 mSv for the lens of the eye.

The recommendations in ICRP Publication 60 are somewhat different. An individual annual effective dose limit of 1 mSv is also set for nonoccupational exposures. There is a proviso that a higher annual limit may be applied, if the annual average over 5 y does not exceed 1 mSv.

### Negligible Individual Dose

In its 1987 Report No. 91, the NCRP defined a Negligible Individual Risk Level for radiation as that level below which efforts to reduce exposure to an individual are not warranted. This concept took cognizance of such factors as the mean and variance of natural background exposure levels, the natural risk for the same health effects, risks to which people are accustomed in life, and the perception of risk levels. Also, the magnitude of the implied dose level and the difficulty of measuring it were considered.

NCRP Report No. 116 defines a negligible individual dose (NID), without a corresponding risk level, as follows:

> The Council ... recommends that an annual effective dose of
> 0.01 mSv be considered a Negligible Individual Dose (NID)
> per source or practice.

ICRP Publication 60 does not make a recommendation on the subject.

## Exposure of Individuals Under 18 Years of Age

Exposure of persons under 18 years of age might be warranted for training or educational purposes. The NCRP states that exposures then be made only under conditions of "high assurance" that the annual effective dose will be maintained at less than 1 mSv, the equivalent dose to the lens of the eye at less than 15 mSv, and that to the hands, feet, and skin to less than 50 mSv (excluding medical and natural background exposures). Such exposures are considered part of the "infrequent" nonoccupational limits given earlier in the section for members of the public.

ICRP Publication 60 makes no special recommendations for persons in this category.

The principal recommendations from NCRP Report No. 116 and ICRP Publication 60 are summarized in Table 14.4. The NCRP report gives additional guidance for situations not discussed here, such as emergency occupational exposures and remedial action levels for natural radiation sources.

### Example

What is the effective dose to a worker who receives uniform, whole-body doses of 8.4 mGy from gamma rays and 1.2 mGy from 80-keV neutrons?

### Solution

The effective dose $E$ is computed from its definition, Eq. (14.4). For uniform, whole-body irradiation, the equivalent dose $H_T$ is the same in every tissue and $E$ is numerically equal to $H_T$ [Eq. (14.5)]. The latter is given by Eq. (14.2). The radiation weighting

**Table 14.4** Exposure Limits from NCRP Report No. 116 and ICRP Publication 60

|  | NCRP-116 | ICRP-60 |
| --- | --- | --- |
| Occupational Exposure |  |  |
| Effective Dose |  |  |
| Annual | 50 mSv | 50 mSv |
| Cumulative | 10 mSv × age (y) | 100 mSv in 5 y |
| Equivalent Dose |  |  |
| Annual | 150 mSv lens of eye; | 150 mSv lens of eye; |
|  | 500 mSv skin, hands, feet | 500 mSv skin, hands, feet |
| Exposure of Public |  |  |
| Effective Dose |  |  |
| Annual | 1 mSv if continuous | 1 mSv; higher if needed, provided |
|  | 5 mSv if infrequent | 5-y annual average ⩽1 mSv |
| Equivalent Dose |  |  |
| Annual | 15 mSv lens of eye; | 15 mSv lens of eye; |
|  | 50 mSv skin, hands, feet | 50 mSv skin, hands, feet |

factors are given in Table 14.1: $w_R = 1$ for the gamma rays and, in this example, $w_R = 10$ for the neutrons. Thus, with the doses expressed in mGy, we find that

$$E = H_T = (1 \times 8.4 + 10 \times 1.2) = 20 \text{ mSv}. \tag{14.9}$$

### Example

During the year, a worker receives 14 mGy externally from uniform, whole-body gamma radiation. In addition, he receives estimated 50-y "committed" doses of 8.0 mGy from internally deposited alpha particles in the lung and 180 mGy from beta particles in the thyroid. (a) What is the effective dose for this worker? (b) How much additional external, uniform, whole-body gamma dose could he receive during the year without technically exceeding the NCRP/ICRP annual limit? (c) Instead of the gamma dose in (b), what additional committed alpha-particle dose to the red bone marrow would exceed the annual effective-dose limit?

### Solution

(a) Using the radiation weighting factors from Table 14.1, we obtain the following equivalent doses for the individual tissues, with the tissue weighting factors from Table 14.2 shown on the right:

$$H_{\text{Lung}} = 8.0 \times 20 = 160 \text{ mSv} \quad (w_T = 0.12) \tag{14.10}$$

$$H_{\text{Thyroid}} = 180 \times 1 = 180 \text{ mSv} \quad (w_T = 0.05) \tag{14.11}$$

$$H_{\text{Whole-body}} = 14 \times 1 = 14 \text{ mSv} \quad (w_T = 1.00). \tag{14.12}$$

The effective dose is, by Eq. (14.4),

$$E = 160 \times 0.12 + 180 \times 0.05 + 14 \times 1 = 42 \text{ mSv}. \tag{14.13}$$

(b) In order not to exceed the annual limit, any additional effective dose must be limited to $50 - 42 = 8$ mSv. Therefore, an additional uniform, whole-body gamma dose of 8 mGy would bring the worker's effective dose to the annual limit of 50 mSv.

(c) We need to compute the dose to the red bone marrow that results in an effective dose of 8 mSv. The weighting factor for this tissue is, from Table 14.2, $w_T = 0.12$. Therefore, the committed equivalent dose to the red bone marrow is limited to $H_T = (8 \text{ mSv})/0.12 = 67$ mSv. Since the radiation weighting factor for alpha particles is 20 (Table 14.1), the limiting average absorbed dose to the red bone marrow is, by Eq. (14.1), $D_{\text{RBM},\alpha} = 67/20 = 3.4$ mGy.

Depending on a one's exposure history, it is possible for the cumulative limit in Table 14.4 to be the limiting factor, rather than the annual limit. The NCRP cumulative limit is 10 mSv times the age of a worker in years. The ICRP cumulative limit is 100 mSv in any 5-y period. (See Problem 15.)

**Table 14.5** Exposure limits from NCRP Report No. 91

| | **NCRP-91** |
|---|---|
| Occupational Exposure | |
| Effective Dose Equivalent | |
|    Annual | 50 mSv |
|    Cumulative | 10 mSv × age (y) guidance |
| Dose Equivalent | |
|    Annual | 150 mSv lens of eye; |
| |    500 mSv all other tissues and organs |

## 14.7
### Occupational Limits in the Dose-Equivalent System

ICRP Publication 60 (1991) and NCRP Report No. 116 (1993) appeared at a time during which an extended and intense review of radiation-protection regulations was being conducted by federal agencies in the United States. Prior to these two new publications, radiation-protection practices were generally administered under the system based on dose equivalent (Section 12.2).

Table 14.5 shows the earlier recommendations for occupational exposures given in NCRP Report No. 91 (1987), the predecessor of Report No. 116. Compared with Table 14.4, the effective dose equivalent in Table 14.5 was superseded in Report No. 116 by the effective dose, $E$. The respective numerical values for the annual and cumulative limits are the same in both reports. The effective dose equivalent is defined like $E$ in Eq. (14.4), with $H_T$ then representing the dose equivalent instead of the equivalent dose.[4] The restrictions on effective dose and effective dose equivalent are employed in both systems in order to limit stochastic effects of radiation. The equivalent-dose limits for individual tissues and organs in Table 14.4 and the corresponding dose-equivalent limits in Table 14.5 were made in order to prevent deterministic effects from occurring.

There is a significant difference in the recommendations of Reports No. 91 and 116 for the limitation of annual intakes. Under Report No. 91 (p. 19), "The Annual Limit on Intake (ALI) is the maximum quantity of a radionuclide that can be taken into the body based on ICRP Reference Man ... each year without the committed effective dose equivalent being in excess of the annual effective dose equivalent limit ... or the committed dose equivalent to any tissue being in excess of the nonstochastic [deterministic] limit." In both systems, the deterministic limits apply whether an individual tissue or organ is exposed selectively or together with other tissues and organs. As we saw in Section 14.6, values of the ARLI in Report No. 116 are determined from an annual 50-y committed effective-dose limit of 20 mSv. The analogous ALI in Report No. 91 are based on a limit of 50 mSv. As a result of this reduction, there are only a few radionuclides that could approach lifetime doses of concern for

---

4  In addition to the conceptual differences between these two quantities (Section 14.4), some radiation and tissue weighting factors were also revised in Report No. 116.

deterministic effects under NCRP Report No. 116. Specific equivalent-dose limits are needed only for the lens of the eye, the skin, hands, and feet, as shown in Table 14.4. (The effective-dose limit protects the skin sufficiently against stochastic effects.)

The newer ICRP/NCRP recommendations were not fully implemented into U.S. regulations by the federal and state reviews, which were completed in the early 1990s. For example, the present rules of the Nuclear Regulatory Commission (NRC) and the Department of Energy (DOE) have continued based on dose equivalent.[5] The basic limits are those shown in Table 14.5, except that there is no lifetime cumulative limit in the federal rules. Also, the 20-mSv annual committed limit was not adopted federally, and so the 500 mSv deterministic limit for "all other tissues and organs" in Table 14.5 is still in effect. Like all parts of the current regulations in the U.S., these decisions followed extensive studies by various organizations, public comments, and reviews of past operating experiences. It was judged that additional changes at the time would result in very little reduction in annual doses, which remain well below the limits, averaging only a few mSv or less. Application of the ALARA principle, which was further emphasized in the revised federal regulations, has played an effective role in keeping occupational exposures low within current ICRP/NCRP recommendations.

*Example*

In the next-to-last example, a worker received external whole-body doses of 8.4 mGy from gamma rays and 1.2 mGy from 80-keV neutrons. (a) Calculate his total effective dose equivalent based on NCRP Report No. 91. The values of the gamma and neutron quality factors are, respectively, $Q = 1$ and $Q = 13$. (b) In addition to these exposures during a year, how much additional internal beta dose ($Q = 1$) to the thyroid alone could be received without technically exceeding the limits in Table 14.5? The thyroid weighting factor is 0.03.

*Solution*

(a) The total effective dose equivalent is, in place of the effective dose, Eq. (14.9),

$$\text{TEDE} = 1 \times 8.4 + 13 \times 1.2 = 24 \text{ mSv.} \tag{14.14}$$

(b) The addition of $50 - 24 = 26$ mSv would reach the stochastic limit in Table 14.5. Given the tissue weighting factor, the thyroid dose equivalent for this contribution would be $(26 \text{ mSv})/(0.03) = 867$ mSv. This amount exceeds the deterministic limit of 500 mSv in Table 14.5. Since the thyroid already has 24 mSv from the external radiation in Part (a), an additional dose equivalent of $500 - 24 = 476$ mSv from the beta irradiation would reach the deterministic limit for the organ. In order to comply with Table 14.5, the maximum additional internal thyroid dose, therefore, could not exceed 476 mGy. (The TEDE for the year is thereby limited to $24 + 0.03 \times 476 = 38$ mSv by the deterministic dose limit to the thyroid.)

---

5   See 10 CFR Part 20 and 10 CFR Part 835 listed in Section 14.11. These laws became effective, respectively, in 1991 for NRC and 1993 for DOE. At this writing, DOE is considering adoption of ICRP-60 *terminology* for occupational radiation exposures. See, for example, Federal Register/Vol. 71, No. 154/Thursday, August 10, 2006/Proposed Rules.

The last example shows that one must pay attention to both the stochastic and the deterministic limits in the dose-equivalent system. Neither should be exceeded, and in specific cases one or the other can be the deciding factor.

## 14.8
### The "2007 ICRP Recommendations"

It was mentioned at the end of Section 14.3 that, at the time of this writing (2007), the ICRP has under consideration a draft of major revisions to its recommendations. The draft consolidates developments made in several ICRP publications subsequent to the 1991 Publication 60 and also introduces new conceptual framework and approaches to radiation protection. In this regard, however, it reaffirms the appropriateness of the numerical limits in Publication 60. As noted in Section 14.5, the total probability coefficients in Table 14.3 have also been revised to somewhat lower values than those shown. The draft recommendations have been posted on the ICRP Web site, where they have received extensive public comment and discussion. We summarize briefly some of the major new aspects.

Under Publication 60, the system of radiation protection is based on three factors: the *justification* of a practice, the *optimization* of protection (essentially, the application of ALARA), and individual *dose limits.* The proposed system would drop justification as part of a radiation-protection program. It would apply to practices already declared as justified by an appropriate authority, such as a government agency. It would also apply to natural radiation sources that are controllable. Very low-level sources and natural sources that are not controllable are to be excluded. Medical exposures are considered in the new recommendations.

The proposed system employs *dose constraints, dose limits,* and *optimization.* There are three numerical dose constraints, shown in Table 14.6, which apply to an individual person in one of three situations. It is expected that regulatory authorities would set limits well below these numerical constraints, which, if exceeded, would be considered an unacceptable result. In addition to the constraints, continued use of ICRP Publication 60 limits for normal operations only is proposed. The last entry in Table 14.6, 0.01 mSv, is the smallest constraint value that should be used for any situation. It is associated with a trivial risk. It serves as a basis for the new concept of exclusion levels of activity concentrations ($Bq\,g^{-1}$) for designated radionuclides, below which radiation protection is not needed. The concept of optimization is generalized to include a broader range of factors than before, such as economic and social considerations.

Some new definitions are introduced. For example, to avoid the confusion between the terms "dose equivalent" and "equivalent dose," the name "weighted dose" is proposed for the latter. Changes are also made in numerical values for some radiation- and tissue-weighting factors. An analytic function is provided for the neutron weighting factor as a function of energy.

The Commission proposes a framework for the protection of non-human species and the environment. Nomenclature, data sets, reference dose models for certain

**Table 14.6** Maximum dose constraints recommended for workers and members of the public from single dominant sources for all types of exposure situations that can be controlled

| Maximum constraint (effective dose, mSv in a year) | Situation to which it applies |
|---|---|
| 100 | In emergency situations, for workers, other than for saving life or preventing serious injury or preventing catastrophic circumstances, and for public evacuation and relocation; and for high levels of controllable existing exposures. There is neither individual nor societal benefit from levels of individual exposure above this constraint. |
| 20 | For situations where there is direct or indirect benefit for exposed individuals, who receive information and training, and monitoring or assessment. It applies into occupational exposure, for countermeasures such as sheltering, iodine prophylaxis in accidents, and for controllable existing exposures such as radon, and for comforters and carers to patients undergoing therapy with radionuclides. |
| 1 | For situations having societal benefit, but without individual direct benefit, and there is no information, no training, and no individual assessment for the exposed individuals in normal situations. |
| 0.01 | Minimum value of any constraint. |

species, and the interpretation of effects will be developed. There is a need for unified, international standards for environmental discharges. Criteria for the management and assessment of environmental impacts of radiation practices will be developed.

No short summary can do justice to the major undertaking of the 2007 Recommendations. The reader is referred to the ICRP Web site (www.icrp.org).

## 14.9
## ICRU Operational Quantities

In a series of reports,[6] the International Commission on Radiation Units and Measurements has addressed the relationship between practical radiation-protection measurements and assessment of compliance with the limits set forth by the ICRP. The basic organ and tissue doses specified in the limits are essentially unmeasurable, but can be estimated from measurements made at appropriate locations in tissue-equivalent phantoms and from calculations. Accordingly, the ICRU has introduced several operational quantities for practical measurements under well-

---

6  See ICRU Report 51 listed in Section 14.11 and earlier reports of the Commission cited in this reference.

defined conditions with explicitly stated approximations. The measurements are intended to give adequate approximations to quantities that can be used for limitation purposes for potentially exposed persons.

The ICRU operational quantities are defined in terms of the older concept of dose equivalent (Section 12.2) at certain locations in a tissue-equivalent sphere, having a diameter of 30 cm. The quality factor for calculating dose equivalent is that given in ICRP Publication 60 (Table 12.2). The ICRU sphere has unit density and a mass composition of 76.2% oxygen, 11.1% carbon, 10.1% hydrogen, and 2.6% nitrogen. Two quantities are defined for area monitoring that link an external radiation field to the effective dose equivalent [see definition following Eq. (14.5)] and the dose equivalent to the skin and to the lens of the eye. Limits for the latter organs usually restrict exposures when only weakly penetrating radiation is present. (Radiation can be characterized as strongly or weakly penetrating on the basis of which dose equivalent is closer to its limiting value.) A third quantity pertains to individual monitoring.

An external radiation field is characterized at a given point $P$ by the particle fluence and its directional and energy distributions there. Given the field at $P$, the ICRU defines an "expanded" field as the uniform field that has everywhere the properties of the actual field at $P$. It further defines the "expanded and aligned" field like the expanded one, except that the fluence is unidirectional. The two quantities for area monitoring are defined in terms of dose equivalents that would occur at specific depths in the ICRU sphere if placed in these two fictitious fields:

1. The ambient dose equivalent, $H^*(d)$, at a point $P$ in a radiation field is the dose equivalent that would be produced by the expanded and aligned field in the ICRU sphere at a depth $d$ on the radius opposing the direction of that field. For strongly penetrating radiation, the ICRU recommends a depth $d = 10$ mm and, for weakly penetrating radiation, $d = 0.07$ mm for the skin and $d = 3$ mm for the eye. (By convention, the ICRU specifies the depths $d$ in mm.) Measurement of $H^*(d)$ generally requires that the radiation field be uniform over the dimensions of the instrument and that the instrument have isotropic response.

2. The directional dose equivalent, $H'(d, \Omega)$, at $P$ is the dose equivalent that would be produced by the expanded field at a depth $d$ on a radius specified by the direction $\Omega$. The same recommendations are made for $d$ as with $H^*(d)$. Measurement of $H'(d, \Omega)$ requires that the radiation field be uniform over the dimensions of the instrument and that the instrument have the required directional response with respect to $\Omega$.

The third operational quantity, which does not involve the ICRU sphere, is defined for individual monitoring:

3. The personal dose equivalent, $H_p(d)$, is the dose equivalent in soft tissue at a depth $d$ below a specified point in the body. For weakly penetrating radiation, depths $d = 0.07$ mm and $d = 3$ mm are recommended for the skin and eye. For strongly penetrating radiation, $d = 10$ mm is to be used. The personal dose equivalent $H_p(d)$ can be measured by a calibrated detector, worn at the surface of the body and covered with the appropriate thickness of tissue-equivalent material. The quantities $H_p(10)$ and $H_p(0.07)$ are, respectively, associated with the regulatory assessments of the *deep* and *shallow* dose equivalents for an individual.

## 14.10
### Probability of Causation

As we have pointed out several times, it is difficult, if not impossible, to attribute a given malignancy in a person to his or her past radiation history. Diseases induced by radiation, from either natural or man-made sources, also occur spontaneously. The concept *of probability of causation* has been introduced to provide an estimate of the probability that a given cancer in a specific tissue or organ was caused by previous exposure to a carcinogen, such as radiation. Although not a part of limits setting, the concept is closely related to the health effects and risk estimates that we have been discussing here.

If $R$ denotes the excess relative risk for the cancer that results from a given radiation dose, then the probability of causation $P$ is defined as

$$P = \frac{R}{1 + R}.$$

(14.15)

This concept does not take other factors into account, such as uncertainties in dose and in the models used to determine $R$.

In the United States, the Congress mandated the use of the probability of causation to evaluate claims of radiation injury from nuclear weapons testing, fallout, and uranium mining. The National Institutes of Health developed tables for cancers in various organs of persons of both sexes who received various doses at different ages.[7]

In 1992, the National Council on Radiation Protection and Measurements issued a statement on the probability of causation, discussing its usefulness and its limitations. The statement and comments about it from a statistician and from a physician are published in *Radiation Research*.[8]

---

7  Report of the NIH *Ad Hoc Working Group to Develop Radioepidemiology Tables*, NIH Publication No. 85-2748, U.S. Government Printing Office, Washington, D.C. (1984).

8  *Rad. Res.* 134, 394-397 (1993).

# 14.11
# Suggested Reading

The World Wide Web is an invaluable source of information on the subjects of this chapter. The reader is referred to the entire comprehensive reports published by the National Council on Radiation Protection and Measurements and the International Commission on Radiological Protection. The documents cover a wide variety of specialized topics dealing with radiation and the practice of radiation protection from all manner of sources. Other organizations that provide publications concerned with radiation protection include the following: American National Standards Institute (ANSI), Food and Drug Administration (FDA), International Atomic Energy Agency (IAEA), International Commission on Radiation Units and Measurements (ICRU), International Labor Organization (ILO), National Academy of Sciences—National Research Council (NAS–NRC), Society of Nuclear Medicine—Medical Internal Radiation Dose (MIRD) Committee, United Nations Scientific Committee on the Effects of Atomic Radiation (UNSCEAR), U.S. Department of Energy (DOE), U.S. Environmental Protection Agency (EPA), and the U.S. Nuclear Regulatory Commission (NRC). In the United States, specific legal radiation-exposure regulations are published in the Federal Register under Title 10, Part 20 of the Code of Federal Regulations.

Several of the works cited in Section 1.6 describe the history of the principal organizations and the development of the radiation-protection concepts, practices, and limits that we have today.

## Other Suggested Reading

1 Clark, R. and Valentin, J., "A History of the International Commission on Radiological Protection," *Health Phys.* **88**, 717–732 (2005).

2 Health Physics Society, "Radiation Risk in Perspective," Position Statement of the Health Physics Society, McLean, VA (Revision Aug. 2004).

3 ICRP Publication 60, *1990 Recommendations of the International Commission on Radiological Protection*, Annals of the ICRP, Vol. 21/1–3 (1991).

4 ICRU Report 51, *Quantities and Units in Radiation Protection*, International Commission on Radiation Units and Measurements, Bethesda, MD (1993). [Also see earlier ICRU Reports cited in this reference.]

5 ICRU Report 66, *Determination of Operational Dose Equivalent Quantities for Neutrons*, J. ICRU, Vol. 1, No. 3 (2001).

6 Jones, C. G., "A Review of the History of U.S. Radiation Protection Regulations, Recommendations, and Standards," *Health Phys.* **88**, 697–716 (2005).

7 Leggett, R. W. and Eckerman, K. F., *Dosimetric Significance of the ICRP's Updated Guidance and Models, 1989–2003, and Implications for U.S. Federal Guidance*, ORNL/TM-2003/207, Oak Ridge National Laboratory, Oak Ridge, TN (2003).

8 NCRP, Proceedings 39th Annual Meeting, April 9–10, 2004. "Radiation Protection at the Beginning of the 21st Century—A Look Forward," *Health Phys.* **87**, 249–318 (2004).

9 NCRP Report No. 91, *Recommendations on Limits for Exposure to Ionizing Radiation*, National Council on Radiation Protection and Measurements, Bethesda, MD (1987).

10 NCRP Report No. 115, *Risk Estimates for Radiation Protection*, National

Council on Radiation Protection and Measurements, Bethesda, MD (1993).

11 NCRP Report No. 116, *Limitation of Exposure to Ionizing Radiation*, National Council on Radiation Protection and Measurements, Bethesda, MD (1993).

12 NCRP Report No. 126, *Uncertainties in Fatal Cancer Risk Estimates used in Radiation Protection*, National Council on Radiation Protection and Measurements, Bethesda, MD (1997).

13 NCRP Report No. 136, *Evaluation of the Linear-Nonthreshold Dose-Response Model for Ionizing Radiation*, National Council on Radiation Protection and Measurements, Bethesda, MD (2001). ["In keeping with previous reviews by the NCRP..., the Council concludes that there is no conclusive evidence on which to reject the assumption of a linear-nonthreshold dose-response relationship for many of the risks attributable to low-level ionizing radiation although additional data are needed...."]

## 14.12
## Problems

1. In earlier years, fluoroscopes were available in stores for inspecting how well shoes fit. What are the benefits and risks from this use of X rays?

2. Welds in metal structures can be inspected with gamma rays (e.g., from $^{137}$Cs) to detect flaws not visible externally. The production, transport, and use of such sources entails exposures of workers and the public to some radiation. Give an example to show how banning the use of gamma rays for this purpose could be of greater detriment to society than the radiation exposures it entails.

3. Discuss risks and benefits associated with the development of nuclear power. What risk and benefit factors are there, apart from the potential health effects of radiation?

4. Discuss the risks and benefits associated with having nuclear submarines.

5. A proposal is made to test a new dental-hygiene procedure for children that is said to have the potential of greatly reducing tooth decay. Dental X rays of several thousand children would have to be made periodically during a 5-y study in order to perfect and evaluate the procedure. Discuss the rationale on which a decision could be made either to implement or to reject the proposal.

6. A gamma source is used for about an hour every day in a laboratory room. If the source, when not in use, is kept in its container in the room, the resulting dose equivalent to persons working there is well within the allowable limit. Alternatively, the source could be stored in the unoccupied basement, two

floors below, with virtually no exposure to personnel. How would the ALARA principle apply in this example?

7. Show that 50 mSv y$^{-1}$ is equivalent to an average rate of 0.025 mSv h$^{-1}$ for 40 h wk$^{-1}$, 50 wk y$^{-1}$.

8. A worker receives uniform, whole-body doses of 0.30 mGy from 100-keV neutrons, 0.19 mGy from 1.5-MeV neutrons, and 4.3 mGy from gamma rays. Calculate the effective dose.

9. A worker has received a committed equivalent dose of 106 mSv to the gonads during a year. What additional, uniform, whole-body external gamma-ray dose could he or she receive without technically exceeding the NCRP annual limit on effective dose?

10. According to NCRP Report No. 116, an equivalent dose of 10 mSv to the lung gives the same effective dose as (choose one or more):
    (a) 10 mSv to the skin
    (b) 1.2 mSv to the whole body
    (c) 24 mSv to the liver
    (d) 10 mSv to the gonads
    (e) 5 mSv to the stomach.

11. (a) What equivalent dose to the thyroid represents the same total detriment to an individual as 5 mSv to the whole body?
    (b) What equivalent dose to the lung represents the same total detriment as 50 mSv to the thyroid?

12. A radiation worker is 31 years of age. What is his cumulative effective-dose limit in mSv according to
    (a) the NCRP?
    (b) the ICRP?

13. A worker receives a lung dose of 6 mGy from alpha radiation from an internally deposited radionuclide plus a 20-mGy uniform, whole-body dose from external gamma radiation.
    (a) What is the equivalent dose to the lung?
    (b) What is the his or her effective dose?

14. A worker has a thyroid dose of 200 mGy from an internally deposited beta emitter and a lung dose of 8 mGy from an internal alpha emitter.
    (a) What is the effective dose?
    (b) What uniform, whole-body dose from thermal neutrons would result in the same effective dose?

15. In the example at the end of Section 14.6, suppose that the worker receives these lung, thyroid, and whole-body exposures while 40 years of age. Furthermore, just as the worker turned 40, his cumulative effective dose was 354 mSv. In order that

this worker's exposure not exceed NCRP limits, what are now the answers to parts (b) and (c) of the example?

16. Calculate the effective dose for an individual who has received the following exposures:

    1 mGy alpha to the lung

    2 mGy thermal neutrons, whole body

    5 mGy gamma, whole body

    200 mGy beta to the thyroid.

17. The annual reference level of intake by inhalation for $^{32}$P is $1 \times 10^7$ Bq.

    (a) What is the committed effective dose per unit activity for the radionuclide?

    (b) If an amount of $^{32}$P equal to the ARLI were ingested on January 15, what would be the effective dose during that calendar year?

18. What is the value of the DRAC for $^{32}$P from the last problem?

19. Ingestion of a certain radionuclide results in a committed effective dose of $5.2 \times 10^{-5}$ mSv Bq$^{-1}$. What is the ARLI?

20. Distinguish between lifetime equivalent dose and committed equivalent dose.

21. As a result of the single intake of $6.3 \times 10^3$ Bq of a radionuclide, a certain organ of the body will receive a dose during the next 50 y of 0.20 mGy from beta particles and 0.15 mGy from alpha particles. The organ has a tissue weighting factor of 0.05.

    (a) Calculate the committed equivalent dose to the organ.

    (b) If this organ is the only tissue irradiated, calculate the committed effective dose.

    (c) What is the ARLI for this route of intake?

22. Critique the probability of causation as a basis for attributing a given case of malignancy to a prior exposure to radiation.

23. If the excess relative risk for a certain cancer is 0.015 at a time 20 y after receiving a dose of 25 mGy, what is the probability of causation?

24. The BEIR-V Report estimated that about 800 extra cancer deaths would be expected as a result of exposing a population of 100,000 persons (of all ages) to a whole-body gamma dose of 0.1 Gy. About 20,000 deaths by cancer would be expected in the population in the absence of radiation. How many persons exposed to 0.1 Gy of gamma radiation would be needed in order to observe such a level of excess cancers at the 95% confidence limit?

**14.13**

**Answers**

8. 11 mSv

9. 29 mGy

13. (a) 140 mSv

(b) 34 mSv

14. (a) 29 mSv

(b) 5.8 mGy

15. 4 mGy gamma,
1.7 mGy alpha

17. (a) $2 \times 10^{-6}$ mSv Bq$^{-1}$

(b) 20 mSv

21. (a) 3.2 mSv

(b) 0.16 mSv

(c) $8 \times 10^{5}$ Bq

23. 0.0148

# 15
# External Radiation Protection

We now describe procedures for limiting the dose received from radiation sources outside the human body. In the next chapter we discuss protection from radionuclides that can enter the body.

## 15.1
### Distance, Time, and Shielding

In principle, one's dose in the vicinity of an external radiation source can be reduced by increasing the distance from the source, by minimizing the time of exposure, and by the use of shielding. Distance is often employed simply and effectively. For example, tongs are used to handle radioactive sources in order to minimize the dose to the hands as well as the rest of the body. Limiting the duration of an exposure significantly is not always feasible, because a certain amount of time is usually required to perform a given task. Sometimes, though, practice runs beforehand without the source can reduce exposure times when an actual job is carried out.

While distance and time factors can be employed advantageously in external radiation protection, shielding provides a more reliable way of limiting personnel exposure by limiting the dose rate. In principle, shielding alone can be used to reduce dose rates to desired levels. In practice, however, the amount of shielding employed will depend on a balancing of practical necessities such as cost and the benefit expected.

In this chapter we describe methods for determining appropriate shielding for the most common kinds of external radiation: gamma rays, X rays from diagnostic and therapeutic machines, beta rays with accompanying bremsstrahlung, and neutrons.

*Atoms, Radiation, and Radiation Protection.* James E. Turner
Copyright © 2007 WILEY-VCH Verlag GmbH & Co. KGaA, Weinheim
ISBN: 978-3-527-40606-7

## 15.2
### Gamma-Ray Shielding

In Section 8.7 we discussed attenuation coefficients and described the transmission of photons through matter under conditions of "good" geometry. The relative intensity $I/I_0$ of monoenergetic photons transmitted without interaction through a shield of thickness $x$ is, from Eq. (8.43),

$$I = I_0 e^{-\mu x}, \tag{15.1}$$

where $\mu$ is the linear attenuation coefficient. If the incident beam is broad, as in Fig. 8.10, then the measured intensity will be greater than that described by Eq. (15.1) because scattered photons will also be detected. Such conditions usually apply to the shields required for protection from gamma-ray sources. The increased transmission of photon intensity over that measured in good geometry can be taken into account by writing

$$I = B I_0 e^{-\mu x}, \tag{15.2}$$

where $B$ is called the buildup factor ($B \geq 1$). For a given shielding material, thickness, photon energy, and source geometry, $B$ can be obtained from measurements or calculations.

Figures 15.1–15.5 show exposure buildup factors for five materials for monoenergetic photons with energies up to 10 MeV from point isotropic sources. The

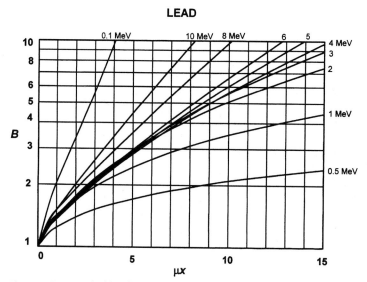

**Fig. 15.1** Exposure buildup factors, $B$, in lead for point sources of monoenergetic photons of energies from 0.1 MeV to 10 MeV as functions of the number of relaxation lengths, $\mu x$.

thickness of a shield for which the photon intensity in a narrow beam is reduced to 1/e of its original value is called the relaxation length. One relaxation length, therefore, is equal to $1/\mu$, the mean free path. The dependence of $B$ in the figures on shield thickness is expressed by its variation with the number of relaxation lengths, $\mu x$. In addition to isotropic point sources, Fig. 15.6 shows an example of buildup factors for a broad, parallel beam of monoenergetic photons normally incident on a uranium slab.

Figures 15.1–15.6 can be used with Eq. (15.2) to calculate the shielding thickness $x$ necessary to reduce gamma-ray intensity from a value $I_0$ to $I$. Since the exponential attenuation factor $e^{-\mu x}$ and the buildup factor $B$ both depend on $x$, which is originally unknown, the appropriate thickness for a given problem usually has to be found by making successive approximations until Eq. (15.2) is satisfied. An initial (low) estimate of the amount of shielding needed can be obtained by solving Eq. (15.2) for $\mu x$ with assumed narrow-beam geometry, that is, with $B = 1$. One can then add some additional shielding and see whether the values of $B$ and the exponential for the new thickness satisfy Eq. (15.2).

Two examples will illustrate gamma-ray shielding calculations.

*Example*
Calculate the thickness of a lead shield needed to reduce the exposure rate 1 m from a 10-Ci point source of $^{42}$K to 2.5 mR h$^{-1}$. The decay scheme of the $\beta^-$ emitter is shown in Fig. 15.7. The daughter $^{42}$Ca is stable.

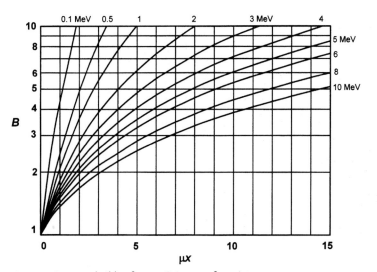

**WATER**

**Fig. 15.2** Exposure buildup factors, $B$, in water for point sources of monoenergetic photons of energies from 0.1 MeV to 10 MeV as functions of the number of relaxation lengths, $\mu x$.

## CONCRETE

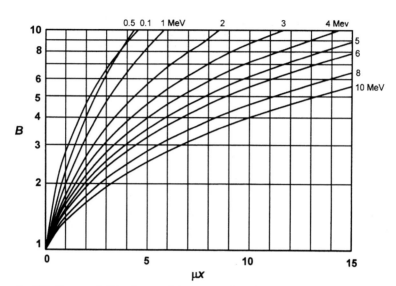

**Fig. 15.3** Exposure buildup factors, *B*, in concrete for point sources of monoenergetic photons of energies from 0.1 MeV to 10 MeV as functions of the number of relaxation lengths, $\mu x$.

## ALUMINUM

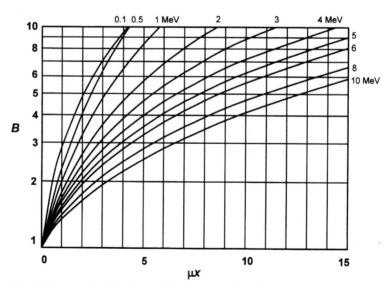

**Fig. 15.4** Exposure buildup factors, *B*, in aluminum for point sources of monoenergetic photons of energies from 0.1 MeV to 10 MeV as functions of the number of relaxation lengths, $\mu x$.

**URANIUM**

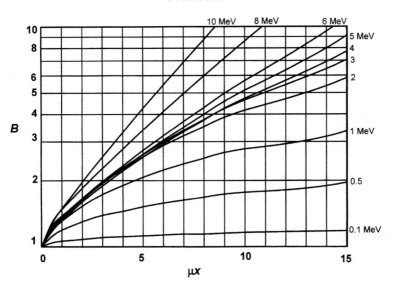

Fig. 15.5 Exposure buildup factors, *B*, in uranium for point sources of monoenergetic photons of energies from 0.1 MeV to 10 MeV as functions of the number of relaxation lengths, $\mu x$.

**URANIUM**

**(Broad Beam)**

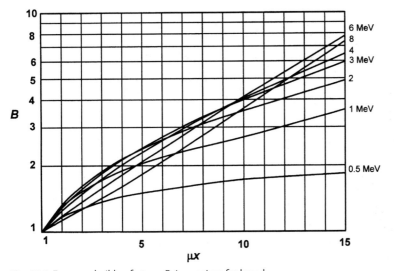

Fig. 15.6 Exposure buildup factors, *B*, in uranium for broad, parallel beams of monoenergetic photons of energies from 0.5 MeV to 8 MeV as functions of the number of relaxation lengths, $\mu x$.

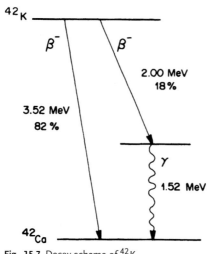

**Fig. 15.7** Decay scheme of $^{42}$K.

*Solution*

With no shielding, the exposure rate at $r = 1$ m is given by Eq. (12.27):

$$\dot{X} = 0.5CE = 0.5 \times 10 \times (0.18 \times 1.52) = 1.37 \text{ R h}^{-1}. \tag{15.3}$$

We make an initial estimate of the shielding required to reduce this to 2.5 mR h$^{-1}$ on the basis of narrow-beam geometry. The number of relaxation lengths $\mu x$ needed would then satisfy the relation

$$e^{-\mu x} = \frac{2.5}{1370} = 1.82 \times 10^{-3}, \tag{15.4}$$

or $\mu x = 6.31$. The energy of the photons emitted by $^{42}$K is 1.52 MeV. We see from Fig. 15.1 (point source) that for photons of this energy in lead and a thickness of 6.31 relaxation lengths, the value of the buildup factor is about 3. To keep the required reduction (15.4) the same when the buildup factor is used, the number of relaxation lengths in the exponential must be increased. The number $y$ of added relaxation lengths that compensates a buildup factor of 3 is given by $e^{-y} = 1/3$, or $y = \ln 3 = 1.10$. Added to the initial value, the estimated shield thickness becomes $6.31 + 1.10 = 7.41$ relaxation lengths. Inspection of Fig. 15.1 shows that the buildup factor has now increased to perhaps 3.5. Thus, a better guess is $y = \ln 3.5 = 1.25$, with an estimated shield thickness of $6.31 + 1.25 = 7.56$, which we round off to 7.6 relaxation lengths. It remains to verify a final solution numerically by trial and error. For $\mu x = 7.6$ in Fig. 15.1, we estimate that $B = 3.6$. The reduction factor with buildup included is then

$$Be^{-\mu x} = 3.6e^{-7.6} = 1.8 \times 10^{-3}, \tag{15.5}$$

which agrees with (15.4) within our degree of precision. From Fig. 8.8, we obtain for the mass attenuation coefficient $\mu/\rho = 0.051$ cm$^2$ g$^{-1}$. With $\rho = 11.4$ g cm$^{-3}$ for lead, we have $\mu = 0.581$ cm$^{-1}$. The required thickness of lead shielding is $x = 7.6/\mu =$

7.6/0.581 = 13 cm. A shield of this thickness can be interposed anywhere between the source and the point of exposure. Usually, shielding is placed close to a source to realize the greatest solid-angle protection.

Until now we have discussed monoenergetic photons. When photons of different energies are present, separate calculations at each energy are usually needed, since the attenuation coefficients and buildup factors are different.

*Example*

A 10-Ci point source of $^{24}$Na is to be stored at the bottom of a pool of water. The radionuclide emits two photons per disintegration with energies 2.75 MeV and 1.37 MeV in decaying by $\beta^-$ emission to stable $^{24}$Mg. How deep must the water be if the exposure rate at a point 6 m directly above the source is not to exceed 20 mR h$^{-1}$? What is the exposure rate at the surface of the water right above the source?

*Solution*

The mass attenuation coefficients for the two photon energies can be obtained from Fig. 8.9. Since we are dealing with water, they are numerically equal to the linear attenuation coefficients. Thus, $\mu_1 = 0.043$ cm$^{-1}$ and $\mu_2 = 0.061$ cm$^{-1}$, respectively, for the 2.75-MeV and 1.37-MeV photons. The approach we use is to consider the harder photons first and find a depth of water that will reduce their exposure rate to a level somewhat below 20 mR h$^{-1}$, and then see what additional exposure rate results from the softer photons. The final depth can be adjusted to make the total 20 mR h$^{-1}$. The exposure rate from 2.75-MeV photons at a distance $d = 6$ m with no shielding is

$$\dot{X}_{2.75} = \frac{0.5CE}{d^2} = \frac{0.5 \times 10 \times 2.75}{6^2} = 0.382 \text{ R h}^{-1}. \tag{15.6}$$

To reduce this level to 20 mR h$^{-1}$ under conditions of good geometry requires a water depth $\mu_1 x$ given by

$$20 = 382e^{-\mu_1 x}, \tag{15.7}$$

or $\mu_1 x = 2.95$ relaxation lengths. From Fig. 15.2, one can see that the buildup factor $B_1$ for 2.75-MeV photons for a water shield of this thickness is about 3.4. The number of relaxation lengths that compensate for this amount of buildup is $y = \ln 3.4 = 1.22$. Adding this amount to the preceding gives an estimate $\mu_1 x = 2.95 + 1.22 = 4.17$. At this depth, the buildup factor has increased to about 4.4, for which the compensating depth added to the first estimate is $y = \ln 4.4 = 1.5$. Therefore, we try an estimated relaxation length of $2.95 + 1.5 = 4.5$. Thus, we obtain for the shielded exposure rate for the more energetic photons,

$$\dot{X}_{2.75} = 4.4 \times 382e^{-4.5} = 19 \text{ mR h}^{-1}. \tag{15.8}$$

For the 1.37-MeV photons, the thickness of this shield in relaxation lengths is larger by the ratio of the attenuation coefficients; that is, $\mu_2 x = 4.5 \times (0.061/0.043) = 6.4$. From Fig. 15.2, the buildup factor for the 1.37-MeV photons is estimated to be, approximately, $B_2 = 12$. The exposure rate at 6 m for these photons without shielding is

$$\dot{X}_{1.37} = \frac{0.5 \times 10 \times 1.37}{6^2} = 0.190 \text{ R h}^{-1}. \tag{15.9}$$

With the shield it is

$$\dot{X}_{1.37} = 12 \times 190 \times e^{-6.4} = 3.8 \text{ mR h}^{-1}. \tag{15.10}$$

The total shielded exposure rate is

$$\dot{X} = \dot{X}_{2.75} + \dot{X}_{1.37} = 19 + 4 = 23 \text{ mR h}^{-1}. \tag{15.11}$$

Some additional thickness is needed. One can proceed simply by incrementing the relaxation length until the desired value of the total exposure rate is obtained. In place of 4.5, for example, one can try $\mu_1 x = 4.7$, for which $B_1 = 4.8$, $\mu_2 x = 6.7$, and $B_2 = 15$. One finds then that

$$\dot{X} = \dot{X}_{2.75} + \dot{X}_{1.37} = 16.7 + 3.5 = 20.2 = 20 \text{ mR h}^{-1}. \tag{15.12}$$

It follows that the required depth of water is $x = 4.7/\mu_1 = 4.7/(0.043 \text{ cm}^{-1}) = 109$ cm. The exposure rate at the surface of the water is $20(600/109)^2 = 610 \text{ mR h}^{-1}$.

In principle, different sets of buildup factors are needed to compute different transmitted quantities, such as dose, exposure, kerma, and energy fluence. However, values given in Figs. 15.1–15.6 specifically for exposure are not very different from those for the others. We shall make no distinction in using the single set of figures for all computations.[1]

## 15.3
### Shielding in X-Ray Installations

X-ray machines have three principal uses—as diagnostic, therapeutic, and non-medical radiographic devices. An X-ray tube is usually housed in a heavy lead casing with an aperture through which the primary, or useful, beam emerges. Typically, the beam passes through metal filters (e.g., Al, Cu) to remove unwanted, less penetrating radiation and is then collimated to reduce its width. The housing, supplied by the manufacturer, must conform to certain specifications in order to limit the leakage radiation that emerges from it during operation. For diagnostic X-ray tubes, regulations require that manufacturers limit the leakage exposure rate at a distance of 1 m from the target of the tube to 0.1 R h$^{-1}$ when operated continuously at its maximum rated current and potential difference.

---

1 Figures 15.1–15.5 are based on Table 6.5.1 from B. Schleien, L. A. Slayback, Jr., and B. K. Birky, *Handbook of Radiological Health*, *3rd Ed.*, Williams and Wilkins (1998).

Figure 15.6 is plotted from data on p. 147, *Radiological Health Handbook, Revised Ed.*, Jan. 1970, Bureau of Radiological Health, Rockville, MD.

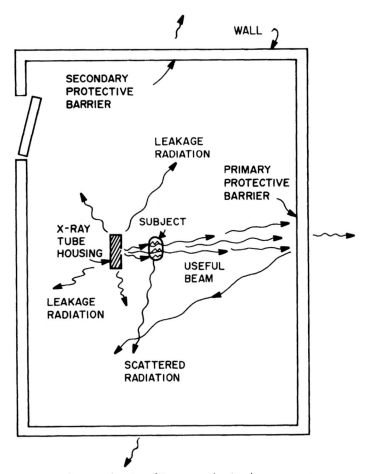

**Fig. 15.8** Schematic plan view of X-ray room showing the different radiation components considered in the design of structural shielding to provide primary and secondary protective barriers.

The shielding provided by the X-ray housing is referred to as source shielding. Additional protection is obtained by the use of structural shielding in an X-ray facility. The basic components of the radiation field considered in the design of structural shielding are shown in Fig. 15.8. A primary protective barrier, such as a lead-lined wall, is fixed in place in any direction in which the useful beam can be pointed. This shield reduces the exposure rate outside the X-ray area in the direction of the primary beam. Locations not in the direct path of the beam are also exposed to photons in two ways. As illustrated in Fig. 15.8, leakage radiation escapes from the housing in all directions. In addition, photons are scattered from exposed objects in the primary beam and from walls, ceilings, and other structures. Secondary protective barriers are needed to reduce exposure rates outside the X-ray area from both leakage and scattered radiation. Sometimes existing struc-

**Table 15.1** Air-Kerma Shielding-Design Goals, *P*, from NCRP Report No. 147

| Controlled Areas | Uncontrolled Areas |
| --- | --- |
| 0.1 mGy wk$^{-1}$ | 0.02 mGy wk$^{-1}$ |

tures, such as concrete walls, provide sufficient secondary barriers; otherwise, additional shielding, such as lead sheets, must be added to them.

Generally, structural shielding has been designed in a manner consistent with limiting the effective dose to an individual outside the X-ray room to 1 mSv wk$^{-1}$ in controlled areas and to 0.1 mSv wk$^{-1}$ in uncontrolled areas. A controlled area is one in which access and occupancy are regulated in conjunction with operation of the facility. Persons working there have special training in radiation protection, and radiation exposures are monitored. In contrast, individuals are free to come and go in uncontrolled areas. These design goals adhere to the annual limits of 50 mSv and 5 mSv for occupational and nonoccupational radiation. Since many instruments used to monitor X radiation are calibrated to measure exposure in roentgen (R), the shielding design objectives that have traditionally been employed are expressed as 0.1 R wk$^{-1}$ and 0.01 R wk$^{-1}$. Numerically, an exposure of 1 R produces an absorbed dose of 8.76 mGy in air (Sect. 12.2). Conversely, a 1-mGy absorbed dose in air is equivalent to 0.114 R.

In 2004 the National Council on Radiation Protection and Measurements issued Report No. 147, *Structural Shielding Design for Medical X-Ray Imaging Facilities*. In this Report, the Council recommends that *air kerma, K* (Sect. 12.10), be the quantity used for making X-ray shielding calculations. It specifies that an instrument reading in R can be divided by 114 to obtain the air kerma in Gy. The recommended design goal for occupational exposure is also revised. The cumulative effective-dose limit (Table 14.4) implies an average annual limit of 10 mSv. *In the design of new facilities*, the Council recommends one-half this value, or 5 mSv y$^{-1}$, and a weekly design goal of $P = 0.1$ mGy air kerma. Using one-half also accomplishes adherence to a monthly equivalent-dose limit of 0.5 mSv to a worker's embryo or fetus. For uncontrolled areas, the recommended design goal is $P = 0.02$ mGy wk$^{-1}$ air kerma, corresponding to the annual 1 mSv shown in Table 14.4. Report No. 147 "is intended for use in planning and designing new facilities and in remodeling existing facilities." Installations designed before publication of Report No. 147 and meeting previous NCRP requirements need not need be reevaluated. The design goals of the Report are summarized in Table 15.1.

Our discussion of X-ray shielding design will be directed toward the levels for individuals given in Table 15.1. In what follows we shall present a prototype example that illustrates how elements of an X-ray shielding calculation can be performed and put together into a final design. As will be pointed out, some important details will be simplified and will not specifically follow NCRP Report No. 147. For the transmission of X rays through different materials, we shall use measured data as traditionally given in terms of the exposure (R). Much of the information available in the literature is presented in this form. Whatever procedures and assumptions are used in making the shielding design, radiation surveys after installation must be performed in order to evaluate the results.

### Design of Primary Protective Barrier

The attenuation of primary X-ray beams through different thicknesses of various shielding materials has been measured experimentally. The data have been plotted to give empirical attenuation curves, which are used to design protective barriers. It is found experimentally that the primary beam intensity transmitted through a shield depends strongly on the peak operating voltage but very little on the filtration of the beam. (The effect of filters on exposure rate is small compared with that of the thicker shields.) In addition, at fixed kVp, the exposure from transmitted photons at a given distance from the X-ray machine is proportional to the time integral of the beam current, usually expressed in milliampere-minutes (mA min). In other words, the total exposure per mA min is virtually independent of the tube operating current itself. These circumstances permit the presentation of X-ray attenuation data for a given shielding material as a family of curves at different kVp values. Measurements are conveniently referred to a distance of 1 m from the target of the tube with different thicknesses of shield interposed.

Attenuation curves measured for lead and concrete at a number of peak voltages (kVp) are shown in Figs. 15.9 and 15.10. The ordinate, $Q$, gives the exposure of the attenuated radiation in $R\,mA^{-1}\,min^{-1}$ at the reference distance of 1 m. The abscissa gives the shield thickness. Figure 15.9 shows, for example, that behind 2 mm of lead, the exposure 1 m from the target of an X-ray machine operating at 150 kVp is $10^{-3}\,R\,mA^{-1}\,min^{-1}$. If the machine is operated with a beam current of 200 mA for 90 s, that is, for $200 \times 1.5 = 300$ mA min, then the exposure at 1 m will be $300 \times 10^{-3} = 0.3$ R behind the 2 mm lead shield. The same exposure results if the tube is operated at 300 mA for 60 s. The exposure at other distances can be obtained by the inverse-square law; for example, the exposure per mA min at 2 m is $10^{-3}/2^2 = 2.5 \times 10^{-4}\,R\,mA^{-1}\,min^{-1}$. The 2 mm of lead shielding can be located anywhere between the X-ray tube and the point of interest.

The amount of shielding needed to provide the primary barrier for an area adjoining an X-ray room can be found from the attenuation curves, once the appropriate value of $Q$ has been determined. In addition to the peak voltage, the value of $Q$ in a specific application will depend on several other circumstances:

1. The weekly design goal, $P$, which is one of the two values in Table 15.1, depending on the area to be protected.
2. The workload, $W$, or weekly amount of use of the X-ray machine, expressed in mA min wk$^{-1}$.
3. The use factor, $U$, or fraction of the workload during which the useful beam is pointed in a direction under consideration.

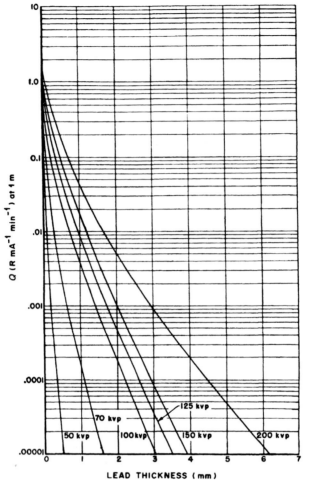

**Fig. 15.9** Attenuation in lead of X rays produced with (peak) potential differences from 50 kVp to 200 kVp. (*National Bureau of Standards Handbook 76,* 1961, Washington, DC.)

4. The occupancy factor, $T$, which takes into account the fraction of the time that an area outside the barrier is likely to be occupied by a given individual. (Average weekly exposure rates may be greater than $P$ in areas not occupied full time by anyone.) In the absence of more specific information, the occupancy factors given in Table 15.2 may be used as guides for shielding design. The allowed average kerma rate in the area is $P/T$ mGy wk$^{-1}$.
5. The distance, $d$, in meters from the target of the tube to the location under consideration. The curves in Figs. 15.9 and

**Table 15.2** Suggested Occupancy Factors, $T^*$

| Description | $T$ |
|---|---|
| Offices, laboratories, and other work areas; children's indoor play areas; X-ray control room | 1 |
| Patient examination and treatment rooms | 1/2 |
| Corridors; patient rooms; staff lounges and rest rooms | 1/5 |
| Corridor door areas | 1/8 |
| Unattended waiting rooms and vending areas; storage areas; public toilets; outdoor areas with seating | 1/20 |
| Attics; stairways; janitor closets; unattended elevators and parking lots; outdoor areas with only transient pedestrians or vehicular traffic | 1/40 |

\* Adapted from NCRP Report No. 147.

15.10 give the value of $Q$ for a distance of 1 m. At other distances, a factor of $d^2$ enters in the evaluation of $Q$.

Since an air kerma of 1 mGy is equivalent to 0.114 R, the relationship between the numerical values of the design goal $P$, involving the unit mGy, and the variable $Q$, involving R, is $Q = 0.114P$. With these considerations, the value of $Q$ can be computed from the formula

$$Q = \frac{0.114 P d^2}{WUT}. \tag{15.13}$$

With $P$ in mGy wk$^{-1}$, $d$ in m, and $W$ in mA min wk$^{-1}$, $Q$ gives the exposure of the transmitted radiation in R mA$^{-1}$ min$^{-1}$ at 1 m. For a given kVp, attenuation curves such as those in Figs. 15.9 and 15.10 can be used to find the shield thickness that results in the design level $P$.

The role of the various factors in Eq. (15.13) is straightforward. When $d = 1$ m, $W = 1$ mA min wk$^{-1}$, and the useful beam is always pointed ($U = 1$) in the direction of an area of full occupancy ($T = 1$), then it follows that $Q$ in R is numerically equal to 0.114$P$. If $U$ and $T$ are not unity, then the weekly kerma in the area can be increased to $P/UT$, which is reflected in a larger value of $Q$ and hence a smaller shield thickness. The factor $d^2$, with $d$ expressed in meters, adjusts $Q$ for locations other than 1 m. Finally, since exposure is proportional to the workload, one divides by $W$ in Eq. (15.13).

*Example*

A diagnostic X-ray machine is operated at 125 kVp and 220 mA for an average of 90 s wk$^{-1}$. Calculate the primary protective barrier thickness if lead or concrete alone were to be used to protect an uncontrolled hallway 15 ft from the tube target (Fig. 15.11). The useful beam is directed horizontally toward the barrier 1/3 of the time and vertically into the ground the rest of the time.

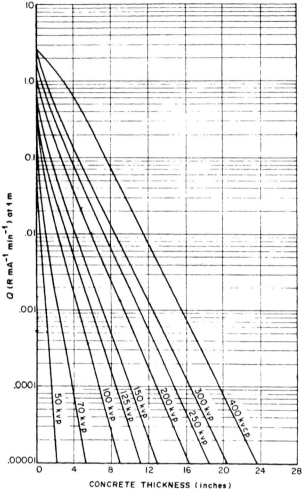

**Fig. 15.10** Attenuation in concrete of X rays produced with (peak) potential differences from 50 kVp to 400 kVp. (*National Bureau of Standards Handbook 76*, 1961, Washington, DC.)

*Solution*

From Table 15.1, $P = 0.02$ mGy wk$^{-1}$ for the hall, which is an uncontrolled area. The distance is $d = 15$ ft $= 4.57$ m, and the workload is $W = 220$ mA $\times$ 1.5 min wk$^{-1}$ = 330 mA min wk$^{-1}$. Note that Eq. (15.13) requires these particular units. The use factor is $U = 1/3$ and, from Table 15.2, the occupancy factor is $T = 1/5$. Equation (15.13) gives

$$Q = \frac{0.114 \times 0.02 \times (4.57)^2}{330 \times \frac{1}{3} \times \frac{1}{5}} = 2.16 \times 10^{-3} \ \text{R mA}^{-1} \text{min}^{-1} \text{ at 1 m.} \qquad (15.14)$$

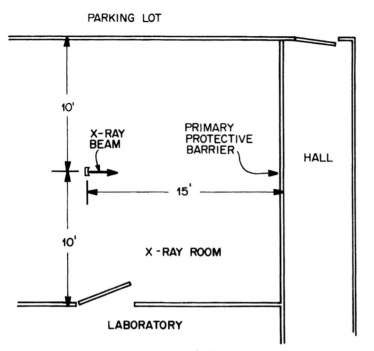

PARKING LOT

10'

X-RAY
BEAM

PRIMARY
PROTECTIVE
BARRIER

HALL

15'

10'

X - RAY  ROOM

LABORATORY

**Fig. 15.11** Schematic plan view of an X-ray facility.

From Fig. 15.9 we find that the needed thickness of lead for 125 kVp is about 1.5 mm. From Fig. 15.10, we estimate the needed thickness of concrete to be about 5.0 in.[2]

Lead is a most effective and practical material for X-ray shielding. On a weight basis, it is considerably lighter than concrete, as the last example illustrates. The lead thickness, expressed as the mass per unit area (density $= 11.4$ g cm$^{-3}$), is $0.15$ cm $\times 11.4$ g cm$^{-3} = 1.7$ g cm$^{-2}$. The concrete thickness (density $= 2.35$ g cm$^{-3}$) is $5.0$ in $\times 2.54$ cm in$^{-1} \times 2.35$ g cm$^{-3} = 30$ g cm$^{-2}$. The ratio of concrete-to-lead thicknesses is $30/1.7 = 18$. A concrete shield covering the same wall area as lead would thus weigh 18 times as much. The principal physical reason for this difference is the much larger photoelectric attenuation coefficient of lead (and other materials of high atomic number) for low-energy photons (cf. Figs. 8.8 and 8.9). The peak X-ray energy in the last example was 125 kVp. At higher peak energies, the difference between lead and concrete, though substantial, becomes less pronounced.

A primary protective barrier can either be erected when a structure is built or it can be provided by adding shielding to an existing structure. The attenuation of many common building materials per g cm$^{-2}$ is approximately the same as that of

---

2 Note. For the computation we have used the distance $d$ from the source to the opposite side of the wall. NCRP Report No. 147 considers that the nearest likely approach of the sensitive organs of a person is not less than 0.3 m beyond the wall.

**Table 15.3** Average Densities of Commercial Building Materials

| Material | Density $(g\,cm^{-3})$ |
|---|---|
| Barytes concrete | 3.6 |
| Brick (soft) | 1.65 |
| Brick (hard) | 2.05 |
| Earth (packed) | 1.5 |
| Granite | 2.65 |
| Lead | 11.4 |
| Lead glass | 6.22 |
| Sand plaster | 1.54 |
| Concrete | 2.35 |
| Steel | 7.8 |
| Tile | 1.9 |

**Table 15.4** Half-Value Layers for X Rays (Broad Beams) in Lead and Concrete

| Peak Voltage (kVp) | HVL Lead (mm) | HVL Concrete (cm) |
|---|---|---|
| 50 | 0.06 | 0.43 |
| 70 | 0.17 | 0.84 |
| 100 | 0.27 | 1.6 |
| 125 | 0.28 | 2.0 |
| 150 | 0.30 | 2.24 |
| 200 | 0.52 | 2.5 |
| 250 | 0.88 | 2.8 |
| 300 | 1.47 | 3.1 |
| 400 | 2.5 | 3.3 |

concrete, which has an average density of 2.35 $g\,cm^{-3}$. Table 15.3 gives the average densities of some common commercial materials. X-ray attenuation in similar materials can be obtained from the curves in Fig. 15.10 for concrete of equivalent thickness. For example, the attenuation provided by 2 in. of tile (average density 1.9 $g\,cm^{-3}$) is equivalent to that of $2(1.9/2.35) = 1.62$ in. of concrete. (If the building materials are of significantly higher atomic number than concrete, this procedure tends to overestimate the amount of shielding needed.) Layers of lead are commonly used with building materials to provide protective barriers. For computing the attenuation it is convenient to have half-value layers for both lead and concrete at different tube operating potentials. These are given in Table 15.4. As stressed in the last paragraph, the shielding properties of concrete and lead for X rays cannot be accurately compared on the basis of the ratio of their densities.

*Example*

If, in the previous example, an existing 3-in. sand plaster wall separates the X-ray room and the hallway in Fig. 15.11, find the thickness of lead that must be added to the wall to provide the primary protective barrier.

*Solution*

We estimated in the last example that 5.0 in of concrete would provide an adequate primary protective barrier. The 3-in sand plaster wall (density $= 1.54$ g cm$^{-3}$) is equivalent to concrete of thickness $3(1.54/2.35) = 2.0$ in. Therefore, additional shielding equivalent to $5.0 - 2.0 = 3.0$ in of concrete is needed. From Table 15.4, 1 HVL of concrete for 125-kVp X rays is 2.0 cm $= 0.79$ in, and so the additional shielding required is $3.0/0.79 = 3.8$ HVLs. Table 15.4 shows that this thickness of lead added to the sand plaster would be $3.8 \times 0.28 = 1.1$ mm.

The use of half-value layers, rather than densities, to compare the shielding of lead and concrete is also only approximate, though more accurate. As an X-ray beam penetrates matter, the softest rays—those photons with the lowest energies—are selectively filtered out, and the beam hardens. The magnitude of the "half-value layer" for incident X rays of a given kVp thus tends to increase with penetration depth. This fact is reflected in the decreasing slopes of the attenuation curves in Fig. 15.9. Numerical values for shielding obtained by the procedures described here when two or more materials are used can depend on the order in which the materials are considered, particularly when the shielding is thick. For instance, for the last example we could use Table 15.4 to find the number of HVLs of concrete alone or lead alone that would provide the needed shielding. From the value of $Q$ in Eq. (15.14) we found that the required thicknesses of lead and concrete were, respectively, 1.5 mm and 5.0 in. From Table 15.4 we find that these thicknesses represent $1.5/0.28 = 5.4$ HVLs for lead and $(5.0 \times 2.54)/2.0 = 6.4$ HVLs for concrete. The estimated number of HVLs needed should ideally be independent of the shielding material, since it is intended to represent the number of factors of 2 by which the exposure rate is to be reduced. However, for the reasons given here, the use of half-value layers for X rays is only an approximate, but useful, concept. The simplified procedures given here for calculating shielding are also only approximate, and conservative judgments are to be applied in their use.

**Design of Secondary Protective Barrier**

As illustrated in Fig. 15.8, the secondary barrier is designed to protect areas not in the line of the useful beam from the leakage and scattered radiation. Physically, these two components of the radiation field can be of quite different quality. Therefore, the shielding requirements are computed separately for each and the final barrier thickness is chosen to be adequate for their sum. Because conditions vary greatly, no single method of calculation is always satisfactory; however, the one presented here can be used as a guide. We assume that the leakage and scattered radiations are isotropic. The use factor for them is then unity ($U = 1$).

**Leakage Radiation** Limits placed on the manufacturer for the maximum allowed leakage radiation from the housing of diagnostic and therapeutic machines are given in terms of the exposure rate $Y$ in $R\,h^{-1}$ at the reference distance of 1 m from the source. As mentioned at the beginning of this section, for instance, the limit $Y = 0.1\ R\,h^{-1}$ applies to diagnostic machines. Given the numerical value of $Y$, the secondary barrier thickness for the leakage radiation is computed as the number of half-value layers needed to restrict the exposure of individuals in other areas to allowed levels. When the facility is operated under constant conditions $t$ min wk$^{-1}$, the weekly exposure to leakage radiation in R at a distance $d$ m from the tube with no structural shielding present does not exceed $(Y/d^2)(t/60) = Yt/(60d^2)$. If $P$ denotes the design goal in mGy wk$^{-1}$ for an area with occupancy factor $T$, then the required reduction $B$ for the leakage X-ray intensity is given by

$$0.114P = B\frac{YtT}{60d^2}. \tag{15.15}$$

(The factor 0.114 converts the number $P$ of mGy wk$^{-1}$ into R wk$^{-1}$, the unit of exposure employed in $Y$.) Solving for $B$ and writing $t = W/I$, where $W$ is the workload in mA min wk$^{-1}$ and $I$ is the average current in mA, we obtain

$$B = \frac{6.84PId^2}{YWT}. \tag{15.16}$$

The number $N$ of half-value layers that reduces the radiation to the factor $B$ of its unshielded value is given by $B = 2^{-N}$, or

$$N = -\frac{\ln B}{\ln 2} = -\frac{\ln B}{0.693}. \tag{15.17}$$

The leakage radiation is filtered and hardened by the lead tube housing. For the shielding estimation, we assume that its penetration is described by the half-value layers given in Table 15.4, which depend primarily on the operating potential of the tube. In addition to other factors, the foregoing derivation also assumes operation of the equipment at a given value of the kVp.

*Example*

The X-ray machine in the last two examples is to be replaced by another diagnostic unit, located in the same position as that shown in Fig. 15.11. The new unit will operate at 200 kVp with an average current of 20 mA and workload of 9200 mA min wk$^{-1}$. Other conditions remain the same as before. How many half-value layers of shielding would be needed to protect the laboratory (a controlled area) from the leakage radiation alone?

*Solution*

We use Eq. (15.16) to compute the needed reduction, $B$. Like Eq. (15.13), it requires that the quantities be expressed in the standard units we have been using. Thus, with $d = 10$ ft $= 3.05$ m, direct substitution into (15.16) gives

$$B = \frac{6.84 \times 0.1 \times 20 \times (3.05)^2}{0.1 \times 9200 \times 1} = 0.138. \tag{15.18}$$

From Eq. (15.17), the number of HVLs is

$$N = -\frac{\ln 0.138}{0.693} = 2.9 \tag{15.19}$$

for protection from the leakage radiation alone. Table 15.4 can be used to find the corresponding thickness of lead or concrete. However, we next consider the scattered radiation before specifying the secondary protective barrier thickness.

**Scattered Radiation** For the purpose of estimating shielding thickness, if the tube operation potential is not more than 500 kVp, then the barrier penetrating capability of the scattered X rays is considered to be the same as that of the useful beam. (Low-energy photons lose a relatively small fraction of their energy in Compton scattering.) Thus, we may use the attenuation curves in Figs. 15.9 and 15.10 for the scattered radiation. Also, as already mentioned, we make the additional assumption that it is isotropic.

The value of $Q$ for the scattered radiation can be determined from a modified form of Eq. (15.13). Measurements show that the exposure rate of scattered X rays at a location 1 m from a scatterer and 90° from the primary-beam direction is about $10^{-3}$ times as large as the incident exposure rate at the scatterer. Therefore, to account for the smaller intensity of the scattered radiation compared with the useful beam, $Q$ in Eq. (15.13) is increased by a factor of 1000. For the scattered radiation, then, we have with $U = 1$ the modified version of Eq. (15.13):

$$Q = \frac{114 P d^2}{WT}. \tag{15.20}$$

As before, $P$ is in mGy wk$^{-1}$, $d$ in m, $W$ in mA min wk$^{-1}$, and $Q$ is the exposure of the transmitted radiation in R mA$^{-1}$ min$^{-1}$ at 1 m.

*Example*
Estimate the number of HVLs that would be needed to shield the laboratory in the last example from the scattered radiation alone.

*Solution*
Applying Eq. (15.20) with $P = 0.1$ mGy wk$^{-1}$ for the controlled area gives

$$Q = \frac{114 \times 0.1 \times (3.05)^2}{9200 \times 1} = 0.012 \text{ R mA}^{-1} \text{ min}^{-1} \text{ at 1 m.} \tag{15.21}$$

Using Fig. 15.9, we estimate the needed thickness of lead shielding to be about 1.6 mm. From Table 15.4, the HVL is 0.52 mm. Therefore, for the scattered radiation alone, the number of HVLs of lead is $1.6/0.52 = 3.1$. Alternatively, using Fig. 15.10 with $Q = 0.012$ indicates that about 5.5 in $= 14$ cm of concrete would be needed. From Table 15.4, the corresponding thickness of concrete is $14/2.5 = 5.6$ HVLs. As discussed at the end of the section on the primary protective barrier, the number of HVLs should, ideally, be independent of the material. We choose the larger, more conservative estimate of 5.6 HVLs for shielding the scattered radiation alone.

Having computed the number of HVLs for the leakage and scattered radiations separately, one must select a secondary protective barrier that is adequate for both together. In our simplified treatment, the following rule, which has been used in the past, will be applied. If the barrier thicknesses for leakage and scattered radiations are found to be within 3 HVLs of one another, then adding 1 HVL to the larger gives a sufficient secondary barrier for both. If the two differ by more than 3 HVLs, then the thicker one alone suffices.

### Example

What thickness of lead must be added to an existing 2.5-in plaster wall between the X-ray room and the laboratory in Fig. 15.11 to provide an adequate secondary protective barrier for the facility considered in the last two examples?

### Solution

We found that 2.9 HVLs would be needed for the leakage alone and 5.6 HVLs for the scattered radiation alone. The difference in the two is 2.7 HVLs. By the rule just cited, since this amount is less than three, we estimate the secondary barrier as $5.6 + 1 = 6.6$ HVLs. The existing wall provides some of the barrier. Its concrete-equivalent thickness is $2.5(1.54/2.35) = 1.64$ in $= 4.16$ cm, as found from the densities in Table 15.3. From Table 15.4, this amount represents $4.16/2.5 = 1.7$ HVLs; and so the amount of additional shielding to be added to the wall is $6.6 - 1.7 = 4.9$ HVLs. The thickness of lead (HVL $= 0.52$ mm) required is $4.9 \times 0.52 = 2.6$ mm.

### NCRP Report No. 147

We have presented examples of how shielding for a diagnostic X-ray facility can be determined. While such procedures, including various modifications, have been commonly applied, the most up-to-date methodology is furnished in the 2004 NCRP Report No. 147 (see Sect. 15.6). We briefly mention some features of the Report in order to supplement the discussions in our examples.

Our approach employed the revised NCRP design goals, based on air kerma (Table 15.1), for new X-ray facilities recommended in the Report. The treatment of workload in the Report underwent a major overhaul. We mentioned previously that, whereas the amount of shielding is a very strong function of kVp, the transmission curves at constant kVp are represented "per mA min" of use. Many facilities perform multiple procedures at different kVp values and/or contain several X-ray tubes. Determining proper workload information for the design of shielding can then become complicated. Data from the literature have been assembled in the Report from surveys that provide workload distributions *per patient* as functions of the kVp operating potentials for a number of facility types (general radiographic, fluoroscopic, chest X-ray, computed tomography, ...). These distributions are folded into tables and graphs used for barrier computations. The data have also been fitted with parameters from which transmission factors and barrier thicknesses can be calculated numerically. Also, the angular dependence of the scattered radiation is not assumed to be isotropic, as in our estimates. We also stressed the need for verification. In this regard, the Report states, "While specific recommendations on shielding design methods are given in this Report, alternative methods may prove equally satisfactory in providing radiation protection. The final assessment of the adequacy of the design and construction of protective shielding can only be based on the post-construction survey performed by a qualified expert."

NCRP Report No. 151, *Structural Shielding Design and Evaluation for Megavoltage X- and Gamma-Ray Radiotherapy Facilities*, was published in 2005. The reader is also referred to the 2003 NCRP Report No. 145, *Radiation Protection in Dentistry* (see Sect. 15.6).

## 15.4
## Protection from Beta Radiation

Beta (including positron) emitters present two potential external radiation hazards, namely, the beta rays themselves and the bremsstrahlung they produce in the source and in adjacent materials. In addition, annihilation photons are always present with positron sources. Beta particles can be stopped in a shield surrounding the source if it is thicker than their range. To minimize bremsstrahlung production, this shield should have low atomic number [cf. Eq. (6.14)]. It, in turn, can be enclosed in another material (preferably of high atomic number) that is thick enough to attenuate the bremsstrahlung intensity to the desired level. For a shielded beta emitter bremsstrahlung may be the only significant external radiation hazard.

The bremsstrahlung shield thickness can be calculated in approximate fashion by the following procedure. Equation (6.14) is used to estimate the radiation yield, letting $T = T_{max}$ be the maximum beta-particle energy. This assumption overestimates the actual bremsstrahlung intensity, because most of the photons have energies much lower than the upper limit $T_{max}$. To roughly compensate, one ignores buildup in the shielding material and uses the linear attenuation coefficient for photons of energy $T_{max}$ to estimate the bremsstrahlung shield thickness. Since the bremsstrahlung spectrum is hardened by passing through the shield, the exposure rate around the source is calculated by using the air absorption coefficient for photons of energy $T_{max}$.

### Example
Design a suitable container for a $3.7 \times 10^{11}$ Bq source of $^{32}$P in a 50-mL aqueous solution, such that the exposure rate at a distance of 1.5 m will not exceed 1 mR h$^{-1}$. $^{32}$P decays to the ground state of $^{32}$S by emission of beta particles with an average energy of 0.70 MeV and a maximum energy of 1.71 MeV.

### Solution
We choose a bottle made of some material, such as polyethylene (density = 0.93 g cm$^{-3}$), with elements of low atomic number to hold the aqueous solution. It should be thick enough to stop the beta particles of maximum energy. From

Fig. 6.4, the range for $T_{max} = 1.71$ MeV is about 0.80 g cm$^{-2}$. The thickness of the polyethylene bottle should therefore be at least $0.80/0.93 = 0.86$ cm. To estimate the bremsstrahlung yield by Eq. (6.14), we need the effective atomic number of the medium in which the beta particles lose their energy. Most of the energy will be lost in the water, a small part being absorbed in the container walls. The effective atomic number for water is

$$Z_{eff} = \frac{2}{18} \times 1 + \frac{16}{18} \times 8 = 7.22. \tag{15.22}$$

The estimated fraction of the beta-particle energy that is converted into bremsstrahlung is, by Eq. (6.14) with $Z = Z_{eff}$ and $T = T_{max}$,

$$Y \cong \frac{6 \times 10^{-4} \times 7.22 \times 1.71}{1 + 6 \times 10^{-4} \times 7.22 \times 1.71} = 7.4 \times 10^{-3}. \tag{15.23}$$

The rate of energy emission by the source of beta particles with an average energy of 0.70 MeV is[3]

$$\dot{E}_{\beta} = 3.7 \times 10^{11} \times 0.70 = 2.59 \times 10^{11} \text{ MeV s}^{-1}. \tag{15.24}$$

The rate of energy emission in the form of bremsstrahlung photons is therefore

$$Y\dot{E}_{\beta} = 7.4 \times 10^{-3} \times 2.59 \times 10^{11} = 1.92 \times 10^{9} \text{ MeV s}^{-1}. \tag{15.25}$$

We next compute the exposure rate from the unshielded bremsstrahlung, treated as coming from a point source at a distance of 1.5 m. Following our procedure, we use the mass energy-absorption coefficient of air for 1.71 MeV photons, which is (from Fig. 8.12) $\mu_{en}/\rho = 0.026$ cm$^2$ g$^{-1}$. Since the intensity at a distance $r = 1.5$ m $= 150$ cm is given by [Eq. (8.61)]

$$I = \frac{Y\dot{E}_{\beta}}{4\pi r^2} = \frac{1.92 \times 10^{9} \text{ MeV s}^{-1}}{4\pi(150 \text{ cm})^2} = 6.79 \times 10^{3} \text{ MeV cm}^{-2} \text{ s}^{-1}, \tag{15.26}$$

the dose rate in air is

$$\dot{D} = \frac{\mu_{en}}{\rho} I = \frac{0.026 \text{ cm}^2}{g} \times \frac{6.79 \times 10^{3} \text{ MeV}}{\text{cm}^2 \text{ s}} = 177 \text{ MeV g}^{-1} \text{s}^{-1}. \tag{15.27}$$

Converting units and remembering that 1 R = 0.0088 Gy in air (Sect. 12.2), we find for the exposure rate

$$\dot{X} = \frac{177 \text{ MeV}}{g s} \times \frac{1.60 \times 10^{-13} \text{ J MeV}^{-1}}{10^{-3} \text{ kg g}^{-1}} \times \frac{1 \text{ Gy}}{\text{J kg}^{-1}} \times \frac{1 \text{ R}}{0.0088 \text{ Gy}} \tag{15.28}$$

$$= 3.22 \times 10^{-6} \text{ R s}^{-1} = 11.6 \text{ mR h}^{-1}. \tag{15.29}$$

Lead is a convenient material for the bremsstrahlung shield. As specified, one ignores buildup and uses the linear attenuation coefficient for photons of energy $T_{max}$ to

3  When the average beta-particle energy is not
known, one can approximate it as one-third
the maximum beta-particle energy (Sect. 3.4).

compute the bremsstrahlung shield thickness. Figure 8.8 gives for 1.71-MeV photons in lead $\mu/\rho = 0.048$ cm$^2$ g$^{-1}$, and so $\mu = 0.048 \times 11.4 = 0.55$ cm$^{-1}$. The thickness $x$ needed to reduce the exposure rate to 1 mR h$^{-1}$ is given by

$$1 = 11.6e^{-0.55x}, \tag{15.30}$$

or $x = 4.5$ cm. A lead container of this thickness could be used to hold the polyethylene bottle. We have ignored any bremsstrahlung shielding that the bottle itself affords.

## 15.5
## Neutron Shielding

Photon shielding design is simplified by a number of factors that do not apply to computations for neutrons. Whereas photon cross sections vary smoothly with atomic number and energy, neutron cross sections can change irregularly from element to element and have complicated resonance structures as functions of energy. In addition, photon cross sections are generally better known than those for neutrons. Elaborate computer codes, using Monte Carlo and other techniques, are available for calculating neutron interactions and transport in a variety of materials. Sometimes circumstances permit useful estimations to be made by simpler means. In this section we present only a general discussion of neutron shielding.

Basically, a neutron shield acts to moderate fast neutrons to thermal energies, principally by elastic scattering, and then absorb them. Most effective in slowing down neutrons are the light elements, particularly hydrogen [cf. Eq. (9.3)]. Many hydrogenous materials, such as water and paraffin, make efficient neutron shields. However, water shields have the disadvantage of needing maintenance; also, evaporation can lead to a potentially dangerous loss of shielding. Paraffin is flammable. Concrete (ordinary or heavy aggregate) or earth is the neutron shielding material of choice in many applications. Often temporary neutron shielding must be provided in experimental areas around a reactor or an accelerator. Movable concrete blocks are convenient for this purpose. One must exercise care to assure that cracks, access ports, and ducts in such shielding do not permit the escape of neutrons. Vertical cracks should be staggered. (Natural sagging of concrete blocks under the force of gravity usually precludes the existence of horizontal cracks.) Generally, surveys are desirable to check temporary neutron shielding before and during extensive use.

Hydrogen captures thermal neutrons through the reaction $^1$H(n,$\gamma$)$^2$H with a cross section of 0.33 barns. Other materials, like cadmium, have a very high (n,$\gamma$) thermal-neutron capture cross section (2450 barns) and are therefore frequently used as neutron absorbers. Hydrogen and cadmium have the disadvantage of emitting energetic (2.22-MeV and 9.05-MeV) capture gamma rays, which might themselves require shielding. Other nuclides, such as $^{10}$B and $^6$Li, capture thermal neutrons through an (n,$\alpha$) reaction without emission of appreciable gamma radiation. In addition to possible health-physics problems that may arise from capture gamma rays, these shields can acquire induced radioactivity through neutron capture or other reactions.

Examples of neutron attenuation in a hydrogenous material are provided by the depth–dose curves in Chapter 12 for monoenergetic neutrons normally incident on tissue slabs. Figure 12.10 for 5-MeV neutrons, for instance, shows that the absorbed dose decreases by an order of magnitude over 30 cm. The energy spectrum of the neutrons changes with the penetration depth as the original 5-MeV neutrons are moderated. The relative number of thermal neutrons at different depths can be seen from the dose curve labeled $E_\gamma$ for the $^1$H$(n,\gamma)^2$H thermal-neutron capture reaction. The thermal-neutron density builds up to a maximum at about 10 cm and thereafter falls off as the total density of neutrons decreases by absorption. In paraffin, the half-value layer for 1-MeV neutrons is about 3.2 cm and that for 5-MeV neutrons is about 6.9 cm.

Neutron shielding can sometimes be estimated by a simple "one-velocity" model that employs neutron removal cross sections. Such shielding must be sufficiently thick and the neutron source energies so distributed that only the most penetrating neutrons in a narrow energy band contribute appreciably to the dose beyond the shield. The neutron dose can then be represented by an exponential function of shield thickness. Conditions must also be such that the slowing-down distance from the most penetrating energies down to 1 MeV is short. In addition, the shield must contain enough hydrogen to assure a short average transport distance from 1 MeV down to thermal energy and the point of absorption. The removal cross sections for various elements are roughly three-quarters of the total cross sections (except $\sim$0.9 for hydrogen). Most measurements of removal cross sections have been made with fission-neutron sources and shields of such a thickness that the principal component of dose arises from source neutrons in the energy range 6–8 MeV. Table 15.5 gives macroscopic removal cross sections, $\Sigma_r$, and attenuation lengths, $1/\Sigma_r$, in some shielding and reactor materials.

**Table 15.5** Macroscopic Neutron Removal Cross Sections and Attenuation Lengths in Several Materials

| Material | Macroscopic Removal Cross Section $\Sigma_r$ (cm$^{-1}$) | Attenuation Length $1/\Sigma_r$ (cm) |
|---|---|---|
| Water | 0.103 | 9.7 |
| Paraffin | 0.106 | 9.4 |
| Iron | 0.1576 | 6.34 |
| Concrete (6% H$_2$O by weight) | 0.089 | 11.3 |
| Graphite (density 1.54 g cm$^{-3}$) | 0.0785 | 12.7 |

*Source*: Data in part from *Protection Against Neutron Radiation*, NCRP Report No. 38, National Council on Radiation Protection and Measurements, Washington, D.C. (1971).

A number of approximate formulas for neutron shielding, based on removal cross sections, have been developed for reactor cores having various shapes and other characteristics. We will not attempt to cover them here. A simple, useful formula is available, however, for radioactive neutron sources.[4] Because of the small intensities compared with fission sources, relatively thin shields are needed. Therefore, the scattered neutrons contribute significantly to the dose outside the shield, and their effect can be represented by a buildup factor $B$. The dose-equivalent rate $\dot{H}$ outside a shield of thickness $T$ at a distance $R$ cm from a point source of strength $S$ neutrons s$^{-1}$ is given by

$$\dot{H} = \frac{BSqe^{-\Sigma_r T}}{4\pi R^2},$$
(15.31)

where $\Sigma_r$ is the removal cross section and $q$ is the dose-equivalent rate per unit neutron fluence rate (e.g., Sv h$^{-1}$ per neutron cm$^{-2}$ s$^{-1}$) for neutrons of the source energy. The factor $q$ can be obtained from Table 12.5; $\dot{H}$ and $q$ will have the same units for dose equivalent. For Po–Be and Po–B sources with a water or paraffin shield at least 20 cm thick, $B \cong 5$.

### Example

Calculate the dose-equivalent rate 1.6 m from an unshielded $3.0 \times 10^{10}$ Bq $^{210}$Po–Be source, which emits $2.05 \times 10^6$ neutrons s$^{-1}$. By what factor is the rate reduced by a 25-cm water shield? What is the dose-equivalent rate behind 50 cm of water?

### Solution

In Eq. (15.31) the presence of the shield introduces the factors $B\exp(-\Sigma_r T)$. For the unshielded source, $\dot{H}_0 = Sq/4\pi R^2$ and $S = 2.05 \times 10^6$ neutrons s$^{-1}$. Table 9.2 shows the average energy to be 4.2 MeV, for which Table 12.5 indicates that about 16 neutrons cm$^{-2}$ s$^{-1}$ give a dose-equivalent rate of 0.025 mSv h$^{-1}$. Therefore we have

$$q = \frac{0.025 \text{ mSv h}^{-1}}{16 \text{ cm}^{-2} \text{ s}^{-1}} = 0.00156 \text{ mSv h}^{-1} \text{ cm}^2 \text{ s};$$
(15.32)

and so

$$\dot{H}_0 = \frac{(2.05 \times 10^6 \text{ s}^{-1})(0.00156 \text{ mSv h}^{-1} \text{ cm}^2 \text{ s})}{4\pi (160 \text{ cm})^2} = 0.0099 \text{ mSv h}^{-1}$$
(15.33)

for the unshielded source. With $\Sigma_r = 0.103$ cm$^{-1}$ from Table 15.5, $B = 5$, and $T = 25$ cm, the dose-equivalent rate is reduced by the factor

$$Be^{-\Sigma_r T} = 5e^{-0.103 \times 25} = 0.38.$$
(15.34)

The rate with 50 cm of water interposed is

$$\dot{H} = \frac{5 \times 2.05 \times 10^6 \times 0.00156}{4\pi (160)^2} e^{-0.103 \times 50} = 2.9 \times 10^{-4} \text{ mSv h}^{-1}.$$
(15.35)

4   Protection Against Neutron Radiation Up to 30 Million Electron Volts, *Handbook 63*, National Bureau of Standards, Washington, DC (1957).

Note that Eq. (15.31) with $B \cong 5$ applies to shields thicker than 20 cm, as was the case here. Note also that one should generally be concerned with gamma-ray shielding where neutrons are present. In this example, however, $^{210}$Po is a weak gamma emitter (0.001%) and decays to stable $^{206}$Pb.

## 15.6
## Suggested Reading

1 Archer, Benjamin R., "Recent History of the Shielding of Medical X-Ray Imaging Facilities," *Health Phys.* **88**, 579–586 (2005). [Review article in 50th Anniversary Issue of *Health Physics*. Describes the recent work of NCRP Scientific Committee 9 (with support of the American Association of Physicists in Medicine), resulting in significant revisions in recommendations for radiation barrier design for medical imaging facilities and NCRP Report 147. Additional perspective into the Report is given.]

2 Brunette, Jeffrey J., Book Review of NCRP Report No. 147, *Health Phys.* **89**, 183 (2005). [Brief but valuable and clear summary of Report No. 147 and important changes from the 1976 NCRP Report No. 49.]

3 Cember, H., *Introduction to Health Physics*, 3rd Ed., McGraw-Hill, New York, NY (1996).

4 Faw, R. E., and Shultis, J. K., *Radiological Assessment*, Prentice Hall, Engelwood Cliffs, NJ (1993).

5 NCRP Report No. 49, *Structural Shielding Design and Evaluation for Medical Use of X Rays and Gamma Rays of Energies up to 10 MeV*, National Council on Radiation Protection and Measurements, Washington, DC (1976). [Predecessor of NCRP Report No. 147.]

6 NCRP Report No. 145, *Radiation Protection in Dentistry*, National Council on Radiation Protection and Measurements, Bethesda, MD (2003).

7 NCRP Report No. 147, *Structural Shielding Design for Medical X-Ray Imaging Facilities*, National Council on Radiation Protection and Measurements, Bethesda, MD (2004).

8 NCRP Report No. 151, *Structural Shielding Design and Evaluation for Megavoltage X- and Gamma-Ray Radiotherapy Facilities*, National Council on Radiation Protection and Measurements, Bethesda, MD (2005).

9 Radiation Safety Information Computational Center (RSICC). This valuable resource on radiation shielding at the Oak Ridge National Laboratory can be consulted on line. The Center collects, analyzes, and distributes computer software and data sets related to radiation transport and safety. Technical assistance is also available.

10 Selman, Joseph, *The Fundamentals of Imaging Physics and Radiobiology*, 9th Ed., Charles C. Thomas, Springfield, IL (2000). [A fine textbook for training X-ray technicians, this work contains much practical information on X-ray production, technology, utilization, and dosimetry.]

11 Shultis, J. Kenneth, and Faw, Richard E., "Radiation Shielding Technology," *Health Physics* **88**, 587–612 (2005). [Review article in 50th Anniversary Issue of *Health Physics*. Presents an historical review and a summary of shielding-calculation techniques. Extensive bibliography.]

## 15.7
## Problems

1. Calculate the thickness of lead shielding needed to reduce the exposure rate 2.5 m from a $5.92 \times 10^{11}$ Bq point source of $^{137}$Cs to $1.0$ mR h$^{-1}$.

2. To what factor of its unshielded value would the exposure rate around the source in the last problem be reduced by 20 cm of concrete shielding (Fig. 15.3) in place of lead in the last problem?

3. A small $1.85 \times 10^{11}$ Bq $^{42}$K source is placed inside an aluminum (Fig. 15.4) pipe (on the axis) having an inside diameter of 1 in. and an outside diameter of 2.5 in. What is the exposure rate opposite the source at a point 6 ft away from the center of the pipe?

4. How thick must a spherical lead container be in order to reduce the exposure rate 1 m from a small $3.70 \times 10^{9}$ Bq $^{24}$Na source to 2.5 mR h$^{-1}$?

5. An unshielded 1 Ci $^{60}$Co source is to be used in a room at the spot $x$ shown in Fig. 15.12. Calculate the thickness $t$ of concrete that is needed to limit the exposure rate to 10 mR h$^{-1}$ outside the wall.

6. A broad, parallel beam of 500-keV photons is normally incident on a uranium sheet (Fig. 15.6) that is 1.5 cm thick. If the exposure rate in front of the sheet is 1.08 mR min$^{-1}$, what is it behind the sheet?

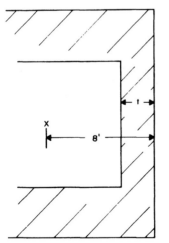

**Fig. 15.12** Diagram of room to be used with an unshielded 1 Ci $^{60}$Co source (Problem 5).

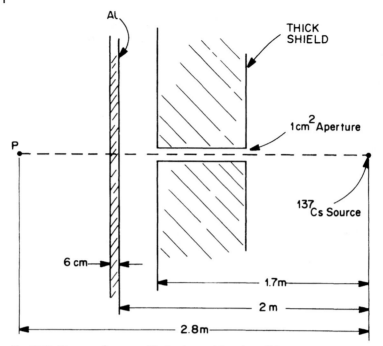

**Fig. 15.13** Diagram of a room with aluminum slab and parallel thick shield used with a 2.96 × 10¹¹ Bq ¹³⁷Cs point source (Problem 8).

7. What thickness of lead shielding is needed around a $7.40 \times 10^{13}$ Bq point source of $^{60}$Co to reduce the exposure rate to $10 \, \mathrm{mR\,h^{-1}}$ at a distance of 50 m?

8. The front of a 6-cm aluminum slab is located 2 m from a $2.96 \times 10^{11}$ Bq $^{137}$Cs point source, as shown in Fig. 15.13. The back of a parallel thick shield, which completely absorbs direct radiation, is 1.7 m from the source. The shield has a cylindrical aperture of area 1 cm² with the source on its axis.
   (a) Calculate the exposure rate at a point $P$ 2.8 m from the source on the cylindrical axis.
   (b) What is the exposure rate at $P$ with the thick shield removed?

9. An $8.33 \times 10^6$ MBq point source of a radioisotope emits a 1.0-MeV gamma photon in 67% of its transformations. This is the only gamma radiation emitted.
   (a) Calculate the thickness of lead shielding needed to reduce the exposure rate to $10.0 \pm 0.5 \, \mathrm{mR\,h^{-1}}$ at a distance of 1.5 m.
   (b) What fraction of the incident photons pass through the shield without having an interaction?

10. A uranium shield (Fig. 15.5) is to be placed around a point source of $^{40}$K. Calculate the thickness of uranium needed to reduce the exposure rate from a $3.7 \times 10^{12}$ Bq source to 10 mR h$^{-1}$ at a distance of 1.5 m. (Ignore the small amount of annihilation radiation.)

11. A $5.18 \times 10^{12}$ Bq point source emits a 1-MeV gamma photon in 67% of its transformations. No other photons are emitted. What thickness of uranium shielding is needed to reduce the exposure rate to 2.5 mR h$^{-1}$ at a point 5 m away?

12. A certain point source emits gamma photons with energies of 1 MeV and 2 MeV. A 2-cm lead shield surrounds the source.
    (a) What is the relative number of 1-MeV photons that pass through the shield without interacting?
    (b) At a certain distance from the unshielded source, the exposure rate due to the 1-MeV photons is 100 mR h$^{-1}$ and that due to the 2-MeV photons is 60 mR h$^{-1}$. What is the total exposure rate at this distance with the shield present?

13. The exposure rate measured at a distance of 82 cm from an unshielded point source of $^{42}$K is 463 mR h$^{-1}$.
    (a) What is the source strength?
    (b) The shielding effect of 1 g cm$^{-2}$ of concrete compared with 1 g cm$^{-2}$ of lead for this source would be less, comparable, or better? Why?

14. The nuclide $^{37}$S emits a 3.09-MeV photon in 90% of its transformations. No other photons are emitted.
    (a) Compute the thickness of uranium shielding needed to reduce the exposure rate of a $3.7 \times 10^{11}$ Bq point source of this nuclide to 5 mR h$^{-1}$ at a distance of 3 m.
    (b) With the shield in place, what is the exposure rate due to the uncollided photons alone?

15. A point source emits gamma photons of energies 0.5 MeV and 1.0 MeV. The total unshielded exposure rate of 760 mR h$^{-1}$ at a certain distance from the source is due to 610 mR h$^{-1}$ from the 0.5-MeV photons and 150 mR h$^{-1}$ from the 1.0-MeV photons.
    (a) Calculate the thickness of lead shielding required to reduce the total exposure rate to 50 mR h$^{-1}$.
    (b) What fraction of the incident photons pass through this lead shield without having an interaction?
    (c) What fraction of the 50 mR h$^{-1}$ is due to these uncollided photons?

16. (a) Why are the semilogarithmic curves in Fig. 15.9 nonlinear?

(b) Why do the slopes decrease with increasing thickness?

(c) Why, on the other hand, do the curves become essentially linear after the initial depths?

17. (a) Why is the leakage radiation from an X-ray machine generally more penetrating than the scattered radiation?

(b) Using the density equivalent of concrete for a material of high atomic number tends to overestimate the amount of shielding needed for X rays. Why?

18. A 200-kVp diagnostic X-ray machine is installed in the position shown in Fig. 15.11. Its average weekly workload of 250 mA min is divided between 150 mA min when the useful beam is pointed horizontally in the direction of the hall and 100 mA min when pointed horizontally in the direction of the unattended parking lot. The tube current is 100 mA.

(a) Determine the thickness of concrete needed for the primary protective barrier for the hall.

(b) Determine the thickness of concrete needed for the primary protective barrier for the parking lot.

19. If a $1\frac{1}{2}$-in. plaster wall exists between the X-ray room and the hall in the last problem (Fig. 15.11), what additional thickness of lead will provide an adequate primary barrier for the hall?

20. (a) If a $2\frac{1}{4}$-in. concrete wall separates the X-ray room and the unattended parking lot in Problem 18 (Fig. 15.11), how much additional lead shielding is needed to make the primary protective barrier?

(b) What changes, if any, should be made in the design of this primary barrier if the parking lot has an attendant 24 h per day?

21. For the attenuation of 300-kVp X rays, 1 cm of concrete is equivalent to what thickness of lead?

22. An 8-in. hard-brick wall of a building separates a public sidewalk from a 200-kVp X-ray machine located inside the building, 9 ft from the outside of the wall. The X-ray machine, which operates at 200 mA, is used an average of 18 minutes per day, 5 days per week. It is directed perpendicularly at this brick wall one-third of the time that it operates. How much lead shielding must be added to the wall to provide an adequate primary barrier?

23. The primary beam from a 250-kVp diagnostic X-ray machine in a doctor's office is directed perpendicularly at a wall that separates the office from a public sidewalk just outside, 12 ft from the X-ray machine. The wall consists of 1.0 in. of hard brick covered by 2.67 in. of sand plaster. The machine is

operated 45 minutes per week with a beam current of 200 mA, and the beam is always directed at this wall. Calculate the thickness of lead that should be added to the wall to provide a proper primary protective barrier.

24. The useful beam of a 150-kVp diagnostic X-ray machine is directed at right angles to a soft-brick wall, which is 5.2 in. thick. An unattended parking lot is located outside the wall, at a distance of 9 ft from the X-ray machine. The machine operates an average of 2.2 minutes per day, 5 days per week, with a beam current of 100 mA. What thickness of lead shielding must be added to the wall to provide an adequate primary protective barrier?

25. The useful beam of a diagnostic X-ray machine is directed 70% of the time at right angles to a 10.1-in. hard-brick wall, which is 4.0 ft away. The machine operates at 250 kVp with a current of 200 mA for an average of 2.5 minutes per day, 5 days per week. An uncontrolled corridor is on the other side of the wall.

    (a) What thickness of lead has to be added to the wall to make an adequate primary barrier for the X rays?
    (b) With the proper shielding in place, procedures are to be changed so that the useful beam is *always* directed toward the wall. Without adding still more shielding, what would be the maximum workload permitted, all other factors remaining the same?

26. The useful beam of a 300-kVp X-ray machine is directed 40% of the time normally at a hard-brick wall 8 ft away. The wall, which is 4.0 in. thick, separates the X-ray room from an uncontrolled hall outside. The X-ray machine is operated with a current of 290 mA, 2 minutes per day, 5 days per week. Calculate the thickness of lead shielding that needs to be added to the brick wall for the primary protective barrier.

27. A hospital uses a 300-kVp X-ray machine with a current of 200 mA for an average of 40 seconds per day, 5 days per week. The beam is always directed normally at a 4-in. hard-brick wall 15 ft away. An elevator shaft is on the other side of the wall. The elevator has an operator and is used for hospital visitors. How much lead shielding must be added to the wall to form an adequate primary protective barrier? (Assume $T = 1/5$).

28. Calculations for a secondary protective barrier at the facility described in the last problem give the following results for the leakage and scattered radiations:

$$B = 0.028$$

$$Q = 0.0060 \text{ R mA}^{-1} \text{min}^{-1} \text{ at 1 m.}$$

If the secondary barrier is to be a wall made of sand plaster, how thick must it be?

29. A 300-kVp diagnostic X-ray machine is separated by a 3-in. soft-brick wall from a realtor's office, 3 ft away. The useful beam can never be directed toward this wall. The machine operates an average of 10 minutes per week at a current of 280 mA. For the secondary protective barrier for this wall, calculate:
    (a) The number of HVLs of concrete needed for shielding the scattered radiation alone.
    (b) The number of HVLs of concrete needed for the leakage radiation alone.
    (c) The number of HVLs of concrete to protect from both scattered and leakage radiations.
    (d) The thickness of lead that needs to be added to the wall for an adequate secondary protective barrier.

30. A diagnostic X-ray machine operates at 300 kVp with a current of 200 mA. It is located 2 ft from the outside of a granite wall, which is 2.0 in. thick. The beam is never directed toward the wall, which separates the X-ray facility from a children's playground outside. The machine is operated an average of 28 minutes per day, 5 days per week.
    (a) Calculate the number of HVLs of concrete that would be needed to shield the playground from the leakage radiation alone, if the granite wall were not there.
    (b) Repeat (a) for the scattered radiation alone.
    (c) What thickness of lead needs to be added to the granite wall to provide a sufficient secondary protective barrier to shield the playground?

31. (a) Calculate the number of half-value layers needed to shield the laboratory from the leakage radiation in Problem 18 (Fig. 15.11).
    (b) Repeat for the scattered radiation.
    (c) If a $\frac{3}{4}$-in. plaster wall separates the X-ray room and the laboratory, how much additional lead shielding is needed to make a secondary protective barrier?

32. Figure 15.14 shows a schematic plan view of an X-ray facility. A diagnostic machine, operated at 150 kVp with a maximum current of 120 mA, is used an average of 22.1 min d$^{-1}$, 5 d wk$^{-1}$. The horizontal beam is always pointed in the direction of the sidewalk. Calculate the thickness of additional lead shielding needed for the primary protective barrier.

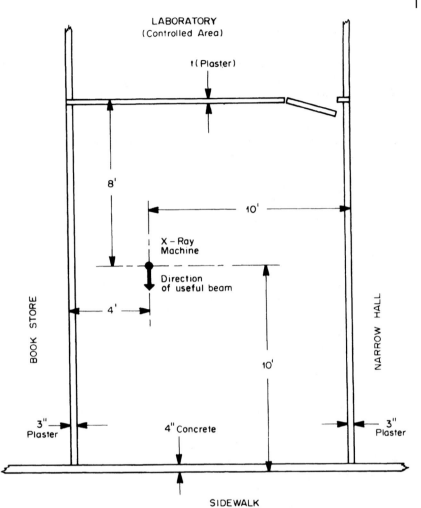

LABORATORY
(Controlled Area)

t (Plaster)

8'

10'

X – Ray
Machine

Direction
of useful beam

BOOK STORE

4'

NARROW HALL

10'

3"
Plaster

4" Concrete

3"
Plaster

SIDEWALK

**Fig. 15.14** Schematic plan view of an X-ray facility (Problem 32).

33. **(a)** Calculate the thickness of additional lead shielding
   needed in the last problem (Fig. 15.14) to make a
   secondary protective barrier for the book store.
   **(b)** Calculate the thickness of additional lead shielding
   needed for the hallway.
34. What thickness $t$ must the plaster wall between the X-ray
   room and the laboratory have in Problem 32 (Fig. 15.14) to
   provide an adequate secondary protective barrier?
35. A law firm is located in an office directly below the 2-in.
   concrete floor of a dentist's office. No other shielding
   separates the two businesses. The dentist's staff has
   routinely operated a 100-kVp X-ray machine with an average

**Fig. 15.15** Small vial containing $^{32}$P in aqueous solution enclosed in an aluminum can (Problem 37).

weekly workload of 100 mA min for the past 4 y. One of the law partners has gone bald in the two years since the firm moved into their present office. Should the dentist fear a law suit? Explain.

36. A 30-mL solution containing $7.4 \times 10^{10}$ Bq of $^{90}$Sr in equilibrium with $^{90}$Y is to be put into a small glass bottle, which will then be placed in a lead container having walls 1.5 cm thick.

   (a) How thick must the walls of the glass bottle be in order to prevent any beta rays from reaching the lead?
   (b) Estimate the bremsstrahlung dose rate in air at a distance of 1.75 m from the center of the lead container.

37. A small vial containing $7.4 \times 10^{12}$ Bq of $^{32}$P in aqueous solution is enclosed in an 8-mm aluminum can as shown in Fig. 15.15. Calculate the thickness of lead shielding needed to reduce the exposure rate to 2.5 mR h$^{-1}$ at a distance 2 m from the can. Treat the vial as a point source and assume that all of the bremsstrahlung is produced by beta particles slowing down in the water.

38. A $1.11 \times 10^9$ Bq source of $^{60}$Co is made by neutron activation of a small piece of cobalt metal ($^{59}$Co, 100% abundant).

   (a) Estimate the energy fluence rate of the bremsstrahlung from the source at a distance of 60 cm. (No shielding is present, and all beta particles stop in the source itself.)
   (b) Estimate the bremsstrahlung dose rate in air at this distance. (This will, of course, be much smaller than the gamma-dose rate from this source.)

39. $^{60}$Co activity can be induced by neutron activation of a sample of natural cobalt ($^{59}$Co, 100% abundant) in a reactor. A small 370-MBq source of $^{60}$Co is made in this fashion. Assume that all of the beta particles emitted are stopped in the cobalt sample itself and that all photons emitted escape from it. For a location 1 m from the source in air, estimate:

   (a) The exposure rate from the $^{60}$Co gamma rays.

(b) The exposure rate from the bremsstrahlung emitted by the stopping beta particles.

40. A small sample of pure sulfur contains $1.96 \times 10^{10}$ Bq of $^{35}$S embedded in it.

(a) Estimate the rate of energy emission from the source in the form of bremsstrahlung.

(b) Estimate the bremsstrahlung dose rate in air 1 m from the source.

(c) What thickness of aluminum would reduce this dose rate by a factor of 100?

41. An $8.51 \times 10^{11}$ Bq source of $^{35}$S is contained in aqueous solution. Assume that the beta particles are stopped by the water and that self-absorption of bremsstrahlung in the solution is negligible.

(a) Estimate (for bremsstrahlung shielding purposes) the dose rate in air due to bremsstrahlung at a distance of 50 cm from the source.

(b) Estimate the fraction to which this dose rate would be reduced by a 0.30-cm sheet of tin.

42. (a) Estimate the dose-equivalent rate at a distance of 80 cm from a $^{210}$Po–B point source that emits $2.2 \times 10^7$ neutrons s$^{-1}$ and is shielded by 30 cm of water.

(b) How thick would the water shield have to be to reduce the dose-equivalent rate to 0.025 mSv h$^{-1}$?

43. (a) What is the maximum number of neutrons s$^{-1}$ that a $^{210}$Po–Be point source can emit if it is to be stored behind 65 cm of paraffin and the dose-equivalent rate is not to exceed 0.10 mSv h$^{-1}$ at a distance of 1 m?

(b) By what factor would a 31-cm shield reduce the dose-equivalent rate?

## 15.8
## Answers

Interaction coefficients and attenuation curves in the text can be read only approximately; therefore, answers may not agree precisely with those given here.

1. 6.4 cm
2. 0.18
4. 10.5 cm
6. 0.010 mR min$^{-1}$
8. (a) 85 mR h$^{-1}$
   (b) 220 mR h$^{-1}$
11. 5.3 cm

12. (a) 0.207
    (b) 61 mR h$^{-1}$
14. (a) 8.1 cm
    (b) 1.6 mR h$^{-1}$
18. (a) 9.0 in.
    (b) 6.0 in.
19. 4.2 mm

**23.** 9.6 mm

**25.** (a) 7.7 mm

(b) 1750 mA min wk$^{-1}$

**28.** 42 cm

**29.** (a) 11

(b) 3.1

(c) 11

(d) 14 mm

**31.** (a) None

(b) None

(c) None

**32.** 2.2 mm

**33.** (a) 2.4 mm

(b) 0.85 mm

**36.** (a) 1.1 g cm$^{-2}$

(b) $9.1 \times 10^{-3}$ mGy h$^{-1}$

**38.** (a) 12.1 MeV cm$^{-2}$ s$^{-1}$

(b) 0.21 $\mu$Gy h$^{-1}$

**39.** (a) 13 mR h$^{-1}$

(b) $2.9 \times 10^{-3}$ mR h$^{-1}$

**40.** (a) $1.53 \times 10^6$ MeV s$^{-1}$

(b) $5.1 \times 10^{-11}$ Gy s$^{-1}$

(c) 11.4 cm

**42.** (a) 0.09 mSv h$^{-1}$

(b) 42 cm

**43.** (a) $1.6 \times 10^9$ s$^{-1}$

(b) 27

# 16
# Internal Dosimetry and Radiation Protection

## 16.1
## Objectives

In this chapter we deal with radionuclides that can enter the body via inhalation, ingestion, wounds, or other means. As with external radiation, procedures have been developed to assess the effective dose and equivalent dose from internal emitters. Particular attention is given to determining the committed equivalent dose to various organs and the committed effective dose that results when an individual incorporates radioactive materials. The limits for all exposures are those recommended by the International Commission on Radiological Protection (ICRP) and the National Council on Radiation Protection and Measurements (NCRP). As described in Chapter 14, secondary limits have also been developed by these two organizations for the constant rate of intake of a given radionuclide that would, by itself, result in the limiting doses for an individual. These annual limits on intake (AL) are derived from assumed models and conditions for calculating internal doses. The ALI, in turn, can be used to derive air, food, and water concentrations as exposure guides for inhalation and ingestion.

This methodology, which is applied to control exposures from internal emitters, is the subject of the present chapter. The calculations employ various metabolic models for "standard," or "reference," individuals under specific given conditions. To this end, the 1975 *Report of the Task Group on Reference Man*, ICRP Publication 23 provided an extensive set of anatomical and metabolic data needed to carry out the formalism. This document has now been superseded by the 2003 ICRP Publication 89, described in the next section. Beginning in 1979, ICRP Publication 30 (including its Supplements) have long served as the foundation for modeling the internal translocation of chemical elements in the body for the computation of organ doses as functions of time following an intake. The formalism includes specific models for the respiratory tract, the gastrointestinal tract, and bone. While subsequent ICRP Publications have revised and improved a great deal of the material in Publication 30, much of the original content is still in current use. In this chapter we shall largely follow ICRP-30 procedures, pointing out where later modifications have been made by the ICRP. In this way, the presentation is kept

*Atoms, Radiation, and Radiation Protection.* James E. Turner
Copyright © 2007 WILEY-VCH Verlag GmbH & Co. KGaA, Weinheim
ISBN: 978-3-527-40606-7

relatively uncomplicated with simpler models, while maintaining the general conceptual framework, which is still employed. We shall be concerned primarily with occupational exposures for the adult male, as represented by reference man, in the numerical calculations to follow in this chapter.

## 16.2
## ICRP Publication 89

The 1975 ICRP Publication 23 on Reference Man was superseded in 2003 by ICRP Publication 89, *Basic Anatomical and Physiological Data for Use in Radiological Protection: Reference Values*. Whereas ICRP-23 concentrated on characteristics for a standard man, ICRP-89 presents data for males and females of six different ages: newborn, 1 y, 5 y, 10 y, 15 y, and adult. The reference values presented for various anatomical and physiological parameters are those for Western Europeans and North Americans, populations for which data are extensive and well documented. Some comparisons are made with several Asian populations. While relatively few individuals in any group will have all characteristics close to the reference values, this concept plays an important role in internal dosimetry. It enables base-line calculations of organ doses to be made for a radionuclide incorporated under a set of very specific, well defined assumptions. The formalism furnishes the basis for analysis in routine monitoring and bioassay programs throughout the world. When applied to special situations, appropriate adjustments of some of the assumptions can be made in order to obtain more realistic internal-dose estimates for a particular individual.

Table 16.1 shows reference values for the masses of some organs and tissues, as an example of some of the extensive anatomical data presented in ICRP-89. A number of mass reference values for the adult male in ICRP-89 are the same as those for the 70-kg reference man in Publication 23. The total mass for the adult male has been revised to 73 kg in ICRP-89. In addition to organ and tissue masses, tables of reference values are provided for a number of other anatomical characteristics. These include height, mass and total-body surface area; red blood-cell and plasma volumes; total mass and distribution of lymphocytes in the body; structural parameters for the eye and the skin; and other properties. Physiological reference values are furnished for metabolic rates, respiratory volumes and capacities (also at different levels of physical activity), daily secretion rates into different regions of the gastrointestinal tract, blood-flow rates through various tissues, and many other characteristics.

As described in the next section, internal-dose assessments include the irradiation of one organ by sources that are located in other organs. Different source- and target-organ masses are sometimes needed. For example, parts of the digestive tract can assume either role, depending on whether they contain a radioactive source or whether they are irradiated from a source located elsewhere. As seen in Table 16.1, for the adult male the mass for the stomach wall (as a target organ) is 150 g and that for the stomach contents (as a source organ) is 250 g.

**Table 16.1** Reference Values for Organ and Tissue Masses (grams) from ICRP Publication 89

| Organ/tissue | Newborn | 1 year | 5 years | 10 years | 15 years M | 15 years F | Adult M | Adult F |
|---|---|---|---|---|---|---|---|---|
| Adipose | 930 | 3,800 | 5,500 | 8,600 | 12,000 | 18,700 | 18,200 | 22,500 |
| Alimentary system | | | | | | | | |
| Stomach wall | 7 | 20 | 50 | 85 | 120 | 120 | 150 | 140 |
| Stomach contents | 40 | 67 | 83 | 117 | 200 | 200 | 250 | 230 |
| Small intestine wall | 30 | 85 | 220 | 370 | 520 | 520 | 650 | 600 |
| Small intestine contents | 56 | 93 | 117 | 163 | 280 | 280 | 350 | 280 |
| Liver | 130 | 330 | 570 | 830 | 1,300 | 1,300 | 1,800 | 1,400 |
| Integumentary system | | | | | | | | |
| Skin | 175 | 350 | 570 | 820 | 2,000 | 1,700 | 3,300 | 2,300 |
| Muscle, skeletal | 800 | 1,900 | 5,600 | 11,000 | 24,000 | 17,000 | 29,000 | 17,500 |
| Respiratory system | | | | | | | | |
| Lung with blood | 60 | 150 | 300 | 500 | 900 | 750 | 1,200 | 950 |
| Lung tissue only | 30 | 80 | 125 | 210 | 330 | 290 | 500 | 420 |
| Spleen | 9.5 | 29 | 50 | 80 | 130 | 130 | 150 | 130 |
| Thymus | 13 | 30 | 30 | 40/30 | 35 | 30 | 25 | 20 |
| Thyroid | 1.3 | 1.8 | 3.4 | 7.9 | 12 | 12 | 20 | 17 |
| Urogenital system | | | | | | | | |
| Kidneys (2) | 25 | 70 | 110 | 180 | 250 | 240 | 310 | 275 |
| Testes (2) | 0.85 | 1.5 | 1.7 | 2 | 16 | | 35 | |
| Ovaries (2) | 0.3 | 0.8 | 2.0 | 3.5 | | 6 | | 11 |
| Uterus | 4.0 | 1.5 | 3 | 4 | | 30 | | 80 |
| Total body | 3,500 | 10,000 | 19,000 | 32,000 | 56,000 | 53,000 | 73,000 | 60,000 |

As discussed in Section 16.8, radiation-transport calculations have been carried out to evaluate doses absorbed in one organ from radiation emitted in others. Of historical interest is the Snyder–Fisher phantom—a mathematical model of reference man used for such computations, performed, for example by Monte Carlo procedures (Section 11.12). The model is depicted in Fig. 16.1. Figure 16.2 is a photograph of a life-sized replica of the mathematical phantom that was fabricated and used to measure quantities to compare with calculations. A radionuclide could be placed in one position in the phantom to simulate a particular source organ, and the doses then measured in other locations.

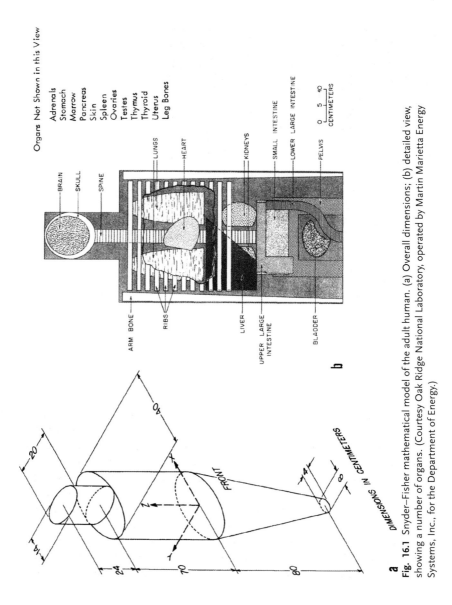

**Fig. 16.1** Snyder–Fisher mathematical model of the adult human. (a) Overall dimensions; (b) detailed view, showing a number of organs. (Courtesy Oak Ridge National Laboratory, operated by Martin Marietta Energy Systems, Inc., for the Department of Energy.)

**Fig. 16.2** Life-size model built to represent Snyder–Fisher
phantom. Various chemical mixtures are used to simulate
tissue composition in each region. Source organs can be filled
with solution of desired radionuclide. (Courtesy Oak Ridge
National Laboratory, operated by Martin Marietta Energy
Systems, Inc., for the Department of Energy.)

## 16.3
## Methodology

The principal methodology that is implemented for internal dosimetry in ICRP-30
can be described with reference to Fig. 16.3. At time $t = 0$ a given amount of a
specific radionuclide (e.g., unit activity) is inhaled or ingested into the body. For
inhalation, the ICRP-30 model of the respiratory system (Section 16.4) is used to
calculate the initial deposition of the radionuclide in various compartments of the
lung and its subsequent retention there as well as its transfer to the body fluids
and the gastrointestinal (GI) tract. For ingestion of the radionuclide or its transfer
from the lung (e.g., by ciliary action and swallowing), the ICRP gastrointestinal-
tract model (Section 16.6) is used to compute the transfer of the radionuclide from
the GI tract to the body fluids as well as its excretion. Thus, by either route of intake,
the radionuclide enters the body fluids, represented by compartment $a$ in Fig. 16.3.

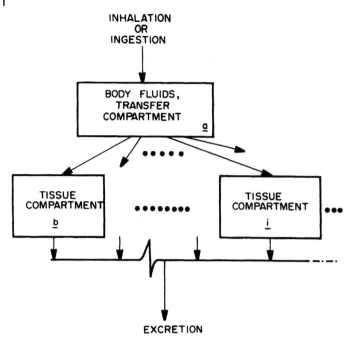

**Fig. 16.3** Mathematical model usually employed for transfer of
a radionuclide from the body fluids (compartment a) to various
organs and tissues and its subsequent excretion. [From data in
*Annals of the ICRP*, Vol. 2, No. 3/4, ICRP Publ. 30, Part I,
International Commission on Radiological Protection, Sutton,
England (1979).]

Compartment $a$ is linked to other compartments, $b, c, \ldots, i, \ldots$, representing various tissues and organs in the body, from which the nuclide can later be excreted. A radionuclide undergoes metabolic clearance from these compartments, as well as radioactive decay. Unless otherwise specified, the metabolic half-life used for the transfer compartment $a$ is 0.25 day. Radioactive transformations that occur in this compartment are assumed to be distributed uniformly in the 70-kg reference man. For simplicity, excretion in the model does not involve transport through compartment $a$, although, realistically, the body fluids are involved. For this reason, the calculated amount of a radionuclide in compartment $a$ at some time after inhalation or ingestion cannot be used as an estimate of the amount of the radionuclide in the body fluids at that time.

After a radionuclide leaves the transfer compartment $a$, the ICRP metabolic model continues to trace its movement through the various tissues and organs in the body, represented by compartments $b, c, \ldots, i, \ldots$, which may be interconnected. Explicit metabolic parameters are given for each radionuclide, its pertinent chemical forms, and all organs and tissues. The activity of the radionuclide in any organ can be calculated explicitly as a function of time after intake. The equivalent-dose rate in the organ from decay of the radionuclide it contains can be computed as a function of time. In addition, if the nuclide emits penetrating radiation, other organs of the body will be irradiated from the decay in a given organ. Using a mathematical model for reference man (Fig. 16.1), the ICRP methodology enables one to calculate the equivalent-dose rates in other organs from disintegrations in a given organ as functions of time. A complete calculation thus provides the equivalent-dose rates in every organ and tissue of the body and the effective-dose rate as functions of time. The committed equivalent doses and the committed effective dose resulting from the initial intake can be evaluated, and the annual reference level of intake (ARLI) and derived reference air concentration (DRAC) determined (Section 14.4). When the original inhaled or ingested nuclide decays into radioactive daughters, contributions of the latter are included in the determination of the ARLI and DRAC.

We now outline explicitly how the calculation of the committed effective dose for a radionuclide in the body can be carried out by using the ICRP metabolic models and reference man. Several concepts will be introduced now and developed further in the sections that follow. By definition, the committed effective dose is given by Eq. (14.7) with $\tau = 50$ y. We write

$$E(50) = \sum_T w_T H_T(50), \tag{16.1}$$

where the $w_T$ are the tissue weighting factors (Table 14.2), the $H_T(50)$ are the committed equivalent doses, given by Eq. (14.6), and the sum extends over all organs and tissues T of the body. As described in the last paragraph, the committed equivalent dose in a given organ is a result of irradiation by sources in both the organ itself and in other organs. For a given organ, considered as a target organ T, one computes and adds the contributions to the committed equivalent dose from all organs, considered as source organs S:

$$H_T(50) = \sum_S U_S \hat{H}(T \leftarrow S). \tag{16.2}$$

Here $\hat{H}(T \leftarrow S)$ denotes the equivalent dose in T (average) per disintegration, or transformation, of the radionuclide in S. The sum extends over all source organs and tissues S of the body, including T. The quantity $U_S$ denotes the number of transformations of the nuclide in S during the 50-y period of the committed equivalent dose.

Having outlined the calculation of the committed effective dose following the intake of a radionuclide, we turn now in the next sections to the individual elements needed for the computations.

## 16.4
### ICRP-30 Dosimetric Model for the Respiratory System

The ICRP-30 model for the respiratory system, shown in Fig. 16.4, is divided into three major parts—the nasal passage (NP), the trachea and bronchial tree (TB), and

| | | Class | | | | | |
|---|---|---|---|---|---|---|---|
| | | **D** | | **W** | | **Y** | |
| Region | Compart-ment | T (d) | F | T (d) | F | T (d) | F |
| N–P ($D_{N-P}$ = 0.30) | a | 0.01 | 0.5 | 0.01 | 0.1 | 0.01 | 0.01 |
| | b | 0.01 | 0.5 | 0.40 | 0.9 | 0.40 | 0.99 |
| T–B ($D_{T-B}$ = 0.08) | c | 0.01 | 0.95 | 0.01 | 0.5 | 0.01 | 0.01 |
| | d | 0.2 | 0.05 | 0.2 | 0.5 | 0.2 | 0.99 |
| P ($D_P$ = 0.25) | e | 0.5 | 0.8 | 50 | 0.15 | 500 | 0.05 |
| | f | n.a. | n.a. | 1.0 | 0.4 | 1.0 | 0.4 |
| | g | n.a. | n.a. | 50 | 0.4 | 500 | 0.4 |
| | h | 0.5 | 0.2 | 50 | 0.05 | 500 | 0.15 |
| L | i | 0.5 | 1.0 | 50 | 1.0 | 1 000 | 0.9 |
| | j | n.a. | n.a. | n.a. | n.a. | ∞ | 0.1 |

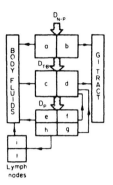

**Fig. 16.4** Compartmental model for the respiratory system. Table gives initial deposition fractions, half-times, and removal fractions for each compartment and class of material for aerosol with an AMAD of 1 $\mu$m (see text). [Reprinted with permission from *Annals of the ICRP*, Vol. 2, No. 3/4, ICRP Publ. 30, Part I, p. ii (Errata), International Commission on Radiological Protection, Sutton, England (1979). Copyright 1979 by ICRP.]

the pulmonary parenchyma (P). In addition, a pulmonary lymphatic system (L) is included for the removal of dust from the lungs. The direct deposition of inhaled material, which occurs in the first three regions, varies with the particle-size distribution of the material. Basic calculations are made for reference man with an assumed log-normal distribution of particle diameters having an assumed activity median aerodynamic diameter (AMAD) of 1 $\mu$m. The fractions $D$ of inhaled material that are initially deposited in the three regions are then assumed to be $D_{NP} = 0.30$, $D_{TB} = 0.08$, and $D_P = 0.25$. A procedure is given for making particle-size corrections for other values of the AMAD. The deposition fractions for other sizes are given in Fig. 16.5. The model is thus applicable to the inhalation of radioactive aerosols, or particulates. Inhalation of a radioactive gas is treated separately (Section 16.11).

The model of the respiratory system describes the initial deposition and subsequent transport of inhaled radioactive aerosols through various compartments of the system and into the body fluids and the GI tract. It has been found that the dose in the NP region can be neglected for most particle sizes. The target tissue assumed for the lung, therefore, is that of the combined TB, P, and L regions, having a total mass of 1000 g. The committed equivalent dose to the lung has two components, one from the radioactive materials residing there and another from photons emitted by materials that are cleared from the lung and transported to other sites in the body.

ICRP Publication 30 classifies inhaled radioactive materials as D, W, or Y (days, weeks, or years), depending on their retention time in the pulmonary region. Class-D materials have a half-time of less than 10 d; W materials, a half-time from 10 d to 100 d; and Y, greater than 100 d. As seen in Fig. 16.4, the four major regions

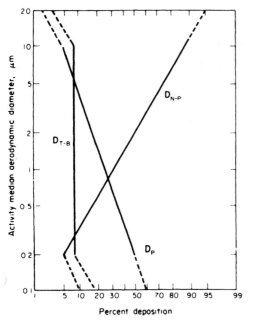

**Fig. 16.5** Deposition fractions in regions of the respiratory model for aerosols of different AMAD (see text) between 0.2 μm and 10 μm. Dashed lines show provisional extensions of curves outside this range. [Reprinted with permission from *Annals of the ICRP*, Vol. 2, No. 3/4, ICRP Publ. 30, Part I, p. i (Errata), International Commission on Radiological Protection, Sutton, England (1979). Copyright 1979 by ICRP.]

of the model are each subdivided into two or four compartments, each associated with a particular pathway of clearance. For the three classes D, W, and Y, the table in Fig. 16.4 gives the half-time $T$ used in each compartment and the fraction $F$ of material that leaves it at that implied rate. Compartments, a, c, and e are associated with the uptake of material from the respiratory system into the body fluids. Compartments b, d, f, and g are associated with the physical transport of particles (e.g., by mucociliary action and swallowing) into the GI tract. Compartment h in the P region provides the pathway to the lymph system L, where some material can be further translocated via i to the body fluids or else retained indefinitely in j. (Compartment j is used only for class-Y materials.)

Given the rate of inhalation $\dot{I}(t)$ of a radionuclide, ten differential equations are used to describe the activities $q_a(t), \ldots, q_j(t)$ in each of the ten compartments shown in Fig. 16.4. The equation describing a, for example, can be written

$$\frac{dq_a}{dt} = \dot{I} D_{NP} F_a - \lambda_a q_a - \lambda_R q_a, \tag{16.3}$$

where $D_{NP} = 0.30$; $F_a$ and $T_a$ are obtained directly from the values of $F$ and $T$ in the table in Fig. 16.4 for class D, W, or Y; $\lambda_a = 0.693/T_a$; and $\lambda_R$ is the radioactive decay constant. The classifications D, W, or Y for different chemical forms of a radionuclide are provided by the ICRP along with the metabolic data. Radioactive daughters, included in the calculations, are assumed to have the same metabolic behavior as the original parent. While Eq. (16.3) involves only a single activity, $q_a$, others are more complicated. For compartment d, for example, we have

$$\frac{dq_d}{dt} = \dot{I} D_{TB} F_d + \lambda_f q_f + \lambda_g q_g - \lambda_d q_d - \lambda_R q_d, \tag{16.4}$$

which couples the activity $q_d$ in d to those in f and g. Given a set of initial conditions, the system of ten linear, coupled, differential equations is solved to obtain the activities in each of the compartments a–j as functions of time. The rate of transfer of the inhaled radionuclide into the body fluids as a function of time is then given by

$$BF(t) = \lambda_a q_a(t) + \lambda_c q_c(t) + \lambda_e q_e(t) + \lambda_i q_i(t). \tag{16.5}$$

Similarly, the rate of transfer into the GI tract is

$$G(t) = \lambda_b q_b(t) + \lambda_d q_d(t). \tag{16.6}$$

The ICRP-30 respiratory-system model thus specifies the deposition, retention, and removal of inhaled materials in various components of the pulmonary–lymph system. It is used to calculate the number of transformations $U$ for the committed equivalent dose to the lung and to calculate source terms for the body fluids and the GI tract.

## 16.5
## ICRP-66 Human Respiratory Tract Model

The ICRP-30 lung model has served well to calculate occupational annual limits on intake (ALI) and to be the foundation for many applied monitoring and control programs and procedures. It has been an extremely valuable tool for the dosimetry of inhaled radionuclides. Nevertheless, there are significant problems in lung dosimetry that the model was not designed or equipped to handle. For example, many radioactive compounds were found to clear from the respiratory system at rates considerably different from those assigned. Also, the lung dose was calculated as an average over the total lung mass, whereas it is relatively rare that the respiratory tract is uniformly irradiated by internally deposited aerosols. Moreover, different tissues of the lung have different radio-sensitivity. Such factors are particularly relevant for inhaled radon daughters or hot particles. In addition, the ICRP-30 worker-oriented lung model lacked the flexibility to be applied generally to members of the public, a matter related to increasing environmental concerns.

These and other considerations, including continued research and a growing body of new information, led the ICRP to review and address lung dosimetry anew. Rather than developing an entirely new model, efforts were aimed at improving and building on the ICRP-30 model to meet an expanded variety of needs. The work resulted in the 1994 Publication 66, *Human Respiratory Tract Model for Radiological Protection*, with the present and considerably more sophisticated ICRP lung model, abbreviated HRTM. The task and outcome are described in ICRP-66. "The objective was a model that would (1) facilitate calculation of biologically meaningful doses; (2) be consistent with morphological, physiological, and radiobiological characteristics of the respiratory tract; (3) incorporate current knowledge; (4) meet all radiation protection needs; (5) be no more sophisticated than necessary to meet dosimetric objectives; (6) be adaptable to development of computer software for the calculation of relevant radiation doses from knowledge of a few readily measured exposure parameters; (7) be equally useful for assessment purposes as for calculating recommended values for limits on uptake, e.g. ALIs; (8) be applicable to all members of the world's population, including specific individuals; (9) allow use of information on the deposition and clearance of specific materials; and (10) consider the influence of smoking, air pollutants, and diseases on the inhalation, deposition, and clearance of radioactive particles from the respiratory tract. These objectives have been largely met within the constraints of the available data."

Some specific major revisions in HRTM can be briefly mentioned as examples. Figure 16.6, taken from Publication 66, shows a considerably more involved, four-region respiratory tract than before. Figure 16.7, which is Table 1 from ICRP-66,

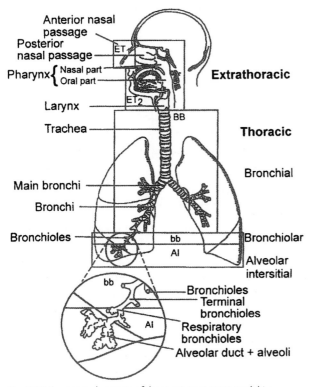

**Fig. 16.6** Anatomical regions of the respiratory tract model in the 1994 ICRP Publication 66. (Courtesy International Commission on Radiological Protection.)

Table 1. Morphology, cytology, histology, function, and structure of the respiratory tract and regions used in the dosimetry model

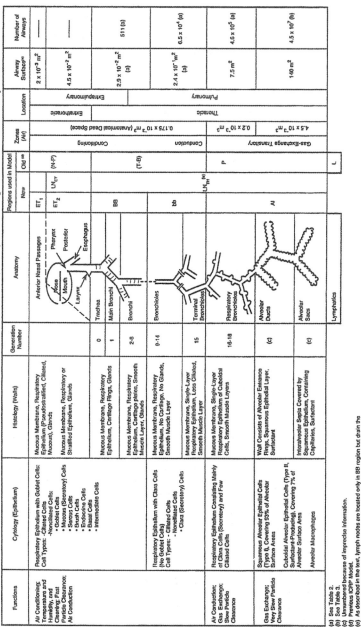

| Functions | Cytology (Epithelium) | Histology (Walls) | Generation Number | Anatomy | Regions used in Model (New / Old[d]) | Zones (Air) | Location | Airway Surface[a] | Number of Airways |
|---|---|---|---|---|---|---|---|---|---|
| Air Conditioning; Temperature and Humidity, and Cleaning; Fast Particle Clearance; Air Conduction | Respiratory Epithelium with Goblet Cells: -Ciliated Cells -Nonciliated Cells: • Goblet Cells • Mucous (Secretory) Cells • Serous Cells • Brush Cells • Endocrine Cells • Basal Cells • Intermediate Cells | Mucous Membrane, Respiratory Epithelium (Pseudostratified, Ciliated, Mucous), Glands | | Anterior Nasal Passages / Nose / Mouth / Pharynx / Posterior | ET₁ (N-P) | Conditioning / 0.175 × 10⁻³ m³ (Anatomical Dead Space) | Extrathoracic | 2 × 10⁻³ m² | — |
| | | Mucous Membrane, Respiratory or Stratified Epithelium, Glands | | Larynx / Esophagus | ET₂ LNₑₜ | | Extrathoracic | 4.5 × 10⁻² m² | — |
| | | Mucous Membrane, Respiratory Epithelium, Cartilage Rings, Glands | 0 | Trachea | BB (T-B) | Conduction / 0.2 × 10⁻³ m³ | Thoracic | 2.9 × 10⁻² m² (a) | 511 (a) |
| | | Mucous Membrane, Respiratory Epithelium, Cartilage plates, Smooth Muscle Layer, Glands | 1 | Main Bronchi | | | | | |
| | | | 2-8 | Bronchi | | | | | |
| Air Conduction; Gas Exchange; Slow Particle Clearance | Respiratory Epithelium with Clara Cells (No Goblet Cells) Cell Types: - Ciliated Cells - Nonciliated Cells • Clara (Secretory) Cells | Mucous Membrane, Respiratory Epithelium, No Cartilage, No Glands, Smooth Muscle Layer | 9-14 | Bronchioles | bb | | | 2.4 × 10⁻³ m² (a) | 6.5 × 10⁴ (a) |
| | | Mucous Membrane, Single-Layer Respiratory Epithelium, Less Ciliated, Smooth Muscle Layer | 15 | Terminal Bronchioles | LNₜₕ (e) | | Thoracic | | |
| | Respiratory Epithelium Consisting Mainly of Clara Cells (Secretory) and Few Ciliated Cells | Mucous Membrane, Single-Layer Respiratory Epithelium of Cuboidal Cells, Smooth Muscle Layers | 16-18 | Respiratory Bronchioles | AI | Gas-Exchange / 4.5 × 10⁻³ m³ | Pulmonary | 7.5 m² | 4.5 × 10⁵ (a) |
| Gas Exchange; Very Slow Particle Clearance | Squamous Alveolar Epithelial Cells (Type II), Covering 93% of Alveolar Surface Area | Wall Consists of Alveolar Entrance Rings, Squamous Epithelial Layer, Surfactant | (c) | Alveolar Ducts | | Transitory | | | |
| | Cuboidal Alveolar Epithelial Cells (Type II), Surfactant-Producing), Covering 7% of Alveolar Surface Area | Interalveolar Septa Covered by Squamous Epithelium, Containing Capillaries, Surfactant | (c) | Alveolar Sacs | | | | 140 m² | 4.5 × 10⁷ (b) |
| | Alveolar Macrophages | | | | L | | | | |
| | | | | Lymphatics | | | | | |

(a) See Table 2.
(b) See Table 3.
(c) Unnumbered because of imprecise information.
(d) Previous ICRP Model.
(e) As described in the text, lymph nodes are located only in BB region but drain the bronchiolar and alveolar-interstitial regions as well as the bronchial region.

**Fig. 16.7** As an illustration of the scope of the HRTM, Table 1 from ICRP-66, shown here, indicates details of the mathematical model used to calculate radiation doses. (Courtesy International Commission on Radiological Protection.)

conveys an idea of the detailed nature of the model. It also shows the relationship to the N-P, T-B, and P regions from the ICRP-30 model. Clearance and absorption compete in each compartment with time-dependent rates in the revised model. Breathing rates, volumes, and other parameters have been revised. The default particle-size AMAD is increased from 1 $\mu$m to 5 $\mu$m. The revised model allows dose calculations for specific target regions containing cells that are considered to be susceptible to cancer induction, in contrast to the mean dose furnished by ICRP-30. The methodology can be applied to males and females, young and old. Effects of smoking and other inhaled pollutants can be included.

## 16.6
## ICRP-30 Dosimetric Model for the Gastrointestinal Tract

The ICRP dosimetric model for the GI tract is shown in Fig. 16.8. Each of the four sections consists of a single compartment: the stomach (ST), small intestine (SI), upper large intestine (ULI), and lower large intestine (LLI). There are two pathways out of the SI. One leads to the ULI and the other to the body fluids, the only route

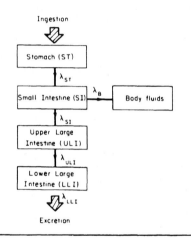

| Section of GI tract | Mass of walls* (g) | Mass of contents* (g) | Mean residence time (d) | $\lambda$ (d$^{-1}$) |
|---|---|---|---|---|
| Stomach (ST) | 150 | 250 | 1/24 | 24 |
| Small Intestine (SI) | 640 | 400 | 4/24 | 6 |
| Upper Large Intestine (ULI) | 210 | 220 | 13/24 | 1.8 |
| Lower Large Intestine (LLI) | 160 | 135 | 24/24 | 1 |

*From ICRP Publication 23 (1975).

**Fig. 16.8** Dosimetric model for the gastrointestinal system. Table gives masses of the sections and their contents and clearance-rate data. [Reprinted with permission from *Annals of the ICRP*, Vol. 2, No. 3/4, ICRP Publ. 30, Part I, p. 33, International Commission on Radiological Protection, Sutton, England (1979). Copyright 1979 by ICRP.]

by which ingested materials are assumed to reach the body fluids. The metabolic rate constants $\lambda_B$ are used by the ICRP for the various chemical elements. The fraction $f_1$ of a stable element that reaches the body fluids after ingestion is given by

$$f_1 = \frac{\lambda_B}{\lambda_{SI} + \lambda_B}. \qquad (16.7)$$

Values of $f_1$ are specified in the metabolic data of the ICRP (Section 16.12).

As with the lung, each of the four compartments in Fig. 16.8 gives rise to a first-order differential equation describing the activity changes. Given initial conditions and the rate of intake $\dot{I}(t)$ into compartment ST, the four equations can be solved for the activities in each section as functions of time. Activity entering the GI tract from the respiratory system is included in $\dot{I}(t)$. Radioactive daughters are included with the parent in the ICRP calculations. The table in Fig. 16.8 gives the assumed masses of the sections and their contents and the clearance rates. The activity transferred to the body fluids as a function of time is given by $\lambda_B q_{SI}(t)$.

When the source organ is a section of the GI tract, the committed equivalent dose is estimated for the mucosal layer of the walls of each section for penetrating and nonpenetrating radiations. Other organs are also irradiated by sources in the contents of the GI tract, and the tract is irradiated by materials located in other parts of the body. The ICRP has compiled values of $U_S$ for various ingested radionuclides and daughters in the sections of the GI tract. It also gives data for sections of the GI tract as target organs with the contents of these sections and other organs of the body as source organs.

In 2007 the ICRP issued its revised dosimetric model of the GI tract, designated HATM. Publication 100, *Human Alimentary Tract Model for Radiological Protection*, treats intakes by children as well as male and female adults, with applicability to both occupational and environmental exposures. Also, like the companion HRTM, the HATM enables dose calculations to be made for specific target regions considered important for cancer induction.

## 16.7
### Organ Activities as Functions of Time

The calculation of activities in various organs and tissues of the body as functions of time after intake is exemplified by computations for the GI tract. We let $\dot{I}(t)$ represent the rate of activity intake (e.g., in Bq s$^{-1}$) of a given radionuclide into the stomach as a function of time $t$. The rate of change of the activity, $q_{ST}(t)$, in the stomach at any time is then given by the equation

$$\frac{dq_{ST}}{dt} = \dot{I}(t) - \lambda_R q_{ST} - \lambda_{ST} q_{ST}. \qquad (16.8)$$

Here, $\lambda_R$ is the radioactive decay constant of the nuclide and $\lambda_{ST}$ is the metabolic rate constant given in Fig. 16.8 ($\lambda_{ST} = 24$ d$^{-1}$). The rate of depletion, $(\lambda_R + \lambda_{ST})q_{ST}$, is governed by the two processes, radioactive decay and biological removal. In general, the effective decay constant for a radionuclide in any tissue or organ compartment with a metabolic removal rate $\lambda_M$ is given by writing

$$\lambda_{Eff} = \lambda_R + \lambda_M. \tag{16.9}$$

In terms of the radiological half-life, $T_R = 0.693/\lambda_R$, and the metabolic half-time, $T_M = 0.693/\lambda_M$, it follows that the effective half-life in the compartment is given by

$$T_{Eff} = \frac{T_R T_M}{T_R + T_M}. \tag{16.10}$$

Note that $T_{Eff}$ is smaller than both $T_R$ and $T_M$.

We treat explicitly the introduction of a single amount of activity $A_0$ of a radionuclide into the stomach at time $t = 0$. Then $\dot{I}(t) = 0$ when $t > 0$ in Eq. (16.8). With the initial condition, $q_{ST}(0) = A_0$, the equation has the solution

$$q_{ST}(t) = A_0 e^{-\lambda_1 t}, \tag{16.11}$$

where $\lambda_1 = \lambda_R + \lambda_{ST}$ has been written to simplify notation. Mathematically, the result (16.11) is the same as Eq. (4.8) for radioactive decay.

The activity $q_{SI}(t)$ in the small intestine satisfies the rate equation (see Fig. 16.8),

$$\frac{dq_{SI}}{dt} = \lambda_{ST} q_{ST} - \lambda_2 q_{SI}, \tag{16.12}$$

where $\lambda_2 = \lambda_R + \lambda_{SI} + \lambda_B$. The rate constants $\lambda_{SI}$ and $\lambda_B$ are defined in Fig. 16.8, and $q_{ST}$ is given by Eq. (16.11). The term $\lambda_{ST} q_{ST}$ represents the metabolic rate of activity transfer from the stomach into the small intestine. (Whereas the depletion rate in a compartment is the result of radioactive decay and metabolic processes, the input rate is governed by biological processes alone.) Writing

$$\frac{dq_{SI}}{dt} + \lambda_2 q_{SI} = \lambda_{ST} A_0 e^{-\lambda_1 t}, \tag{16.13}$$

we multiply both sides by the integrating factor $e^{\lambda_2 t}$ in order to solve for $q_{SI}$.[1]

$$e^{\lambda_2 t}\left(\frac{dq_{SI}}{dt} + \lambda_2 q_{SI}\right) \equiv \frac{d}{dt}(q_{SI} e^{\lambda_2 t}) = \lambda_{ST} A_0 e^{(\lambda_2 - \lambda_1)t}. \tag{16.14}$$

Integration gives

$$q_{SI} e^{\lambda_2 t} = \frac{A_0 \lambda_{ST}}{\lambda_2 - \lambda_1} e^{(\lambda_2 - \lambda_1)t} + c_1, \tag{16.15}$$

1   Equation (16.12) is mathematically equivalent to Eq. (4.31), which was solved earlier for serial radioactive decay. We show the use of an integrating factor here, because it is a convenient device for continuing to the next compartments.

where $c_1$ is the constant of integration. With the initial condition, $q_{SI}(0) = 0$, we find $c_1 = -A_0\lambda_{ST}/(\lambda_2 - \lambda_1)$. Thus,

$$q_{SI}e^{\lambda_2 t} = \frac{A_0\lambda_{ST}}{\lambda_2 - \lambda_1}[e^{(\lambda_2 - \lambda_1)t} - 1], \tag{16.16}$$

or

$$q_{SI}(t) = \frac{A_0\lambda_{ST}}{\lambda_2 - \lambda_1}[e^{-\lambda_1 t} - e^{-\lambda_2 t}]. \tag{16.17}✳}$$

[This solution is similar to Eq. (4.40).]

For the activity in the upper large intestine,

$$\frac{dq_{ULI}}{dt} = \lambda_{SI}q_{SI} - \lambda_3 q_{ULI}, \tag{16.18}$$

where $\lambda_3 = \lambda_R + \lambda_{ULI}$. Using the integrating factor $e^{\lambda_3 t}$ and Eq. (16.17) for $q_{SI}$, we write in place of (16.18)

$$e^{\lambda_3 t}\left(\frac{dq_{ULI}}{dt} + \lambda_3 q_{ULI}\right) = \frac{A_0\lambda_{SI}\lambda_{ST}}{\lambda_2 - \lambda_1}[e^{(\lambda_3 - \lambda_1)t} - e^{(\lambda_3 - \lambda_2)t}]. \tag{16.19}$$

Integrating both sides yields

$$q_{ULI}e^{\lambda_3 t} = \frac{A_0\lambda_{SI}\lambda_{ST}}{\lambda_2 - \lambda_1}\left[\frac{e^{(\lambda_3 - \lambda_1)t}}{\lambda_3 - \lambda_1} - \frac{e^{(\lambda_3 - \lambda_2)t}}{\lambda_3 - \lambda_2}\right] + c_2, \tag{16.20}$$

where $c_2$ is the constant of integration. With the initial condition, $q_{ULI}(0) = 0$, it follows that

$$c_2 = -\frac{A_0\lambda_{SI}\lambda_{ST}}{\lambda_2 - \lambda_1}\left[\frac{1}{\lambda_3 - \lambda_1} - \frac{1}{\lambda_3 - \lambda_2}\right] = \frac{A_0\lambda_{SI}\lambda_{ST}}{(\lambda_3 - \lambda_2)(\lambda_3 - \lambda_1)}. \tag{16.21}$$

Substituting this value for $c_2$ and multiplying both sides of Eq. (16.20) by $e^{-\lambda_3 t}$, we obtain

$$q_{ULI}(t) = A_0\lambda_{SI}\lambda_{ST}\left[\frac{e^{-\lambda_1 t}}{(\lambda_2 - \lambda_1)(\lambda_3 - \lambda_1)} + \frac{e^{-\lambda_2 t}}{(\lambda_3 - \lambda_2)(\lambda_1 - \lambda_2)}\right.$$

$$\left. + \frac{e^{-\lambda_3 t}}{(\lambda_3 - \lambda_2)(\lambda_3 - \lambda_1)}\right]. \tag{16.22}✳$$

We could proceed in similar fashion to solve for the activity in the lower large intestine as a function of time, but will not do so.

The general decay and growth functions, such as those described by Eqs. (16.11), (16.17), and (16.22), are often referred to as the Bateman equations. They date from the early days of study with the naturally occurring radioactive decay chains.[2]

2   H. Bateman, *Proc. Cambridge Philos. Soc.* **15**, 423 (1910).

*Example*

(a) What is the effective half-life of $^{198}$Au in the stomach? (b) What is the effective decay constant there?

*Solution*

(a) The effective half-life is given by Eq. (16.10). From Appendix D, the radiological half-life is $T_R = 2.70$ d. From Fig. 16.8, we find that the metabolic half-time is $T_{ST} = 0.693/\lambda_{ST} = 0.693/(24\ \text{d}^{-1}) = 0.0289$ d. Equation (16.10) gives

$$T_{Eff} = \frac{T_R T_{ST}}{T_R + T_{ST}} = \frac{2.70 \times 0.0289}{2.70 + 0.0289} = 0.0286\ \text{d}. \tag{16.23}$$

(b) The effective decay constant is

$$\lambda_{Eff} = \frac{0.693}{T_{Eff}} = \frac{0.693}{0.0286\ \text{d}} = 24.2\ \text{d}^{-1}. \tag{16.24}$$

One can also calculate $\lambda_{Eff}$ as the sum of the radiological and metabolic constants. One has $\lambda_R = 0.693/(2.70\ \text{d}) = 0.257\ \text{d}^{-1}$ and, from Fig. 16.8, $\lambda_{ST} = 24\ \text{d}^{-1}$, giving $\lambda_{Eff} = \lambda_R + \lambda_{ST} = 24.3\ \text{d}^{-1}$.

*Example*

For a single ingestion of $^{198}$Au at time $t = 0$, calculate the fractions of the initial activity in the stomach, small intestine, and upper large intestine as functions of time. In Eq. (16.7), $f_1 = 0.1$.

*Solution*

We evaluate the rate constants for Eqs. (16.11), (16.17), and (16.22). It is convenient to express time in days. From the last example, the radioactive decay constant is $\lambda_R = (0.257\ \text{d}^{-1})\ [(1/24)\ \text{d}\,\text{h}^{-1}] = 0.0107\ \text{h}^{-1}$. With the help of Fig. 16.8 and the given value, $f_1 = 0.1$, we find

$$\lambda_B = \frac{f_1 \lambda_{SI}}{1 - f_1} = \frac{0.1 \times (6/24)\ \text{h}^{-1}}{1 - 0.1} = 0.0278\ \text{h}^{-1}. \tag{16.25} \text{✱}$$

Thus, with time in hours,

$$\lambda_1 = \lambda_R + \lambda_{ST} = 0.0107 + \frac{24}{24} = 1.01\ \text{h}^{-1}. \tag{16.26}$$

$$\lambda_2 = \lambda_R + \lambda_{SI} + \lambda_B = 0.0107 + \frac{6}{24} + 0.0278 = 0.289\ \text{h}^{-1}. \tag{16.27}$$

$$\lambda_3 = \lambda_R + \lambda_{ULI} = 0.0107 + \frac{1.8}{24} = 0.0857\ \text{h}^{-1}. \tag{16.28}$$

The fraction of the ingested activity in the stomach after $t$ hours is, from Eq. (16.11),

$$\frac{q_{ST}(t)}{A_0} = e^{-\lambda_1 t} = e^{-1.01t}. \tag{16.29}$$

The fraction in the small intestine is given by Eq. (16.17), with $\lambda_2 - \lambda_1 = 0.289 - 1.01 = -0.721$ h$^{-1}$ and $\lambda_{ST} = 1$ h$^{-1}$:

$$\frac{q_{SI}(t)}{A_0} = \frac{1}{-0.721}(e^{-1.01t} - e^{-0.289t}) = 1.39(e^{-0.289t} - e^{-1.01t}). \tag{16.30}$$

Additionally, for Eq. (16.22) we need $\lambda_{SI} = 6/24 = 0.250$ h$^{-1}$, $\lambda_3 - \lambda_1 = 0.0857 - 1.01 = -0.924$ h$^{-1}$, and $\lambda_3 - \lambda_2 = 0.0857 - 0.289 = -0.203$ h$^{-1}$. Thus,

$$\frac{q_{ULI}(t)}{A_0} = (0.250)(1)\left[\frac{e^{-1.01t}}{(-0.721)(-0.924)} + \frac{e^{-0.289t}}{(-0.203)(0.721)}\right. \tag{16.31}$$

$$\left. + \frac{e^{-0.0857t}}{(-0.203)(-0.924)}\right]$$

$$= 0.375e^{-1.01t} - 1.71e^{-0.289t} + 1.33e^{-0.0857t}. \tag{16.32}$$

As a check, we note that $q_{ULI}(0) = -0.005$, which, to within roundoff, is the required initial condition, $q_{ULI}(0) = 0$.

The fractional activities, (16.29), (16.30), and (16.32), are plotted in Fig. 16.9. With the 1-h mean metabolic residence time in the stomach, the fraction of the radionuclide there drops rapidly below unity following ingestion. Equation (16.26) shows that radioactive decay adds only 1% to the clearance rate from the stomach. The activity in the small intestine builds up to a maximum at 1.75 h; that in the upper large intestine reaches a maximum at 7 h. About 70% of the originally ingested activity is in these three compartments at this time. After about 24 h, the fractional activity in the ULI is 0.17. Very little additional activity enters the ULI then, and so $q_{ULI}(t)$ declines exponentially thereafter, governed principally by the third term in Eq. (16.32). The

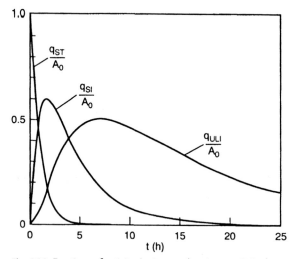

**Fig. 16.9** Fractions of activity $A_0$, ingested at time $t = 0$, in the stomach (ST), small intestine (SI), and upper large intestine (ULI) as functions of time. See Eqs. (16.29), (16.30), and (16.32) in the example in text.

rate $\lambda_3 = 0.0857$ h$^{-1}$ = 2.06 d$^{-1}$ reflects the combined effect of the biological removal and radioactive decay.

Returning to Fig. 16.3, we next set up equations to describe the transport of activity from the body fluids (compartment $a$) to a tissue compartment $b$ and the computation of the activity in $b$ as a function of time. We let $\dot{I}(t)$ represent the rate at which a radionuclide enters the body fluids at time $t$ after its inhalation or ingestion into the body. If $q_a(t)$ is the activity of the radionuclide in compartment $a$, $\lambda_R$ is its radioactive decay constant, and $\lambda_a$ is the metabolic clearance rate of that element from $a$, we then have

$$\frac{dq_a(t)}{dt} = \dot{I}(t) - \lambda_R q_a(t) - \lambda_a q_a(t). \tag{16.33}$$

If $b$ represents the fraction of the element that goes to compartment $b$ when it leaves $a$, then the activity $q_b(t)$ of the radionuclide in $b$ at time $t$ satisfies the equation

$$\frac{dq_b(t)}{dt} = b\lambda_a q_a(t) - \lambda_R q_b(t) - \lambda_b q_b(t), \tag{16.34}$$

where $\lambda_b$ is the metabolic clearance rate from $b$. Similar equations describe the activities $q_i(t)$ of the radionuclide in other tissue compartments. With given initial conditions and specific metabolic data, the equations can be solved for the activities $q_i(t)$ in the various organs and tissues. The 50-y integrals of the $q_i(t)$ then give the numbers of transformations $U_S$ [Eq. (16.2)] in the source organs or tissues $i$.

We treat a single amount of activity, $A_0 = q_a(0)$, of a radionuclide introduced instantaneously into the transfer compartment $a$ at time $t = 0$ and consider a single tissue compartment, $b$. Equations (16.33) and (16.34) can be written

$$\frac{dq_a}{dt} = -(\lambda_R + \lambda_a)q_a \tag{16.35}$$

and

$$\frac{dq_b}{dt} = b\lambda_a q_a - (\lambda_R + \lambda_b)q_b. \tag{16.36}$$

Like Eq. (16.11), having the same initial condition, we write for the solution of Eq. (16.35)

$$q_a(t) = A_0 e^{-(\lambda_R + \lambda_a)t}. \tag{16.37}$$

Equation (16.36) is mathematically the same as Eq. (16.12) with the same initial condition, $q_b(0) = 0$. Analogous to Eq. (16.17), the solution to (16.36) is

$$q_b(t) = \frac{A_0 b \lambda_a}{\lambda_b - \lambda_a} [e^{-(\lambda_R + \lambda_a)t} - e^{-(\lambda_R + \lambda_b)t}]. \tag{16.38}$$

The activity in $b$ thus builds up from its initial value of zero to a maximum from which it decreases thereafter, like the activity $q_{SI}$ in Fig. 16.9.

We continue in the next section with the computation of quantities needed to determine the $\hat{H}(T \leftarrow S)$ in Eq. (16.2).

## 16.8

### Specific Absorbed Fraction, Specific Effective Energy, and Committed Quantities

The specific equivalent dose, $\hat{H}(T \leftarrow S)$, in a target organ T due to the radiation emitted per transformation by a radionuclide in organ S (which can be the same as T), appears in Eq. (16.2). For particular choices of T, S, and radiation type, one can obtain this quantity from knowledge of the fraction of the energy emitted in S that is absorbed in T. This absorbed fraction (AF) is designated $AF(T \leftarrow S)_R$, where R denotes the radiation type. Dividing the absorbed fraction by the mass $M_T$ of the target organ gives the specific absorbed fraction, $AF(T \leftarrow S)_R/M_T$. The ICRP has published tables of specific absorbed fractions, expressed in $g^{-1}$, for use in computing the committed quantities given by Eqs. (16.1) and (16.2).

Table 16.2 shows an example of specific absorbed fractions in a number of target organs for photons of several energies emitted from the thyroid as source organ. The table illustrates the effect of the decrease in the linear attenuation coefficient (Section 8.7) with increasing photon energy over the range considered. The specific absorbed fractions for the thyroid as target decrease with the greater probability of escape of the higher-energy photons from the source organ. The same holds true for the total body. In contrast, the greater penetrability of the higher-energy photons leads to an increase in the specific absorbed fractions in tissues outside the thyroid. Since 0.010-MeV photons are so strongly absorbed, the increase is dramatic at 0.100 MeV.

For most organs it is assumed that the energies of all alpha and beta particles are absorbed in the source organ. (The exceptions, which we shall not go into, are mineral bone and the contents of the gastrointestinal tract.) Thus $AF(T \leftarrow S)_i = 0$ unless T and S are the same when $i$ denotes alpha or beta radiation. Then also $AF(T \leftarrow S)_i = 1$.

The absorbed fractions for gamma rays cannot be evaluated in a simple way. Their values depend in a complicated fashion on the photon energy; the size, den-

**Table 16.2** Specific Absorbed Fraction ($g^{-1}$) of Photon Energy in Several Target Organs and Tissues for Monoenergetic Photon Source in Thyroid (from ICRP Publication 23)

| Target | Photon Energy (MeV) | | |
| --- | --- | --- | --- |
| | 0.010 | 0.100 | 1.00 |
| Stomach wall | 2.07 E-25 | 1.90 E-07 | 4.62 E-07 |
| Small intestines plus contents | 4.58 E-35 | 1.97 E-08 | 1.38 E-07 |
| Lungs | 1.52 E-13 | 3.67 E-06 | 3.83 E-06 |
| Ovaries | 2.33 E-23 | 1.09 E-08 | 9.62 E-08 |
| Red marrow | 2.68 E-09 | 4.87 E-06 | 2.57 E-06 |
| Testes | 2.48 E-28 | 7.87 E-10 | 2.46 E-08 |
| Thyroid | 4.29 E-02 | 1.44 E-03 | 1.54 E-03 |
| Total body | 1.43 E-05 | 4.71 E-06 | 4.26 E-06 |

sity, and relative positions of the source and target organs; and on the specific intervening tissues. The Medical Internal Radiation Dose (MIRD) Committee of the Society of Nuclear Medicine has made extensive calculations of the specific absorbed fractions (absorbed fraction per gram of target) for a number of source and target organs in reference man. Monte Carlo techniques are employed in which the transport of many individual photons through the body is carried out by computer codes and the resulting data compiled to obtain the specific absorbed fractions. Calculations have been performed both for monoenergetic photons and for the spectra of photons emitted by a number of radionuclides.

In general, a radionuclide in S emits several kinds of radiation R, with yields $Y_R$ and average energies $E_R$. Multiplication of $Y_R E_R$ by the specific absorbed fraction gives the average absorbed dose in T per transformation in S contributed by the radiation R. Expressing the energies in MeV, the ICRP defines the specific effective energy (SEE) imparted per gram of tissue in a target organ T from the emission of a specified radiation R in a source organ S per transformation as follows:

$$\text{SEE } (T \leftarrow S)_R \equiv \frac{\text{AF } (T \leftarrow S)_R}{M_T} Y_R E_R w_R \text{ MeV g}^{-1}. \tag{16.39}$$ ✳

This amount of energy absorbed per gram, weighted by the factor $w_R$, thus represents the contribution of radiation of type R emitted per transformation of a radionuclide in S to the equivalent dose in T. To express this contribution in sieverts, we multiply the SEE by the factor $(1.60 \times 10^{-13} \text{ J MeV}^{-1})/(10^{-3} \text{ kg g}^{-1}) = 1.60 \times 10^{-10} \text{ Sv (MeV g}^{-1})^{-1}$. Summing over all types of radiation emitted by the radionuclide, we obtain[3]

$$\hat{H}(T \leftarrow S) = 1.6 \times 10^{-10} \sum_R \text{SEE } (T \leftarrow S)_R \text{ Sv}. \tag{16.40}$$

Returning to Eq. (16.2), we multiply by the number of transformations $U_S$ in 50 y and sum over all organs S to obtain

$$H_T(50) = 1.6 \times 10^{-10} \sum_S U_S \sum_R \text{SEE } (T \leftarrow S)_R \text{ Sv} \tag{16.41}$$ ✳

for the committed equivalent dose in T. Finally, the committed effective dose, Eq. (16.1), can be written

$$E(50) = 1.6 \times 10^{-10} \sum_T w_T \sum_S U_S \sum_R \text{SEE } (T \leftarrow S)_R \text{ Sv}. \tag{16.42}$$

If there are several radionuclides $j$ in the body, then one adds the individual contributions given by Eq. (16.42):

$$E(50) = 1.6 \times 10^{-10} \sum_T w_T \sum_S \sum_j \left[ U_S \sum_R \text{SEE } (T \leftarrow S)_R \right]_j \text{ Sv}. \tag{16.43}$$

3  Generally, dosimetric results are expressed with no more than two significant figures for internal dosimetry. The ARLI and DRAC are given to only one significant figure.

*Example*

What is the contribution of the beta radiation from $^{131}$I in the thyroid to the specific effective energy values for various target organs? The mass of the thyroid is 20 g. What will be the effect on the value of the SEE when the photons from $^{131}$I are included?

*Solution*

In this example the source organ is the thyroid and the only type of radiation R that we are to consider initially is beta rays. Because their range is small compared with the size of the 20-g thyroid, we assume that the beta particles are completely absorbed in the source organ. Therefore, AF(other organs ← thyroid)$_{131_{I(\beta^-)}}$ = 0 and the corresponding contributions to the SEE(other organs ← thyroid)$_{131_{I(\beta^-)}}$ = 0. (The absorbed fractions for the other target organs are not zero for the gamma rays emitted by $^{131}$I in the thyroid. The SEE for the gamma photons are discussed below.) It remains to compute the SEE for the beta rays in the thyroid itself as the target organ. The various factors that enter Eq. (16.39) are determined as follows. The thyroid mass $M_T = 20$ g is given. The absorbed fraction AF(thyroid ← thyroid) = 1, and $w_R = 1$ (Table 14.1). The yields $Y_R$ and energies $E_R$ per transformation can be obtained from Appendix D. (We assume, when not given, that the mean beta-particle energy is one-third the maximum.) Thus we obtain from Eq. (16.39)

$$\text{SEE (thyroid} \leftarrow \text{thyroid)}_{131_{I(\beta^-)}} = \frac{1}{20}(0.006 \times 0.269 + 0.89 \times 0.192$$

$$+ 0.07 \times 0.097 + 0.02 \times 0.069)$$

$$= 0.009 \text{ MeV g}^{-1} \tag{16.44}$$

per transformation. Including the photons from $^{131}$I will make all of the SEE(other organs ← thyroid) $\neq$ 0. Since most of the photon energy emitted inside the small thyroid will escape from the organ, the specific effective energy SEE(thyroid ← thyroid)$_{131_{I(\gamma)}}$ is very much smaller than that of the $\beta^-$. It turns out that

$$\text{SEE(thyroid} \leftarrow \text{thyroid)}_{131_I} = \text{SEE(thyroid} \leftarrow \text{thyroid)}_{131_{I(\beta^-)}}$$

$$+ \text{SEE(thyroid} \leftarrow \text{thyroid)}_{131_{I(\gamma)}}$$

$$= 0.010 \text{ MeV g}^{-1} \tag{16.45}$$

per transformation. As described in the next paragraph, this is the correct value as obtained from the detailed calculations.

Complete, detailed decay-scheme data were used to calculate the specific absorbed fractions and specific effective energies used in the ICRP-30 methodology. As an example, in place of the simple decay scheme from Appendix D that we used for $^{131}$I in the example just given, the ICRP calculations used the complex decay data shown in Fig. 16.10. The yields and average energies are given for five modes of $\beta^-$ decay, a sixth mode ($\beta^-$5) contributing less than 0.1% to $\sum Y_R E_R$ for the beta particles. A total of nine gamma photons are included, along with five internal-conversion electrons and the $K_{\alpha 1}$ and $K_{\alpha 2}$ daughter xenon X rays. For the particular example chosen here, the beta-particle contribution [Eq. (16.44)] turns out to

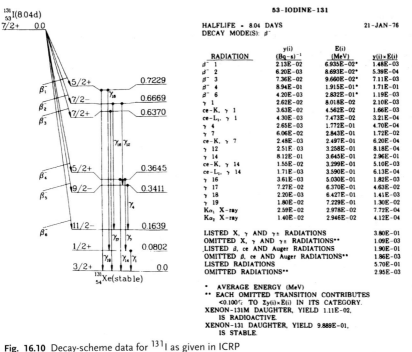

$^{131}_{53}I(8.04d)$

$7/2+$   0.0

53-IODINE-131

HALFLIFE = 8.04 DAYS                          21-JAN-76
DECAY MODE(S): β⁻

| RADIATION | $y(i)$ $(Bq-s)^{-1}$ | $E(i)$ (MeV) | $y(i) \times E(i)$ |
|---|---|---|---|
| β⁻ 1 | 2.13E-02 | 6.935E-02* | 1.48E-03 |
| β⁻ 2 | 6.20E-03 | 8.693E-02* | 5.39E-04 |
| β⁻ 3 | 7.36E-02 | 9.660E-02* | 7.11E-03 |
| β⁻ 4 | 8.94E-01 | 1.915E-01* | 1.71E-01 |
| β⁻ 6 | 4.20E-03 | 2.832E-01* | 1.19E-03 |
| γ 1 | 2.62E-02 | 8.018E-02 | 2.10E-03 |
| ce-K, γ 1 | 3.63E-02 | 4.562E-02 | 1.66E-03 |
| ce-L₁, γ 1 | 4.30E-03 | 7.473E-02 | 3.21E-04 |
| γ 4 | 2.65E-03 | 1.772E-01 | 4.70E-04 |
| γ 7 | 6.06E-02 | 2.843E-01 | 1.72E-02 |
| ce-K, γ 7 | 2.48E-03 | 2.497E-01 | 6.20E-04 |
| γ 12 | 2.51E-03 | 3.258E-01 | 8.18E-04 |
| γ 14 | 8.12E-01 | 3.645E-01 | 2.96E-01 |
| ce-K, γ 14 | 1.55E-03 | 3.299E-01 | 5.10E-04 |
| ce-L₁, γ 14 | 1.71E-03 | 3.590E-01 | 6.13E-04 |
| γ 16 | 3.61E-03 | 5.030E-01 | 1.82E-03 |
| γ 17 | 7.27E-02 | 6.370E-01 | 4.63E-02 |
| γ 18 | 2.20E-03 | 6.427E-01 | 1.41E-03 |
| γ 19 | 1.80E-02 | 7.229E-01 | 1.30E-02 |
| Kα₁ X-ray | 2.59E-02 | 2.978E-02 | 7.72E-04 |
| Kα₂ X-ray | 1.40E-02 | 2.946E-02 | 4.12E-04 |

| | | |
|---|---|---|
| LISTED X, γ AND γ± RADIATIONS | | 3.80E-01 |
| OMITTED X, γ AND γ± RADIATIONS** | | 1.09E-03 |
| LISTED β, ce AND Auger RADIATIONS | | 1.90E-01 |
| OMITTED β, ce AND Auger RADIATIONS** | | 1.86E-03 |
| LISTED RADIATIONS | | 5.70E-01 |
| OMITTED RADIATIONS** | | 2.95E-03 |

\* AVERAGE ENERGY (MeV)
\*\* EACH OMITTED TRANSITION CONTRIBUTES
   <0.100% TO Σy(i)×E(i) IN ITS CATEGORY.
XENON-131M DAUGHTER, YIELD 1.11E-02,
   IS RADIOACTIVE.
XENON-131 DAUGHTER, YIELD 9.889E-01,
   IS STABLE.

**Fig. 16.10** Decay-scheme data for $^{131}I$ as given in ICRP Publication 38. [Reprinted with permission from *Radionuclide Transformations*, ICRP Publ. 38, p. 453, International Commission on Radiological Protection, Sutton, England (1983). Copyright 1983 by ICRP.]

be close to the total value [Eq. (16.45)] obtained from the detailed scheme shown in Fig. 16.10. The close agreement results because (1) we were not far off in estimating $\sum Y_R E_R$ from Appendix D for the beta particles plus conversion electrons and (2) most of the photon energy escapes from the small thyroid. As is more often the case, however, self-absorption of photons in an organ is important and simple estimates are not reliable. When the source and target organs are different, then the detailed Monte Carlo calculations offer the only feasible way of obtaining all of the needed specific absorbed fractions reliably. As mentioned in connection with Fig. 16.2, some absorbed-fraction computations have been checked experimentally.

Today, detailed decay data for more than 800 radionuclides are maintained in computer files by the Dosimetry Research Group at the Oak Ridge National Laboratory.[4] The data base has been assembled over the years in conjunction with Publications 30 and 38 and subsequent work of the ICRP and publications of the MIRD Committee of the Society of Nuclear Medicine, mentioned earlier. An associated program, which can be run on a personal computer, calculates age-dependent

4  K. F. Eckerman, R. J. Westfall, J. C. Ryan, and M. Cristy, *Nuclear Decay Data Files of the Dosimetry Research Group*, Report    ORNL/TM-12350, Oak Ridge National Laboratory, Oak Ridge, Tenn. (Dec., 1993).

specific effective energies for a newborn and children of ages 1 y, 5 y, 10 y, and 15 y as well as a 58-kg adult female and the 70-kg adult reference man. The reader is referred to Section 16.13 for additional information on decay-scheme data for internal dosimetry.

## 16.9
### Number of Transformations in Source Organs over 50 Y

The final step needed in the methodology for obtaining the committed quantities is the determination of the numbers of transformations $U_S$ that occur in source organs and tissues S during the 50 y following the intake of a radionuclide. We present some detailed calculations for a two-compartment model, consisting of the body fluids (transfer compartment $a$) and a single additional compartment $b$.

After the intake of an initial activity $A_0$ of a radionuclide into compartment $a$ at time $t = 0$ (with no prior activity present), Eqs. (16.37) and (16.38) describe the time-dependent activities in $a$ and $b$. We obtain the number of transformations $U_a$ and $U_b$ of the nuclide in the two compartments from time $t = 0$ to any subsequent time $T$ by integrating the activities:

$$U_a(T) = \int_0^T q_a(t)\, dt = -\frac{A_0}{\lambda_R + \lambda_a} e^{-(\lambda_R + \lambda_a)t}\Big|_0^T \tag{16.46}$$

$$= \frac{A_0}{\lambda_R + \lambda_a}(1 - e^{-(\lambda_R + \lambda_a)T}) \tag{16.47}$$

and

$$U_b(T) = \int_0^T q_b(t)\, dt \tag{16.48}$$

$$= \frac{b\lambda_a A_0}{\lambda_b - \lambda_a}\left(\frac{1 - e^{-(\lambda_R + \lambda_a)T}}{\lambda_R + \lambda_a} - \frac{1 - e^{-(\lambda_R + \lambda_b)T}}{\lambda_R + \lambda_b}\right). \tag{16.49}$$

The numbers of transformations $U_a$ and $U_b$ in compartments $a$ and $b$ for a radionuclide with decay constant $\lambda_R$ can thus be evaluated explicitly for the committed equivalent dose ($T = 50$ y), given the metabolic parameters $\lambda_a$, $\lambda_b$, and $b$.

*Example*
Use the two-compartment model just described for a radionuclide having a half-life of 0.430 d. The metabolic half-life in the body fluids (compartment $a$) is 0.25 d, the fraction $b = 0.22$ of the nuclide goes to organ $b$ when it leaves $a$, the metabolic half-life in $b$ is 9.8 y, the organ weighting factor for $b$ is $w_T = 0.05$, and the mass of $b$ is 144 g. Calculate the number of transformations $U_a$ and $U_b$ of the radionuclide in the two compartments during the 50 y following the single entrance of 1 Bq of the radionuclide into compartment $a$ at time $t = 0$. If the radionuclide emits only beta particles with an average energy $E = 0.255$ MeV, what are the resulting committed equivalent

doses for organ $b$ and for $a$ (the total body)? What is the committed effective dose? (Neglect any doses that occur between intake and entry into compartment $a$).

*Solution*

We use Eqs. (16.47) and (16.49) to calculate $U_a$ and $U_b$. From the given radioactive and metabolic half-lives, we have the following decay rates:

$$\lambda_R = \frac{0.693}{0.43} = 1.6 \text{ d}^{-1}, \tag{16.50}$$

$$\lambda_a = \frac{0.693}{0.25} = 2.8 \text{ d}^{-1}, \tag{16.51}$$

$$\lambda_b = \frac{0.693}{9.8 \times 365} = 1.9 \times 10^{-4} \text{ d}^{-1}. \tag{16.52}$$

(We shall express time in days and retain only two significant figures in the computations.) The time period is $T = 50 \times 365 = 1.8 \times 10^4$ d, and so the exponential terms in Eqs. (16.47) and (16.49) are negligible compared with unity. The initial activity in compartment $a$ is

$$A_0 = 1 \text{ Bq} = 1 \text{ s}^{-1} \times 86{,}400 \text{ s d}^{-1} = 8.6 \times 10^4 \text{ d}^{-1}. \tag{16.53}$$

Using (16.47), we find

$$U_a = \frac{8.6 \times 10^4 \text{ d}^{-1}}{(1.6 + 2.8) \text{ d}^{-1}} = 2.0 \times 10^4. \tag{16.54}$$

Using Eq. (16.49) (with $\lambda_b$ neglected compared with $\lambda_R$), we obtain

$$U_b = \frac{b\lambda_a A_0}{\lambda_b - \lambda_a} \left( \frac{\lambda_b - \lambda_a}{(\lambda_R + \lambda_a)(\lambda_R + \lambda_b)} \right) \tag{16.55}$$

$$= \frac{0.22 \times 2.8 \times 8.6 \times 10^4}{(1.6 + 2.8)(1.6)} = 7.5 \times 10^3. \tag{16.56}$$

The committed equivalent dose for each tissue or organ is given by Eq. (16.41) with the specific effective energy calculated from Eq. (16.39). As with the earlier example (Section 16.8) involving $^{131}$I beta particles, $AF = 1$ when the source and target are the same and $AF = 0$ otherwise. Also, $w_R = 1$ and $YE = 0.26$. Transformations in compartment $a$ can be considered to be uniformly distributed over the whole body, having the mass $M = 70{,}000$ g. From Eq. (16.39) we have for the whole body (WB)

$$SEE(WB \leftarrow WB) = \frac{1}{70{,}000}(0.26 \times 1) = 3.7 \times 10^{-6} \text{ MeV g}^{-1} \tag{16.57}$$

per transformation. For compartment $b$, having the given mass 144 g,

$$SEE(b \leftarrow b) = \frac{1}{140}(0.26 \times 1) = 1.9 \times 10^{-3} \text{ MeV g}^{-1} \tag{16.58}$$

per transformation. The committed equivalent doses are, by Eq. (16.41),

$$H_{WB}(50) = 1.6 \times 10^{-10} U_a SEE(WB \leftarrow WB) \tag{16.59}$$

$$= 1.6 \times 10^{-10} \times 2.0 \times 10^4 \times 3.7 \times 10^{-6} = 1.2 \times 10^{-11} \text{ Sv} \tag{16.60}$$

and

$$H_b(50) = 1.6 \times 10^{-10} U_b \text{SEE}(b \leftarrow b) \tag{16.61}$$

$$= 1.6 \times 10^{-10} \times 7.5 \times 10^3 \times 1.9 \times 10^{-3} = 2.3 \times 10^{-9} \text{ Sv}. \tag{16.62}$$

The last number represents the committed equivalent dose in organ $b$ that is delivered by radionuclides in $b$ after they have left $a$. Since organ $b$ is also irradiated as a part of the whole body by radionuclides in $a$, the total committed equivalent dose in $b$ is the sum of (16.62) and (16.60). The later contribution, however, is negligible. The committed effective dose [Eq. (16.42)] is the sum of (16.60) and (16.62), weighted by the given factor $w_T = 0.05$. Thus, $E(50) = 1.2 \times 10^{-11} + 0.05 \times 2.3 \times 10^{-9} = 1.3 \times 10^{-10}$ Sv.

In ICRP Publication 30, the committed dose equivalent per unit intake, the annual limit on intake (ALI), and the derived air concentration (DAC) all refer to the intake of a specified radionuclide alone. If the radionuclide decays into radioactive daughters, then these are also included in the calculations of committed equivalent dose. The computations include either specific metabolic data for the daughters or the assumption that they follow the transport of the parent. Values of the number of transformations $U_S$ in source organs S for a radionuclide are computed together with values $U_S'$, $U_S''$, and so forth, for the daughter radionuclides that build up in the body during the 50 y following intake of the parent.

With the adoption of ICRP Publication 60, the Commission instituted the use of the committed effective dose in place of the committed effective dose equivalent. These are fundamentally different quantities. The ICRP also defined the ALI with reference to a committed effective dose of 20 mSv in place of a committed effective dose equivalent of 50 mSv. In addition, a new set of tissue weighting factors was adopted. These changes and the 1994 ICRP-66 revision of the lung model necessitated new calculations of the relationships between intakes and resulting committed organ equivalent doses and individual committed effective doses. The new information is presented in the 1995 ICRP Publication 68, *Dose Coefficients for Intakes of Radionuclides by Workers*. The complete revision of the ICRP-30 protocols, which is in progress, will take into account the newer anatomical and physiological data (ICRP-89) and continuing development of biokinetic models. *Dose coefficient* is defined in ICRP-68 as "the committed tissue equivalent dose per unit acute intake $h_T(\tau)$ or committed effective dose per unit acute intake $e(\tau)$, where $\tau$ is the time period in years over which the dose is calculated (e.g., $e(50)$)." The dose coefficient has the units Sv Bq$^{-1}$. The quantity is also referred to as a *dose conversion factor* (DCF). The acute intake of an amount of activity (by inhalation or ingestion) multiplied by the DCF gives the committed effective dose. For the 50-y committed effective dose of 20 mSv, the annual limit on intake is thus given in terms of the dose coefficient $e(50)$ by

$$\text{ALI} = \frac{0.020 \text{ Sv}}{e(50) \text{ Sv Bq}^{-1}} = \frac{0.020}{e(50)} \text{Bq}. \tag{16.63}$$

**Table 16.3** Effective Dose Coefficients (Sv Bq$^{-1}$) for Inhalation and Ingestion, $e_{inh}$ (50) and $e_{ing}$ (50), from ICRP Publication 68

| Nuclide | Type$^{\dagger}$ | $f_1$ | Inhalation | | $f_1$ | Ingestion |
|---|---|---|---|---|---|---|
| | | | $e_{inh}$ (50) (1 $\mu$m AMAD) | $e_{inh}$ (50) (5 $\mu$m AMAD) | | $e_{ing}$ (50) |
| $^{90}$Sr | F | 0.300 | $2.4 \times 10^{-8}$ | $3.0 \times 10^{-8}$ | 0.300 | $2.8 \times 10^{-8}$ |
| | S | 0.010 | $1.5 \times 10^{-7}$ | $7.7 \times 10^{-8}$ | 0.010 | $2.7 \times 10^{-9}$ |
| $^{131}$I | F | 1.000 | $7.6 \times 10^{-9}$ | $1.1 \times 10^{-8}$ | 1.000 | $2.2 \times 10^{-8}$ |
| $^{137}$Cs | F | 1.000 | $4.8 \times 10^{-9}$ | $6.7 \times 10^{-9}$ | 1.000 | $1.3 \times 10^{-8}$ |
| $^{226}$Ra | M | 0.200 | $1.6 \times 10^{-5}$ | $1.2 \times 10^{-5}$ | 0.200 | $2.8 \times 10^{-7}$+ |

$^{\dagger}$ Rate of absorption into blood from respiratory tract: F = fast, M = moderate, S = slow.

Table 16.3 shows a sample of dose coefficients for inhalation (1 $\mu$m and 5 $\mu$m AMAD) and for ingestion from Publication 68. Additional data given in ICRP-68 include $f_1$ values for inhalation and ingestion of various compounds, lung clearances, and the treatment of gases and vapors. Whereas the deposition of particulates in the respiratory tract is assumed to be determined by the size distribution, the situation is different for a gas or vapor. In the latter case, the fate of the inhaled radionuclide depends on its specific chemical form.

## 16.10
## Dosimetric Model for Bone

A great deal of effort has been devoted to studying the intake, deposition, and retention of radionuclides in the skeleton and its substructures. We shall make only a few remarks here about bone dosimetry.

ICRP Publication 30 states,

> The cells at carcinogenic risk in the skeleton have been identified as the haematopoietic stem cells of marrow, and among the osteogenic cells, particularly those on endosteal surfaces, and certain epithelial cells close to bone surfaces (ICRP Publication 11). The haematopoietic stem cells in adults are assumed to be randomly distributed predominantly throughout the haematopoietic marrow within trabecular bone (ICRP Publication 11). Therefore, dose equivalent to those cells is calculated as the average over the tissue which entirely fills the cavities within trabecular bone. For the osteogenic tissue on endosteal surfaces and epithelium on bone surfaces the Commission recommends that dose equivalent should be calculated as an average over tissue up to a distance of 10 $\mu$m from the relevant bone surfaces (para. 47, ICRP Publication 26).

The two principal target tissues for bone dosimetry are thus the active red bone marrow and cells near the bone surfaces. Except for gamma emitters, the source tissues in bone are cortical and trabecular bone. The ICRP gives specific details for estimating the number of transformations in trabecular and cortical bone and the needed absorbed fractions. Some inhaled or ingested elements are assumed to become distributed throughout the bone volume, while others are assumed to attach on bone surfaces. Generally, an alkaline-earth radionuclide with a radioactive half-life greater than 15 days belongs to the former group, while the shorter lived elements reside on the bone surfaces.

The alkaline-earth elements, which include calcium and strontium, have received much attention in bone dosimetry. A special task group of the ICRP was formed to study their metabolism in man. The extensive human data and experience with radium in the bone was mentioned in Section 13.7. Bone-seeking radionuclides, which also include those of strontium and plutonium, are considered dangerous because they irradiate the sensitive cells of the marrow. They all produce cancer in laboratory animals at sufficiently high levels of exposure.

Bone dosimetry is a continuing area of active research. In addition to its role in radiation protection for internal emitters, it is often an important consideration in radiation therapy. It then needs to be tailored toward the specific patient. In some treatments, the dose to the active (red) bone marrow is the limiting factor. ICRP Publication 89 includes extensive reference values for the human skeleton. These are the age- and gender-dependent compilations from the 1995 ICRP Publication 70, *Basic Anatomical and Physiological Data for Use in Radiological Protection: The Skeleton*. This document updated and extended the earlier data on ICRP reference man.

## 16.11
### ICRP-30 Dosimetric Model for Submersion in a Radioactive Gas Cloud

The last important model to be described is that for submersion in a cloud of radioactive gas. In this case the organs of the body can be irradiated by gas that is outside the body, absorbed in the body's tissues, and contained in the lungs.

To see how these sources limit the exposure of a radiation worker, ICRP-30 treats a cloud of infinite extent with a constant, uniform concentration $C$ Bq m$^{-3}$ of a gaseous radionuclide. For a person submerged in the cloud, one needs to consider (a) the equivalent-dose rate $\dot{H}_E$ to any tissue from external radiation, (b) the rate $\dot{H}_A$ to the tissue from the gas absorbed internally in the body, and (c) the rate $\dot{H}_L$ to the lung from the gas contained in it. We discuss each of these in turn.

For irradiation from outside the body, we let $s$ represent the equivalent-dose rate in sieverts per hour (Sv h$^{-1}$) in the air per Bq g$^{-1}$ of air. The rate in the air from $C$ Bq m$^{-3}$ is $Cs/\rho_A$, where $\rho_A$ is the density of air ($\sim 1300$ g m$^{-3}$). The equivalent-dose rate at the body surface of a person submerged in the cloud is then given by $Csk/\rho_A$, where $k$ is the ratio of the mass stopping powers of tissue and air ($k \cong 1$). Therefore, one can express the equivalent-dose rate in a small volume of tissue in the body by writing

$$\dot{H}_E = \frac{C s k g_E}{\rho_A} \ \text{Sv}\,\text{h}^{-1}, \tag{16.64}$$

where $g_E$ is a geometrical factor that allows for shielding by intervening tissues. For alpha particles and low-energy beta particles, such as those of tritium, $g_E = 0$. These radiations cannot penetrate to the lens of the eye (at a depth of 3 mm) or to the basal layer of the epidermis (at a depth of 70 $\mu$m). For most other beta emitters and for low-energy photons, $g_E \cong 0.5$ near the body surfaces and approaches zero with increasing depth. For high-energy photons, $g_E \cong 1$ throughout the body.

For irradiation from gas absorbed in the body, the ICRP considers a prolonged exposure to the cloud, which results in equilibrium concentrations of the gas in the air and in tissue. The concentration $C_T$ of gas in the tissue is then given by

$$C_T = \frac{\delta C}{\rho_T} \ \text{Bq}\,\text{g}^{-1}, \tag{16.65}$$

where $\rho_T$ is the density of tissue ($\sim 10^6$ g m$^{-3}$) and $\delta$ is the solubility of the gas in tissue, expressed as the volume of gas in equilibrium with a unit volume of tissue at atmospheric pressure. The solubility increases with the atomic weight of the gas, varying in water at body temperature from $\sim 0.02$ for hydrogen to $\sim 0.1$ for xenon. For adipose tissue the values may be larger by a factor of 3–20. For the equivalent-dose rate in tissue from absorbed gas, the ICRP writes

$$\dot{H}_A = \frac{s \delta C g_A}{\rho_T} \ \text{Sv}\,\text{h}^{-1}, \tag{16.66}$$

where $g_A$ is another geometric factor, depending on the size of a person and the range of the radiation. For alpha and beta particles and low-energy photons, $g_A \cong 1$ for tissues deep inside the body and $g_A \cong 0.5$ for tissues at the surface. For energetic photons $g_A \ll 1$.

The equivalent-dose rate in the lung from the gas it contains can be written

$$\dot{H}_L = \frac{s C V_L g_L}{M_L} \ \text{Sv}\,\text{h}^{-1}, \tag{16.67}$$

where $V_L$ is the average volume of air in the lungs ($\sim 3 \times 10^{-3}$ m$^3$), $M_L$ is the mass of the lungs (1000 g), and $g_L$ is a geometrical factor ($\cong 1$ for alpha and beta particles and low-energy photons and decreasing with increasing photon energy).

The three rates (16.64), (16.66), and (16.67) can be applied in the following way. For tritium, $\dot{H}_E = 0$ for all relevant tissues of the body, because this nuclide emits only low-energy beta particles. The ratio of the equivalent-dose rate in any tissue from absorbed gas to the rate in the lung from the gas it contains is

$$\frac{\dot{H}_A}{\dot{H}_L} = \frac{\delta g_A M_L}{V_L g_L \rho_T}. \tag{16.68}$$

With $g_A \cong g_L \cong 1$ for tritium, substitution of values just given leads to the ratio

$$\frac{\dot{H}_A}{\dot{H}_L} \cong \frac{\delta \times 1 \times 1000 \text{ g}}{3 \times 10^{-3} \text{ m}^3 \times 1 \times 10^6 \text{ g m}^{-3}} = \frac{\delta}{3}. \tag{16.69}$$

The value of $\delta$ for tritium is $\sim 0.02$ for aqueous tissues and $\sim 0.05$ for adipose tissues. It follows that the equivalent-dose rate to the lung from the gas contained in it, which is some 60–150 times greater than the rate to any tissue from absorbed gas, is the limiting factor for submersion in a cloud of elemental tritium. (The limit for tritiated water, which is much smaller than that for elemental tritium, restricts most practical cases of exposure to tritium.)

Many noble gases emit photons and energetic beta particles. Then, for tissues near the surface of the body, $g_E \cong 0.5$. From Eqs. (16.64) and (16.67) one has

$$\frac{\dot{H}_E}{\dot{H}_L} = \frac{k g_E M_L}{\rho_A V_L g_L} \cong \frac{1 \times 0.5 \times 1000}{1300 \times 3 \times 10^{-3} g_L} \gtrsim 130, \tag{16.70}$$

where the inequality reflects the fact that $g_L \leq 1$. Thus the equivalent-dose rate to tissues near the body surfaces is more than 130 times $\dot{H}_L$. From (16.64) and (16.66) we find

$$\frac{\dot{H}_E}{\dot{H}_A} = \frac{k g_E \rho_T}{\rho_A \delta g_A} \cong \frac{1 \times 0.5 \times 10^6}{1300 \delta g_A} \tag{16.71}$$

$$\cong \frac{400}{\delta g_A} \gtrsim \frac{400}{\delta}, \tag{16.72}$$

since $g_A \leq 1$. Since $\delta \lesssim 2$ always,

$$\frac{\dot{H}_E}{\dot{H}_A} \gtrsim 200. \tag{16.73}$$

For these noble gases, the external radiation will be the limiting factor for a person submerged in a cloud.

Conversion factors are used by the ICRP to apply the results obtained for infinite clouds to exposure in rooms of sizes from 100 m³ to 1000 m³.

## 16.12
### Selected ICRP-30 Metabolic Data for Reference Man

In this section we provide a small sample of metabolic data for reference man for several radionuclides. The ICRP-30 fractions $f_1$ for these nuclides are the same as the current values shown in Table 16.3. Information such as this can be used with the ICRP-30 models as presented in this chapter.

## Hydrogen

Reference-man data:

| | |
|---|---:|
| Hydrogen content of the body | 7,000 g |
| of soft tissue | 6,300 g |
| Daily intake of hydrogen | 350 g |
| Water content of the body | 42,000 g |
| Daily intake of water, including water of oxidation | 3,000 g. |

Water makes up about 80% of the mass of some soft tissues. As discussed in the last section, exposure to *elemental* tritium is limited by the equivalent dose from tritium in the lung. In contrast, tritiated water that is inhaled, ingested, or absorbed through the skin is assumed to become instantaneously and uniformly distributed throughout all the soft tissues of the body. While some tritium from tritiated water can become organically bound in the body, the ICRP assumed a single-exponential retention function for the body, based on tritiated water alone, with a biological half-life of 10 d. The fraction of tritium, taken in as tritiated water at time $t = 0$, retained at time $t$ days is given by

$$R(t) = e^{-0.693t/10}. \tag{16.74}$$

If the body contains $q$ Bq, then the concentration in soft tissue (mass 63,000 g) is $q/63{,}000$ Bq g$^{-1}$.

## Strontium

Reference-man data:

| | |
|---|---|
| Strontium content of the body | 0.32 g |
| of the skeleton | 0.32 g |
| of soft tissues | 3.3 mg |
| Daily intake in food and fluids | 1.9 mg. |

Based on human and animal data, the ICRP-30 used $f_1 = 0.3$ for soluble salts and 0.01 for $SrTiO_3$ as the fractional uptake of ingested strontium by the body fluids [Eq. (16.7)]. For inhalation, soluble compounds were assumed to be in class D and $SrTiO_3$ in class Y (Section 16.4). As discussed in Section 16.10, strontium isotopes $^{90}Sr$, $^{85}Sr$, and $^{89}Sr$, having half-lives greater than 15 d, were assumed to be distributed uniformly in the volume of mineral bone. In contrast, other strontium isotopes, with shorter half-lives, were assumed to be distributed uniformly over bone surfaces. The detailed metabolic model was used to estimate the number of transformations in soft tissue, cortical bone, and trabecular bone during the 50 y following the introduction of unit activity into the transfer compartment of the body (Fig. 16.3).

**Iodine**

Reference-man data:

| | |
|---|---|
| Iodine content of the body | 11.0 mg |
| of the thyroid | 10.0 mg |
| Daily intake in food and fluids | 0.2 mg. |

Iodine is absorbed rapidly and almost completely from the gut, and so $f_1 = 1$ was used. All data indicate that compounds of iodine belong to inhalation class D. When iodine enters the transfer compartment, the fraction 0.3 was assumed to be taken up by the thyroid and the rest excreted directly. Iodine in the thyroid was assigned a biological half-life of 120 d, leaving the gland as organic iodine. The organic iodine became distributed uniformly in the other tissues of the body with a half-life of 12 d. One-tenth of this organic iodine was then assumed to be excreted while the rest returned to the transfer compartment.

**Cesium**

Reference-man data:

| | |
|---|---|
| Cesium content of the body | 1.5 mg |
| of muscle | 0.57 mg |
| of bone | 0.16 mg |
| Daily intake in food and fluids | 10.0 $\mu$g. |

Cesium compounds are usually rapidly and almost completely absorbed in the GI tract, and so $f_1 = 1$. They were assigned to inhalation class D. A two-compartment retention function was used for cesium:

$$R(t) = ae^{-0.693t/T_1} + (1-a)e^{-0.693t/T_2}. \tag{16.75}$$

When the element enters the transfer compartment, the fraction $a = 0.1$ was transferred to one tissue compartment (Fig. 16.3) and retained there with a metabolic half-life $T_1 = 2$ d; the remainder, $1 - a = 0.9$, was transferred to another tissue compartment and kept there with a half-life $T_2 = 110$ d. The cesium in both of these compartments was assumed to be distributed uniformly throughout the body.

**Radium**

Reference-man data:

| | |
|---|---|
| Radium content of the body | 31.0 pg |
| of the skeleton | 27.0 pg |
| Daily intake in food and fluids | 2.3 pg. |

Available data lead to the choices $f_1 = 0.2$ and inhalation class W for all commonly occurring compounds of radium. A comprehensive retention model for radium in adults was used to calculate the numbers of transformations in soft tissue, cortical bone, and trabecular bone for obtaining the committed doses per unit intake.

# 16.13
## Suggested Reading

A great deal of information is available through the World Wide Web on radionuclide decay data, specific absorbed fractions, specific effective energies, and other important parameters needed for internal dosimetry. Publications of the International Commission on Radiological Protection (ICRP) and the Medical Internal Radiation Dose (MIRD) Committee are particularly relevant. Some are included below.

1 Cember, H., *Introduction to Health Physics*, 3rd Ed., McGraw-Hill, New York, NY (1996). [Chapter 8 provides an excellent coverage of radiation safety guides. It includes an extensive description of ICRP internal-dosimetry methodology up to the time. There is an appendix with data on reference man and one giving specific absorbed fractions for photons of different energies, both based on ICRP Publication 23 on reference man.]

2 Endo, A., Yamaguchi, Y., and Eckerman, K. F., *Nuclear Decay Data for Dosimetry Calculation: Revised Data of ICRP Publication 38*, JAERI 1347, Japan Atomic Energy Research Institute, Tokai-mura, Japan (2005).

3 Endo, A., *Nuclear Decay Data for Dosimetry Calculation: Revised Data of ICRP Publication 38—Supplement to JAERI 1347*, Japan Atomic Energy Research Institute, Tokai-mura, Japan (2005).

4 ICRP Publication 23, *Task Group Report on Reference Man*, Pergamon Press, Oxford, England (1975).

5 ICRP Publication 30, *Limits for Intakes of Radionuclides by Workers*, Pergamon Press, Elmsford, NY (1979). [Publication 30 appears as a series of parts and supplements.]

6 ICRP Publication 60, *1990 Recommendations of the International Commission on Radiological Protection*, Pergamon Press, Elmsford, NY (1991).

7 ICRP Publication 66, *Human Respiratory Tract Model for Radiological Protection*, Annals of the ICRP **24**, Nos. 1–3 (1994).

8 ICRP Publication 68, *Dose Coefficients for Intakes of Radionuclides by Workers*, Annals of the ICRP **24**, No. 4 (1995). [Dose coefficients and other data based on ICRP-60 limits and revised lung model of ICRP-66. ICRP-68 replaces ICRP-61, which employed ICRP-30 lung model.]

9 ICRP Publication 70, *Basic Anatomical and Physiological Data for Use in Radiological Protection: the Skeleton*, Annals of the ICRP **25**, No. 2 (1995).

10 ICRP Publication 89, *Basic Anatomical and Physiological Data for Use in Radiological Protection: Reference Values*, Annals of the ICRP **32**, Nos. 3–4 (2003).

11 ICRP Publication 100, *Human Alimentary Tract Model for Radiological Protection*, Annals of the ICRP (2007).

12 MIRD Committee Pamphlets. [The Medical Internal Radiation Dose (MIRD) Committee of the Society of Nuclear Medicine publishes pamphlets and other information, available on line. Targeted for 2006 is the revision of the 1989 MIRD Radionuclide Data and Decay Scheme book.]

13 NCRP Report No. 116, *Limitation of Exposure to Ionizing Radiation*, National Council on Radiation Protection and Measurements, Bethesda, MD (1993).

## 16.14
## Problems

1. What is reference man and what is his role in internal dosimetry?

2. The annual limit on intake for $^{32}$P by ingestion is $8 \times 10^6$ Bq. What is the committed effective dose per unit activity of ingested $^{32}$P?

3. The ALI for inhalation of $^{235}$U aerosols having retention times of the order of weeks in the pulmonary region is $1 \times 10^4$ Bq. What is the corresponding derived air concentration?

4. A single inhalation of $10^6$ Bq of a certain radionuclide results in a committed effective dose of 6.1 mSv.
   (a) What is the ALI?
   (b) What is the DAC?

5. In what ways is the limitation of the annual effective dose different for external and internal radiation when the ALI concept is applied?

6. As a result of the single intake of $6.3 \times 10^3$ Bq of a radionuclide, a certain organ of the body will receive doses of 0.20 mGy from low-energy beta particles and 0.05 mGy from alpha particles during the next 50 y. These are the only radiations emitted, and this organ is the only one that receives appreciable irradiation as a result of the intake. The organ has a weighting factor, $w_T = 0.05$.
   (a) What is the committed equivalent dose to the organ?
   (b) What is the committed effective dose to the individual?
   (c) What is the ALI for this route of intake?

7. (a) What fraction of inhaled aerosols with an AMAD of 1 $\mu$m is assumed to be deposited in the trachea and bronchial tree in the ICRP-30 model of the respiratory system?
   (b) What fraction is exhaled?

8. Write a differential equation, analogous to Eq. (16.3), that describes the rate of change $\dot{q}_b(i)$ of the activity in compartment b of the respiratory-system model (Fig. 16.4) for a rate $\dot{I}(t)$ of inhalation.

9. Given the inhalation rate $\dot{I}(t)$, write two differential equations that describe the rates of change $\dot{q}_i(t)$ and $\dot{q}_j(t)$ of activity in the lymph-node compartments of the respiratory-system model in Fig. 16.4.

10. An activity of 1000 Bq of a class-W aerosol (AMAD $= 1$ $\mu$m) is inhaled in a single intake.

(a) According to the ICRP-30 model for the respiratory system, what is the initial activity deposited in compartment e?

(b) If the radionuclide has a physical half-life of 32 d, what fraction of the original activity in compartment e will still be there after 1 wk?

11. (a) Use the data in Fig. 16.4 to find numerical values for the following quantities for a class-Y aerosol with AMAD $= 1 \ \mu$m: $D_{TB}$, $F_d$, $\lambda_f$, $\lambda_g$, and $\lambda_d$.

(b) If the activity of a class-W aerosol in compartment e of the ICRP-30 lung model is $10^3$ Bq, what is the rate of transfer of activity from compartment e to the body fluids?

12. Consider the inhalation of $^{239}$Pu as a class-W aerosol with an AMAD of 1 $\mu$m.

(a) If 50 Bq of $^{239}$Pu is inhaled, how much activity is deposited in compartment b of the ICRP-30 lung model?

(b) What fraction of $^{239}$Pu, deposited in compartment b at time $t = 0$, clears out of b in the first 24 h?

13. At a certain time following inhalation of a class-W aerosol, the activities in compartments b and d of the ICRP-30 respiratory-system model are, respectively, $7.8 \times 10^4$ Bq and $1.5 \times 10^4$ Bq. What is the rate of transfer of activity to the gastrointestinal tract?

14. In the dosimetric model for the GI tract (Fig. 16.8), show that $\lambda_B$ can be estimated from $f_1$, the fraction of a stable element that reaches the body fluids after ingestion, by writing $\lambda_B = f_1 \lambda_{SI}/(1 - f_1)$.

15. At a certain time following ingestion of a radionuclide, the activity in the contents of the small intestine is $2.80 \times 10^5$ Bq. If the fraction of the stable element that reaches the body fluids after ingestion is 0.41, what is the rate of transfer of activity from the small intestine to the body fluids?

16. What is the effective half-life of a radionuclide in the body-fluid transfer compartment, if its radioactive half-life is 8 h?

17. An activity of $5 \times 10^6$ Bq of a radioisotope, having a half-life of 2 d, enters the body fluids. How much activity remains in this compartment at the end of 4 d?

18. A worker with a burden of $^{131}$I in his thyroid was monitored in a whole-body counter under fixed conditions of geometry and counting time. The net number of gamma counts (background is subtracted) measured from the thyroid was initially 1129. The net numbers of counts at this and three subsequent times are shown here. From these data, determine the metabolic half-life of the iodine in the thyroid.

| Time (d) | Net Counts |
|----------|-----------|
| 0 | 1129 |
| 7 | 589 |
| 14 | 314 |
| 21 | 156 |

19. A drug, containing $4.27 \times 10^5$ Bq of $^{86}$Y, is ingested ($f_1 = 1 \times 10^{-4}$) at time $t = 0$. After 10 h, what is the activity of the $^{86}$Y in
    (a) the stomach,
    (b) the small intestine,
    (c) the upper large intestine?

20. What is the rate of transfer of activity from the GI tract into the body fluids at time $t = 10$ h in the last problem?

21. At what time after ingestion does the activity in the small intestine reach a maximum in Problem 19?

22. What is the equivalent dose to the kidneys of an adult male per transformation from a source in the lungs that emits a single 500-keV photon per transformation, if the specific effective energy (kidney ← lung) is $5.82 \times 10^{-9}$ MeV g$^{-1}$?

23. What is the absorbed fraction (kidney ← lung) in the last problem? (Mass of the kidneys is given in Table 16.1.)

24. The specific absorbed fraction for irradiation of the red bone marrow by 200-keV photons from a source in the liver is $4.64 \times 10^{-6}$ g$^{-1}$. Calculate the specific effective energy for the liver (source organ) and red marrow (target tissue) for a gamma source in the liver that emits only a 200-keV photon in 85% of its transformations.

25. What is the committed equivalent dose to the red marrow from the source in the liver in the last problem if $2.23 \times 10^{15}$ transformations occur in the liver over a 50-y period?

26. What are the specific absorbed fractions for various target organs for the pure beta emitter $^{14}$C in the liver (mass = 1800 g) as source organ?

27. What are the corresponding values of the specific effective energies in the last problem?

28. A certain radionuclide emits a 1-MeV photon in 62% of its transformations. Use Table 16.2 to compute the equivalent dose delivered to the lungs by these photons as a result of $10^6$ transformations of the nuclide, located in the thyroid.

29. The specific absorbed fraction for the testes for 1-MeV photons emitted in the thyroid is $2.46 \times 10^{-8}$ g$^{-1}$ (Table 16.2).

(a) What is the specific effective energy SEE(testes ← thyroid) for a nuclide in the thyroid that emits a 1-MeV photon in 30% of its transformations, other radiations being negligible?

(b) How much total energy is imparted to the testes (Table 16.1, mass = 35 g) as a result of $10^9$ transformations of the nuclide in the thyroid?

30. A radionuclide in the lungs (mass = 1000 g) emits a 1-MeV photon in 72% of its transformations. That is the only radiation that reaches the thyroid (mass = 20 g). The absorbed fraction, AF(thyroid ← lungs), for 1-MeV photons is $9.4 \times 10^{-5}$.

(a) Calculate the SEE(thyroid ← lungs) for this nuclide.

(b) What equivalent dose does the thyroid receive from $10^8$ transformations of the nuclide in the lungs?

31. For a 0.5-MeV photon source in the lungs, the absorbed fraction for the liver (mass = 1800 g) is AF(liver ← lungs) = 0.0147. A nuclide in the lungs emits a single 0.5-MeV photon in 70% of its transformations. This is the only radiation that reaches the lungs.

(a) Calculate SEE(liver ← lungs) for this nuclide.

(b) Calculate the equivalent dose to the liver per transformation of the nuclide in the lungs.

32. A source organ S in the body contains a radionuclide that emits a 0.80-MeV gamma photon in 90% and a 1.47-MeV photon in 48% of its transformations. The corresponding absorbed fractions for a target organ T, having a mass of 310 g, are, respectively, AF = $4.4 \times 10^{-6}$ and AF = $1.8 \times 10^{-6}$. Organ T is irradiated only by these photons.

(a) Calculate SEE(T ← S) for this case.

(b) What is the equivalent dose in T as a result of $10^{14}$ transformations of the nuclide in S?

(c) If the nuclide has a radiological half-life of 2.0 y and a metabolic half-life of 6.0 y in S, how long does it take for the activity in S to decrease by a factor of 10?

33. A radionuclide emits a 5.80-MeV alpha particle in 60% of its transformations and a 5.60-MeV alpha particle followed by a 0.20-MeV gamma photon in 40% of its transformations. These are the only radiations emitted. A worker has a burden of $4.1 \times 10^6$ Bq of this nuclide in his lungs.

(a) What is the equivalent-dose rate to the worker's lungs (mass = 1000 g) from the alpha radiation?

(b) What is the equivalent-dose rate to the spleen (mass = 180 g) if AF(spleen ← lungs) = $1.47 \times 10^{-3}$ for the photons?

(c) For a total of $10^{12}$ transformations of this nuclide in the lung, how much does the resulting equivalent dose to the spleen contribute to the worker's effective dose? (See Table 14.2. Spleen $w_T = 0.05/10 = 0.005$.)

34. Show that Eq. (16.49) follows from (16.38).

35. By letting $\lambda_a \to \infty$ in Eq. (16.49), show that the second term represents the number of transformations that would have occurred in compartment $b$, had the material been transferred to it instantaneously. (The first term, therefore, represents the effect on $U_b$ of the finite residence time in compartment $a$.)

36. Use the two-compartment model described in Section 16.9. An activity of $10^6$ Bq of a radionuclide, having a half-life of 18 h, enters compartment $a$ (body fluids). The fraction that goes to organ $b$ when it leaves $a$ is 0.30, and the metabolic half-life in $b$ is 2 d. Calculate the number of transformations in compartments $a$ and $b$ during the two days after the radionuclide enters $a$.

37. Assume that the radionuclide in the last problem is an alpha or low-energy beta emitter with a stable daughter. What fraction of the committed equivalent dose is delivered to organ $b$ in the 2 d after entry of the radionuclide into $a$?

38. (a) Repeat Problem 36 for a radionuclide that has a radioactive half-life of 90 y and a metabolic half-life in compartment $b$ of 40 y.

    (b) How many transformations occur in compartments $a$ and $b$ over the 50 y of the committed equivalent dose?

39. In the last problem, assume that the radionuclide emits beta particles of average energy 49.5 keV and no other radiation ($^{14}C$). Assume, further, that compartment $b$ is the lung (mass $= 1000$ g), which is the only organ of the body that receives appreciable radiation over 50 y.

    (a) Calculate the committed equivalent dose to the lung.

    (b) Calculate the committed effective dose.

    (c) What is the ALI for this nuclide (for the route of intake that occurred)?

40. Use the two-compartment dose model described in Section 16.7. A single intake of a radionuclide, having a physical half-life of 170 d, is made into compartment $a$ (body fluids, metabolic half-life $= 0.25$ d) at time $t = 0$. Six-tenths of the activity that is transferred out of $a$ goes into compartment $b$, where the metabolic half-life is 236 d.

    (a) What fraction of the initial activity is left in compartment $a$ at time $t = 21$ h?

    (b) What fraction of the initial activity is located in compartment $b$ at $t = 200$ d?

(c) If $3 \times 10^8$ disintegrations occurred in compartment $b$ between $t = 0$ and $t = 200$ d, how much activity was initially taken into compartment $a$?

41. Use the two-compartment model described in Section 16.6 for ingestion of a nuclide that emits a single 5.5-MeV alpha particle per transformation and has a radioactive half-life of 8.6 d. No other radiation is emitted. The metabolic half-life in the body fluids (compartment $a$) is 0.25 d, and that in compartment $b$ is 3.5 d. The fraction of the radionuclide that goes to $b$ when it leaves $a$ is 0.80. Assume that 100% of the ingested nuclide goes directly into the body fluids and that compartment $b$ receives a negligible dose as a result of the activity in $a$. The mass associated with compartment $b$ is 1800 g.
   (a) Calculate the specific effective energy, SEE $(b \leftarrow b)$.
   (b) Calculate the number of transformations in compartment $b$ over the 50-y period following ingestion of 1 Bq of activity at time $t = 0$.
   (c) What is the committed equivalent dose in $b$ per Bq of ingestion?
   (d) If the organ weighting factor for $b$ is 0.05, determine the ALI for ingestion for this radionuclide.

42. Why is bone dosimetry of particular importance?

43. What are the target tissues for bone?

44. Why is the dosimetric model for submersion in a radioactive gas cloud different from the model for the respiratory system?

45. Calculate the equivalent-dose rate in air in $Sv\,h^{-1}$ due to 1 Bq of $^{14}C$ per gram of air.

46. Estimate the equivalent-dose rate at the surface of the skin of a person immersed in air (at STP) containing $2.4 \times 10^3$ $Bq\,m^{-3}$ of $^{14}CO_2$.

47. Calculate the equivalent-dose rate in a large air volume (at STP) that contains a uniform distribution of $2 \times 10^3$ $Bq\,m^{-3}$ of $^{137}Cs$.

48. Why is the DRAC for tritiated water so much smaller than that for elemental tritium?

49. (a) What activity of tritium, distributed uniformly in the soft tissue of the body (reference man), would result in an equivalent-dose rate of 0.05 $Sv\,y^{-1}$?
   (b) What would be the total mass of tritium in the soft tissue?

50. Estimate the time it takes for the body to expel by normal processes 95% of the tritium ingested in a single intake of tritiated water. Would the retention time be affected by increasing the intake of liquids?

51. What fraction of the iodine in the total body of reference man is in the thyroid?

## 16.15

## Answers

2. $2.5 \times 10^{-6}$ mSv Bq$^{-1}$

4. (a) $3 \times 10^6$ Bq
   (b) $1 \times 10^3$ Bq m$^{-3}$

6. (a) 1.2 mSv
   (b) 0.060 mSv
   (c) $2 \times 10^6$ Bq

7. (a) 0.08
   (b) 0.37

9. $\dot{q}_i = F_i \lambda_h q_h - \lambda_i q_i - \lambda_R q_i$
   $\dot{q}_j = F_j \lambda_h q_h - \lambda_R q_j$

10. (a) 37.5 Bq
    (b) 0.78

11. (b) 14 Bq d$^{-1}$

13. $1.9 \times 10^5$ Bq d$^{-1}$

17. 19 Bq

18. $\sim$ 100 d

22. $9.3 \times 10^{-19}$ Sv

25. 0.28 Sv

27. $2.8 \times 10^{-5}$ MeV g$^{-1}$ for liver; 0 for others

28. $3.8 \times 10^{-10}$ Sv

29. (a) $7.4 \times 10^{-9}$ MeV g$^{-1}$
    (b) 260 MeV

33. (a) 6.5 Sv d$^{-1}$

   (b) $3.6 \times 10^{-5}$ Sv d$^{-1}$
   (c) $5 \times 10^{-7}$ Sv

36. $2.3 \times 10^{10}$; $1.3 \times 10^{10}$

38. (a) $3.1 \times 10^{10}$
       $4.2 \times 10^{10}$
    (b) $3.1 \times 10^{10}$
       $2.7 \times 10^{14}$

39. (a) 2.1 Sv
    (b) 260 mSv
    (c) $8 \times 10^4$ Bq

41. (a) 0.061 MeV g$^{-1}$
    (b) $2.4 \times 10^5$
    (c) $2.4 \times 10^{-6}$ Sv
    (d) $2 \times 10^5$ Bq

45. $2.9 \times 10^{-8}$ Sv h$^{-1}$

46. $5.4 \times 10^{-8}$ Sv h$^{-1}$
    at surface;
    $2.7 \times 10^{-8}$ Sv h$^{-1}$
    superficial layer

47. $7.3 \times 10^{-7}$ Sv h$^{-1}$ total;
    $5.0 \times 10^{-7}$ Sv h$^{-1}$ $\gamma$
    alone ($4\pi$ geometry)

50. 43 d; yes

51. 0.91

# Appendix A
# Physical Constants

Planck's constant, $h = 6.6261 \times 10^{-34}$ J s
$$\hbar = h/2\pi = 1.05457 \times 10^{-34} \text{ J s}$$
Electron charge, $e = -1.6022 \times 10^{-19}$ C $= -4.8033 \times 10^{-10}$ esu
Velocity of light in vacuum, $c = 2.997925 \times 10^{8}$ m s$^{-1}$
Avogadro's number, $N_0 = 6.0221 \times 10^{23}$ mole$^{-1}$
Molar volume at STP (0°C, 760 torr) $= 22.414$ L
Density of air at STP (0°C, 760 torr) $= 1.293$ kg m$^{-3}$
$$= 1.293 \times 10^{-3} \text{ g cm}^{-3}$$
Rydberg constant, $R_\infty = 1.09737 \times 10^{7}$ m$^{-1}$
First Bohr orbit radius in hydrogen, $a_0 = 5.2918 \times 10^{-11}$ m
Ratio proton and electron masses $= 1836.15$
Electron mass, $m = 0.00054858$ AMU $= 0.51100$ MeV $= 9.1094 \times 10^{-31}$ kg
Proton mass $= 1.0073$ AMU $= 938.27$ MeV $= 1.6726 \times 10^{-27}$ kg
H atom mass $= 1.0078$ AMU $= 938.77$ MeV $= 1.6735 \times 10^{-27}$ kg
Neutron mass $= 1.0087$ AMU $= 939.57$ MeV $= 1.6749 \times 10^{-27}$ kg
Alpha-particle mass $= 4.0015$ AMU $= 3727.4$ MeV $= 6.6447 \times 10^{-27}$ kg
Boltzmann's constant, $k = 1.3807 \times 10^{-23}$ J K$^{-1}$

Principal source: E. R. Cohen and B. N. Taylor, "The Fundamental Physical Constants," *Physics Today*, **56** (No. 8), pp. BG6-BG13, August (2003). Updated and available online at http://www.physicstoday.org/guide/fundcon.html

The metric (SI) system of units is summarized in Robert A. Nelson's "Guide for Metric Practice," available at http://www.physicstoday.org/guide/metric.html See also National Institute of Standards and Technology, Physical Reference Data, http://www.physics.nist.gov/cuu/Constants

# Appendix B
# Units and Conversion Factors

$1 \text{ cm} = 10^4 \ \mu\text{m} = 10^8 \text{Å}$

$1 \text{ in.} = 2.54 \text{ cm (exactly)}$

$1 \text{ barn} = 10^{-24} \text{ cm}^2$

$1 \text{ L} = 1 \text{ dm}^3 = 10^{-3} \text{ m}^3$

$1 \text{ dyne} = 1 \text{ g cm s}^{-2} = 10^{-5} \text{ kg m s}^{-2} = 10^{-5} \text{ N}$

$1 \text{ kg} = 2.205 \text{ lb}$

$1 \text{ erg} = 1 \text{ dyne cm} = 1 \text{ g cm}^2 \text{ s}^{-2} = 1 \text{ esu}^2 \text{ cm}^{-1}$

$1 \text{ J} = 1 \text{ N m} = 1 \text{ kg m}^2 \text{ s}^{-2} = 1.11265 \times 10^{-10} \text{ C}^2 \text{ m}^{-1}$

$10^7 \text{ erg} = 1 \text{ J}$

$1 \text{ eV} = 1.6022 \times 10^{-12} \text{ erg} = 1.6022 \times 10^{-19} \text{ J}$

$1 \text{ AMU} = 931.49 \text{ MeV} = 1.6605 \times 10^{-27} \text{ kg}$

$1 \text{ gram calorie} = 4.186 \text{ J}$

$1 \text{ W} = 1 \text{ J s}^{-1} = 1 \text{ V A}$

$1 \text{ statvolt} = 299.8 \text{ V}$

$1 \text{ esu} = 3.336 \times 10^{-10} \text{ C}$

$1 \text{ A} = 1 \text{ C s}^{-1}$

$1 \text{ C} = 1 \text{ V F}$

$1 \text{ Ci} = 3.7 \times 10^{10} \text{ s}^{-1} = 3.7 \times 10^{10} \text{ Bq (exactly)}$

$1 \text{ R} = 2.58 \times 10^{-4} \text{ C kg}^{-1} \text{ air } (= 1 \text{ esu cm}^{-3} \text{ air at STP})$

$1 \text{ rad} = 100 \text{ erg g}^{-1} = 0.01 \text{ Gy}$

$1 \text{ Gy} = 1 \text{ J kg}^{-1} = 100 \text{ rad}$

$1 \text{ Sv} = 100 \text{ rem}$

$0°\text{C} = 273 \text{ K}$

$1 \text{ atmosphere} = 760 \text{ mm Hg} = 760 \text{ torr} = 101.3 \text{ kPa}$

$1 \text{ day} = 86,400 \text{ s}$

$1 \text{ yr} = 365 \text{ days} = 3.1536 \times 10^7 \text{ s}$

$1 \text{ radian} = 57.30°$

*Atoms, Radiation, and Radiation Protection.* James E. Turner
Copyright © 2007 WILEY-VCH Verlag GmbH & Co. KGaA, Weinheim
ISBN: 978-3-527-40606-7

# Appendix C
# Some Basic Formulas of Physics (MKS and CGS Units)

## Classical Mechanics

Momentum = mass × velocity, $p = mv$
    units: $kg\,m\,s^{-1}$; $g\,cm\,s^{-1}$
Kinetic energy, $T = \frac{1}{2}mv^2 = p^2/2m$
    units: $1\,J = 1\,kg\,m^2\,s^{-2}$; $1\,erg = 1\,g\,cm^2\,s^{-2}$
Force = mass × acceleration, $F = ma$
    units: $1\,N = 1\,kg\,m\,s^{-2}$; $1\,dyne = 1\,g\,cm\,s^{-2}$
Work = force × distance = change in energy
    units: $1\,J = 1\,N\,m = 1\,kg\,m^2\,s^{-2}$; $1\,erg = 1\,dyne\,cm = 1\,g\,cm^2\,s^{-2}$
Impulse = force × time = change in momentum, $I = Ft = \Delta p$
    units: $1\,N\,s = 1\,kg\,m\,s^{-1}$; $1\,dyne\,s = 1\,g\,cm\,s^{-1}$
Angular momentum, uniform circular motion, $L = mvr$
    units: $1\,kg\,m^2\,s^{-1} = 1\,J\,s$; $1\,g\,cm^2\,s^{-1} = 1\,erg\,s$
Centripetal acceleration, uniform circular motion, $a = v^2/r$
    units: $m\,s^{-2}$; $cm\,s^{-2}$

## Relativistic Mechanics (units same as in classical mechanics)

Relativistic quantities:
    $v$ = speed of object
    $c$ = speed of light in vacuum
    $\beta = v/c$, dimensionless, $0 \le \beta < 1$
    $\gamma = 1/\sqrt{1-\beta^2}$, dimensionless, $1 \le \gamma < \infty$
Rest energy, $E_0 = mc^2$, $m$ = rest mass
Relativistic mass, $m/\sqrt{1-\beta^2} = \gamma m$
Total energy, $E_T = mc^2/\sqrt{1-\beta^2} = \gamma mc^2$
Kinetic energy = total energy—rest energy,

$$T = E_T - E_0 = mc^2(\gamma - 1) = mc^2\left(\frac{1}{\sqrt{1-\beta^2}} - 1\right)$$

Momentum, $p = \gamma mv = mv/\sqrt{1-\beta^2}$

*Atoms, Radiation, and Radiation Protection.* James E. Turner
Copyright © 2007 WILEY-VCH Verlag GmbH & Co. KGaA, Weinheim
ISBN: 978-3-527-40606-7

Relationship between energy and momentum,

$$E_T^2 = p^2 c^2 + m^2 c^4 = (mc^2 + T)^2$$

## Electromagnetic Theory

Force $F$ between two point charges, $q_1$ and $q_2$, at separation $r$ (Coulomb's law) in vacuum,

MKS: $F = k_0 q_1 q_2 / r^2$, $q_1$, $q_2$ in C, $r$ in m, $F$ in N, and $k_0 = 8.98755 \times 10^9$ N m$^2$ C$^{-2}$ ($= 1/(4\pi\varepsilon_0)$) in terms of permittivity constant $\varepsilon_0$)

CGS: $F = q_1 q_2 / r^2$, $q_1$, $q_2$ in esu (statcoulombs), $r$ in cm, and $F$ in dynes

Potential energy of two point charges at separation $r$ in vacuum,

MKS: PE $= k_0 q_1 q_2 / r$

CGS: PE $= q_1 q_2 / r$

Electric field strength (force per unit charge), $F/q$

units: 1 N C$^{-1}$ $= 1$ V m$^{-1}$; 1 dyne esu$^{-1}$ $= 1$ statvolt cm$^{-1}$

Change in potential energy $\Delta E$ of charge $q$ moved through potential difference $V$, $\Delta E = qV$. With $q =$ number of electron charges and $V$ in volts, $\Delta E$ is in electron volts (eV) of energy, by definition of the eV.

Capacitance, $Q/V$

units: 1 F $= 1$ C/V

Current, $I = Q/t$ (charge per unit time)

units: 1 A $= 1$ C s$^{-1}$

Power, $P = VI$ (potential difference $\times$ current)

units: 1 W $= 1$ V A $= 1$ J s$^{-1}$

Relationship between wavelength $\lambda$ and frequency $\nu$ of light in vacuum (speed of light $= c$), $\lambda\nu = c$

## Quantum Mechanics

de Broglie wavelength, $\lambda = h/p = h/\gamma mv$

Photon energy, $E = h\nu$

Photon momentum, $p = E/c = h\nu/c$

Bohr quantization condition for angular momentum, $L = n\hbar$

Bohr energy levels, $E = -13.6 Z^2 / n^2$ eV

Uncertainty relations, $\Delta p_x \Delta x \gtrsim \hbar$, $\Delta E \Delta t \gtrsim \hbar$.

# Appendix D
# Selected Data on Nuclides

| Nuclide | Natural Abundance (%) | Mass Difference $\Delta = M - A$ (MeV) (at. mass − at. mass No.) | Type of Decay | Half-Life | Major Radiations, Energies (MeV), and Frequency per Disintegration (%) |
|---|---|---|---|---|---|
| $^1_0$n | — | 8.0714 | $\beta^-$ | 12 min | $\beta^-$: 0.78 max |
| $^1_1$H | 99.985 | 7.2890 | — | — | — |
| $^2_1$H | 0.015 | 13.1359 | — | — | — |
| $^3_1$H | — | 14.9500 | $\beta^-$ | 12.3 y | $\beta^-$: 0.0186 max, no $\gamma$ |
| $^3_2$He | 0.00013 | 14.9313 | — | — | — |
| $^4_2$He | 99.99+ | 2.4248 | — | — | — |
| $^6_3$Li | 7.42 | 14.088 | — | — | — |
| $^7_3$Li | 92.58 | 14.907 | — | — | — |
| $^7_4$Be | — | 15.769 | EC | 53.3 d | $\gamma$: 0.478 (10.3%) |
| $^{10}_5$B | 19.7 | 12.052 | — | — | — |
| $^{11}_6$C | — | 10.648 | $\beta^+$ 99 + %<br>EC 0.2% | 20.38 min | $\beta^+$: 0.960 max (avg 0.386)<br>$\gamma$: 0.511 (200%, $\gamma^\pm$) |
| $^{12}_6$C | 98.892 | 0 | — | — | — |
| $^{14}_6$C | — | 3.0198 | $\beta^-$ | 5730 y | $\beta^-$: 0.156 max (avg 0.0495), no $\gamma$ |
| $^{14}_7$N | 99.635 | 2.8637 | — | — | — |
| $^{15}_7$N | 0.356 | 0.100 | — | — | — |
| $^{16}_8$O | 99.759 | -4.7366 | — | — | — |
| $^{17}_8$O | 0.037 | -0.808 | — | — | — |
| $^{22}_{10}$Ne | 8.82 | -8.025 | — | — | — |

| Nuclide | Natural Abundance (%) | Mass Difference $\Delta = M - A$ (MeV) (at. mass – at. mass No.) | Type of Decay | Half-Life | Major Radiations, Energies (MeV), and Frequency per Disintegration (%) |
|---|---|---|---|---|---|
| $^{22}_{11}$Na | — | −5.182 | $\beta^+$ 89.8% <br> EC 10.2% | 2.602 y | $\beta^+$: 0.545 max (avg 0.215) <br> $\gamma$: 0.511 (180%, $\gamma^\pm$), 1.275 (100%), Ne X rays |
| $^{23}_{11}$Na | 100. | −9.528 | — | — | — |
| $^{24}_{11}$Na | — | −8.418 | $\beta^-$ | 15.00 h | $\beta^-$: 1.390 max (avg 0.554) <br> $\gamma$: 1.369 (100%), 2.754 (100%) |
| $^{24}_{12}$Mg | 78.60 | −13.933 | — | — | — |
| $^{26}_{12}$Mg | 11.3 | −16.214 | — | — | — |
| $^{26}_{13}$Al | — | −12.211 | $\beta^+$ 81.8% <br> EC 18.2% | $7.16 \times 10^5$ y | $\beta^+$: 1.174 max (avg 0.544) <br> $\gamma$: 0.511 (164%, $\gamma^\pm$), 1.130 (2.5%), 1.809 (100%), Mg X rays |
| $^{26m}_{13}$Al | — | −11.982 | $\beta^+$ | 6.4 s | $\beta^+$: 3.21 max <br> $\gamma$: 0.511 (200%, $\gamma^\pm$) |
| $^{32}_{15}$P | — | −24.303 | $\beta^-$ | 14.29 d | $\beta^-$: 1.710 max (avg 0.695) <br> No $\gamma$ |
| $^{32}_{16}$S | 95.0 | −26.013 | — | — | — |
| $^{35}_{16}$S | — | −28.847 | $\beta^-$ | 87.44 d | $\beta^-$: 0.167 max (avg 0.0488) No $\gamma$ |
| $^{37}_{16}$S | — | −27.0 | $\beta^-$ | 5.06 min | $\beta^-$: 1.6 max (90%) <br> 4.7 max (10%) <br> $\gamma$: 3.09 (90%) |
| $^{38}_{16}$S | — | −26.8 | $\beta^-$ | 2.87 h | $\beta^-$: 1.1 max (95%), 3.0 max (5%) <br> $\gamma$: 1.88 (95%) <br> Daughter radiations from $^{38}$Cl |

| Nuclide | Natural Abundance (%) | Mass Difference $\Delta = M - A$ (MeV) (at. mass − at. mass No.) | Type of Decay | Half-Life | Major Radiations, Energies (MeV), and Frequency per Disintegration (%) |
|---|---|---|---|---|---|
| $^{35}_{17}$Cl | 75.53 | −29.015 | — | — | — |
| $^{37}_{17}$Cl | 24.47 | −31.765 | — | — | — |
| $^{35}_{18}$Ar | — | −23.05 | $\beta^+$ | 1.8 s | $\beta^+$: 4.94 max<br>$\gamma$: 0.511 (200%, $\gamma^{\pm}$), 1.22 (5%), 1.76 (2%) |
| $^{37}_{18}$Ar | — | −30.951 | EC | 35.02 d | $\gamma$: Cl X rays |
| $^{40}_{19}$K | 0.0118 | −33.533 | $\beta^-$ 89%<br>EC 11% | $1.28 \times 10^9$ y | $\beta^-$: 1.312 max (avg 0.585)<br>$\gamma$: 1.461 (11%), Ar X rays |
| $^{42}_{19}$K | — | −35.02 | $\beta^-$ | 12.36 h | $\beta^-$: 1.996 max (avg 0.822) (18%),<br>3.521 max (avg 1.564) (82%)<br>$\gamma$: 1.525 (17.9%) |
| $^{55}_{25}$Mn | 100 | −57.705 | — | — | — |
| $^{55}_{26}$Fe | — | −57.474 | EC | 2.70 y | $\gamma$: Mn X rays |
| $^{59}_{26}$Fe | — | −60.660 | $\beta^-$ | 45.53 d | $\beta^-$: 0.131 max (avg 0.036) (1.3%)<br>0.273 max (avg 0.081) (45.3%)<br>0.465 max (avg 0.149) (53.2%)<br>1.565 max (0.2%)<br>$\gamma$: 0.143 (1.0%), 0.192 (3.0%),<br>1.099 (56.1%), 1.292 (43.6%) |
| $^{57}_{27}$Co | — | −59.339 | EC | 270.9 d | $\gamma$: 0.014 (9%), 0.122 (86%),<br>0.136 (11%), Fe X rays |
| $^{60}_{27}$Co | — | −61.651 | $\beta^-$ | 5.271 y | $\beta^-$: 0.318 max (avg 0.096) (99.92%)<br>1.491 max (0.08%)<br>$\gamma$: 1.173 (99.98%), 1.332 (99.90%) |

| Nuclide | Natural Abundance (%) | Mass Difference $\Delta = M - A$ (MeV) (at. mass – at. mass No.) | Type of Decay | Half-Life | Major Radiations, Energies (MeV), and Frequency per Disintegration (%) |
|---|---|---|---|---|---|
| $^{57}_{28}$Ni | — | –56.10 | EC 60% $\beta^+$ 40% | 36.08 h | $\beta^+$ : 0.843 max $\gamma$ : 0.127 (13%), 0.511 (80%, $\gamma^{\pm}$), 1.38 (78%), 1.76 (7%), 1.92 (15%), Co X rays |
| $^{60}_{28}$Ni | 26.16 | –64.471 | — | — | — |
| $^{65}_{29}$Cu | 30.9 | –67.27 | — | — | — |
| $^{65}_{30}$Zn | — | –65.92 | EC 98.5% $\beta^+$ 1.5% | 243.9 d | $\beta^+$ : 0.330 max (avg 0.143) $\gamma$ : 0.511 (3.0%, $\gamma^{\pm}$), 1.116 (51%), Cu X rays $e^-$ : 1.107 |
| $^{85}_{36}$Kr | — | –81.48 | $\beta^-$ | 10.72 y | $\beta^-$ : 0.687 max (avg 0.251) $\gamma$ : 0.514 (0.43%) |
| $^{84}_{37}$Rb | — | –79.753 | EC 70% $\beta^+$ 26% $\beta^-$ 4% | 32.77 d | $\beta^+$ : 1.658 max (avg 0.756) (14%), 0.777 max (avg 0.338) (12%), $\beta^-$ : 0.890 max (avg 0.331) $\gamma$ : 0.511 (52%, $\gamma^{\pm}$), 0.882 (71%) Kr X rays |
| $^{90}_{38}$Sr | — | –85.95 | $\beta^-$ | 29.12 y | $\beta^-$ : 0.546 max (avg 0.196), no $\gamma$ Daughter radiations from $^{90}$Y |
| $^{85}_{39}$Y | — | –77.79 | $\beta^+$ 70% EC 30% | 5.0 h | $\beta^+$ : 2.24 max $\gamma$ : 0.231 (13%), 0.511 (140%, $\gamma^{\pm}$), 0.770 (8%), 2.16 (9%), Sr X rays $e^-$ : 0.215 |

| Nuclide | Natural Abundance (%) | Mass Difference $\Delta = M - A$ (MeV) (at. mass − at. mass No.) | Type of Decay | Half-Life | Major Radiations, Energies (MeV), and Frequency per Disintegration (%) |
|---|---|---|---|---|---|
| $^{86}_{39}$Y | — | −79.23 | EC 67%<br>$\beta^+$ 33% | 14.74 h | $\beta^+$ : 3.17 max (1.8%) and others; avg $\beta^+$ 0.759<br>$\gamma$ : 0.443 (17%), 0.511 (66%, $\gamma^\pm$), 0.628 (33%), 0.703 (15%), 0.777 (22%), 1.077 (83%), 1.153 (31%), 1.854 (17%), 1.925 (21%), Sr X rays |
| $^{90}_{39}$Y | — | −86.50 | $\beta^-$ | 64.0 h | $\beta^-$ : 2.284 max (avg 0.935), no $\gamma$ |
| $^{99}_{43}$Tc | — | −87.33 | $\beta^-$ | $2.12 \times 10^5$ y | $\beta^-$ : 0.294 max (avg 0.101), no $\gamma$ |
| $^{99m}_{43}$Tc | — | −87.18 | IT | 6.02 h | $\gamma$ : 0.140 (89%), Tc X rays<br>$e^-$ : 0.002, 0.119 |
| $^{103}_{45}$Rh | 100 | −88.014 | — | — | — |
| $^{103m}_{45}$Rh | — | −87.974 | IT | 56.12 min | $\gamma$ : 0.040 (0.07%), Rh X rays<br>$e^-$ : 0.017, 0.037, 0.039 |
| $^{103}_{46}$Pd | — | −87.46 | EC | 16.96 d | $\gamma$ : 0.062 (0.001%), 0.295 (0.003%), 0.357 (0.022%), 0.497 (0.004%), Rh X rays<br>Daughter radiations from $^{103m}$Rh |
| $^{113}_{48}$Cd | 12.26 | −89.041 | — | — | — |
| $^{114}_{48}$Cd | 28.86 | −90.018 | — | — | — |
| $^{115}_{49}$In | 95.70 | −89.54 | $\beta^-$ | $5.1 \times 10^{15}$ y | $\beta^-$ : 0.494 max (avg 0.152), no $\gamma$ |
| $^{116m}_{49}$In | — | −88.14 | $\beta^-$ | 54.15 min | $\beta^-$ : 1.007 max<br>$\gamma$ : 0.417 (29%), 0.819 (11%), 1.097 (56%), 1.294 (84%), 1.507 (10%), 2.112 (16%) |

| Nuclide | Natural Abundance (%) | Mass Difference $\Delta = M - A$ (MeV) (at. mass − at. mass No.) | Type of Decay | Half-Life | Major Radiations, Energies (MeV), and Frequency per Disintegration (%) |
|---|---|---|---|---|---|
| $^{126}_{52}$Te | 18.71 | −90.05 | — | — | — |
| $^{126}_{53}$I | — | −87.90 | EC 60.2% $\beta^-$ 36.5% $\beta^+$ 3.3% | 13.02 d | $\beta^-$: 1.251 max (8%) and others; avg $\beta^-$ 0.290 $\beta^+$: 1.130 max (avg 0.508) $\gamma$: 0.341 (39%), 0.491 (3%) 0.511 (6.6%, $\gamma^\pm$), 0.666 (33%), 0.754 (4%), Te X rays |
| $^{131}_{53}$I | — | −87.441 | $\beta^-$ | 8.05 d | $\beta^-$: 0.248 max (avg 0.069) (2%) 0.334 max (avg 0.097) (7%) 0.606 max (avg 0.192) (89%) 0.807 max (0.6%) $\gamma$: 0.080 (2.6%), 0.284 (6.1%), 0.364 (81%), 0.637 (7.3%), 0.723 (1.8%) $e^-$: 0.046, 0.330 Daughter radiations from $^{131m}$Xe |
| $^{126}_{54}$Xe | 0.090 | −89.15 | — | — | — |
| $^{137}_{55}$Cs | — | −86.9 | $\beta^-$ | 30.0 y | $\beta^-$: 1.174 max (5%) 0.512 max (avg 0.173) (95%) $\gamma$: 0.662 (85%, $^{137m}$Ba), Ba X rays $e^-$: 0.624, 0.656 |
| $^{137}_{56}$Ba | 11.3 | −88.0 | — | — | — |
| $^{137m}_{56}$Ba | — | −87.4 | IT | 2.55 min | $\gamma$: 0.662 (90%), Ba X rays $e^-$: 0.624, 0.656 |

| Nuclide | Natural Abundance (%) | Mass Difference $\Delta = M - A$ (MeV) (at. mass − at. mass No.) | Type of Decay | Half-Life | Major Radiations, Energies (MeV), and Frequency per Disintegration (%) |
|---|---|---|---|---|---|
| $^{191}_{76}$Os | — | −36.4 | $\beta^-$ | 15.4 d | $\beta^-$: 0.142 max (avg 0.038) <br> $\gamma$: 0.129 (26%), Ir X rays <br> $e^-$: 0.029, 0.031, 0.039, 0.053 <br> Daughter radiations from $^{191m}$Ir included above |
| $^{191m}_{77}$Ir | — | −36.5 | IT | 4.94 s | $\gamma$: 0.129 (26%), Ir X rays <br> $e^-$: 0.029, 0.031, 0.039, 0.053 |
| $^{197}_{79}$Au | 100 | −31.17 | — | — | — |
| $^{198}_{79}$Au | — | −29.59 | $\beta^-$ | 2.696 d | $\beta^-$: 0.225 max (avg 0.079) (1%) <br> 0.961 max (avg 0.315) (99%) <br> $\gamma$: 0.412 (96%), 0.676 (1%), 1.088 (0.2%), Hg X rays <br> $e^-$: 0.329, 0.398 |
| $^{198}_{80}$Hg | 10.2 | −30.97 | — | — | — |
| $^{203}_{80}$Hg | — | −25.26 | $\beta^-$ | 46.60 d | $\beta^-$: 0.213 max (avg 0.058) <br> $\gamma$: 0.279 (82%), Tl X rays |
| $^{205}_{81}$Tl | 70.50 | −23.81 | — | — | — |
| $^{205}_{82}$Pb | — | −23.77 | EC | $1.43 \times 10^7$ y | $\gamma$: Tl L X rays |
| $^{206}_{82}$Pb | 25.1 | −23.79 | — | — | — |

| Nuclide | Natural Abundance (%) | Mass Difference $\Delta = M - A$ (MeV) (at. mass − at. mass No.) | Type of Decay | Half-Life | Major Radiations, Energies (MeV), and Frequency per Disintegration (%) |
|---|---|---|---|---|---|
| $^{210}_{82}$Pb | — | −14.73 | $\beta^-$ | 22.3 y | $\beta^-$ : 0.017 max (avg 0.004) (80%) 0.063 max (avg 0.016) (20%) $\gamma$ : 0.047 (4%), Bi X rays $e^-$ : 0.030, 0.043 Daughter radiations from $^{210}$Bi, $^{210}$Po |
| $^{214}_{82}$Pb | — | −0.15 | $\beta^-$ | 26.8 min | $\beta^-$ : 1.024 max (6%) and others; avg $\beta^-$ 0.219 $\gamma$ : 0.053 (1%), 0.242 (7%), 0.295 (19%), 0.352 (37%), 0.786 (1%) $e^-$ : 0.037, 0.049 Daughter radiations from $^{214}$Bi, $^{214}$Po |
| $^{210}_{83}$Bi | — | −14.79 | $\beta^-$ | 5.012 d | $\beta^-$ : 1.162 max (avg 0.389), no $\gamma$ |
| $^{214}_{83}$Bi | — | −1.19 | $\beta^-$ | 19.9 min | $\beta^-$ : 3.27 max (18%) and others; avg $\beta^-$ 0.638 $\gamma$ : 0.609 (46%), 1.120 (15%), 1.765 (15%) |
| $^{210}_{84}$Po | — | −15.95 | $\alpha$ | 138.38 d | $\alpha$ : 5.305 (100%) $\gamma$ : 0.802 (0.0011%) |
| $^{214}_{84}$Po | — | −4.47 | $\alpha$ | 164 $\mu$s | $\alpha$ : 7.69 $\gamma$ : 0.800 (0.010%) |

| Nuclide | Natural Abundance (%) | Mass Difference $\Delta = M - A$ (MeV) (at. mass − at. mass No.) | Type of Decay | Half-Life | Major Radiations, Energies (MeV), and Frequency per Disintegration (%) |
|---|---|---|---|---|---|
| $^{218}_{84}$Po | | 8.38 | $\alpha$ (99.98%) $\beta^-$ (0.02%) | 3.05 min | $\alpha$: 6.003 <br> $\beta^-$: 0.256 max <br> Daughter radiations from $^{214}$Pb, $^{214}$Bi, $^{214}$Po |
| $^{219}_{86}$Rn | — | 8.85 | $\alpha$ | 3.96 s | $\alpha$: 6.82 (80%), 6.55 (12%), 6.42 (7%) <br> $\gamma$: 0.271 (10%), 0.402 (7%) <br> Po X rays |
| $^{220}_{86}$Rn | — | 10.61 | $\alpha$ | 55.6 s | $\alpha$: 6.29 <br> $\gamma$: 0.55 (0.07%) <br> Daughter radiations from $^{216}$Po |
| $^{222}_{86}$Rn | — | 16.39 | $\alpha$ | 3.8235 d | $\alpha$: 5.490 <br> $\gamma$: 0.510 (0.078%) <br> Daughter radiations from $^{218}$Po, $^{214}$Pb, $^{214}$Bi, $^{214}$Po |
| $^{226}_{88}$Ra | — | 23.69 | $\alpha$ | 1600 y | $\alpha$: 4.602 (5.5%), 4.785 (94.4%) <br> $\gamma$: 0.186 (3.3%), 0.262 (0.005%) <br> Rn X rays <br> $e^-$: 0.088, 0.170 <br> Daughter radiations from $^{222}$Rn, $^{218}$Po, $^{214}$Pb, $^{214}$Bi, $^{214}$Po |
| $^{235}_{92}$U | 0.720 | 40.93 | $\alpha$ | $7.038 \times 10^8$ y | $\alpha$: 4.218 (6%), 4.326 (5%), 4.366 (18%), 4.400 (56%), 4.417 (2%), 4.505 (2%), 4.558 (4%), 4.599 (5%) |

| Nuclide | Natural Abundance (%) | Mass Difference $\Delta = M - A$ (MeV) (at. mass – at. mass No.) | Type of Decay | Half-Life | Major Radiations, Energies (MeV), and Frequency per Disintegration (%) |
|---|---|---|---|---|---|
| | | | | | $\gamma$: 0.109 (2%), 0.144 (11%), 0.163 (5%), 0.186 (54%), 0.202 (1%), 0.205 (5%), Th X rays<br>$e^-$: 0.016, 0.020<br>Daughter radiations from $^{231}$Th, etc. |
| $^{238}_{92}$U | 99.276 | 47.33 | $\alpha$ | $4.468 \times 10^9$ y | $\alpha$: 4.149 (23%), 4.198 (77%)<br>$\gamma$: 0.050 (0.070%), Th X rays<br>$e^-$: 0.030, 0.033, 0.046, 0.050<br>Daughter radiations from $^{234}$Th, $^{234m}$Pa, etc. |
| $^{239}_{94}$Pu | — | 48.60 | $\alpha$ | 24.065 y | $\alpha$: 5.105 (11%), 5.143 (15%), 5.156 (74%)<br>$\gamma$: 0.010 (0.001%), 0.039 (0.006%), 0.052 (0.021%), 0.099 (0.001%), 0.129 (0.006%), 0.375 (0.002%), U X rays<br>$e^-$: 0.009, 0.013, 0.018, 0.022, 0.031, 0.034, 0.047<br>Daughter radiations from $^{235}$U, etc. |

*Source:* Most of the decay data were obtained with the utility code, DEXRAX: K. F. Eckerman, R. J. Westfall, J. C. Ryman, and M. Cristy, *Nuclear Decay Data Files of the Dosimetry Research Group*, ORNL/TM-12350, Oak Ridge National Laboratory, Oak Ridge, Tenn. (December 1993). See also K. F. Eckerman, R. J. Westfall, J. C. Ryman, and M. Cristy, "Availability of Nuclear Decay Data in Electronic Form, Including Beta Spectra not Previously Published," *Health Phys.* **67**, 338–345 (1994). The author is grateful to Dr. Eckerman for his substantial help. Values of the mass differences, $\Delta$, are from the *Radiological Health Handbook*, U.S. Public Health Service Publ. No. 2016, Bureau of Radiological Health, Rockville, Md. (1970).

# Appendix E
# Statistical Derivations

## Binomial Distribution

### Mean

The mean value $\mu$ of the binomial distribution is defined by Eq. (11.15):

$$\mu \equiv \sum_{n=0}^{N} nP_n = \sum_{n=0}^{N} n\binom{N}{n} p^n q^{N-n}. \tag{E.1}$$

To evaluate this sum, we first use the binomial expansion to write, for an arbitrary (continuous) variable $x$,

$$(px + q)^N = \sum_{n=0}^{N} \binom{N}{n} p^n x^n q^{N-n} = \sum_{n=0}^{N} x^n P_n. \tag{E.2}$$

Differentiation with respect to $x$ gives

$$Np(px + q)^{N-1} = \sum_{n=0}^{N} nx^{n-1} P_n. \tag{E.3}$$

Letting $x = 1$ and remembering that $p + q = 1$ gives

$$Np = \sum_{n=0}^{N} nP_n \equiv \mu. \tag{E.4}$$

### Standard Deviation

The variance is defined by Eq. (11.17):

$$\sigma^2 \equiv \sum_{n=0}^{N} (n - \mu)^2 P_n. \tag{E.5}$$

*Atoms, Radiation, and Radiation Protection.* James E. Turner
Copyright © 2007 WILEY-VCH Verlag GmbH & Co. KGaA, Weinheim
ISBN: 978-3-527-40606-7

This definition implies that

$$\sigma^2 = \sum_{n=0}^{N} \left( n^2 P_n - 2\mu n P_n + \mu^2 P_n \right) \tag{E.6}$$

$$= \sum_{n=0}^{N} n^2 P_n - 2\mu \sum_{n=0}^{N} n P_n + \mu^2 \sum_{n=0}^{N} P_n. \tag{E.7}$$

The first summation gives the expected value of $n^2$, the square of the number of disintegrations. From Eq. (E.4) it follows that the second term is $-2\mu^2$. The sum in the last term is unity [Eq. (11.14)]. Thus, we can write in place of Eq. (E.7)

$$\sigma^2 = \sum_{n=0}^{N} n^2 P_n - 2\mu^2 + \mu^2 = \sum_{n=0}^{N} n^2 P_n - \mu^2. \tag{E.8}$$

We have previously evaluated $\mu$ [Eq. (E.4)]; it remains to find the sum involving $n^2$. To this end, we differentiate both sides of Eq. (E.3) with respect to $x$:

$$N(N-1)p^2(px+q)^{N-2} = \sum_{n=0}^{N} n(n-1)x^{n-2} P_n. \tag{E.9}$$

Letting $x = 1$ with $p + q = 1$, as before, implies that

$$N(N-1)p^2 = \sum_{n=0}^{N} n(n-1) P_n \tag{E.10}$$

$$= \sum_{n=0}^{N} n^2 P_n - \sum_{n=0}^{N} n P_n = \sum_{n=0}^{N} n^2 P_n - \mu. \tag{E.11}$$

Thus,

$$\sum_{n=0}^{N} n^2 P_n = N(N-1)p^2 + \mu. \tag{E.12}$$

Substituting this result into Eq. (E.8) and remembering that $\mu = Np$, we find that

$$\sigma^2 = N(N-1)p^2 + Np - N^2 p^2 = Np(1-p) = Npq. \tag{E.13}$$

The standard deviation of the binomial distribution is therefore

$$\sigma = \sqrt{Npq}. \tag{E.14}$$

### Poisson Distribution

As stated at the beginning of Section 11.5, we consider the binomial distribution when $N \gg 1$, $N \gg n$, and $p \ll 1$. Under these conditions, the binomial coefficient in Eq. (11.13) is approximated well by Eq. (11.29). Also, the last factor in Eq. (11.13) can be written

$$q^{N-n} \cong q^N = (1-p)^N. \tag{E.15}$$

Using the binomial expansion for the last expression, we then have

$$q^{N-n} = 1 - Np + \frac{N(N-1)}{2!}p^2 - \cdots \tag{E.16}$$

$$\cong 1 - Np + \frac{(Np)^2}{2!} - \cdots = e^{-Np}. \tag{E.17}$$

Substitution of Eqs. (11.29) and (E.17) into (11.13) gives

$$P_n = \frac{N^n}{n!}p^n e^{-Np} = \frac{(Np)^n}{n!}e^{-Np}, \tag{E.18}$$

which is the Poisson distribution, with parameter $Np$.

### Normalization

The distribution (E.18) is normalized to unity when summed over *all* non-negative integers $n$:

$$\sum_{n=0}^{\infty} P_n = e^{-Np} \sum_{n=0}^{\infty} \frac{(Np)^n}{n!} = e^{-Np} e^{Np} = 1. \tag{E.19}$$

For the binomial distribution, $P_n = 0$ when $n > N$. As seen from Eq. (E.18), the terms in the Poisson distribution are never exactly zero.

### Mean

The mean value of $n$ can be found from Eq. (E.18). With some manipulation of the summing index $n$, we write

$$\mu \equiv e^{-Np} \sum_{n=0}^{\infty} \frac{n(Np)^n}{n!} = e^{-Np} \sum_{n=1}^{\infty} \frac{n(Np)^n}{n!} \tag{E.20}$$

$$= e^{-Np} \sum_{n=1}^{\infty} \frac{(Np)^n}{(n-1)!} = e^{-Np} Np \sum_{n=1}^{\infty} \frac{(Np)^{n-1}}{(n-1)!} \tag{E.21}$$

$$= e^{-Np} Np \sum_{n=0}^{\infty} \frac{(Np)^n}{n!} = e^{-Np} Np e^{Np} = Np. \tag{E.22}$$

The mean of the Poisson distribution is thus identical to that of the binomial distribution. We write in place of Eq. (E.18) the usual form

$$P_n = \frac{\mu^n e^{-\mu}}{n!}. \tag{E.23}$$

## Standard Deviation

The variance of the Poisson distribution is given by Eq. (E.8), with the $P_n$ defined by Eq. (E.23). As before, it remains to find the expected value of $n^2$. Again, manipulating the index of summation $n$, we write, using Eq. (E.23),

$$\sum_{n=0}^{\infty} n^2 P_n \equiv e^{-\mu} \sum_{n=0}^{\infty} \frac{n^2 \mu^n}{n!} = e^{-\mu} \sum_{n=1}^{\infty} \frac{n^2 \mu^n}{n!} \tag{E.24}$$

$$= e^{-\mu} \mu \sum_{n=1}^{\infty} \frac{n \mu^{n-1}}{(n-1)!} = e^{-\mu} \mu \sum_{n=0}^{\infty} \frac{(n+1)\mu^n}{n!} \tag{E.25}$$

$$= e^{-\mu} \mu \sum_{n=0}^{\infty} \left( \frac{n\mu^n}{n!} + \frac{\mu^n}{n!} \right) = \mu(\mu+1) = \mu^2 + \mu. \tag{E.26}$$

Substitution of this result into Eq. (E.8) gives for the variance

$$\sigma^2 = \mu^2 + \mu - \mu^2 = \mu. \tag{E.27}$$

We obtain the important result that the standard deviation of the Poisson distribution is equal to the square root of the mean:

$$\sigma = \sqrt{\mu}. \tag{E.28}$$

## Normal Distribution

We begin with Eq. (E.23) for the Poisson $P_n$ and assume that $\mu$ is large. We also assume that the $P_n$ are appreciably different from zero only over a range of values of $n$ about the mean such that $|n - \mu| \ll \mu$. That is, the distribution of the $P_n$ is relatively narrow about $\mu$; and both $\mu$ and $n$ are large. We change variables by writing $x = n - \mu$. Equation (E.23) can then be written

$$P_x = \frac{\mu^{\mu+x} e^{-\mu}}{(\mu+x)!} = \frac{\mu^\mu \mu^x e^{-\mu}}{\mu!(\mu+1)(\mu+2)\cdots(\mu+x)}, \tag{E.29}$$

with $|x| \ll \mu$. We can approximate the factorial term for large $\mu$ by means of the Stirling formula,

$$\mu! = \sqrt{2\pi \mu} \, \mu^\mu e^{-\mu}, \tag{E.30}$$

giving

$$P_x = \frac{\mu^x}{\sqrt{2\pi \mu}(\mu+1)(\mu+2)\cdots(\mu+x)} \tag{E.31}$$

$$= \frac{1}{\sqrt{2\pi \mu}\left(1+\dfrac{1}{\mu}\right)\left(1+\dfrac{2}{\mu}\right)\cdots\left(1+\dfrac{x}{\mu}\right)}. \tag{E.32}$$

Since, for small $y$, $e^y \cong 1 + y$, the series of factors in the denominator can be rewritten ($\mu$ is large) to give

$$P_x = \frac{1}{\sqrt{2\pi\,\mu}e^{1/\mu}e^{2/\mu}\cdots e^{x/\mu}} = \frac{1}{\sqrt{2\pi\,\mu}}e^{-(1+2+\cdots+x)/\mu}. \qquad \text{(E.33)}$$

The sum of the first $x$ positive integers, as they appear in the exponent, is $x(1 + x)/2 = (x^2 + x)/2 \cong x^2/2$, where $x$ has been neglected compared with $x^2$. Thus, we find that

$$P_x = \frac{1}{\sqrt{2\pi\,\mu}}e^{-x^2/2\mu}. \qquad \text{(E.34)}$$

This function, which is symmetric in $x$, represents an approximation to the Poisson distribution. The normal distribution is obtained when we replace the Poisson standard deviation $\sqrt{\mu}$ by an independent parameter $\sigma$ and let $x$ be a continuous random variable with mean value $\mu$ (not necessarily zero). We then write for the probability density in $x$ ($-\infty < x < \infty$) the normal distribution

$$f(x) = \frac{1}{\sqrt{2\pi}\sigma}e^{-(x-\mu)^2/2\sigma^2}, \qquad \text{(E.35)}$$

with $\sigma^2 > 0$. It can be shown that this density function is normalized (i.e., its integral over all $x$ is unity) and that its mean and standard deviation are, respectively, $\mu$ and $\sigma$. The probability that the value of $x$ lies between $x$ and $x + dx$ is $f(x)\,dx$. Whereas the Poisson distribution has the single parameter $\mu$, the normal distribution is characterized by the two independent parameters, $\mu$ and $\sigma$.

## Error Propagation

We determine the standard deviation of a quantity $Q(x, y)$ that depends on two independent, random variables $x$ and $y$. A sample of $N$ measurements of the variables yields pairs of values, $x_i$ and $y_i$, with $i = 1, 2, \ldots, N$. For the sample one can compute the means, $\bar{x}$ and $\bar{y}$; the standard deviations, $\sigma_x$ and $\sigma_y$; and the values $Q_i = Q(x_i, y_i)$. We assume that the scatter of the $x_i$ and $y_i$ about their means is small. We can then write a power-series expansion for the $Q_i$ about the point $(\bar{x}, \bar{y})$, keeping only the first powers. Thus,

$$Q_i = Q(x_i, y_i) \cong Q(\bar{x}, \bar{y}) + \frac{\partial Q}{\partial x}(x_i - \bar{x}) + \frac{\partial Q}{\partial y}(y_i - \bar{y}), \qquad \text{(E.36)}$$

where the partial derivatives are evaluated at $x = \bar{x}$ and $y = \bar{y}$. The mean value of $Q_i$ is simply

$$\bar{Q} \equiv \frac{1}{N}\sum_{i=1}^{N}Q_i = \frac{1}{N}\sum_{i=1}^{N}Q(\bar{x}, \bar{y}) = \frac{1}{N}NQ(\bar{x}, \bar{y}) = Q(\bar{x}, \bar{y}), \qquad \text{(E.37)}$$

since the sums of the $x_i - \bar{x}$ and $y_i - \bar{y}$ over all $i$ in Eq. (E.36) are zero, by definition of the mean values. Thus, the mean value of $Q$ is the value of the function $Q(x, y)$ calculated at $x = \bar{x}$ and $y = \bar{y}$.

The variance of the $Q_i$ is given by

$$\sigma_Q^2 = \frac{1}{N} \sum_{i=1}^{N} (Q_i - \bar{Q})^2.$$ 

(E.38)

Applying Eq. (E.36) with $\bar{Q} = Q(\bar{x}, \bar{y})$, we find that

$$\sigma_Q^2 = \frac{1}{N} \sum_{i=1}^{N} \left[ \frac{\partial Q}{\partial x}(x_i - \bar{x}) + \frac{\partial Q}{\partial y}(y_i - \bar{y}) \right]^2$$ 

(E.39)

$$= \left( \frac{\partial Q}{\partial x} \right)^2 \frac{1}{N} \sum_{i=1}^{N} (x_i - \bar{x})^2 + \left( \frac{\partial Q}{\partial y} \right)^2 \frac{1}{N} \sum_{i=1}^{N} (y_i - \bar{y})^2$$

$$+ 2 \left( \frac{\partial Q}{\partial x} \right) \left( \frac{\partial Q}{\partial y} \right) \frac{1}{N} \sum_{i=1}^{N} (x_i - \bar{x})(y_i - \bar{y}).$$ 

(E.40)

The last term, called the *covariance* of $x$ and $y$, vanishes for large $N$ if the values of $x$ and $y$ are uncorrelated. (The factors $y_i - \bar{y}$ and $x_i - \bar{x}$ are then just as likely to be positive as negative, and the covariance also decreases as $1/N$). We are left with the first two terms, involving the variances of the $x_i$ and $y_i$:

$$\sigma_Q^2 = \left( \frac{\partial Q}{\partial x} \right)^2 \sigma_x^2 + \left( \frac{\partial Q}{\partial y} \right)^2 \sigma_y^2.$$ 

(E.41)

This is one form of the error propagation formula, which is easily generalized to a function $Q$ of any number of independent random variables.

# Index

*Atoms, Radiation, and Radiation Protection.* James E. Turner
Copyright © 2007 WILEY-VCH Verlag GmbH & Co. KGaA, Weinheim
ISBN: 978-3-527-40606-7

# PERIODIC TABLE OF

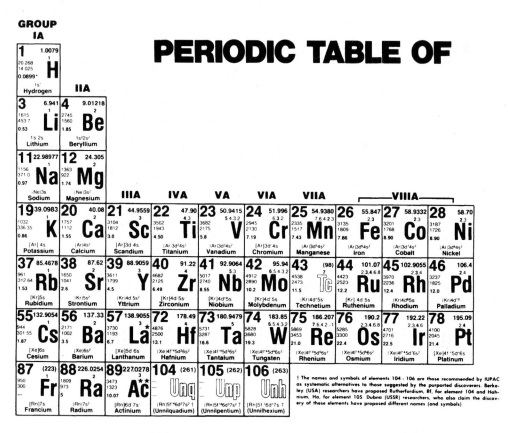

† The names and symbols of elements 104 - 106 are those recommended by IUPAC as systematic alternatives to those suggested by the purported discoverers. Berkeley (USA) researchers have proposed Rutherfordium, Rf, for element 104 and Hahnium, Ha, for element 105. Dubna (USSR) researchers, who also claim the discovery of these elements have proposed different names (and symbols)

KEY

ATOMIC NUMBER
ATOMIC WEIGHT
BOILING POINT, K
OXIDATION STATES (Bold most stable)
MELTING POINT, K
SYMBOL (1)
DENSITY at 300 K (g/cm³)
ELECTRON CONFIGURATION
NAME

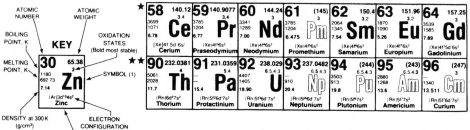